Industrial Engineering Terminology

ANSI Standard Z94.0-1982

Industrial Engineering Terminology

A revision, consolidation, and redesignation of ANSI Z94 Index and ANSI Z94.1-12

An American National Standard
Approved December 9, 1982

Published by the
Institute of Industrial Engineers
Industrial Engineering & Management Press

Distributed in cooperation with
Wiley-Interscience, a division of John Wiley & Sons, Inc.

Included in this volume are abbreviations, symbols, acronyms, functional designations, and letter combinations that are not official, nor are they comprehensive or definitive. They are intended only as a convenient reference supplement when so many aspects of industrial engineering are changing and expanding.

ISBN 0-471-80270-0

Additional copies may be obtained by contacting:
Publication Sales
Industrial Engineering and Management Press
Institute of Industrial Engineers
25 Technology Park/Atlanta
Norcross, Georgia 30092
(404) 449-0460

Quantity discounts available.

Contents

INTRODUCTION

William J. Jaffe, Engr. Sc. D., P.E., Chairman

American National Standards Committee Z94

Makeup of the Z94.0 Standard

To preserve a closer relation to the original document issued in 1972, Z94.0 is the connotation assigned to this new single-volume combination of the revision of Z94 plus the five new documents in the INDUSTRIAL ENGINEERING TERMINOLOGY. The 1972 document consisted of thirteen individually issued documents:

Z94.1 BIOMECHANICS
Z94.2 COST ENGINEERING
Z94.3 DATA PROCESSING AND SYSTEMS DESIGN
Z94.4 DISTRIBUTION AND MARKETING
Z94.5 ENGINEERING ECONOMY
Z94.6 FACILITY PLANNING (including layout, maintenance, materials handling, etc.)
Z94.7 MATERIALS PROCESSING (including tool engineering, foundry operations, automation, plastics, etc.)
Z94.8 APPLIED MATHEMATICS (including operations research, management science, statistical quality control, statistical and mathematical applications, etc.)
Z94.9 ORGANIZATION PLANNING AND THEORY
Z94.10 PRODUCTION PLANNING AND CONTROL (including inventory control, routing, scheduling, dispatching, etc.)
Z94.11 APPLIED PSYCHOLOGY (including psychological, sociological, and human factors, job evaluation, wage incentives, personnel practices, etc.)
Z94.12 WORK MEASUREMENT AND METHODS
— 1972 INDEX TO Z94.1-Z94.12

In accordance with ANSI practice of periodic examination, all these volumes were reexamined. Eleven of these were completely revised and updated and five new ones added. In a sense, the added terminologies — viz., HUMAN FACTORS, WAGE AND SALARY ADMINISTRATION, ANTHROPOMETRY, SAFETY, OCCUPATIONAL HEALTH AND MEDICINE — are really expansions of APPLIED PSYCHOLOGY, the twelfth of the originally issued terminologies, which appears, for the most part, in its original state in Section 11.

At an early meeting (1967) of Z94, the main and the working committees adopted the breakdown of the industrial engineering field as advocated by the American Institute of Industrial Engineering's AIIE RESEARCH ABSTRACTS, which listed APPLIED PSYCHOLOGY and included in this field such disciplines as biomechanics, human factors, wage incentives, safety, etc. Although this heading offered great opportunities, it also presented problems. One of the most glaring of the problems was the questionable inclusion of biomechanics under an APPLIED PSYCHOLOGY heading — since BIOMECHANICS = BIOLOGY + MECHANICS. Hence, the committee gave biomechanics its own category (Z94.1). Again, there was no listing of distribution and marketing, a most important aspect — since this field "consists of everything that comes before the design of a product as well as all that comes after its manufacture." Hence, this field was added (Z94.4) in the original edition. In the present terminology, HUMAN FACTORS, WAGE AND SALARY ADMINISTRATION, ANTHROPOMETRY, SAFETY, and OCCUPATIONAL HEALTH AND MEDICINE — which by now have been developed far enough to indicate individual disciplines of their own — have been removed from the APPLIED PSYCHOLOGY category and comprise Sections 13 through 17, respectively, in this present one-volume combined INDUSTRIAL ENGINEERING TERMINOLOGY.

Human factors and ergonomics (terms often used synonymously) were long recognized especially by the Gilbreths in the early pioneering days of industrial engineering, when they emphasized, among other factors, the worker and the workplace. Then, too, especially during and after World War II, the Office of Naval Research, as well as the National Space agencies, paid special deference to this aspect. Anthropometry was another field to which attention was paid: the Gilbreths, again, were among the first to take note of a worker's physical dimensions. OSHA, for example, has led to increased attention to safety and occupational health. Wage and salary administration, with the increased regulations of government and labor unions, must now command its own special category. Hence, all these disciplines have been removed from the APPLIED PSYCHOLOGY category and expanded on their own. However, the APPLIED PSYCHOLOGY category has been retained because it still provides and will continue, in the future, to be a source of many of the advancements that will undoubtedly be part of industrial engineering. Hence, except for some additions and deletions, the old Z94.11 appears in Section 11.

These studied actions should clearly indicate the contemporary aspects of these standards and should also bear witness to the dynamic aspects of the field. Too many — both lay and professional — fail to understand this and believe that, in the physical and chemical sciences, for example, standards remain fixed and permanent while all other fields have illusory and flexible norms. Too many still mistakenly believe that "one cubic centimeter of water weighs one gram," that the unit of force (or weight) is no different from a unit of mass — the fact that these are measured in "pounds" is ample "proof" for them — and that SI is a new name given to the cgs system they "learned" in their elementary education programs.

Because industrial engineering is so concerned with "people," this TERMINOLOGY has taken cognizance of contemporary standards as illustrated in the revised and enlarged definition that the AIIE has established:

> Industrial Engineering is concerned with the design, improvement, and installation of integrated systems of people, materials, equipment and energy. It draws upon specialized knowledge and skill in the mathematical, physical, and social sciences together with the principles and methods of engineering analysis and design to specify, predict, and evaluate the results to be obtained from such systems.

Standardized job titles no longer carry such terms as *foreman* and/or *forelady*, nor do they use *foreperson*, but rather there has been a complete change of connotation — e.g., *supervisor* (often bringing the need for distinguishing this from (say) *superintendent*. However, there need not be a complete revolution every time the term *man* is used, in an attempt not to disregard the "other half of the population". Although *chairperson* seems to have been accepted, any attempt to alter such terms as *manpower, manhours, management,* etc., would make these terms almost entirely unintelligible. Frankly, the term *management* has nothing to do with the male of the species, but has the same root as the word *menagerie*, being derived from the Italian *maneggie* from *maneggiare*, referring to the control of horses. Again, going even further back in its etymology, there is the Latin *manus* (hand). Hence, manufacture, manpower, manhours, etc., take on a greater meaning. Again, to avoid any tinge of sexism, many are led to the grammatical error of using *they* third person plural when the third person, singular is meant. This TERMINOLOGY has recognized the problem and, after discussions, has attempted to treat the problem rationally. (As for the third person, singular pronoun, this TERMINOLOGY may, however, go to the extent of using he/she, her/him, etc.)

Of course, Z94 has also taken cognizance of SI, a system that is the accepted and legal standard in *all* countries but one or two. The USA first declared the metric system as its only *legitimate* measuring system — and also signed the International Treaty of the Metre — over a hundred years ago, and only recently renewed its commitment to SI. It is now over a dozen years that it has been the policy, established by the ASME COUNCIL, to "encourage and guide the use or inclusion of SI units as appropriate in Codes and Standards and other ASME publications" (ASME ORIENTATION AND GUIDE FOR USE OF SI [METRIC] UNITS).

Although this present edition of the INDUSTRIAL ENGINEERING TERMINOLOGY takes cognizance of the contemporary trends in IE, it is only appropriate to mention those leaders who, while not directly associated with this new edition, did lay the basis for the TERMINOLOGY. Z94 has been in existence long enough so that some of these early sponsors have passed: Col. Clarence E. Davies (whose early association was with H. L. Gantt but whose later work was involved, for more than half a century, with ASME at the highest level), Professor David B. Porter, who also worked with Gantt, drew the first Gantt Progress Chart, and established the first collegiate course in motion study well over half a century ago, Henry N. Muller, who was a main force behind the negotiating committee for the establishment of the TERMINOLOGY project, as well as President of ASME, and Dr. Harold B. Maynard (MTM founder and author of ASME's 1955 publication, INDUSTRIAL ENGINEERING TERMINOLOGY). These well known leaders, together with AIIE Emeritus Executive Director Jack Jericho, Professor Delmar W. Karger (Emeritus Dean and Ford Foundation Professor at Rensselaer Polytechnic Institute), and Dr. William J. Jaffe (Emeritus Distinguished Professor, Newark College of Engineering, New Jersey Institute of Technology), were involved in establishing the INDUSTRIAL ENGINEERING TERMINOLOGY project.

As in the previous edition, Dr. Jaffe has served as Chairman of Z94 and took upon himself the entire task of compiling the INDEX, coordinating all the terminologies, and editing of the entire manuscript. (Corrections and suggestions, however, may be addressed to the new Chairman of Z94 Dr. Robert L. Williams, Professor and Chairman, Industrial and Systems Engineering, Ohio University, Athens, Ohio 45701.)

Purpose and Scope

The Committee originally was the direct result of the suggestion that The American Society of Mechanical Engineers publish a revision of its 1955 publication, INDUSTRIAL ENGINEERING TERMINOLOGY (ASME Standard No. 106) [1] — as well as an attempt at expansion of the INDUSTRIAL ENGINEERING TERMINOLOGY MANUAL [2], which the American Institute of Industrial Engineers published in 1965. Both publications were heavily oriented toward work standards and measurement — a natural consequence, since *Industrial Engineering* (IE), early in its development from *Scientific Management,* took on and expanded this phase as one of its basic fields. It is of interest, too, to note historically, that in 1952, the Society for the Advancement of Management issued a GLOSSARY OF TERMS USED IN METHODS, TIME STUDY AND WAGE INCENTIVES [3], which, as its very title suggested, was limited in scope.

ASME's publication received excellent support from the profession, but it was apparent that any further reprinting — without taking cognizance of the rapid developments in the application of statistical, mathematical, psychological, physiological, and sociological aspects to IE — would certainly hamper its utility. In bringing out a revision, there were many reasons for ASME proceeding on its own. Essentially, the publication was initially prepared by ASME. Moreover, ASME's role in the field was an historical one, going back to the last century when it was the sole rostrum for the field: in fact, Person [4], Taylor's great advocate, indicated Towne's THE ENGINEER AS ECONOMIST [5] paper for ASME as a "convenient bench mark for the beginning of this . . . movement." Moreover, as early as 1920, the ASME officially adopted [6] the recommendation "... that Industrial Engineering is a major subject for consideration by the Society and shall be placed on par with all major technical subjects". More specifically, on the other hand, in 1948, the American Institute of Industrial Engineers was established with IE as its principal interest.

Hence, a joint venture by both ASME and AIIE was welcomed by all. This co-sponsorship was effectively executed through the offices of the Standards Department of ASME, which coordinated all activities.

To fulfill the objective of encompassing as many aspects of the IE field as possible, invitations were extended to all societies, governmental, and quasi-governmental agencies, trade associations, and labor and industrial organizations dealing with any phase of IE. The following is a listing of those which accepted membership — along with the names of their present representatives on the Z94 Committee.

AMERICAN NATIONAL STANDARDS COMMITTEE Z94

Industrial Engineering Terminology

W. J. Jaffe, Chairman   D. W. Karger, Vice Chairman   I. A. Delloff, Secretary

Standards Committee

AMERICAN ARBITRATION ASSOCIATION

H. Gittler, P.E., Vice President (Manufacturing), Gilford Instrument Labs., Incorporated, Oberlin, Ohio

AMERICAN ASSOCIATION OF COST ENGINEERS

R. E. Templeton, M. W. Kellogg Company, Houston, Texas

AMERICAN FEDERATION OF LABOR AND CONGRESS OF INDUSTRIAL ORGANIZATIONS

J. L. Zalusky, American Federation of Labor and Congress of Industrial Organizations, Washington, D.C.

R. Oswald, Alternate, American Federation of Labor and Congress of Industrial Organizations, Washington, D.C.

AMERICAN INSTITUTE OF CHEMICAL ENGINEERS

G. C. Thomas, American Institute of Chemical Engineers, Newport Beach, California

AMERICAN PRODUCTION AND INVENTORY CONTROL SOCIETY, INC.

T. Wallace, T. F. Wallace & Company, Cincinnati, Ohio

AMERICAN SOCIETY FOR ENGINEERING EDUCATION

Prof. B. W. Niebel, The Pennsylvania State University, University Park, Pennsylvania

AMERICAN SOCIETY OF MECHANICAL ENGINEERS, THE
Dr. W. J. Jaffe, P.E., Chairman, New Jersey Institute of Technology, Newark, New Jersey
Dr. J. Mihalasky, P.E., New Jersey Institute of Technology, Newark, New Jersey

ASSOCIATION OF CONSULTING MANAGEMENT ENGINEERS, INCORPORATED
P. W. Shay, Association of Consulting Management Engineers, Incorporated, New York, New York

GENERAL SERVICES ADMINISTRATION
J. W. Gilbert, General Services Administration, Washington, D.C.

HUMAN FACTORS SOCIETY OF AMERICA
B. Crist, Eastman Kodak Company, Rochester, New York

INDUSTRIAL MANAGEMENT SOCIETY OF AMERICA
J. E. Scott, Stonybrook, New York

INDUSTRIAL RELATIONS RESEARCH ASSOCIATION
Dr. W. Gomberg, University of Pennsylvania, Philadelphia, Pennsylvania

INDUSTRIAL TRUCK ASSOCIATION, THE
E. R. Sirotak, Towmotor Corporation, Mentor, Ohio

INSTITUTE OF INDUSTRIAL ENGINEERS
Prof. D. W. Karger, P.E., Vice Chairman, Rensselaer Polytechnic Institute, Troy, New York
Dr. R. L. Williams, Ohio University, Athens, Ohio
I. A. Delloff, P.E., Alternate, Secretary, Motivation and Training Programs, Fairlawn, New Jersey

INSTITUTE OF MANAGEMENT SCIENCES, THE
Dr. J. Monahan, Boston University, Boston, Massachusetts

INTERNATIONAL MATERIAL MANAGEMENT SOCIETY
W. G. Rheim, Cincinnati Technical College, Cincinnati, Ohio

MASSACHUSETTS DEPARTMENT OF LABOR AND INDUSTRIES
E. F. Stokham, Department of Labor and Industries, North Andover, Massachusetts

SOCIETY FOR THE ADVANCEMENT OF MANAGEMENT
Prof. D. Beach, (resigned), Rensselaer Polytechnic Institute, Troy, New York
C. McCurry, Alternate, Robertshaw Fulton Company, Knoxville, Tennessee

SOCIETY OF MANUFACTURING ENGINEERS
Vacant

UNITED STATES DEPARTMENT OF COMMERCE
Vacant

History of the Z94 Committee

Z94 Project, INDUSTRIAL ENGINEERING TERMINOLOGY, was originally organized in 1966 under the procedures of the American National Standards Institute (formerly known as American Standards Institute and then U.S. Standards Institute). After a lengthy period of preliminary discussions, the project was organized as a joint venture between the American Society of Mechanical Engineers and the American Institute of Industrial Engineers (now IIE) — with Dr. William J. Jaffe, Professor (now Distinguished Professor) of Industrial Engineering, Newark College of Engineering (now NJ Institute of Technology), as Chairman, and Prof. Delmar W. Karger (then Dean and now) Ford Foundation Professor of Management, School of Management, Rensselaer Polytechnic Institute, as Vice Chairman. Dr. Jaffe represented ASME and Prof. Karger AIIE. Originally the Secretary of the Committee came from the Society for Advancement of Management, but he was replaced by Mr. Irving A. Delloff (formerly of the Presto Lock Co.), an AIIE Alternate.

At its first working session, the Z94 Committee, in its anxiety to allow full and authoritative investigation into as many aspects of IE as possible, decided to take advantage of the expertise existing in the various fields of its cooperating members. Using the breakdown advocated by the AIIE RESEARCH ABSTRACTS — with one modification and one addition — twelve working subcommittees were formed. Essentially the ABSTRACTS had ten divisions, but the Committee added Marketing and Distribution and removed Biomechanics from the Applied Psychology Section to form its own entity. The Committee accepted this action unanimously.

It was felt that Marketing and Distribution, which had no role in the ABSTRACTS, was entitled to a separate status, because, as one Committee member expressed it, "Marketing and Distribution was concerned with what takes place before a product goes on a production line as well as with what takes place after it comes off." Moreover, even in the early pioneering days when the work of Taylor, the Gilbreths, and Gantt was beginning, Dexter Kimball [7] insisted that "distribution" deserved particular attention, and ASME itself devoted a distinct division to its pursuit. In fact, the famous TEN YEAR REPORTS of ASME on management development have contained separate sections on this most important aspect. In addition, the application of mathematical and statistical methods to marketing problems pervade the literature of operations research and management science. Finally, from the economic and industrial practice points of view, the cost aspect is still not too far from the old "fifty-nine cents of the dollar" figure that became so famous as long ago as the Franklin Roosevelt era.

As for Biomechanics, which was included under the Applied Psychology Section of the ABSTRACTS, this, in the opinion of the experts who were in the field, was incorrect. The practitioners insisted that "Biomechanics" was precisely what the name implied — viz. namely Biology plus Mechanics. This notion was further reinforced by the Biomechanics Division of ASME.

In revising Z94-1972, the ASME and AIIE suggested that Z94.0 INDUSTRIAL ENGINEERING TERMINOLOGY appear in a single volume instead of the individual ones of the earlier Standard. However, to retain the advantages of separating the disciplines — see below (DEFINITIONS) — it was decided that all of the twelve previously issued single volumes (Z94.1: BIOMECHANICS; Z94.2: COST ENGINEERING, etc.) now appear, in their revised forms as Sections 1, 2, etc., respectively, in the new edition. Hence, Sections 1 through 12 here are the revisions of the 1972 sections.

As noted above (MAKEUP OF THE Z94.0 STANDARD), the APPLIED PSYCHOLOGY Section (Z94.11-1972) was further "dissected," and the fields of HUMAN FACTORS, WAGE AND SALARY ADMINISTRATION, ANTHROPOMETRY, SAFETY, and OCCUPATIONAL HEALTH AND MEDICINE were removed, expanded, and given sections of their own: these new standards appear as Sections 13, 14, 15, 16, and 17, respectively.

Thus, seventeen subcommittees were formed — each headed by an expert in the field, who, in turn, appointed subcommittee members who were experts in their respective fields. These subcommittees worked diligently — with no compensation other than the professional realization that all would benefit from their labors. The members of these seventeen working committees are listed individually in the respective sections along with their terminologies. The chairmen of the subcommittees are:

BIOMECHANICS

Dr. Allan B. Drosin, Chairman
    Formerly of the Biomechanics Program, The Center for Safety, New York University Medical Center

COST ENGINEERING

Dr. Carl Wolf, P.E., Chairman
    Professor, Management Engineering, N.J. Institute of Technology

DATA PROCESSING AND SYSTEMS DESIGN

Dr. Victor B. Godin, P.E., Chairman
    Associate Professor, Business Administration, Northeastern University

DISTRIBUTION AND MARKETING

Richard A. Forberg, Co-chairman
    Retired Director, Management Systems Division, Procter & Gamble Company
George D. Montillon, Co-chairman
    Retired Manager, Product Development, Management Systems Division, Procter & Gamble Company

ENGINEERING ECONOMY

Dr. Gerald Fleischer, Co-chairman
Professor, Industrial Engineering, University of Southern California
Dr. John Mihalasky, P.E., Co-chairman
Professor, Industrial Engineering, N.J. Institute of Technology

FACILITY PLANNING

Edward R. Sirotak, Chairman
Chief Engineer, Towmotor Corporation

MATERIALS PROCESSING

Prof. Benjamin W. Niebel, Chairman
Emeritus Head and Professor, Industrial and Management Systems Engineering, The Pennsylvania State University

APPLIED MATHEMATICS

Thomas Calvin, Chairman
International Business Machines

ORGANIZATION PLANNING AND THEORY

Prof. Delmar W. Karger, P.E., Co-chairman
Emeritus Dean and Ford Foundation Professor, Management, Rensselaer Polytechnic Institute; Professor, University of West Florida's; Consultant
Prof. Dale S. Beach, Co-chairman (resigned)
Professor, Management, Rensselaer Polytechnic Institute

PRODUCTION PLANNING AND CONTROL

Thomas F. Wallace, Chairman
T. F. Wallace & Company

APPLIED PSYCHOLOGY

Bertram Gottlieb, P.E., Chairman
Bertram Gottlieb Associates

WORK MEASUREMENT AND METHODS

Dr. Robert L. Williams, Chairman
Professor and Chairman, Industrial and Systems Engineering, Ohio University

HUMAN FACTORS

Brian Crist, Chairman
Human Factors Division, Eastman Kodak

WAGE AND SALARY ADMINISTRATION

Prof. James L. Rigassio, P.E., Chairman
Chairman, Industrial and Management Engineering, N.J. Institute of Technology

ANTHROPOMETRY

Charles Mauro, Chairman
Charles Mauro Associates

SAFETY

Dr. John Mihalasky, P.E., Chairman
Professor, Industrial and Management Engineering, N.J. Institute of Technology

OCCUPATIONAL HEALTH AND MEDICINE

Edward Bernacki, M.D., Chairman
Corporate Medical Director, United Technologies

Definitions

The standard definitions for each section were derived by the respective Subcommittee. Upon adoption by the Subcommittee, the definitions were next submitted to all other subcommittees of Z94 for examination and suggestions. Revisions were often made, and the definitions adopted by the respective subcommittees were approved and adopted by the Z94 Standards Committee. Although the subcommittee members were chosen for their expertise, it was obvious that some definitions cut across all fields of IE, despite any attempt at dividing the entire IE discipline. Thus, for example, *planning* could not be the sole possession of the Organization Planning Subcommittee, but could be found in other sections as well (e.g., Production Planning and Control). Again, the term *control* was a matter of concern to the Applied Mathematics Subcommittee (e.g., Statistical Quality Control) as well as the Cost Engineering Subcommittee. It was natural that, over the years, many supposedly common terms were used differently among the various divisions of the field. This should not be surprising, since even in the quantitative aspects the letter "c" could refer to *c-charts* (plotting number of defects) and to sampling plans where it represented the *acceptance number*. What was really surprising was the fact that in relatively few instances were there any real disagreements. However, in some *very few* instances, the term had become so ingrained that it was necessary to give both definitions. Again, in some cases, the "application" of the term was allowed to stand along with the definition. Thus, the standard definition of *Linear Programming* (LP) is naturally found in the Applied Mathematics Section, but the Production Planning and Control Subcommittee's term was also allowed to stand since this was not so much a definition as an example of the use of LP in Distribution and Marketing.

Again, there was naturally the question as to the exact place where a definition belonged: this was especially true in cases where more than one subcommittee used the term. This may not be as simple as it sounds, for it is not limited to aspects where the same term is used in an exact sense. Consider, for example, the term *Industrial Engineering,* which might properly belong in any or all sections. However, because of the historical — as well as the continuing — association with *Motion and Time Study,* the term is found in the Work Measurement and Methods Section. In all instances, the reader should bear in mind that the fundamental definition belongs to the basic subcommittee, and, in the event that another is given, the less general application is limited to use in a particular aspect of IE (e.g., the Distribution and Marketing Subcommittee's application of *Linear Programming* — an Applied Mathematics Subcommittee term — to a marketing problem). However, a cross-reference between Z94 volumes is noted.

It must be admitted that this problem of determining where a definition belonged could have been easily avoided by grouping all definitions together — with no consideration as to a subdivision of the entire IE field. As tempting as that may have been, in view of the possible boundary questions, the purpose of breaking up the IE field into specific subdivisions and the resulting separate sections was for the convenience of the practitioner. It would undoubtedly prove quite inconvenient, for example, for an industrial engineer engaged in, say, labor negotiations, to carry with him to the bargaining table, an overall composite of IE terms ranging from Applied Mathematics to Facility Planning. Hence, the breakdown into seventeen separate sections. It must be admited that even this is not without its impending problems in view of the fact that IE terms cut across the fictitious boundaries. To overcome this, an overall Index was prepared by the Chairman, and, it is hoped this will prove useful in finding definitions in any specific category.

Although the basic responsibilities for the definitions are those of the particular subcommittee, in particular, and the Z94 Committee, in general, it is likely that the reader, particularly the practitioner, for whom these definitions are of more than academic interest, will have suggestions as to omissions, revisions, and additions. Such suggestions are not only invited, but solicited with advance appreciation. It will be helpful if these discerning readers and practitioners will contact Dr. Williams, the Z94 Chairman, Industrial and Systems Engineering, Ohio University, Athens, Ohio 45701.

The Industrial Engineering Field

The task of determining the exact limits of the IE field was not an easy one, despite the Committee's acceptance of the official AIIE definition:

> Industrial Engineering is concerned with the design, improvement, and installation of integrated systems of people, materials, equipment, and energy. It draws upon specialized knowledge and skill in the mathematical, physical, and social sciences, together with the principles and methods of engineering analysis and design, to specify, predict and evaluate the results obtained from such systems.

Official recognition is, of course, exceedingly helpful, but in spite of all admonitions of strict adherence and interpretation, the field is a living and dynamic one. Even more important, terms are used by people and are at their mercy.

There is, perhaps, as far as the industrial engineer and his predecessors are concerned, no better example of the changing field than the statement made by the pioneer, Hoxie [8], who is probably best remembered as the one who was among the first to define the relationship between scientific management and organized labor, and who insisted that "... time and motion study, therefore, must be regarded as the chief cornerstone." Although hardly anyone would deny this as being the historical pattern, some today would seriously question so basic a role ascribed to this aspect of the field. Even the terms themselves — *time and motion study* and *scientific management* — are seldom used by practitioners at present. In fact, these are terms that are foreign to many of today's students, who may be unaware of the historical roots of IE.

If a pragmatic beginning is sought for INDUSTRIAL ENGINEERING, it can be found in the pioneering work of Taylor in his attempt to answer the question: WHAT IS A FAIR DAY'S WORK? Taylor was fond of quoting President Theodore Roosevelt who insisted that "The conservation of our national resources is only preliminary to the larger question of national efficiency." Almost a hundred years ago, Taylor noted:

> We can see our forests vanishing, our water-powers going to waste, our soil being carried by floods into the sea; and the end of our coal and iron is in sight. But our larger wastes of human effort, which go on every day through such of our acts are as blundering, ill-directed, or inefficient . . .

Hence, Taylor looked for the competently, well-trained employee and employer. "The principal object of management," he asserted, "should be to secure the maximum prosperity for the employer, coupled with the maximum prosperity of each employee." Initiative and incentive of the worker was only half the problem, management was the other half. Hence, he developed four principles: develop a science for each element of the employee's work to replace the old rule of thumb method; scientifically select, train, teach, and develop the worker; heartily cooperate so that all work is accomplished in accordance with the principles of the science developed; an almost equal division of work and responsibility between management and workers.

Early in his career it was Taylor's need to discover "some rule or law which would enable a foreman to know in advance how much of any kind of heavy laboring work one who was well-suited to the job ought to do in a day: that is, to study the tiring effect of heavy labor on a first-class man." Research on earlier work, indicated that two types of experiments had been made. Physiologists studied the endurance of the human animal; engineers, determined to discover what part of a horse-power a man-power was, had men lifting loads by turning winches, lifting weights, etc. Workers were studied by Taylor, and data were translated into foot-pounds. Using the concept of the product of force and distance could obviously not yield a relationship between foot-pounds of work and the tiring effect on the worker. (Clearly, a worker, standing still but holding a weight in his hands, was, according to the physicists, not working.) Convinced that there was a relationship somewhere in his data, Taylor assigned the task to Carl Barth, who derived a law, confined to that class of work in which the limit of a worker's capacity is reached because of being tired. This was the law of heavy laboring where the worker's strength is exerted by pushing, pulling, grasping. The worker could be under load only part of the day. Thus when pig iron was handled (with a pig weighing 92 pounds), the worker could be under load only 43 percent of the day, and free of load for 53 percent. Handling half a pig, the worker could be under load 58 percent of the day. As the load lightens, the time under load increases, until the load is so light, the worker can carry the load the entire day! At this point the law of endurance ceases to operate, and some other law must be found to indicate a worker's capacity for work.

Taylor, influenced by his early teachers — especially the celebrated American mathematics text writer, Wentworth — relied on the stopwatch. Hence, the pioneering work took place in what became known as TIME STUDY. The Gilbreths, insisting on emphasizing the METHOD of work, set the goal as THE ONE BEST WAY; this became known as MOTION STUDY.

This schism has, over the years, subsided — but Motion and Time Study marks the beginning of Industrial Engineering as we know it.

It is altogether possible that the educators themselves — few, if any, of whom identify their respective departments as teaching *scientific management* — have added to the confusion of all, including students. Educators are prone to identify their IE departments as teaching solely *industrial engineering*. In fact, they often couple the IE with numerous other connotations. Thus, there are departments of Industrial Engineering and . . . Operations Research, Management Science, Management Engineering, Systems Engineering, Administrative Engineering, Engineering Management, to mention a few. To complicate matters further, they also teach, in addition to courses with the aforementioned titles, those in Industrial Management, Production Management, etc.

The IE educators are not alone in this endeavor to search for more encompassing appelations. The practitioners, themselves, have also engaged in this practice. Many, schooled in IE techniques, have identified themselves by titles quite similar to those that the educators have adopted: thus, they use the names *Management Consultant, Management Engineer, Systems Engineer,* and so on. It must be admitted, however, that the adoption of the term *Engineer,* of course, is limited to those licensed by the respective state boards. Many of these, in fact, simply use the term *Professional Engineer,* without further definition of their specialties. Naturally, their respective societies have followed suit, and these practitioners are found in divisions of the "founding" engineering societies, as well as in the "management," the "operations research," the "system" societies, in addition to the basic society for the IE's. In all of these societies, there has been, from time to time, strong movements for name changes. The reasons, like those in the educational institutions, have been manifold: the field is dynamic and not static, the splinter societies are attracting a great number of members merely by stressing these new fields, the name no longer identifies what the practitioner does, etc.

There is a sufficient history for this — in other fields as well. Probably the best example is the oldest of the engineering societies' field — Civil Engineering. In many of the American schools of engineering, the "classical" title *Civil Engineering Department* is no more. If the name has not been changed completely, it has taken on companion titles — Sanitary Engineering, Urban Planning, Environmental Engineering, and others. Again, this is understandable, for the history of engineering shows a pattern of fragmentation, consolidation, and then further fragmentation and consolidation. Thus, in the United States, the earliest of the engineering societies, the American Institute of Civil Engineers, owes its foundation (1852) to those who felt that engineering had applications other than those with which the military was concerned. Moreover, this great interest and concern for nonmilitary (or civil) engineering was not limited to this country. The British Institution of Civil Engineers adopted quite early the famous definition of Thomas Tredgold (1882) and included on its membership certificates [9]:

> A society for the general advancement of mechanical science and more particularly for promoting the acquisition of that species of knowledge which constitutes the profession of civil engineer, being the art of directing the great sources of power in nature for the use and convenience of man.

Hence, what began as a seeming fragmentation resulted in an all-encompassing objective.

Yet this was followed by more fragmentation. The engineers with specific interest in mining, for example, broke away and formed their own society which, incidentally, followed the fragmentation by a consolidation with the metallurgists to establish the American Institute of Mining and Metallurgical Engineers. The nonmilitary group soon experienced further fragmentation when the engineer with interests in machinery formed The American Society of Mechanical Engineers. Soon the electrical machine group broke away to form the American Institute of Electrical Engineers — only to experience similar fragmentation and consolidation with the radio engineers and the electronics group. Hence, the breaking away of the group of engineers and others who rallied around Frederick W. Taylor — himself a former ASME president — repeated the cycle that had become so common in engineering. In the first place, the "break" was never clear-cut. Although a few left ASME to join the Taylor Society, many retained dual membership — a practice that was continued when the Taylor Society combined with the Society of Industrial Engineers to form the Society for (the) Advancement of Management. In fact, when the AIIE came on the scene in 1948, many maintained triple membership.

This process continued as more and more specialized interest groups appeared — Quality Control, Operations Research, Management Science, Systems, etc. The IE practitioner, like the IE educator, often views these newer groups with forebodings. Unsure that his discipline has taken sufficient cognizance of this development, and fearful that many, in pursuit of this newer field, will foresake the older, he seeks to identify the old with the new. Hence, there is often a coupling of titles — Industrial and Management Engineering, Industrial Engineering and Operations Research, Industrial Engineering and Management Science, and, more recently, Industrial and Systems Engineering. However, this search for the "best of all possible worlds" is not without its dangers, as well as its confusion.

Consider the use of the "systems" term. A "system," to quote Johnson, Kast, and Rosenzweig, authors of the THEORY AND MANAGEMENT OF SYSTEMS [10], is an "organized or complex whole; an assemblage or combination of things or parts forming a complex of unitary whole." Thus, the data processing engineer, the electronics engineer, the civil as well as the geographer, the clergyman, the anatomist, to mention but a few — can lay claim to data processing systems, communications systems, transportation systems, religious systems, anatomical systems, etc. The same is true of the industrial engineer. The quoted authors insist that, in input-output language, managers are essentially the agents who convert the inputs of disorganized resources of man (now identified as people), money and machinery, into outputs of useful goods and services. Thus management is decision systems, communication systems, control systems, etc.

This is not something new. In fact, it is almost impossible to discuss IE without reference to its historical roots, *scientific management* and Taylor. Here, too, matters seem a bit hazy, for even in reference to *management* itself, a babel of terms has persisted and continue to plague us. Even among English-speaking countries, there is confusion about the meaning of the term and its use.

Although the articulate may carefully distinguish between *management, rationalization, rationalisation, organization, organisation, administration,* the terms seem to be used almost interchangeably [11]. Although *management* has Latin roots (from Italian *maneggiare* = to handle horses; from Latin *manus* = hand), none of the other languages especially the Romance languages — use any form of the term. They advocate: *organisation scientifique* (thus, for example, *Comité International de l'Organisation Scientifique*), *organizacao racional de trabalho, racionalizacion del travajo, organizzazione del lavore, Wirtschaflichkeit, Betriebslehre, Betriebsführung,* etc. Thus, the noun in the title of Fayol's classic, *ADMINISTRATION INDUSTRIELLE ET GENERALE,* was first translated as *administration* by Conbrough (1929) and later *management* by Storr (1949). As a matter of fact, Urwick, not only noted the correctness of the Storr translation, but he also noted, unhappily, that the term *management* is popularly associated with definitions connotating cajolery and trickery on the one hand, but used more authoritatively, as an activity, a body of knowledge, a group of persons, etc. Just as there were problems of translating the title of Fayol's title into English, there were problems in the reverse process — viz., translating Taylor's title into French by Le Chatelier, who used the term *l'organisation scientifique du travail,* which essentially is the French equivalent for a labor union. As for the German term, it is not a happy choice, for *Betriebsführung* is related to the German verb *treiben* — to drive.

The direct association of *management* with *engineering* — particularly in the United States — came early in the history of ASME, and when Taylor spoke of *management* — particularly *shop management* — the concern was a matter that had occupied the attention of the earlier ASME presidents. In fact, R. H. Thurston [12] the first ASME president, in his inaugural address, included *social economy* among the "Objects of the Society." Not only did H. R. Towne, in that "bench mark of scientific management," THE ENGINEER AS ECONOMIST, [5] include "shop management" as a field of interest, but, in an ASME session, substituting for President Coleman Sellars, [6] he suggested "industrial or economic questions" because ASME had, among its membership, "more than the Civil Engineers or the Mining Engineers men who are managers of labor, who are either owners or representatives of owners, and who, therefore, control capital." And, it is of special interest to note that (1920) the ASME council approved and adopted the *FIRST REPORT OF THE COMMITTEE OF AIMS AND ORGANIZATION* [6] (L. C. Marburg, Chairman), which included the statement: . . . that Industrial Engineering is a major subject for consideration by the Society and shall be placed on par with all major subjects. In his COST AND PRODUCTION HANDBOOK, L. P. Alford [13], codifier of management laws and principles, included under the all-inclusive term *Industrial Engineering,* the "laws" of Taylor, Fayol, and Alford, himself. Little wonder, then, that ASME in 1955, as an outgrowth of its Management Division's Work Standardization Committee (1948), sponsored an IE TERMINOLOGY, which, itself, has had its influence on this present TERMINOLOGY.

Although it must be noted that the ASME never did establish an IE division, its IE partisans were mainly housed in its Management Division, again indicating some sort of concomittance between *IE* and *Management*. Nor was this the only instance. As already noted [14], some ASME members joined with some non-ASME members to form a group — at first informal, but later formal — known as the Society to Promote the Science of Management, which, after Taylor's death in 1915, became known as the Taylor Society. This society, in 1936, merged with the Society of Industrial Engineers — a group which, in 1917, probably as a measure for helping the war effort, was formed from the Western Efficiency Society (1910) — to establish the Society for (the) Advancement of Management. SAM attracted a wide spectrum of members — some with ASME's Management Division, some with the National Association of Cost Accountants, some with the National Personnel Association, some unattached — to form an overall umbrella for most of the management practitioners. SAM's interests can be readily perceived from its *GLOSSARY OF TERMS USED IN METHODS, TIME STUDY AND WAGE INCENTIVES,* already noted. Among its other interests were problems relating to budgets and costing in production and distribution, performance in factories, offices, and stores, as well as operations in both profit and nonprofit institutions. In addition there were numerous other organizations specializing in all of these plus many more aspects of IE. Finally, in 1948, the

American Institute of Industrial Engineers was established: its objectives were many and compelling, but perhaps the most important was the professional recognition of IE.

The growth of AIIE has tended to emphasize the engineering aspects of the field, and, at the same time, has caused a continual re-examination and re-appraisal of the old concept of equating *IE* with *Management.* To date, no clear cut definitions have been recognized as far as the lay public is concerned, even though specific moves have been made by the professionals in identifying the field and acquainting the public with such notions. However, in many instances there have been concerted efforts to broaden the limits so as to encompass as much territory as possible. This has been especially true as new concepts and ideas have come into being: for example, operations research, bioengineering, systems engineering, to mention but a few. Many practitioners have found that such acquisition can be accomplished by considering themselves not only as adherents to an IE practice, but as advisers to management, in general. Many of the professional societies, fearful of either an exodus of members from their societies, or a failure to attract the new adherents who might seek independent societies devoted entirely to these "new frontiers" have attempted to solve the dilemma by establishing divisions in which specific interests may be pursued. Some societies have, in fact, considered name changes. This concern about territorial limits as well as the accompanying titles has long been a problem of the colleges which, of course, are among the first to sense the rise of the new branches.

Probably, of all the colleges, the most concerned have been the engineering colleges, in general, and the IE departments, in particular. Although accreditation is a major factor in all recognized colleges, it is particularly acute in the engineering college because of its relation to professional registration, a matter which is less critical in the liberal arts and the business colleges. In fact, a whole system of "common law" definitions have come about with *management* (without an identifying adjective) as the property of the business school, and *management* (accompanied by such terms as *Engineering* and/or *Industrial Engineering*) as the concern of the engineering school, and *management* (with the single adjective *industrial*) as the subject matter of the technological institute. There is, of course, no hard and fast rule, but "IE" has been taken as belonging solely to the engineering school. However, this, despite the AIIE definition, did not solve the boundary problem.

Historically [15], the first IE department was established at The Pennsylvania State — then, College, but now University, with Hugo Diemer at the head. He had been recommended by Taylor, who, in turn, had been asked by Beaver, a former governor of Pennsylvania and President of the Penn State Board of Trustees, to suggest a "man who could teach mechanical engineering from the standpoint of production rather than machine performance." (1907). It is interesting to note that Diemer had, since 1905, been teaching courses in *shop management,* at the University of Kansas [16]. (The University of Pennsylvania, on the other hand, can also be credited with pioneering with its Wharton School, which had been established [1881] on the recommendation of James Wharton, in the belief that "management should be separated from ownership and developed as a profession.") Because of the "scientific management" tradition associated with Taylor, it was clear that the engineering school's program included *management* courses. In fact, the management term appeared frequently in describing the courses and the departments, because many shied away from the IE term since it was associated by some with "efficiency" practitioners. And so, for a while, many departments named themselves with such titles as Administrative Engineering, Engineering Administration, Industrial Administration, Manufacturing Engineering, Manufacturing Management, Management Engineering, Commercial Engineering, etc. However, a basic group retained the IE appelation, which, eventually with accreditation programs, became the dominating title, especially with the establishment of AIIE and its promotion of a professional status.

Despite the title, *management* courses formed a most important part of the IE curriculum, and these attracted many students less technically oriented but anxious for positions in industry with high status and high wages. AIIE's attempt to solidify the professional engineering aspects, provided an external impetus, while the development of courses in mathematical and statistical applications served as an internal catalyst in strengthening the IE program. Soon the titles *Operations Research, Management Science,* and *Systems Engineering,* were added to the IE name.

These supplementary titles — used by the educators and the practitioners alike — have certainly not clarified matters, and both students and non-students are not too clear about the domain of IE. The defense, if any is needed, is quite simple: The IE field is broad and can and does encompass all such disciplines. Moreover, as each new vista appears, it attracts many, and so the consolidation — fragmentation cycle, noted above, begins all over again. Thus, in the World War II period, it was the application of statistical methods (especially in the field of quality and inspection), next came the more sophisticated mathematical and team approach (with operations research and management science), then the systems concept. In all probability, the next phase will be the application of biology and the life sciences.

IE, Management, etc.

Through all these metamorphoses and connotations, one aspect seems to have prevailed from the beginning. This is MANAGEMENT, for IE begins, as we know it, with Taylor — "the Father of Scientific Management." All of the pioneers were, in one way or another, concerned with it: Church (*The Science and Practice of Management*), Alford (*Laws of Management Applied to Manufacturing*), Cooke (*Spirit and Social Significance of Scientific Management*), Diemer (*Factory Organization and Administration*), the Gilbreths (*Primer of Scientific Management, The Psychology of Management,* etc.), Gantt (*Industrial Leadership,* etc.), Kimball (*Plant Management*), et alia.

However, it must be admitted that no single definition of *management* has been accepted. The famous ASME Report of 1912 [17] dealt with the "new element," but did not offer any specific definition of *management, scientific* or otherwise. In fact, the very term *scientific management* was seemingly concocted by a distinguished group — Gantt, Gilbreth, Emerson, and others — to aid Brandeis in the celebrated Eastern Rate Case Hearings (1910), even though Taylor had used the term in a limited sense earlier (1903). The adjective was probably meant to indicate that *management* could be studied and practiced deliberately and systematically, and that it was not the result of capricious, rule-of-thumb, seat-of-pants, meanderings. However, the term (*scientific management*) has had limited use during approximately the last quarter of a century. On the other hand, the term, *management,* has been used extensively both as a noun and as an adjective especially in all fields associated with IE.

Numerous definitions of *management* have appeared, and even the briefest examination of a few of these will show the wide diversity the matter has aroused even among the professionals. [18] Thus,

> Management is a broad term and covers almost all factors of an enterprise. [19]
>
> Management proper is the function in industry concerned in the execution of policy, within the limits set up by administration and the employment of the organization for the particular objectives set before it. [20]
>
> Management is the science of applied human effort . . . [21]
>
> Management is involved in the avoidance of waste in any human effort. [22]
>
> Management — the development of people, not the direction of things. [23]
>
> Management is the art and science of organizing, preparing and directing the human effort applied to control the forces and utilize the materials of nature for the benefit of mankind. [24]
>
> Management is the art of getting work done. [25]
>
> To manage is to forecast and plan, to organize, to command, to coordinate and control. [26]

Of course, there are others, but the definition given in this Z94 Standard (see Z94.9 Organization Planning and Theory) is:

> Management:
>
> 1. The process of utilizing material and human resources to accomplish designated objectives. It involves the activities of planning, organizing, directing, coordinating, and controlling.
> 2. That group of people who perform the functions described in 1, above.

This definition is basic. Certainly, it can be restated in many other ways, but retaining these fundamental concepts. Thus, it can be seen to be

> . . . the process by which the execution of a given purpose is put into operation and supervised.

Or, again, in another way,

> Management may be defined as the process by means of which the purposes and objectives of a particular human group are determined, clarified, and effectuated.

Thus, Prof. D. W. Karger can paraphrase his School of Management (Rensselaer Polytechnic Institute) Long Range Planning Document to say:

> Management refers to the systematic organization, allocation, and application of economic and human resources to bring about a controlled change; the role of management in these proceedings is one of organization and coordination of human effort in exploiting resources for the improvement of the organization involved.

With all these diverse "definitions" among the experts, it is not surprising to find that *management* takes on a wide variety of meanings among the laymen (including *cajolery, trickery,* and *totalitarianism*). However, even among the "experts" two other words — *administration* and *organization* — have often been used interchangeably with it, especially *administration*. The interchangeable use, among many, of *management* and *administration* has been noted in this Z94 Standard, which also calls attention to a distinction (concerned with the goal-establishing, policy-formulation part of *management*). This concept is consistent with Sheldon and Spriegel, who point out some very interesting aspects of the two terms. According to Spriegel [27], business circles consider *management* the more inclusive term, while governmental agencies consider *administration* the more inclusive. To the business world, *administration* is usually equated with that part of *management* usually described as *top management*. Sheldon [20] puts it quite succinctly in considering *administration* as "the function in industry concerned with the determination of corporate policy."

Whether it is because, as Dionysius has said, "education is contact with manners," or, as Shakespeare has insisted, namely "that which we call a rose, by any other name would smell as sweet," the engineering college, in general, and the IE Department in particular, have adopted a great variety of names — not only to describe their courses, but to describe the IE Department itself: Thus, some colleges have a "Department of Industrial and Management Engineering," a "Department of Industrial Engineering and Management Science," a "Department of Industrial Engineering and Operations Research," a "Department of Industrial and Systems Engineering," and so on. Others, of course, shy away from the IE connotation completely, and choose such names as a "Department of Administrative Engineering," a "Department of Engineering Administration," etc. Theoretically, *Industrial Management,* a term that had for years been used to describe some of the Industrial Engineering Department courses concerned with application of *management* to industrial problems, has been exiled to a strange limbo — a field that combines *management* with *business,* and "outside" the "true" engineering curriculum. It is paradoxical that it has been adopted by some of our foremost technological institutes — with great success. More often than not, the departure of *industrial management* from current IE curricula arises from a fear of a misunderstanding by engineering accreditation groups, as well as an inhibition on the part of purists who sense some contamination. On the other hand, in full realization of the need of such knowledge on the part of its students, quite frequently, the IE Department has adopted the material under some other name — probably being careful to add the term *engineering* to the title: e.g. *Management Engineering, Administrative Engineering.*

Because of the heavy mathematical orientation, *Operations Research* (OR) and *Management Science* (MS) have been quickly embraced as titles by the IE Department. In fact, so enthusiastic has been the adoption that many have given these equal billing with IE. (This has not been without problems, for the broad base of OR, for example, has caused it, in at least one great American engineering college, to move the subject from the IE Department to the Civil Engineering Department. In another, it was moved to the School of Commerce.) The breadth of the OR field allowed the IE Department to assume this discipline among its courses, while, on the other hand, OR provided a potent weapon for attacking IE problems.

The strength of OR can be readily observed from the definition given by Churchman, Ackoff, and Arnoff in their pioneering text [28]:

> Operations research in the most general sense can be characterized as the application of scientific methods, techniques, and tools to problems involving the operation of systems so as to provide those in control of the operations with the optimum solution of the problem.

With heavy emphasis on the multi-disciplined team and with a heavy orientation to a mathematical and statistical approach, OR, originally applied to military matters (hence its name), soon was useful in civilian matters. And the industrial engineer especially became interested in its application to management problems, particularly when they could be studied by mathematical models with optimization as a prime objective. Weinwurm [29] and others felt the "need for a thorough and detailed analysis of the decision making process . . . as a cause for the development of a management science" — with two approaches possible. One approach was "rationalistic with emphasis on building models," while the other insisted that the human element could not be neglected and took special cognizance of the fact that decisions were made by human beings — with the consequent result of a great need for the incorporation of "behavioristic and organization concepts." An integrated theory of organization (Simon) [30] soon took place.

Thus, *IE,* whose relationship with *Management* continued, could not possibly overlook the development of OR and MS. And, hence, its association with these fields developed. But, this IE-Management association did not stop here. In more recent years management has come to be viewed in terms of *systems*. A *system* can be defined as [10]

> . . . an organized or complex whole; an assemblage or combination of things or parts forming a complex of unitary whole.

The organization is, in this sense, a system wherein certain inputs are transformed to certain outputs — i.e. *Management* is a system that converts men, money, materials and machines into goods and services. Moreover, management can thus be viewed as control systems, decision systems, and communications systems — domains claimed by managers and their IE advisers.

Little wonder, then, that IE, whose official definition includes such concepts as *systems, mathematical, physical,* and *social sciences, engineering analysis and design,* can take on, along with *management,* a term which has been with IE since its beginning, such appellations as *OR, MS* and *Systems.* Admittedly, the IE field is broad — exceedingly broad — and further and further horizons may be expected, and the consolidation-fragmentation cycle will, in all probability, continue. It is hopeful that this Z94.0 IE TERMINOLOGY will — like the others that preceded and are bound to follow it — become a chart for delineating and defining the field.

References and Notes

[1] Industrial Engineering Terminology, ASME Standard #106, 1955.

[2] "Industrial Engineering Terminology Manual," *Journal of Industrial Engineering,* AIIE, XVI, No. 6. Nov.-Dec. 1965.

[3] *Glossary of Terms Used in Methods, Time Study and Wage Incentive,* SAM 1952.

[4] H. S. Person. "The Origin and Nature of Scientific Management," *Scientific Management in American Industry,* Taylor Society, Harper, 1929, p. 5.

[5] H. R. Towne. "The Engineer as Economist," *Transactions ASME, VII,* 1886, pp. 428-432.

[6] W. J. Jaffe, L. P. Alford. *The Evolution of Modern Industrial Management,* New York, New York University Press, 1957, pp. 29-34.

[7] D. Kimball. "Another Side of Efficiency Engineering," *American Machinist,* XXXV, pp. 263-265, Aug. 10, 1911.

[8] R. F. Hoxie. "Scientific Management and Labor Welfare," *Journal of Political Economy,* XXIV, 1916, p. 838. (Cf. "Taylor said many years ago: 'Time study is by far the most important element in scientific management.'" q. by A. H. Mogensen, *Common Sense Applied to Motion and Time Study,* N.Y. McGraw-Hill, 1932, p. 7.)

[9] A. C. Humphreys. *Lecture Notes on the Business Features of Engineering Practice,* Stevens Institute of Technology, Hoboken, 3rd Ed. 1916, p. 7.

[10] R. A. Johnson, F. E. Kast, and J. E. Rosenzweig, *The Theory and Management of Systems,* N.Y. McGraw-Hill, 1963, p. 4.

[11] L. M. Gilbreth and W. J. Jaffe. "Management's Past — A Guide to its Future," *Fifty Years Progress in Management 1910-1960,* ASME, 1960, p. 6.

[12] R. H. Thurston. "President's Inaugural Address," *Transactions ASME,* I, 1880, pp. 14-29.

[13] L. P. Alford (editor). *Cost and Production Handbook,* New York, Ronald, 1934.

[14] W. J. Jaffe. op. cit 111-117.

[15] W. J. Jaffe. op. cit 286-288.

[16] C. W. Lytle. "Collegiate Course for Management," *Professional Society for Promotion of Engineering Education,* XXXIX, 1936, pp. 806-839.

[17] "The Present State of the Art of Industrial Management, ASME's 1912 Report," *Fifty Years Progress in Management 1910-1960,* pp. 293-323.

[18] E. H. Anderson and G. T. Schwenning. *The Science of Organization Production,* New York, Wiley, 1938, pp. 9-29.

[19] R. H. Lansburgh. *Industrial Management,* N.Y., Wiley, 1920, 2nd Ed. p. 4.

[20] O. Sheldon. *The Philosophy of Management,* London Pitman, 1923, pp. 101 ff.

[21] A. H. Church. *The Science and Practice of Management,* New York Engineering Magazine Co. 1914, p. 282.

[22] S. Webb. *The Works Manager Today,* New York, Longmans, Green 1918, p. 5.

[23] B. W. Niebel. Letter to author.

[24] ASME Management Division 1921.

[25] L. P. Alford. "Scientific Industrial Management," Paper No. 341, *World Engineering Congress, Management Section,* Tokyo, Oct. 30, 1929.

[26] H. Fayol. *General and Industrial Management,* Storrs translation, London Pitman, 1949, pp. 5-6.

[27] W. P. Spriegel. *Industrial Management,* New York, Wiley, 1955, 5th Ed. pp. 1, 8-10.

[28] C. W. Churchman, R. L. Ackoff, and E. L. Arnoff. *Introduction to Operations Research,* New York, Wiley, 1957, Chapter 1.

[29] E. H. Weinwurm. "Management Science," *Fifty Years Progress in Management, 1910-1960,* ASME 1960, pp. 90-95.

[30] H. A. Simon. *Administrative Behavior,* New York, Macmillian, 147.

Preface. In some sections — e.g., Section 8 Applied Mathematics — many terms are followed by a code number which refers to a specific source listed in the Bibliography.

How to Use this Terminology Book

1. The *Terminology* consists of seventeen sections: (1) *Biomechanics,* (2) *Cost Engineering,* (3) *Data Processing and Systems Design,* (4) *Distribution and Marketing,* (5) *Engineering Economy,* (6) *Facility Planning,* (7) *Materials Processing,* (8) *Applied Mathematics,* (9) *Organization Planning and Theory,* (10) *Production Planning and Control,* (11) *Applied Psychology,* (12) *Work Measurement and Methods,* (13) *Human Factors,* (14) *Wage and Salary Administration,* (15) *Anthropometry,* (16) *Safety,* and (17) *Occupational Health and Medicine*. All terms in each of the respective sections are listed alphabetically. (However, if the term begins with a symbol e.g., "λ-Criterion — the key word will be "LAMBDA" — look under the word the symbol represents.)
2. In addition to the seventeen sections, there is an *OVERALL INDEX* — containing references to ALL the terms in all of the sections. Hence, if you are not certain in which section you can find the term you seek, consult the *OVERALL INDEX*. All terms in the *OVERALL INDEX* are listed under the "KWIK Index System."
3. The "KWIK Index System" allows you to find any term — especially terms containing more than one word — without knowing how an indexer would "break up" the term. In other words, you do not have to know whether the term will be listed in a straightforward fashion or with the main classification word first. E.g., you do not have to guess whether the indexer used "Expected Work Pace," "Work Pace, Expected," or "Pace, Expected Work." You will find this particular term listed more than once:

   EXPECTED work pace
   expected WORK pace
   expected work PACE

   Thus, you will find the term under "E," under "W," and under "P," with the key words of "EXPECTED," "WORK," and/or "PACE."
4. The *OVERALL INDEX* will also list the section in which the desired term is to be found. There will be no need for page numbers since, it should be remembered, all terms in each of the volumes are listed alphabetically.
5. In some sections — e.g., Section 7 Materials Processing — there is appended, at the end of a definition, a key description as to the aspect of the overall subject to which the term belongs (e.g., "automation," "forging," "metal fastening," etc.).
6. Although many terms owe their definition to conferences among the various Z94.0 subcommittees, some definitions come from accepted sources (books, societies, etc.). Hence, a Bibliography usually follows the Preface. In some sections — e.g., Section 8 Applied Mathematics — many terms are followed by a code number which refers to a specific source listed in the Bibliography.

Biomechanics

Z94.1

1

The original (Z94.1 BIOMECHANICS-1972) edition of this section was prepared for use mostly by industrial engineers and key management personnel. However, it has become useful to many in other disciplines. Hence, this revision has been prepared with full awareness of such expanded use.

Terms from related fields of science, engineering, medicine, psychology and physiology have been included. Adequate and meaningful definitions have been provided, using the simplest but most effective language.

The work of the Subcommittee was made considerably easier because of the accomplishments of the contributors to the 1972 edition: Dr. E. R. Tichauer and members of the Biomechanics Staff (G. Schulz and M. Miller) and the participants in the U.S. Public Health Program at New York University (I. Nathan, H. Gage. R. Jarvik, and R. Wolkenberg) and Dr. J. Sarno, Medical Consultant.

However, the present Subcommittee, distinguished in its own right, made contributions which can, in no way, be diminished by the use of the earlier document as a basis:

CHAIRMAN:

Allan B. Drosin, Ph.D.
Formerly, Division of Biomechanics, Center for Safety, New York University Medical Center

MEMBERS:

Joseph G. Benton, M.D. (deceased)
Late Professor and Chairman, Rehabilitation Medicine, State University of New York, Downstate Medical Center

Erwin Beretsky, M.D.
Formerly, Good Samaritan Hospital, Suffern, N.Y.

William B. Blesser, M.E.E., P.E.
Professor and Chairman, Bioengineering, Polytechnic Institute of New York

Albert H. Burstein, Ph.D.
Director, Biomechanics, Hospital for Special Surgery, Cornell University Medical Center

George Van B. Cochran, M.D.
Director, Biomechanics Research, Helen Hayes Hospital

Myron Cohen, Ph.D., P.E.
Director, R & D Hospital Products, Cheesebrough Pond, formerly, Director, Mechanical Engineering Laboratory, Stevens Institute of Technology

Willibald Nagler, M.D.
Professor and Chairman, Physical Medicine and Rehabilitation, New York Hospital, Cornell University Medical Center

James A. Nicholas, M.D.
Director, Sports Medicine and Athletic Trauma, Lenox Hill Hospital

Edward Peizer, Ph.D.
Assistant Director, Veterans' Administration Prosthetics Center, New York

Morris H. Shamos, Ph.D.
Vice President and Chief Scientific Officer, Technicon, Inc., and Professor, Physics, New York University

Richard Skalak, Ph.D.
Finch Professor, Engineering Mechanics, Columbia University

Lawrence Slote, Engr.Sc.D.
Professor and Chairman, Occupational Health and Safety, New York University

Philip D. Wilson, M.D.
Surgeon-in-Chief, Hospital for Special Surgery, Cornell University Medical Center

abdominal wall, The covering of the abdominal cavity composed of skin, fatty tissue, muscles and fibrous structures. Extends from the rib cage to the pelvis. The muscles of the abdominal wall assist in respiration and in the maintenance of posture. The workplace should not mechanically press against the abdominal wall.

abduction, Deflection of a limb away from a sagittal plane (q.v.). A basic element of motions inventory. (See adduction.)

adbuctors, Muscles which, by contraction, effect abduction of a limb. E.g., the deltoid (q.v.) abducts the arm, and the gluteus medius abducts the leg, moving the limb away from the body.

acceleration, A measure of the time rate of change of velocity. Also a vector quantity which possesses magnitude, direction, sense, and point of application, in a manner similar to velocity.

acceleration signature, A recording showing the specific pattern of acceleration and deceleration of an anatomical reference point (q.v.) in the performance of a task. The signature is an index of performance effectiveness as well as the health of the individual. (See signature, velocity signature.)

accelerometry, Quantitative measurement of acceleration and deceleration of the entire body or its members in the performance of a task. Direct measurements or double differentiation of displacement in space of an anatomical reference point are used.

acclimatization, Physiological, emotional and behavioral adjustment to changes in the environment. Proper performance depends on adequate acclimatization to the workplace including significant mechanical features, such as seat height, lighting, etc. Physiologically, heat, cold, humidity and light are important.

acidosis, Often caused by insufficient respiration or excessive muscular activity leading to high concentration of acid metabolic by-products. Work rates should be set to avoid protracted periods of acidosis which are associated with fatigue and inefficient energy use. Acidosis is also observed as a result of disease such as diabetes. (See alkalosis.)

acoustic impedance, The ratio between a pressure perturbation and the particle velocity caused by a sound. The unit of measurement is the acoustic ohm.

acoustic stimulus, Sound signal which indicates that an individual is to initiate, maintain or terminate an activity.

acromion, A bony outcrop of the shoulder blade which forms a joint with the collarbone and is attached to the deltoid and trapezius muscles. It can be felt through the skin in front of the shoulder. It is used as an anatomical reference point (q.v.) in anthropometry (q.v.).

action potential, Electric activity produced in nerve, muscle or other excitable tissue during activity. Once the threshold of stimulation is reached, the action potential is triggered and the "all or none law" (q.v.) applies. Most electrophysiological recordings are produced by conditioned signals evoked originally by action potentials, e.g., electrocardiograms, electromyograms.

activating receptors, End-organs of nerves which trigger a specific response reaction upon receipt of a stimulus. (See sensory end organs.)

adduction, Movement of a limb towards a sagittal plane (q.v.). (See abduction.)

adductors, Muscles which by contraction effect adduction of a limb. E.g., pectoralis major adducts the arm and the adductor magnus adducts the thigh.

aerobic metabolism, Physiological "combustion" of body fuels with oxygen in muscle during exertion. This is a highly efficient form of energy release available for long periods of time in light and medium work. During heavy work aerobic muscle metabolism changes to anerobic (q.v.) muscle metabolism after relatively short periods of time, and after initial, all out effort.

afferent, Conveying a stimulus or a body fluid inwardly towards a biological processing station. (See efferent.)

airway, The path that air of other respiratory gases take in going from the mouth or nostrils, to the alveoli.

airway resistance, The resistance, usually measured as a pressure drop, which must be overcome for air to flow through the airway.

alkalosis, A pathological condition in the body of excessive base or below normal acids. One occupationally important cause is hyperventilation (q.v.), which may result from anxiety, drugs or disease or from bad environmental factors such as high temperature, high concentration of solvents or too rapid or heavy a work rate. Physiological response is evidenced in respiratory rate, cardiac rate and kidney function, or it may cause fainting.

all or none law, Nerve fibers and associated muscle fibers respond entirely to a stimulus or do not respond at all. In muscle, strength of contraction is governed by the number of fibers stimulated and the frequency of stimulation; it is not determined by the individual fiber, which, once stimulated, cannot be stopped halfway in its contractile element, but is limited by its elasticity.

anaerobic metabolism, When energy demand of a task exceeds — due to inadequate supply of oxygen in a muscle — the energy available by aerobic metabolism (q.v.), then a state of "oxygen debt" exists and body fats are "burned" without the use of oxygen. This is a low efficiency process because the anaerobic combustion products must be removed from the body by a secondary aerobic process proceeding after the activity has finished. Hence, man pants after heavy exercise. Anaerobic activity should be minimized in task design. There must be a payback for any anaerobic effort.

analytic workplace design, Based on established biomechanical and behavioral concepts including the known operating characteristics of man. Produces a workplace situation well within the range of human capacity and does not generally require modification, improvement, or preliminary experimental "mock-ups."

anatomical position, Subject standing erect against a wall with feet parallel and touching, arms adducted (q.v.) and supinated (q.v.), with palm faced forward. Used as a reference posture in anatomical description.

anatomical reference point, A prominent structure or feature in anatomy which can be located and described by visual inspection or palpation of the body surface. Movements and postures are often defined by description of the relative position and displacement of various anatomical reference points with respect to each other. Also known as anatomical landmark.

anatomy, The discipline dealing with the geometrical and topographical description of the structures of the body. It includes description of the structures such as bones, muscles, blood vessels and nerves, including the shape and dimensions (gross anatomy) as well as description of the finer features of human tissues visible only through the microscope (histology). Since structure determines function, a knowledge of anatomy as the structural basis of human performance is indispensable for the proper understanding of human function. It is both static and dynamic.

anatomy of function, A subdiscipline of kinesiology (q.v.) which describes changes in the configuration of limbs during the performance of a task. Different from "functional anatomy" (q.v.) which relates to physiological function.

angle of abduction, Angle between the longitudinal axis of a limb and a sagittal plane (q.v.).

angular acceleration, The time rate of change of the angular velocity.

angular velocity, The time rate of change of the angular position of a line.

anisotropic, Indicating nonuniform directional physical properties. Body tissues such as bone are anisotropic. The strength of bone differs in different directions. Muscle also has highly directional properties.

anoxia, Lack of oxygen in the blood stream or tissue cell. Although the result of many mechanisms, the ultimate effect frequently is defective function of many sensory and motor functions. Produced environmentally by high altitude, chemically by pollutants and drugs, or locally by overexertion. May result in death.

antagonist, A muscle opposing the action of another muscle. An active antagonist is essential for control and stability of action by a prime mover (q.v.). E.g., the biceps and pronator teres are antagonists in pronation and supination.

anthropometric tables, An arrangement of tabulations and dimensional drawings stating human body measurements. Good quality tables stress data in ranges of dimensions ascribed to certain percentiles of the population rather than listing of mean values. Anthropometric tables are essential to analytical workplace design (q.v.). (See Anthropometry.)

anthropometry, The study of people in terms of their physical dimensions. It is the formal name for the technique used to express quantitatively the form of the human body. (See Anthropometry.)

anthropomorphic, Shaped like man or man's limbs. Mechanical manipulators and bionic devices often have anthropomorphic features. Cosmetic prostheses are anthropomorphic.

applied anatomy, A subdiscipline of anatomy dealing with problems involving physical (as opposed to physiological) function of body systems. The term is sometimes also used to indicate the application of anatomical principles to specific fields of human activity such as surgical anatomy, diagnostic anatomy, anatomy of work.

aponeurosis, A fibrous sheet of connective tissues which serves as attachment of muscles or muscle fibers at origin or insertion on bone; an expanded tendon. It is normally pressure resistant and is a preferred load-bearing site. E.g., the seat back should support the aponeurosis of the latissimus dorsi (q.v.). It does not have the elastic properties of the muscles attached to it.

arm-tool aggregate, The arm holding and manipulating a tool, acting with the tool as an integral biomechanical unit. In estimating physiological work, the arm-tool aggregate mass must be considered.

arthritis, Inflammation of joint structures. Such inflammation may or may not cause pain, and may result in joint destruction and/or limitation of motion. Important cause of partial occupational disability, manipulative limitations, and reduced capacity for head scanning, and increases with aging.

articulation, Commonly known as a joint, e.g., knee, elbow. Refers to the junction of bones and ligaments which allow body motion. Thus, the wrist is the articulation between forearm and hand. (See joint.)

asthenia, Lack or loss of strength and energy. When pertaining to somatotype (body build), refers to condition of very slender, feeble-appearing person. (See ectomorph.)

atrophy, Wasting away of body tissues or organs from disuse, poor nourishment, or from disease, leaving decreased mass of muscle.

available motions inventory, The nature and quality of motions which are available from man in the performance of a specific task. It may be limited by physiological or anatomical deficiencies of individuals.

axis of rotation, The true line about which angular motion takes place at any instant. Not necessarily identical with anatomical axis of symmetry of a limb nor necessarily fixed. Thus, forearm rotates about an axis which extends obliquely from lateral side of elbow to a point between the little finger and ring finger. The elbow joint has a fixed axis maintained by circular joint surfaces, but the knee has a moving axis as its cam-shaped surfaces articulate. Axis of rotation of tools should be aligned with true limb axis of rotation. Systems of predetermined motion times often specify such axes incorrectly. Almost all joints move through multicentric axes.

axis of thrust, Line along which thrust can be transmitted safely. In the forearm it coincides with the longitudinal axis of the radius. Tools should be designed to align with this axis. Ulnar or radial deviation which produces misalignment causes bending stress acting on the wrist.

basic grasp, One of five fundamental means of prehension (q.v.). All activities of grasping and manipulation involve the basic grasps or combinations of them. They are: contact grasp, instrument grasp, power grasp, tripodal grasp and wraparound grasp (q.v.).

behavioral disorders, Behavioral patterns differing from what is accepted as societal norm. They may or may not be indicative of neurotic or psychotic conditions. Behavioral disorders may affect integration of individual with group in team effort.

behavioral dynamics, The behavioral operating characteristics of individuals and groups to the extent that they are conditioned by external working environment (q.v.) and/or interaction between individuals as well as groups of individuals at work. (See sociotaxis.)

behavioral sciences, Those sciences which attempt to explain or predict the behavior of an individual or group of individuals.

benchmark, A physiological or biological reference value against which performance is compared. E.g., "normal" heartbeat is 60-80 beats per minute; "normal" walking speed in industrial work is 2 miles per hour; a "normal" electrocardiogram has an acceptable shape; "normal" acceleration signatures (q.v.) of limit movements are characteristically recognizable. Often used in rating. The effectiveness of evaluative procedures in biomechanics often depends on the intelligent selection of bench marks.

bicipital, Relating to the biceps muscle, or the groove in which its tendon lies.

bifurcation, A division into two parts or branches; the bifurcations of the arterial tree.

biocontrol system, A mechanical system controlled by biological signals; e.g., electromyographically controlled prostheses; e.g., a device alerting truck drivers who are about to fall asleep by means of a bell operated by encephalographic patterns. A BIOCONTROL SYSTEM should make use of signals of unambiguous meaning which cannot be accidentally generated.

biomechanical hypothesis, An *a priori* assumption about the operating characteristics of man in a work situation based on functional anatomy. Thus, it might be hypothesized that a task which requires ulnar deviation will limit the rotation of the wrist.

biomechanical profile, Electromyographic and biomechanical data recorded simultaneously during motion against a resistance. Included are displacement signature (q.v.), which indicates range and pattern of motion; velocity signature (q.v.), which provides an index of strength and speed of motion; and acceleration signature (q.v.), which shows the quality of motion and is an index of control over precision. Electromyogram of muscle masses involved gives an index of their sequencing and coordination. The profile permits objective evaluation of changes in functional capacity resulting from modifications of man-equipment interfaces. (See electromyography and biomechanics.)

biomechanics, The study of the human body as a system operating under two sets of laws: the laws of Newtonian mechanics and the biological laws of life.

bionic device, Man-made device resembling a biological structure in form, mode of operation and function, e.g., space suits, manipulators for radioactive devices, flap-wing aircraft.

biotaxis, Contact with living things. One of the ecological stress vectors. Includes both contact with fellow man, i.e., sociotaxis (q.v.) and with other organisms, e.g., microbiotaxis (q.v.).

bitrochal seat, Seat designed to fit bitrochanteric width.

bitrochanteric width, Anthropometric term roughly synonymous with hip breadth. This dimension is obtained by measurement with calipers applied to the greater trochanters of both femurs in standing position. Of importance in seating design.

blood pressure, The pressure exerted by blood against containing vascular elements, tissues or the heart chamber, measured in millimeters of mercury above atmospheric pressure. In biomechanics the product of systolic blood pressure and pulse rate, under nonstressful conditions, considered a good index of physiological work stress and/or the mechanical output of the heart. Together with the pulse beat, it is used as an index of fitness in work. Blood pressure is dependent on the energy of the heart action, the elasticity of the walls of the arteries, and the volume and viscosity of the blood. The maximum (systoltic) pressure occurs near the end of the strike of the heart. The minimum (diastolic) pressure occurs late in the ventricular diastole. Note: systolic is derived from the Greek *(systole)* meaning *drawing together* or *contraction* and refers to the contraction or period of contraction of the heart, especially that of the ventricle. Diastolic is from the Greek *(diastole)* meaning drawing asunder or expansion and refers to the dilation (or dilation period) of the heart, especially of the ventricles, which coincides with the interval between the second and first heart sounds.

body-load aggregate, In lifting and manual materials handling, the combined weight of the load manipulated and the body segments involved in the task.

brachialis muscle, Short, strong muscle originating at lower end of humerus (q.v.) and inserting into ulna (q.v.). Operates at mechanical advantage, powerful flexor of forearm, employed when lifting.

brightness, The amount of light emitted by, or reflected from, a surface in a specified direction. Measured as light intensity per unit projected area. Also referred to as luminance.

brightness contrast, The relationship between the brightness of an object, B_o, and the brightness of the background, B_b. It is expressed as the ratio of the difference to the background $\frac{B_o - B_b}{B_b}$ and was previously measured in foot-lamberts (q.v.).

candela (cd), The luminous intensity, in the direction of the normal, of a blackbody surface 1/100,000 m^2 in area, at the temperature of solidification of platinum (2042 K) under a pressure of 101,325 N/m^2. (See candle.)

candle, (new unit), 1/60 of the intensity of 1 cm^2 (100 mm^2) of a blackbody radiator at the temperature of solidification of platinum (2042 K). To understand this, a brief history may be in order: the first light standard was the STANDARD CANDLE, that was a candle of spermacel of which six weighed one pound with the candle burning at a rate of 6 grains per hour. Since this spermacel candle was not of sufficiently constant illuminating power to be of much scientific value, other standards were used.

In the USA, the HARCOURT PENTANE LAMP, in which air was drawn over the pentane liquid, with the mixture burned in a standard burner and the flame adjusted to a definite height, with corrections made when atmospheric conditions varied from normal, became the standard. The illuminating power of such a lamp was equal to about 10 standard candles and was accepted as an INTERNATIONAL STANDARD (defined as illuminating power equal to 1/10 of that of the Harcourt Lamp).

In Germany, the HEFNER LAMP (constructed after a standard pattern and burning amyl acetate) was the standard and equal to 0.9 International Candles.

In France, the CARCEL LAMP in which colza oil was

burned was the standard, and the Carcel Standard was 9.62 International Candles.

For practical purposes, to measure the "quantity of light," the INTENSITY OF ILLUMINATION produced by a standard candle at a distance of 1 m was adopted and called a candle-meter. Analagously, there was the foot candle, with 1 candle-meter = 0.093 foot candles.

candlepower, Light intensity of a light source. Also "candle" and a "lumen/steradian." A source of one candle intensity at the center of a sphere emits a total radian flux of 12.57 (= 4π) lumens. The total surface of the sphere is 12.57 steradians; thus one candle provides an intensity of 1 lumen/steradian.

capitulum of humerus, A smooth hemispherical protuberance at the distal end of the humerus (q.v.) articulating with the head of the radius. Irritation caused by pressure between the capitulum and head of the radius is called tennis elbow. (See radiohumeral joint.)

carbon dioxide, The gaseous product of oxidation or aerobic metabolism (q.v.) CO_2 (chemical symbol) is transported in the plasma of venous blood to the lungs for discharge. Its partial pressure, p CO^2, controls many body functions including respiratory rate and cardiac rate.

cardiac rate, Heart beats per minute (BPM). The heart can change its output by increasing its stroke volume or its cardiac rate, where the latter is the most effective and most efficient method. Heart rate varies widely with age, sex, environmental stress, state of health, etc. Therefore, the term "normal" heart rate is meaningless. However, ratio of working cardiac rate over resting cardiac rate is one of the measures of work tolerance and work stress.

carpal tunnel, A passage in the wrist through which important blood vessels and nerves pass to the hand from the forearm. Ulnar or radial deviation cause misalignment of the carpal tunnel and irritation of structures passing through it. (See carpal tunnel syndrome.)

carpal tunnel syndrome, A common affliction of assembly workers caused by compression of the median nerve in the carpal tunnel. Often associated with tingling, pain or numbness in the thumb and first three fingers. Reduces manipulative skills, particularly if thumb is involved, and frequently reduces work output.

cartilaginous plates, Flattened masses of cartilage which act as supporting structures. Intervertebral discs (q.v.) are cartilaginous plates.

cerebellum, The "small brain" located below the cerebral hemispheres. It exerts a regulatory influence of muscular activity and is concerned with balance and posture. It is a subdivision of the mentencephalen, which grows upward and outward from the axis of the brain stem to hide the more inferior portions not covered by the cerebral hemispheres.

cerebrum, The antero-superior part of the brain which governs voluntarily coordinated activities of the body, e.g., sensory-motor functions and associative activities. It is the main portion of the brain, occupying the upper part of the cranial cavity. Its two heispheres, united by the corpus callosum (also known as commisura magna cerebri), form the largest part of the human nervous system.

chemoreceptor, An end organ of the nervous system which triggers neural signaling activity in response to chemical stimuli, e.g., taste buds, olfactory end organs.

chemotaxis, Relating to contact with a chemical environment. A primary consideration of environmental health, e.g., air pollution. One of the ecological stress vectors.

chronocyclegram, The photographic record made by a chronocyclegraph, (q.v.).

chronocyclegraph, An instrument for measuring the pathway taken by a body reference point or an object during the performance of a task as part of the work cycle. It consists of a flickering light attached to an anatomical reference point (q.v.) and photographic recording of the moving light. The chronocyclegraph enables an investigator to determine exact speed, relative speeds and directions as well as changes in motion such as may be caused by fatigue.

clavicle, The collarbone. The sole bony connection between the arm and the trunk. It is easily fractured in falls broken by an extended arm.

conditioned reflex, Patterned reflex response to an external stimulus developed as a result of either habituation or training. Easier to learn than to unlearn, e.g., stop on red, go on green. (See simple reflex.)

conductive deafness, Impaired hearing resulting from interference with, or injury to, the mechanisms for the conduction of sound waves into the inner ear. This transmission takes place either through the external auditory canal and the structures of the middle ear (air conduction) or through vibrations applied to the bones of the skull (bone conduction). This type of hearing loss can be contrasted with Boilermaker's Deafness which is due to perceptive deafness (q.v.).

cones, End organs within the retina (q.v.) involved in perception of color.

contractile tissue, Tissue which is capable of shortening in response to physiological or environmental stimuli. E.g., muscle.

contact grasp, Action of the hand when pushing, e.g., a coin over a flat surface. One of the five basic grasps of the hand. Not a grasp in the literal sense. The index finger is generally used. Overuse may contribute to high stress concentration on base of distal phalanx and physical damage.

coronal plane, Any plane which divides the body into anterior and posterior portions. It is perpendicular to the sigittal and transverse planes (q.v.) and is also known as the frontal plane.

coronoid fossa, A depression in the lower end of the front of the humerus in which the coronoid process of the ulna lies when the arm is bent.

correlative kinesiology, That branch of kinesiology (q.v.) which quantitatively relates myoelectric activity (q.v.) with resultant movement. The basis of rational workplace design (q.v.) for minimal fatigue.

covert lifting task, An operation which may not involve the handling of a load but which exhibits all of the biomechanical characteristics ascribed to an overt lifting task (q.v.). Its principal characteristic resides in the existence of a bending moment acting on the vertebral column for reasons other than load lifting, e.g., excessive heel height may cause postural changes which impose bending moments on the lumbar spine. Side-stepping may impose a lateral bending moment. Reaching forward constitutes a covert lifting task because torque of the arm at the shoulder joint is transferred to the vertebral column. (See lifting task.)

cumulative pathogenesis, The development of trauma or work strain from repeated or continuous application of work stress which for short periods of time or single applications would not be harmful. An important factor to consider in workplace and tool design.

deceleration, Rate of slowing down a motion. A negative acceleration. Very significant in biomechanics in the control of end points of a motion. Adequate coordination and voluntary control requires accurate deceleration. Many drugs (e.g., muscle relaxants) interfere with deceleration control and thus destroy coordination.

deltoid, The muscle of the shoulder, responsible for extending the arm sideways, and for swinging the arm at the shoulder. Overuse of the deltoid muscle may cause fatigue or pain in the shoulder and unwarranted fear of heart disease.

demanded motions inventory, The nature and quality of motions which are required to perform a specific task. The demanded motions inventory may be limited by constraints imposed by the working environment including machinery and tools. Also used to denote the nature and sequence of motions demanded by a specific task design. (See available motion inventory.)

depth perception, The ability to judge the distance to objects and spatial relationships between objects located at different distances. Depends on eye coordination and the use of parallactic observation. Depth is sensed by comparison of images from each eye and by muscle tension which develops as eyes converge on object. Increased background lighting, rather than object brightness, improves depth perception. (See visual acuity.)

diaphysis, The central portion or shaft of a long bone.

distal, Away from the central axis of the body; opposite of proximal.

Doppler effect, An apparent signal frequency change caused by relative motion of transmitter and observer. Characteristically, a light color change or sound pitch change. Sometimes used in work measurement to investigate motion patterns. Current devices are not always reliable in the response ranges of interest.

dynamic moment, A force-distance relationship dealing with dynamic (resulting from inertia and acceleration) forces as opposed to static (resulting from gravitational accelerations only) forces. An example of dynamic moment is that experienced during a lifting task. At the instant of lift-off (load moves off platform), the acceleration of the load is caused by the dynamic moment on the lumbo-sacral joint. The consequences of this phenomenon are not always clearly predictable.

dynamic work, Muscle work performed when one end of a muscle moves with respect to the other end and external movement is produced. (See isometric work.)

dynamometer, Apparatus for measuring force or work output external to a subject. Often used to compare external output with associated physiological phenomena (electromyography, spirometry, etc.) to assess physiological work efficiency.

echography, The use of ultrasonic energy for imaging internal organs and tissues. Same as sonography. (See ultrasonics.)

ecological stress vector, Any vector in the environment producing a physiological response in man. Such response may or may not be pathological. Major stress vectors are: climatotaxis, biotaxis, mechanotaxis, chemotaxis, particulotaxis (q.v.).

ectomorph, One of the somatotypes (q.v.). Having a fragile, thin linear body build with poorly developed musculature. (See endomorph, mesomorph.) *Syn:* hyposthenic.

effective temperature, A measure of "warmth as related to comfort". Combines environmental factors, himidity, true temperature, and air velocity to provide a scale of subjective comfort. Two experimental scales of the American Society of Heating and Ventilating Engineers pertain to normally-dressed and partially-dressed people. A comfortable effective temperature is significant in maintaining high productivity and good quality control.

efferent, Conveying away from a biological processing station, specifically by conveying nervous impulses from a neural center to a muscle. (See afferent.)

elastic limit of tissues, The level of physical deformation caused by the application of a force beyond which tissue damage occurs and beyond which the tissue will not return to its original function when the force is removed.

electrocardiograph, A recording galvanometer which produces an analog representation of the elctrical activity of the conductive tissues of the heart, i.e., an electrocardiogram (ECG or EKG). Useful in the diagnosis of heart disease and assessment of work stress in health and disease.

electrode, A device used to sense the electric potential of body tissues or to transmit such a potential. It can be surface-mounted as a skin electrode or internally mounted as a needle electrode. An effective electrode minimizes contact resistance and junction potentials. (Used in myography, cardiography, and encephalography.)

electromyographic kinesiology, The analysis and evaluation of human motion patterns by means of electromyography. Useful in establishing optimal motion patterns through workplace and tool design.

electromyography, The technique of recording the electrical potential generated by muscle activity. Surface myography will provide a measure of muscular sequence and physiological effort associated with a particular task.

electrophysiological apparatus, Research and evaluation devices which sense bioelectrical signals from heart, muscles, brain, etc. The sensed signal is amplified and displayed in the form of an analogue or digital readout. Examples: electromyography, electrocardiography, polygraph, (q.v.).

electro-silence, Absence of a measurable action potential in a biological structure. A zero recorder pen deflection should ideally occur during electromyography of a noncontracting muscle. In practice, true electro-silence cannot be observed unless random noise signals are filtered out.

empirical workplace design, Design based on operating characteristics and mechanical needs of equipment or on previous experience with similar work situations. Must normally proceed through a number of "improvements" to adapt to the needs of the worker.

endochthon, Innate, originating from within. In hand-operation of a lever, the power is endochthonous to the operator. (See exochton.)

endomorph, One of the somatotypes. Stocky, soft and round in body build with a tendency towards obesity. (See ectomorph, mesomorph.) *Syn:* hypersthenic.

environmental engineering, Application of engineering principles to create and maintain surroundings which are favorable for the sustenance of life.

environmental inputs, Often used as synonym for ecological stress vector (q.v.). More accurately, the following factors to which the worker responds either physiologically or behaviorally; economic, social, managerial, mechanical, climatic, psychological. The resulting outputs are economical (in the form of production), emotional (in the form of employee behavior), and physiological (in the form of employee health and fatigue).

environmental stress vector, Synonym for ecological stress vector (q.v.).

epicondyle, A raised area (bump) on the condyle of a bone, from which muscles originate. E.g., the flexor muscles of the fingers originate from the medial epicondyle of the humerus.

epicondylitis, Inflammation or infection in the general area of an epicondyle, e.g., tennis elbow.

epiphysis, The ends of a long bone. Joint surfaces are always part of the epiphysis.

ergonomics, The application of a body of knowledge (life sciences, physical sciences, engineering, etc.) dealing with the interactions between man and his total working environment, such as atmosphere, heat, light and sound, as well as all tools and equipment of the workplace.

ergonomic analysis, Application of the principles of ergonomics to study in detail all of the specific elements which are pertinent to a man-equipment or man-task interface.

etiology, The study of causes of disease. In biomechanics, etiology of trauma or work strain is complicated because the site of such trauma is often remote from the locus of work stress. Industrial hygiene experience is a valuable aid to determining causes of workplace-generated trauma.

exochthon, Outside of and not inherent to a body. Operation of pneumatic or electrical tools requires management by the operator of power which is exochthonous to his system, i.e., not generated by him. (See endochthon.)

extensor muscle, A muscle which when active increases the angle between limb segments, e.g. the muscles which straighten the knee or elbow, open the hand or straighten the back.

extensor retinaculum, Also known as transverse dorsal ligament. A membranous band of fibers at the back of the hand which forms a tunnel through which the extensor tendons of the fingers pass. The retinaculum acts as a guide and prevents the tendons from bowstringing when the wrist is hyperextended.

extensor tendon, Connecting structure between an extensor muscle and the bone into which it inserts. Examples are the hard, longitudinal tendons found on the back of the hand when the fingers are fully extended. (See tendon, extensor muscle, insertion.)

external mechanical environment, The man-made physical environment, e.g., equipment, tools, machine controls, clothing. *Ant:* internal (bio) mechanical environment (q.v.).

external working environment, The environment external to the body at the workplace. Of particular importance in ergonomics because of the effect of surroundings on work performance. It includes immediate environment such as tool handles, clothing, temperature, humidity, pressure, and composition of atmosphere. (See Gilbrethian variables, internal environment.)

extrinsic, Anatomical term referring to a component, usually a muscle, that originates outside of the structure on which it acts, e.g. relating to vision, the extrinsic muscles of the eye are located in the orbit and move the eyeballs. The intrinsic muscles (q.v.) are located within the eyeball and operate the iris and lens. *Ant:* intrinsic.

eye scanning, Binocular scanning of the visual field by movement of the eyeballs alone and without head rotation through use of the extrinsic muscles of the eye. Eye scanning is a rapid and easy process, but can be effectively performed only with a central visual cone of 60 degrees or loss of depth perception will result. (See head scanning.)

fatigue, That *physiological* state characterized by a lessened capacity for work and/or the inability to maintain consistent work quality levels. May be due to intensive mental or physical activity or depletion of physiological reserves.

feedback system, A control system in which information from the controlled element is returned to the controlling element. The controlling element then modulates its requirements of the controlled element according to performance requirements. In biomechanics, muscle coordination by visual or tactile signals are examples of feedback systems. Improper performance by a worker may be caused by interference in his feedback system (disease) or improper operation of his sensors (optical illusion). Homeostatic systems are examples of physiological feedback systems. (See homeostasis.)

femur, The long bone of the upper leg (thigh). The upper end (head) of the femur articulates with the pelvis to form the hip joint, the lower end (condyles) articulates with the tibia to form the knee joint. The hip and knee joints often are the site of arthritis (q.v.), which may make walking painful and limited (elderly workers).

fibula, The smaller bone of the lower leg (shank). Extending from just below the knee to the ankle joint, it is lateral and posterior to the tibia (q.v.), which is the main load-bearing bone. The fibula is attached to the tibia by the interosseus membrane and provides a site for muscle attachment. The lower end (lateral malleolus) forms the lateral side of the ankle joint itself, is essential to ankle stability, and often is injured by twisting forces.

fifth lumbar vertebra, The lowest vertebral element in the lumbar region, in the small of the back immediately above the sacrum. Because the flexible lumbar section is adjacent to the rigid sacral and pelvic structure, the fifth lumbar vertebra and adjacent discs are subjected to high stress, especially in lifting tasks. Trauma due to excessive stress at this site can cause lower back pain and pain in the legs. (See lumbosacral angle and herniated disc.)

fixed linkage mechanism, Linkage formed between skeletal elements of man and machine in a man-machine system. Movement of any link produces movement in other links, e.g., machine operator and drill press. Configuration of fixed linkage systems determines biomechanical efficiency of worker, e.g., seat height at drill press.

flexion, Movement at a joint whereby the angle between two bones connected at a joint is changed in a manner tending to approximate the body segments it connects. Motion of this type is produced by contraction of flexor muscles.

flexor muscle, A muscle which, when contracting, changes the angle between two limb or body segments in a manner tending to approximate the segments it connects. The principal flexor of the elbow is the brachialis muscle. Flexors of the fingers and the wrist are the large muscles of the forearm originating at the elbow. (See extensor muscle.)

flexor retinaculum, Tough fibrous ligament forming the ceiling of the carpal tunnel (q.v.). Also called transverse carpal ligament, this structure prevents bowstringing of the flexor tendons at the wrist.

flicker fusion, The tendency of an oscillating or flickering sensory input signal to be observed as continuous. At a certain threshold frequency (typically above 6 Hz for vision), the perceiving apparatus, (e.g., eye, ear, tactile or other sensory areas) is no longer able to separate the individual pulses and instead perceives one continuous signal. This effect makes motion pictures seem to run smoothly. Flicker fusion caused, for instance, by low frequency fans or objects on fast conveyor belts passing before a light source, may lead to errors such as seeing a moving saw blade apparently at rest, and leading to accidents. (See stroboscope.)

foot candle, An obsolete measure of the intensity of illumination of a surface, or, the flux density of light incident on a surface. A flux of 1 lumen per square foot of surface is defined as 1 foot candle of illumination. It is no measure of the light reflected from a surface and, therefore, no "index of visibility" of an object to the worker. (See brightness contrast, visual acuity, reflectance, foot-lambert).

foot-lambert, An obsolete measure of the illumination flux density reflected from a surface. Used in measurement of the effectiveness of workplace lighting, the foot-lambert is the only accurate index of light actually perceived by the eye: equals $1/\pi$ candles per square foot. (See reflectance.)

foot-pounds of torque, A popular measure of torque. (See torque.)

force platform (plate), See reactance platform.

force-time, Applied force multiplied by time of application. Used to quantify isometric work (q.v.) and has the dimensions of linear impulse, for short time duration phenomenon.

functional anatomy, Study of the body and its component parts, taking into account those structural features related directly to physiological function. E.g. functional anatomy of elbow defines relative flexor force developed by biceps and brachialis at various degrees of flexion.

frequency, The number of repeated occurrences in a given time interval. The frequency of electrical or mechanical activity is usually specified as cycles per second (c.p.s.) or hertz (Hz.).

gait analysis, Study of human kinetics involved in walking or any other form of ambulation. May utilize videotape or stop-motion photography, electromyography (q.v.), force platform (q.v.), foot contact switches, goniometers (q.v.) or accelerometers (q.v.). Used in evaluation and design of ramps, lower limb prostheses, and load-carrying situations, and in defining neurological or structural abnormalities of gait. Factors examined include muscular activity and coordination, ground reaction forces, joint and limb segment positions, forces and accelerations, and postural changes.

gangrene, A condition in which a localized area of body tissue dies and decays. Gangrene occurs in many forms, but primarily refers to death or necrosis of tissue, as caused by loss of blood flow to a specific area (e.g. arterial injury or occlusion, frostbite, etc.), infection by bacteria (e.g., gas gangrene associated with traumatic wounds), or both. Diabetic workers are more susceptible to gangrene than others due to circulatory insufficiency associated with the disease. Also, paraplegic workers are subject to decubitus ulcers over the buttocks associated with lack of sensation which predisposes to unrelieved sitting pressures occluding capillary circulation long enough to cause tissue death.

gestalt vision, Visual perception and retention in memory of an object as a geometric shape, e.g., an erect octagon means "stop" in international traffic markers. Interference with gestalt vision, such as glare which obscures shape, leads to errors.

Gilbrethian "systems concept", The original notion calling for matching of the internal physiological and biomechanical environment of the worker with his external physiological and mechanical environment. First postulated by Frank B. and Lillian M. Gilbreth in 1911 by their setting forth comprehensive lists of interacting variables affecting the worker, his environment, and his motions requirements. (See Gilbrethian variables.)

Gilbrethian variables, System of three sets of variables intrinsic to every task. They are variables of the worker relating to anatomic and psychological factors: variables of environment including physical and economic factors: and variables of motion including effort and movement patterns used as a basic tool in work system analysis and design.

glenoid cavity, Glenoid cavity is used commonly with reference to the socket of the shoulder joint formed by the scapula (q.v.) to receive the humerus (q.v.). *Syn:* joint socket.

gluteus maximus, Large muscle of the rump which acts across the hip joint and extends the thigh, i.e., brings the thigh from the position occupied in sitting to the position in standing. It is an important muscle for walking, lifting and standing. Electromyogram of the gluteus maximus is a good index of the relative muscle activity during lifting.

goniometer, An instrument by which angles of joint flexion are measured, either statically or during movement of the joint. Important in the evaluation of industrial workplace layouts and in the analysis of the body in motion. (See reflex goniometry.)

grasp reflex, A reflex which is both a basic and conditioned reflex. It is initiated by stimulation of the median nerve (q.v.)

sensory feedback area in the hand. Surgeons need a well-conditioned grasp reflex so that when surgical tools are being transferred, the surgeon can concentrate on the patient. However, in the machine shop people have lost fingers by reflexively grasping moving saw blades, etc. (See prehensile.)

grooving, Designing a tool with grooves on the handle to accommodate the fingers of the user. A bad practice because of the great variation in the size of workers' hands. Grooving interferes with sensory feedback. Intense pain may be caused by grooving to the arthritic hand.

hamate bone, One of the wrist bones which possesses a hook projecting toward the palm of the hand on the side opposite the thumb. This hook provides one attachment of the covering of the carpal tunnel (q.v.) flexor retinaculum (q.v.) through which the flexor tendons of the fingers pass; the hook helps hold these tendons in place. (See palmar arch, ulnar nerve.)

hamstrings, Muscles located at the back of the thigh which are employed in bending the knee and erecting the trunk. They are important in lifting tasks where bent knees are employed and in tasks requiring operation of controls with the leg.

head of radius, Proximal end of radius bone (q.v.) which is in contact with the humerus (q.v.) at the elbow joint. It provides the primary rotational element of the elbow joint is biomechanically significant in the operation of many hand tools.

head scanning, Scanning of the visual field through head movement combined with eye movement. Head scanning is normally a slow process which, when performed too frequently or rapidly, may induce giddiness and lead to errors in visual judgment. Good workplace design minimizes the need for head scanning. (See eye scanning.)

heat acclimatization, Physiological adjustment to temperature changes of the external working environment (q.v.). Such adjustments include change in heart rate, respiratory rate, transpiration and perspiration and peripheral circulation. The effect of these changes is to maintain a constant core-temperature and other physiologic parameters essential to the well-being and efficient work performance of the individual.

heat stroke, Illness caused by exposure to excessively high temperatures resulting in overload and breakdown of the physiological control system permitting acclimatization. Symptoms are headache, dizziness, confusion, hot dry skin, collapse and coma, high fever, and increased pulse rate. Heat stroke may occur where work space is not properly cooled and ventilated and especially where work situations involve intense heat, i.e., steel mills, roasters, etc. Prolonged and repeated exposure can cause serious brain damage and/or death. Clothing (such as impermeable synthetics) which retard adequate heat dissipation can cause heat stroke. In other cases, symptoms of prostration similar to shock with collapse, cool, damp skin and subnormal temperature may occur.

hernia, Also known as rupture. Failure or weakness in the wall which contains an organ or other structure or the protrusion of the organ through the area of weakness. Hernia of the groin, in which intestines may protrude through a mechanically weak area in the abdominal wall, is industrially significant because of the work limitations which it imposes. Good workplace layout and careful attention to posture will minimize restrictions imposed on the worker.

herniated disc, Also known as "slipped disc", this term commonly refers to herniation of the nucleus pulposis or protrusion of the interior soft part of the intervertebral disc through the outer, fibrous containing layer which fails following degeneration or excessive stress. The condition may cause pressure on spinal nerves, with resulting low back and sciatic pain down the leg, disability and often partial paralysis. Low back trauma and/or overloading has a causal relationship.

hinge joint, A single degree of freedom joint such as that encountered at the elbow (humero-ulnar joint) where motion in only one plane is possible. The elbow, as a whole, has two degrees of freedom due to the humero-radial and radio-ulnar joints which permit rotation.

homeostasis, The state of equilibrium or constancy of physiological conditions in the living body, e.g., chemical equilibrium, fluid content, blood pressure. Also used to describe the physiological mechanism by which equilibrium is maintained in the body, e.g., regulation of blood pressure, blood chemical factors, body temperature, etc.

homologous motion, A motion produced by one set of muscles which subsitutes for a motion which could be produced by another set of muscles. In workplace design, the muscles required for a particular task can be chosen by the workplace layout. Thus, a lateral humeral rotation (q.v.) or its homologous motion, a shoulder swing, is exercised depending on seat height or table height. Homologous motions which have the same output may produce substantially different work stress.

human engineering, Application of the results of experimental psychology to the design of work situations. The objective is to optimize human performance in the monitoring and operation of tools and equipment. Sometimes referred to as Human Factors Engineering. (See ergonomics; also Human Factors.)

humeral rotation, Rotation of the arm about the long axis of the humerus (q.v.) as contrasted with circumduction or shoulder swing. Humeral rotation with forearm flexed produces wrist displacement in the plane of the forearm, probably with less physiological work than would be required to achieve the same position by shoulder abduction.

humerus, The bone of the upper arm which starts at the shoulder joint, head of humerus articulates with glenoid cavity, q.v.) and ends in the humeral condyles within the elbow joint. Muscles which move the upper arm, forearm and hand are attached to this bone.

hyperventilation, "Over breathing" (a state in which there is an increased amount of air entering the pulmonary alveoli). Prolonged, rapid and/or deep breathing which leads to excessive expulsion of residual CO_2 from the blood via the lungs. Conducive to alkalosis (q.v.). Often produced by excessive physical work load and/or heat stress or apprehension.

hypoxia, Synonym anoxia, condition in which there is an oxygen deficiency in organs and tissues, i.e., less than normal amount. Mechanisms which cause hypoxia include insufficient partial pressure of oxygen in air, (i.e., high-altitude) insufficient pulmonary ventilation, or exposure to toxic gases (i.e., carbon monoxide) or chemicals. Impeded circulation, caused for example, by pressure of a tool on the hand, can reduce oxygen delivery to tissues resulting in a local area of hypoxia.

idiopathic, Refers to a disease of unknown cause, especially one arising in the body without an identifiable extrinsic cause. Diseases caused by toxic industrial agents tended to be considered idiopathic until causes were found.

iliac crest, The upper rounded border of the lateral aspect of the pelvic bone above the hip joint. No muscles cross the iliac crest and it lies immediately below the skin. It is an important anatomical reference point (q.v.) because it can be felt through the skin. Seat backrests should clear the iliac crest.

impairment, A dysfunction in man in which a body part (or parts) is incapable of performance within established and accepted standards of "normal" performance.

improvement approach to workplace design, See workplace layout. *Syn:* empirical workplace design.

index of work tolerance, Any measure indicative of the length of time during which an individual can perform a specific task with necessary efficiency while at the same time experiencing desirable levels of physiological and emotional well-being.

industrial hygiene, That science and art devoted to the recognition, evaluation and control of those environmental factors or stresses, arising in or from the workplace, which may cause sickness, impaired health and well-being, or significant discomfort and inefficiency among workers. (Adapted from *Fundamentals of Industrial Hygiene,* National Safety Council.)

infrared photography, A technique whereby only the infrared radiation of the spectrum is photographed by using infrared sensitive film and infrared transmitting filters. Infrared radiation varies with the temperature of the radiator. Therefore, infrared photography is used to indicate local temperature variations in human subjects which can be caused by local inflammation, ischemia, or peripheral vasoconstriction. Useful in evaluating effects of hand tools and hand-operated machinery on soft tissues. (See thermograph.)

input variables of external environment, Six motivational vectors in human performance; economic, social, managerial, biomechanical, climatic, and behavioral.

insertion, The anatomic point of attachment of a muscle to a bone which moves when the muscle contracts. The insertion is usually via a tendon or aponeurosis (q.v.). Insertion is at distal (q.v.) end of muscle. (See origin.)

integrated surface myogram, Recording of the level of electrical activity (firing rate) associated with muscle contraction. Sensed by electrodes mounted on the skin and bracketing the belly of the muscle under consideration. The ISM represents a summation of all the number of potential changes generated at any instant by the contraction of the muscle mass. It is proportionate to the number of fibers contracting at any time. It indicates coordination, sequencing and level of muscular activity involved in specific body maneuvers. Important index of work stress, status of training and aptitude for the performance of a task.

internal (bio)mechanical environment, The muscles, bones and tissues of the body, all of which are subject to the same Newtonian force as external objects in their interaction with other bodies and natural forces. When designing for the body one must consider the forces that the internal mechanical environment must withstand.

interphalangeal joints, The finger or the toe joints. The thumb has one interphalangeal joint, the fingers have two interphalangeal joints each.

intervertebral discs, Fibro-cartilaginous pads which separate the bodies of the vertebrae. Bending of the spine requires deformation of the discs which are susceptible to injury during incorrect lifting.

intrinsic muscle, Muscles located within an anatomical entity and contributing to its function, e.g., intrinsic muscles of the hand used for the fine manipulation, as opposed to extrinsic muscles which lie in the forearm but move fingers and wrist. Intrinsic muscles are smaller and weaker and fatigue earlier than extrinsic muscles.

ischial tuberosity, A rounded projection of the ischium (q.v.). It is a point of attachment for several muscles involved in moving the femur (q.v.) and the knee. It can be affected by improper design of chairs and by situations involving trauma to the pelvic region. When seated, pressure is borne at the site of the ischial tuberosities. Chair design should provide support to the pressure projection of the ischial tuberosity through the skin of the buttocks.

ischemia, Lack of blood flow. Loss of sufficient replacements to maintain normal metabolism (q.v.) in the cells. Caused by blockage in the circulatory system or failure of the cardiac system. Blockage may be by internal biological agents such as arterial wall deposits or by external environmental agents such as poorly designed tools or workplace which press against arteries and occlude them. Depending on the degree of ischemia, numbness, fatigue and tingling may be evidenced in the limbs. At the workplace, loss of precision in manipulation may lead to reduced efficiency, poor quality and possibility of accidents.

ischium, The lowest component bone of the hip complex. Important biomechanically as the main support of the body in seating. (See ischial tuberosity.)

isometric work, Referring to a state of muscular contraction without movement. Although no work in the "physics" sense is done, physiologic work (energy utilization and heat production) occurs. In isometric exercise, muscles are tightened against immovable objects. In work measurement isometric muscular contractions must be considered as a major factor of task severity. (See dynamic work, negative work, work.)

joint, Vernacular term for articulation (q.v.) between two bones which may permit motion in one or more planes. They may become the sites for work-induced trauma (see tennis elbow, etc.) or other disorders (e.g., arthritis).

kinematic chain, A combination of body segments connected by joints (q.v.) which, operating together, provide a wide range of motion for the distal (q.v.) element. A single joint only allows rotation, but kinematic chains by combining joints, enable translatory motion to result from the rotary motions of the limb segments. Familiarity with the separate rotary motions and their limitations is necessary for comprehension of the characteristics of the resultant motion. By combining joints whose axes are not parallel, the kinematic chain enables a person to reach every point within his span of reach (q.v.).

kinesiology, The study of human movement in terms of anatomy of function.

kinesiometer, A device which measures parameters of motion such as displacement, velocity and acceleration, either as vector sums, or resolved into component parts. Used to evaluate the effect of changes in the workplace such as chair height, lighting, task elements, drugs, antihistamines, etc., on the motions patterns of the worker.

knee switch, A control device which is operated by lateral movement of the knee when seated, requiring operation of the abductor muscles of the leg. In the seated position the leg cannot rotate but the abductors operate with strength, precision and control. Knee switches are used to control sewing machines, etc.

lactic acid, A product of anaerobic metabolism (q.v.) of muscle which is cleared from the muscle tissues after the activity which produced it has ceased. Presence of lactic acid causes fatigue and muscular pain and contributes to condition of acidosis (q.v.). Lactic acid is removed from the body by a subsequent aerobic metabolism (q.v.). Work schedules should provide periods of recovery following heavy exertion.

lambert surface, An illuminated reflecting surface whose brightness appears equal at any angle of observation. A theoretical ideal for a workplace surface.

lateral displacement, Movement of a limb or body segment away from the mid-sagittal plane (q.v.). Movement of legs (side-stepping) or arms sideways are examples of lateral displacement. (See medial displacement.)

lateral transfer, Personnel reassignment in a company organization to an equivalent position in a collateral subdivision. When lateral transfers are actively sought by employees, organizational or environmental problems should be suspected. Minor discomforts such as sore backs or elbows may be the cause of requests for lateral transfers.

latissimus dorsi, A large flat muscle of the back which originates from the spine of the lower back and inserts into the humerus (q.v.) at the armpit. It adducts the upper arm, and when the elbow is abducted, it rotates the arm medially. It is actively used in operating equipment such as the drill press where a downward pull by the arm is required.

lesion, A wound, injury, or unnatural change in tissue texture. Describes local manifestation of disease whether traumatic or not.

lifting task, Any industrial task which applies a moment to the vertebral column. It includes one or more of the following elements: sagittal bending moment, lateral bending moment, torsional moment, inertial moment, isometric component, negative component, dynamic component, and frequency of lift.

lifting torque, The most reliable measure of lifting stress. The product of load and distance from a fulcrum within the vertebral column created by a lifting task. This torque must be balanced by an opposing one in the musculoskeletal system.

ligament, A tough fibrous band or loop of tissue which connects bones and supports body tissues. The geometry of joint movement depends on the arrangement of the ligaments. The biomechanist must be familiar with arrangement of ligaments in designing workplace and tasks for the industrial worker.

light flux, Rate at which a source emits light energy (evaluated in terms of visual effect) and expressed in lumens (lm) (q.v.).

light intensity, See foot-lambert, candle power.

light task, Any job which requires a physiological output rate of from 0.01 to 0.025 horsepower. This range assumes that general working conditions are sufficient to maintain an 18 percent to 20 percent rate of worker efficiency (the metabolic conversion of fuel into useful energy). This definition is applicable only when large muscle groups are used.

limb-load aggregate, The total load lifted to be considered in biomechanical task analysis. Where the load is small, the physiological effort is primarily dependent on limb weight. Such an effort is a covert lifting task (q.v.). (See arm-tool aggregate.)

linkage, See fixed linkage mechanism.

locomotion, Movement of the body or limbs or an anatomical reference point (q.v.) from one place to another.

locomotor system, Anatomic structure used in locomotion (q.v.) consisting of musculoskeletal system (q.v.) interacting with nervous system. The performance of any motor task involves the use of the locomotor system.

lordosis, Curvature in the sagittal plane (q.v.) of the cervical and lumbar regions of the spine. Normal lordosis is a prerequisite of safe materials handling. Lordotic configuration changes when stress equivalent (q.v.) of loads handled exceeds safe limits. Bone disease may also affect lordosis. In the industrial environment lordotic changes caused by improper heel heights may cause discomfort and inefficiency in workers, especially women.

lordotactic, Mechanically leaning against the lumbar spine.

lumbar spine, Lowest section of the spinal column or vertebral column immediately above the sacrum (q.v.). Located in the small of the back and consisting of five large lumbar vertebrae (q.v.), it is a highly stressed area in work situations and in supporting the body structure.

lumbar vertebrae, The five vertebrae located in the lumbar spine. Largest in cross section and strongest of the vertebrae.

lumbosacral angle, Angle between the back of the lumbar spine and the sacrum. Correct angle is essential to postural integrity and safety and affects walking speed. Frequently distorted by improper shoe heel height and poorly designed backs of working chairs.

lumbosacral joint, The joint between the fifth lumbar vertebra and the sacrum. Often the site of spinal trauma because of large moments imposed by lifting tasks.

lumen, The unit of luminous flux or power emitted by a source of illumination. (See candlepower, luminous flux.) More precisely, the LUMEN (lm) is equal to the luminous flux (q.v.) through a unit solid angle (steradian) from a point source of an intensity of one candela. Again it is the flux on a unit surface, all points of which are at a unit distance from a point source of one candela.

lumen of the transverse canal, Hole in the vertebral column formed by the junction of semicircular notches of successive vertebrae thru which nerves emerge. Injury from the stress of poor lifting technique, from mechanical impact or from biological changes of aging may cause compression of the nerve, thereby causing partial or total disability, paralysis and pain.

luminous flux, The total visible energy emitted by a source per unit of time is called the "total luminous flux from the source." The unit of flux, the LUMEN (lm) is the flux emitted in a unit solid angle (steradian) by a point source of one candela luminous intensity. A uniform point source of one candela intensity thus emits 4π lumens.

man-environment interface, Region of contact of man with his environment in the performance of a work task. It can include areas of exposure between the worker and the parameters which affect output performance levels such as climate, temperature, and illumination.

man-equipment interface, Areas of physical or perceptual contact between man and equipment. The design characteristics of the man-equipment interface determine the transfer of information and motor skill. Poorly designed interfaces with the work place, machines, tools, fixtures and the workpiece may lead to localized trauma (e.g., calluses) or fatigue.

man-equipment-task system, System where the motor skills of man are matched to those demanded by equipment in such a manner as to facilitate the effective performance of a task. Input-output relations in these systems (i.e., efficiency) are specific and are functions of task design.

manipulation, To operate through the use of the hands. Most industrial tasks involve motor skills of manipulation.

manipulative skill, The ability to move objects with reference to location in space, speed of movement and control of movement. Manipulative skill depends on intrinsic muscles (q.v.) of the hand and eye-hand coordination.

man-task system, Simple system involving man performing a given task, usually involving only perception and motor skills.

mechanical analogue, Mechanical model or representation of another type of system (i.e., physiological, anatomical, etc.) which responds to inputs with outputs corresponding to those in the real system.

mechanoreceptor, Sensory end organs receptive only to mechanical stimuli, e.g., pressure sensors in the skin. Impairment of mechanoreceptors can lead to degraded quality of work and increased risk at the workplace because of insufficient sensory feedback (q.v.). Control in manipulation of tools or objects depends on effective mechanoreceptors. (See sensory end organs.)

mechanotaxis, Contact with a mechanical environment consisting of forces (pressure, moment), vibration, etc. One of the ecological stress vectors. Improper design of the mechanotactic interface may lead to instantaneous trauma, cumulative pathogenesis, or death.

medial displacement, Movement of a limb or body segment toward the mid-sagittal plane (q.v.). Bringing the legs or arms together are examples of medial displacement. (See lateral displacement.)

medial popliteal nerve, Important nerve at the back of the leg in the hollow of the knee. (See popliteal region.) Pressure on the medial popliteal nerve caused by poor seat design may cause the leg to "go to sleep". *Syn:* tibial nerve.

median nerve, A major nerve controlling the flexor muscles (q.v.) of the wrist and hand. Tool handles and other objects to be grasped should make good contact with the sensory feedback (q.v.) area of this nerve located in the palmar surface of the thumb, index, middle, and part of ring finger.

mesomorph, Athletic type of body build featuring good musculature, broad shoulders and a flat abdomen. The mesomorph body type is most desirable for heavy labor tasks. Statistically more prone to heart disease.

metabolic cost, The amount of heat or energy consumed as the result of a specific activity, generally stated in calorie (q.v.) units. Can be measured by spirometry (q.v.) or respiratory gas exchange. A helpful measure of task severity. (See metabolism, light work.)

metabolic rate, Rate of energy consumption required for performance of body activities. Basal metabolic rate is the minimum rate at which vital body functions can be maintained. Units are in calories per unit time. Metabolic rate can be determined indirectly by oxygen uptake or carbon dioxide discharge and is a good index of work strain of steady state work when work is medium-heavy or heavy, and large muscle masses are involved.

metabolism, The physiological combustion process. Conversion of foodstuffs into body tissue, energy, and waste products. A continuous complex process influenced by the energy demands of the body as well as its state of health, temperature, emotion, etc. (See aerobic metabolism, metabolic rate, metabolic cost.)

metacarpal bones, The five bones in the hand which connect the carpal (wrist) bones with the phalanges (knuckles). The metacarpals form the structural base of the palm of the hand.

microbiotaxis, Contact with microorganisms. Exposure to an environment of certain microbes. One of the ecological stress vectors. A form of biotaxis (q.v.).

micromotion film, Motion picture made of a specific job or class or work. Each component of the task is portrayed on a single frame for independent study. With such films, it is possible to break down complex motion patterns into their most basic elements for therblig determination and efficiency analysis. In biomechanics this technique is used to identify short duration acceleration patterns and reflex responses, e.g., during lifting, which are too fast for visual observation.

mid-sagittal plane, A reference plane formed by bisecting the human anatomy so as to have a right and left aspect. Human motor function can be described in terms of movement relative to the mid-sagittal plane. (See transverse and coronal planes.)

moment or moment of force (torque), The effectiveness of a force to produce rotation about an axis measured by the product of the force and perpendicular distance from the line of force to the axis. (SI unit is Nm.) Popularly it refers to the twisting or turning effect of a force. Moments, like forces, have both internal and external effects upon the body in which they are acting. The external effect of a moment on a body is to change or attempt to change the angular or rotational velocity of the body. The internal effects of moments are to cause a state of strain. A moment is necessary to produce angular acceleration. (See torque.)

moment concept, The concept based on theoretical and experimental bases that lifting stress depends on the bending moment exerted at susceptible points of the vertebral column rather than depending on weight alone.

moment of inertia, The inertial resistance to a change in rotational motion which must be overcome by a torque (q.v.), in order to produce angular acceleration. It is a function of the distribution of mass about the axis of rotation.

motions inventory, See available motions inventory, demanded motions inventory.

motions pathway, The locus of travel of an anatomical landmark, often a "knuckle," in moving from one point of the workplace to another. It includes the elemental increments of "reach" and "move" as defined by the various predetermined motion time systems (q.v.) and can also include such motor therbligs as "position," "examine," "hold," etc.

motion time system, Abbreviated MTM. (See predetermined motion time system.)

motor nerve, A nerve also known as an efferent nerve which conducts impulses from the brain or spinal cord to muscle. (See sensory nerve.)

musculoskeletal system, The combined system of muscles and bones which comprise the internal biomechanical environment.

myography, See electromyography.

navicular bone, Synonym for scaphoid (q.v.) bone.

negative work, Work performed against gravity as in the lowering of an object or a body segment. Its muscular effort mainly stresses the control of the load rather than the power required in lifting. Its effort is usually rated at from 1/3 to 1/2 of that of positive work (q.v.). One of the elements of a lifting task (q.v.).

neuropathy, Pathology of a nerve trunk of varied etiology. Pain, paralysis, muscle atrophy, numbness, and loss of reflexes may occur in areas supplied by the involved nerve. Neuropathy may be due to injury, exposure to harmful chemicals, infection, compression of a nerve, or poisoning by heavy metals. It may also result from a generalized disease such as diabetes.

nonpositive tactile stimuli, The cessation of feedback signal from the tactile sensors to the brain. For example, if one is sliding an object along a surface toward a hole without looking at it, a nonpositive tactile stimulus will be obtained when the object falls into the hole indicating that the task has been successfully completed.

nystagmogram, A recording of the quick rhythmic oscillations of the eyeballs usually known as saccadic eye movements. These oscillations are typically involuntary and can be either horizontal, vertical or rotary. The recordings are a good indicator of the effects of rotation on postural controls as maintained by the semicircular canals and also on the effectiveness of eye scanning (q.v.) as opposed to head scanning (q.v.).

occupational ecology, Study of the worker, his environment and the interaction of worker with environment. The occupational ecologist is interested in matching man and environment for optimal ergonomic efficiency (q.v.) and minimal disturbance to the environment.

occupational medicine, The medical specialty concerned with the epidemiology, prevention and treatment of disease commonly observed in industrial environments. These include the diseases and/or accidents which are primarily associated with particular occupations.

oculogram, Recording of overall eye movement patterns, obtained by electromyographic observation of the rectus lateralis muscles of both eyes. Instantaneous readout of the position of sight focus within the visual field can be shown if instruments are properly calibrated.

olecranon fossa, A depression in the back of the lower end of the humerus in which the ulna (q.v.) bone rests when the arm is straight.

olecranon process, The elbow bone. A part of the ulna which makes up the bony point of the elbow, best felt when the arm is bent. An important anatomical reference point (q.v.).

origin (muscle), The anatomic point of attachment of the nonmoving end of a muscle to a bone. When the muscle contracts, it brings an associated body segment toward the origin. The origin is at the proximal (q.v.) end of the muscle.

orthoaxis, The true anatomical axis about which a limb rotates as opposed to the assumed axis. The assumed axis is usually the most obvious or geometric one, while the orthoaxis is less evident and can only be referenced by the use of anatomical landmarks.

orthocentre, The instantaneous anatomical center about which a limb rotates. It varies with the angle of rotation and can fall outside the physical limits of the joint involved.

orthosis, A device applied externally or internally, to control or enhance human limb movement, or prevent bone deformity. May take the form of splints, or self-help devices for the disabled.

oscillograph, An electrical recording device which yields a permanent trace of an electrical signal. The recording may be produced by pen and ink on standard paper, pressure stylus on waxed paper, heated stylus on heat-sensitive paper, or a light beam on light-sensitive paper. A good oscillographic recorder will not distort the signal which is being recorded. Used in electromyography (q.v.), cardiography (q.v.), etc.

overload principle, The principle which states that when a system is saturated by the requirements of a performance task, it fails to function entirely. Two pertinent examples of this principle are found in the functioning of muscle groups under excess load and the ability to process information when the information load becomes too large. (See perceptual overload.)

overt lifting task, Lifting and manipulation of substantial loads as opposed to the covert lifting task (q.v.). The stresses which are generated are resisted by the vertebral column and the erector muscles of the trunk and back.

oxygen debt, Quantitative expression of the amount of oxygen necessary for physiological combustion and elimination of the by-products of anaerobic metabolism (q.v.).

palmar arch, Blood vessel in the palm of the hand from which the arteries supplying blood to the fingers are branched. Pressure against the palmar arch by poorly designed tool handles may cause ischemia (q.v.) of the fingers and loss of tactile sensation (q.v.) and precision of movement.

particulotaxis, Contact with particulate matter. One of the ecological stress vectors. Particulotaxis at the workplace has caused numerous industrial diseases such as silicosis, which is commonly called potter's consumption or stonemason's disease, berylliosis, which results from exposure to beryllia dust, and black lung disease found among coal miners. Ventilation, electrostatic precipitators, proper work clothes, masks, etc., can effectively abate the dangers of particulotaxis.

pathocumulus, Trauma resulting from repetitive application of work stress.

pathological process, Any process which causes temporary or permanent changes in physiological function or anatomical structure resulting in a state of disease. Pathological processes may be triggered or exacerbated by poorly designed man-task interfaces.

percentile threshold, Relates to the collective visual acuity of a group, stating the percentile probability of detection of a small object. Object size is expressed by its subtended angle at the eye. An important parameter in visual inspection for quality control. (See visual acuity.)

perceptive deafness, Deafness from conditions involving the cochlear structures or auditory nerve. Persons so affected may be deaf only to high frequency sounds, or, alternatively, only to low frequencies. Perceptive deafness is considered to be occupational since it often occurs in persons working in noisy environments. Also called boilermakers' deafness. (See conductive deafness.)

perceptual overload, Input of too much information. The condition in which a person receives too much sensory information. His nervous system, on becoming saturated, is unable to process or understand the information received. The result of a perceptual overload is no reaction or response. (See overload principle.)

peripheral hemodynamics, The study of blood flow phenomena in the outer regions of the body. Because work tasks involve the extremities, knowledge of peripheral hemodynamics is necessary to understand the interaction of internal and external environments (q.v.).

phalanx, Colloquially known as the knuckle, any of the long bones of the fingers or toes. Frequently used as anatomical reference points (q.v.) in work analysis.

pharmacokinesis, Motions caused by the administration of drugs.

photopic vision, Commonly called daylight vision — a function of the cones of the retina which exhibit a high excitation threshold. Photopic vision permits color vision and perception of fine detail. (See scotopic vision.)

photoreceptors, Sensory end organs (q.v.) of the optic nerve. Elements of the retina called rods and cones (q.v.). They contain biochemicals which operate in the interpretation of perceived light. The industrial designer should be aware of the differences in these end organs. Rods are black/white discriminating, cones are color sensitive. Rods are effective in the peripheral visual field and color perception is limited to the central visual field.

phrenic nerve, Motor nerve to the diaphragm, originating from the vertebral column in the neck and transversing the thoracic cavity. Important in diaphragmatic respiration. Can be irritated in the neck by a poorly designed seat harness. Repeated pressure surges in the abdominal cavity caused by movement of the back required by improper seat height are transmitted to the diaphragm and may cause pain to be projected to the shoulder.

physiological optimal alignment, Alignment of the principal axis of rotation of a task with the orthoaxis of a limb. The motion of a fixed optimal alignment is false, in that each worker and each task may have a different optimal position.

physiological response, Any of the body's reactions (hormonal, electrochemical, muscular, or nervous, etc.) to any internal or external stimulus. Physiological response to external stress is synonymous with work strain (q.v.).

physiology, The study of the biological, biochemical and biophysical functions of living organisms. It is one of the basic disciplines in the practice of biomechanics.

piezoelectric effect, A property exhibited by certain crystalline substances in which the application of mechanical stress produces a redistribution of electric charge, and the application of a voltage produces a mechanical deformation. This property is exhibited by bone and other tissue.

pinch grasp, Also known as key grasp. One of the lesser used grips of the hand. Involves the thumb and the near side of the index finger at the second phalanx (q.v.). Not a natural grip, it must be learned and requires high levels of manual dexterity and tactile discrimination. Used to apply large forces to small objects because of the strength of the thumb.

pivot joint, A joint in which motion is limited to rotation about an axis perpendicular to the contact surface. The atlas and axis, the two uppermost vertebrae, form such a joint which accommodates rotation of the head. Attempts at movement other than pure rotation at such joints may be dangerous or cause discomfort. Proper positioning of the atlas-axis pivot joint is biomechanically significant for proper head scanning (q.v.).

plane of work, The principal plane in which motions occur at the workplace. Typically, assembly on a work surface makes that work surface the plane of work. Workplace layout should orient surfaces with the plane of work such that motions are guided by the surfaces. The best place of work is one which permits lateral transport of workpiece by humeral rotation (q.v.). The best plane of work is not necessarily horizontal.

plantar flexion, The motion about the ankle joint which raises the heel from the ground and points the toe, e.g., standing on toes, operating a gas pedal.

play for position, One of the fundamental motions or therbligs at the workplace. The pre-positioning of an object for subsequent operation. If the worker searches to find the correct position, he is inefficient. Guides should be used where feasible.

pneumotachograph, An instrument used to record the rate of breathing and pulmonary ventilation (q.v.). Design by Wolff is commonly used in work physiology, especially in agriculture, forestry, and foundry work.

polygraph, Multi-channel strip chart recorder of voltages producing analogue tracings with respect to a time base of physiological events (e.g., sweat rate, heart rate, respiratory rate, etc.) simultaneously. It is generally used for work physiological measurements but its popularity is derived from its association with lie-detecting tests.

popliteal clearance, Distance between the front of the seating surface and the popliteal line. This should be about 5″ in good seat design to prevent pressure on the popliteal artery.

popliteal height, Distance from the sole of the foot to the popliteal line. In good seating design, the distance between the floor and the highest point of the seat is approximately 2 inches less than the popliteal height.

popliteal line, The crease at the back of the leg in the hollow of the knee when lower leg is flexed. Important anatomical landmark.

popliteal region, The area of the leg at the back of the knee. A significant area for consideration in the design of seats. Pressures in the popliteal region from the front edge of a seat must be eliminated for good workplace design. Region contains nerves which, if pressed, cause numbness in the leg. (See popliteal clearance, popliteal height.)

positive click adjustment, A method of adjustment which provides the proper amount of motor-sensory feedback to the operator. Positive click refers to incorporation of momentary "stops" with an associated audible or tactile "click" within operations involving motions such as dial turning, pushing buttons, etc. One advantage of positive click adjustment in workplace design is its potential for use by persons suffering from localized sensory deficits; i.e., blindness, deafness, etc. It also replaces graded information with attributive information and reduces the need for judgment.

positive work, Work performed by a person in applying a force through a distance. Muscular effort consists of generating the energy to perform work in the true engineering sense. work stress (q.v.) and strain (q.v.) (muscular shortening) are necesary corollaries of positive work. It is one of the elements of a lifting task (q.v.). (See negative work, isometric work.)

postprandial, Literally after meal. Its importance in the industrial workplace is in understanding the effects of digestion upon work performance. Basically, the consumption of food in substantial quantities causes a shift in the flow of blood away from the skeletal muscles and peripheral organs and toward the viscera, severely limiting metabolic activity of the skeletal muscles. Also, after heavy meals a lower level of alertness and a greater accident proneness may exist. Indication of pre- or postprandial state of subject should be included in work-task study.

power grip, One of five basic grasps of the hand. It consists of the fingers wrapping around the gripped object with the thumb placed against it. This allows use of the strong opposing muscles of the thumb. Used in hammering operations with special pliers. The tool designer should be aware of the strength available from this grip.

predetermined motion time system, Any scheme useful in the prediction of performance times of industrial work tasks. It analyzes all motions into elemental components whose unit times have been computed according to such factors as length, degree of muscle control required, precision, strength, etc. Time standards of several of these are used as bench mark (q.v.) levels for normal performance in biomechanics and work physiology. For these purposes they are divided into systems: derived from taxonomy and kinesiology; nontaxonomic and kinesiological; taxonomic but nonkinesiological.

prehensile, Adapted for taking hold. Usually refers to grasping motions of the hand. The prehensile and manipulative (q.v.) ability varies considerably with levels of training and cultural and social background. Performance of workers on a given task may differ for these reasons. (See grasp reflex.)

prehension, Refers to the ability to grasp or take hold. (See prehensile.)

pressure sensor, One of the sensory end organs (q.v.) responsive to a pressure stimulus. Important for tactile control (q.v.). Feedback from pressure sensors between fingers enables the worker to moderate his grasp. Interference with pressure sensor function, as with heavy gloves or shoes, will impair touch perception and may cause faulty grasp or insufficient foot pedal control.

pressure vector, Descriptive term in hand tool or equipment usage. Describes direction and magnitude of pressure (stress) exerted against the body at man-equipment interfaces. Large pressure vectors may interfere with blood supply or irritate nerves.

prime movers, The muscles which produce or maintain a specific motion or posture. Also called agonists, they are assisted by synergists (q.v.) and opposed by antagonists (q.v.). Prime mover activities may be preceded by action of a trigger muscle initiating the movement. Knowledge of the prime movers is essential in workplace design and in electromyographic evaluation of task severity.

pronation, Rotation of the forearm in a direction to face the palm downward when the forearm is horizontal or backward when the body is in anatomical position (q.v.). An important element of industrial demanded motions inventory (q.v.), it is performed by muscles whose efficiency is a function of arm position. (See range of pronation and supination.)

proprioception, The sensing of one's location relative to the external environment. An important sense for maintenance of balance and for orienting one's self for performing work tasks. An impaired proprioceptive sense can cause industrial accidents or faulty performance where controls are located outside the visual field.

prosthesis, The replacement of a body part. The replacement may be for a limb, a blood vessel, an organ, or a skeletal member. It may be partially or completely nonfunctional when external, although it is usually cosmetic. (See orthosis.)

proximal, Describing that part of a limb which is closest to the point of attachment. The elbow is proximal to the wrist which is proximal to the fingers. Generally refers to distance from trunk.

psychosomatic response, A physical or physiological response resulting from emotional stimulus. Often a defensive reaction. Important to understand in the practice of human engineering (q.v.) and in modifying sociotaxis (q.v.) in ergonomic analysis (q.v.).

pulmonary ventilation, Commonly referred to as breathing. Volume of respiratory gases passing in and out of the lungs per unit time. Used in spirometry (q.v.) as an index of work stress (q.v.).

pulse rate ratio, The ratio of pulse rate after or during exercise to pulse rate at rest. Typically used as a measure of heart function, the pulse ratio is also used as a measure of work stress.

queuing, Commonly known as bottleneck formation. Bunching of events or objects during sequential flow where the rate of advance of each object is variable about a constant average rate. The variation may be produced by the frequency distribution of manipulative performance times. Changes in posture, moment of inertia of the moving limb, or angle of abduction (q.v.) can cause individual task rate changes which may disrupt a production line with queueing problems, i.e., bottlenecks. Improper plane of work or seat design may also be implicated. Idle time, the reverse of queueing, may occur. Biomechanical analysis of posture is desirable whenever a variable rate human processor feeds or is fed by a constant rate mechanical processor or another variable rate human processor.

radial deviation, Movement of the hand which deflects its longitudinal axis toward the radius (q.v.). It causes the head of radius (q.v.) to press against the capitulum of humerus (q.v.) and may lead to irritation known as "tennis elbow." Tool design should minimize radial deviation. Strength of grasp is diminished in radial deviation.

radial nerve, One of the main nerves of the arm providing motor stimuli to extensor muscles of the forearm, wrist and fingers. Receives sensory feedback from back of forearm and hand.

radiocarpal joint, Commonly known as the wrist, this joint provides all movements between the radius-ulna and hand. Because of its flexibility, the radiocarpal joint is often abused in adapting to the use of improperly designed tools.

radiohumeral joint, Part of the elbow which is thrust bearing. The site of tennis elbow (q.v.). Composed of the head of radius (q.v.) and capitulum of humerus (q.v.). (See radial deviation.)

radio-ulnar joints, The mechanism whereby the radius and ulna rotate about one another during pronation (q.v.) and supination (q.v.). At the elbow, the head of radius pivots in a notch in the ulna and is contained by an encircling annular ligament, thus forming a journal bearing. The bones are maintained in relative position by a ligamentous sheet connecting their central sections. During pronation and supination the distal end of the radius glides around the head of the ulna and rotates on its own axis. Workplace design should assure proper wrist and hand alignment during supination and pronation to avoid trauma. (See radial deviation, tennis elbow.)

radius, The long bone of the forearm in line with the thumb. It is the active element in the forearm during pronation (q.v.) and supination (q.v.). It also provides the forearm connection of the wrist joint. (See radiocarpal joint, radiohumeral joint, ulna.)

range of pronation and supination, The total useful rotational capability of the wrist about the longitudinal axis of the forearm. Very important in industrial tasks involving screw threads, twisting motions, etc. Range (approximately 180 degrees) is greatest if no other deflection (ulnar or radial deviation) occurs at the same time.

rational workplace design, Application, *a priori,* of principles of anatomy, physiology, and systems engineering, in optimizing task and performance efficiency. (See analytic workplace design, improvement approach, and empirical workplace design.)

Raynaud's disease, Constriction of the blood vessels of the hand from cold temperature, emotion, or unknown causes. Afflicts women predominantly and affects both hands simultaneously. Hands become cold, blue and numb and lose fine prehensile (q.v.) ability. On recovery hands become red accompanied by burning sensation. Easily confused with one-sided numbness and tingling caused by poor tool design and resulting pressure. (See scalenus anticus syndrome.)

reactance platform, A balance system for measuring forces or accelerations of a supported body. In biomechanics the force platform is used to measure the forces associated with movement and the accelerations associated with these forces. Shifts in the center of gravity of the total body can also be determined. Used in the measurement of work stress resulting from unusual materials handling and in gait analysis (q.v.).

reactions inventory, The available physiological, kinesiological or psychological responses of an individual to a stimulus or set of stimuli. Reactions will vary with time and depend on the condition (preparedness, fatigue, etc.) of the individual.

rectus abdominis, Long strap-like muscle originating in the pubic region and running vertically to insert in the lower ribs and sternum. By contracting, it flexes the vertebral column and compresses the abdominal wall. It functions as the antagonist (q.v.) to the sacrospinalis. The rectus abdominis, by contracting can break the force of a blow to protect the abdominal viscera. However, workplace design should avoid repeated impact of the rectus abdominis.

referred pain, Pain sensed at a place distant from its true origin (e.g., pain originating in stomach may be felt high in the lumbar region; e.g., pressure on the front of the abdomen may produce pain in the shoulder). Unawareness of referred pain patterns is frequent cause of incorrect identification of loci of work stresses.

reflectance, The ratio of the reflected incident light which on striking a surface is reflected. Although lighting at the workplace is important, reflectance is more significant because it provides the visual input to the worker. Reflectance is expressed as a percentage of incident flux density. (See foot-lambert.)

reflex, See simple reflex.

reflex goniometry, Quantification of reflexive response by measurement of angular displacement, velocity and/or acceleration of a limb about a joint. Used in the evaluation of drug effects as related to safety at the workplace.

rehabilitation, Treatment to restore a loss of capacity by retraining, medication, corrective exercise, or substitution of new faculties to replace those lost or damaged. Includes physical, emotional and technological rehabilitation.

respiration, Commonly called breathing. The composite process of internal respiration wherein an exchange of gases between circulatory fluids and cells occurs and external respiration which covers the gas exchange between blood and the air in the lungs. Respiration must occur to sustain life and must be adequate for efficient workplace function, i.e., the task should not interfere with normal breathing. Inadequate internal respiration is one of the causes of muscular fatigue. (See anozia, ischemia, anaerobic metabolism.)

respiratory quotient, The ratio of carbon dioxide volume exhaled to volume of oxygen up-take. Respiratory quotient varies considerably with type of foodstuff being metabolized and is increased by hyperventilation and acidosis. In steady work, the respiratory quotient is an index of work strain.

resting metabolism, Physiological combustion process usually expressed in units of oxygen consumed per unit time in a person at rest while seated or standing in normal position at the workplace. Increase in metabolic rate (q.v.) during specific activities above resting metabolism is the metabolic cost (q.v.) of the activity.

retina, The inner liner of the eye which contains optic end organs. Composed of rods and cones (q.v.) photosensitive cells, it translates light energy into nervous impulses.

rods, Photosensitive cells of the retina which are specifically reactive to dim light. They are black-and-white sensitive only, and their proper function is required in poorly illuminated or night-vision applications. (See scotopic vision.)

rotator cuff, A group of muscles surrounding the shoulder joint. Because of the many components of available movement at the shoulder, it is susceptible to self-generated trauma caused by actions of the muscles of the rotator cuff. In the industrial setting, rapid reversals of movement in the shoulder should be minimized to avoid trauma or fatigue which may result.

sacrospinalis muscle, The long muscle of the back with origin (q.v.) in the sacrum and insertion (q.v.) in the thoracic and lumbar vertebrae. Its primary function is to keep the spine erect. A key muscle to be monitored during electromyographic studies of whole body lifting tasks since its response is characteristic of the total effort being expended. Operates at a poor mechanical advantage and should not be overstressed in task design or damage to vertebral column or spinal nerves (q.v.) may result.

sacrum, A triangular, slightly curved bone at the base of the spine. The joint between sacrum and pelvis and between sacrum and lumbar spine are frequent sites of lower back trauma.

sagittal plane, A plane from back to front vertically dividing the body into right and left portions. Important in anthropometric definitions. Mid-sagittal plane is a sagittal plane symmetrically dividing the body.

saphenous veins, Two major superficial veins in the leg. Long saphenous vein passes approximately from the front of the medial ankle to the groin. Short saphenous vein located at back of leg passes from ankle to the popliteal region (q.v.). Both veins are frequently sites of varicosity. Good chair design eliminates pressure against short saphenous vein, thus avoiding interference with venous return and minimizing risk of irritation and possible general inflammation of vein.

scalenus anticus syndrome, Numbness, pain and tingling in fingers and hand, often in the area of ulnar nerve (q.v.) distribution. Differs from Raynaud's disease (q.v.) in that it may appear on one side only. Caused by a squeezing of large nerves and blood vessels supplying the arm between the scalenus muscles in the neck or sometimes between scalenus muscle and ancillary rib. Occupationally significant because it may manifest itself in some persons only when the arm is abducted or when weights are lifted.

scaphoid, One of the bones of the hand. Located at the base of the palm on the thumb side. It bears against the radius (q.v.) and is directly on the axis of thrust from the hand to the arm. Its proximal surface is an important moving element of the wrist joint involved in flexion, extension, adduction and abduction. Because of its location, the scaphoid is vulnerable to disability through fracture from impact and thrust loads. Also the name of a bone in the tarsal region of the foot. (See scaphoid turbercle.)

scaphoid tubercle, Protuberance of scaphoid bone (q.v.). Important as one of the anchor points of the flexor retinaculum (q.v.). An important anatomical reference point (q.v.) for locating the wrist joint.

scapula, The shoulder blade. Large, triangular flat bone; it forms the socket for the proximal joint of the humerus and constitutes at the same time the place of the large muscle of the shoulder and back. This bone positions and stabilizes the shoulder joint. Backrests of chairs should clear the lowest point of the scapula. A bony ridge on the scapula serves as an important anatomical reference point (q.v.).

scotopic vision, Night vision. Permits, in essence, only black, white and gray discrimination. Scotopic vision appears at a level of below one candela per square meter of illumination. Because only the rods are involved, the visual angle under scotopic conditions is much wider, and one can see objects peripherally. Because of the absence of color clues in scotopic vision, brightness contrast between object and background must be sufficient to maintain visibility and avoid accidents.

semilunar notch, A wide but shallow crescent-shaped indentation found on the upper end of the ulna. It articulates with the trochlea (q.v.) of the humerus and forms the bearing of the hinge joint during extension and flexion of the forearm. The orthoaxis of flexion of the forearm passes through the center of curvature of the semilunar notch.

semispinalis capitis, Muscle of the neck, antagonist to sternomastoid muscle (q.v.). Important element in head scanning (q.v.) and postural control.

sensory end organs, Receptor organs of the sensory nerves located in the skin or other tissues. Each end organ can sense only a specific type of stimulus. Primary stimuli are heat, cold, or pressure, each requiring different end organs. Knowledge of end organ distribution is of importance to the safety engineer. For example, there are a few heat receptors on the outer surface of the forearm, so that the skin may be severely burned before heat is sensed.

sensory feedback, Signals perceived by sense organs (e.g., eye, ear) to indicate quality or level of performance of an event triggered by voluntary action. On the basis of sensory feedback information, decisions may be made permitting or not permitting an event to run its course; enhancing or decreasing activity levels.

sensory nerve, A nerve (also called afferent nerve) which conducts stimuli from sensory end organs (q.v.) which change physical information (such as pressure, temperature, light) into nerve impulses, to the spinal cord or the brain where cognition occurs and/or the stimulus is converted to a motor signal for transmission via a motor nerve (q.v.).

sesamoid bone, Small rounded bone enclosed and/or enveloped by a tendon between tendon and joint structures (e.g., kneecap) and may protect joint (e.g., sesamoid bone at base of big toe). In optimal posture the center of mass of the body is located above the sesamoid bone of the big toe.

signature, Record of a characteristic pattern by a subject (e.g., the electrocardiogram is a signature). Motions pathways (q.v.) or their time derivatives which are repeatable are signatures indicative of an individual's motor skills, state of health or fatigue. (See velocity signature, acceleration signature, chronocyclegraph.)

simple reflex, Automatic motor reaction to an external stimulus where the pathway from sensory end organ (q.v.) to muscle short-circuits the brain and passes through the spinal cord only. The knee jerk and the Achilles tendon reflex are examples of simple reflexes which maintain integrity of posture. (See conditioned reflex.)

skeletal configuration, The general arrangement of body parts, such as muscles and bones, which may be necessary in the performance of tasks. Skeletal configuration largely determines the biomechanical efficiency of performance.

sociotaxis, Contact with fellow man either individually or in groups under circumstances which triggers behavioral reactions. One of the ecological stress vectors. Improper design of sociotactic interfaces may cause emotional work stress, while good sociotactic interfacing will promote emotional well-being and high motivation at work.

somatotyping, System for classifying body builds by grouping them into distinctive types. These classifications are loosely correlated with personality traits, and may be helpful in identifying proneness to disease and accident. (See ectomorph, endomorph, mesomorph.)

sonography, The use of ultrasonic energy for imaging internal organs and tissues. Same as echography. (See ultrasonics.)

span of reach, Hypothetical boundaries of normal reach based on the assumption that it is easy to reach any point within a sphere formed by the rotation of the tip of the index finger around the shoulder joint, using the extended arm as a radius in front of the coronal plane (q.v.). However, there are many points within the so-called "span of reach" which are definitely "out of reach" due to limitations of freedom of skeletal configurations (q.v.).

spasm, Involuntary contraction of isolated bundles of muscle or muscle groups. Generally caused by local chemical imbalance resulting from fatigue, local ischemia or trauma or other reasons. Cannot be consciously controlled and is a potential cause of accidents in the workplace. Must not be confused with spasticity (q.v.). Because the conditions causing spasm can be induced by enforced postural rigidity or by poorly designed tools, the industrial engineer must be aware of the etiology (q.v.) of spasm.

spasticity, Heightened reactivity of affected muscle groups. Results in increased resistance to passive stretch and increased reaction of tendon reflexes. Caused by interference with cerebral or cortical control of musculature. Do not confuse with spasm (q.v.). Important in evaluation of disabled workers in rehabilitative effort. (See simple reflexes.)

specific acoustic impedance, A material property that is equal to the product of the density and the speed of sound in that material. A sound wave will be reflected from the interface of two materials when each has a different specific acoustic impedance.

spike, Recording of an isolated electrophysiological event of less than 25 millisecond duration. It appears as a sharp V or inverted V.

spinal nerve, One of the 31 pairs of nerves originating from the spinal cord. It leaves the vertebral column (q.v.) by two roots and contains both motor and sensory components (q.v.). At the point of exit between two vertebrae the spinal nerve is susceptible to compression trauma which may cause pain, paralysis or numbness at points distal (q.v.) to the site of the lesion.

spine, Also called the backbone. The vertebral column (q.v.) which is the central skeletal structure of the body. Also a sharp projection of any bone.

spinous process of the vertebra, A bony rearward extension of the vertebra. It provides extended surfaces for muscle attachment, and for stabilizing adjoining vertebral segments. The spinous process is that part of the vertebra which can be felt through the skin at the back. Often used as anatomical reference points (q.v.). The spinous process of the seventh cervical vertebra is an example of an important anatomical reference point.

spirometry, Measurement of gas volumes moved during breathing. Used as an index of work stress (q.v.). Also used in indirect inference of metabolic rates (q.v.). In industry, a portable gasometer or a pneumotachograph (q.v.) is commonly used.

static work, *Syn:* isometric work (q.v.).

stereoscopic acuity, Ability to sense three-dimensional aspect of physical space. A function of independent image inputs, i.e., viewing with two eyes. Stereoscopic acuity also depends on brightness contrast and distance from observed objects. (See depth perception.)

sternocleidomastoid muscles, A pair of muscles connecting the breastbone and collar bones to the mastoid processes of the lower skull behind the ears, which provide support for the head. When operating together, the right and left sternocleidomastoids pull the head forward and the back of the skull downward, and when operating singly each turns the head to the opposite side. They oppose the semispinalis (q.v.) muscles and stabilize the head. In the workplace the worker's head position should be near vertical to minimize activity of the semispinal and sternocleidomastoid muscles. The sternocleidomastoid is also functionally important in head scanning (q.v.).

strain, (1) In mechanics, a measure of deformation, either elongation, contraction or angular (shear) deformation. (2) In medicine, any injury involving overextension, compression or twisting of a muscle, ligament or joint. (See also work strain.)

strain propagation, Internal transmission of the response reaction in the body to external stress. Propagation may be by means of mechanical or biological processes such as hormonal, neural (nerves), or circulatory mechanisms. Because of strain propagation resultant trauma is often evidenced at locations remote from the point of stress application, e.g., scalenus anticus syndrome (q.v.).

strain synthesis, Artificial simulation of work strain by combination of work stresses.

stress, (1) In mechanics, a force per unit are either external or internal to the body. Normal stresses are tensile or compressive. Tangential stresses are called shear stresses. (2) In medicine, any force or other stimulus applied to or acting on an individual such as excessive lifting force, excessive noise or emotional upset. (See work stress.)

stress equivalent, In biomechanics, the quantitative relationship between physiological outputs or stresses of different dimensions generated by work stress or work strain. Thus, both isometric work with the dimensions of linear impulse, and dynamic work with conventional dimensions of work, can be compared through the resultant work strain expressed in terms of metabolic activity which under such circumstances serves as the stress equivalent. In lifting, light bulky objects exert the same moment as heavy dense objects and the recorded height of the integrated electromyogram (q.v.) is the stress equivalent of each task expressed in foot-pounds of torque on the spine per millivolt of mean myogram height.

stress transmittal, Mode of transfer of external force from the man-equipment interface (q.v.) into distant points of the body. If the level of stress transmittal is low, then excessive work stress is manifested by superficial lesion (q.v.) as blisters or calluses. If the level of transmittal is high, then cumulative pathogenesis (q.v.) may produce pathological lesions in anatomical structure, remote from the point of work stress application, e.g., tennis elbow caused by poor machine control design; e.g., broken collarbone caused by breaking a fall with outstretched hand.

stroboscope, An optical instrument for observing moving bodies by making them visible intermittently. If the motion is cyclic, such as rotation or vibration, an optical illusion is created of the moving body being stationary or moving at a slower rate. (See flicker fusion.)

supination, Rotation of the forearm about its own longitudinal axis bringing the thumb side of the hand from a position next to the body to one away from it when the arm is in anatomical position (q.v.). Supination tends to turn the palm upward when the elbow is flexed 90 degrees and the forearm is horizontal and forward. Supination is an important element of available motions inventory (q.v.) for industrial application, particularly where tools such as screwdrivers are used. Efficiency in supination depends on arm position. Workplace design should provide for elbow flexion at 90 degrees. (See pronation, range of pronation and supination.)

synapse, Junction of a nerve cell with another. In the simplest case of a reflex arc the synapse joins the sensory (receptor) fiber with the motor fiber to stimulate muscle action. (See simple reflex, sensory nerve, motor nerve.)

synergist, A muscle which acts to assist a prime mover (q.v.) in performing a specific action. A muscle may act as synergist in one action and as antagonist (q.v.) in another action to the same prime mover. Work tasks should be designed to use synergists effectively.

syndrome, A complex of symptoms which identifies a physiological or pathological condition. (See scalenus anticus syndrome, carpal tunnel syndrome, etc.)

synovial fluid, Fluid which provides lubrication to the joints. Secreted by membranes of the joint and contained by surrounding tissues, it is very viscous and an effective lubricant. An excess or lack of fluids resulting from work stress trauma or disease will cause swelling and pain in the joint and may limit its range of motion. (Also known as synovia.)

synovial structures, All of those elements concerned with generation and/or use of synovial fluid (q.v.). Generally located at joints or in tendon sheaths wherein the tendon is lubricated by synovial fluid. When synovial structure is damaged, a state of disease may develop, e.g., tenosynovitis (q.v.), tendonitis, synovitis, or arthritis.

tactile control, Control regulated through tactile sensory feedback (q.v.). It may be reinforced by designing equipment controls to incorporate distinctive texture patterns (e.g., knurling) or shapes to prevent error. Other means of tactile control are the initiation of movement with a vibratory cue and termination through a positive stop. (See tactile sensation, tactile stimulus.)

tactile sense, The sense of touch. Experienced through four types of mechanoreceptors (q.v.) located throughout the body. Significant source of sensory feedback (q.v.). Improper clothing (thick gloves) may impair tactile sensation and cause inaccurate discrimination of tactile signals. (See tactile control, tactile stimuli.)

tactile stimulus, An input originating in the touch mechanoreceptors (q.v.) as sensory end organs (q.v.). The input is converted into a nerve signal for afferent transmission along the sensory nerves (q.v.) to the brain or spinal cord, whence an efferent signal is transmitted along a motor nerve (q.v.) causing muscular reaction. Tactile stimuli in the workplace replace other cognitive judgment and are conducive to quick error-free control action. (See tactile sensation, tactile control.)

task load, A measure of task severity. Expressed as a stress equivalent (q.v.) which is biomechanically a more accurate measure of task load than measure of metabolic activity, heart rate or pulmonary ventilation.

tendon, Fibrous end section of muscle. It attaches to bone at the area of application of tensile force. When its cross section is small, stresses in the tendon are high, particularly because the total force of many muscle fibers is applied at the single terminal tendon. The site of many industrial diseases caused by trauma, biomechanically improper movements, or failure of lubrication, the tendon must be protected in tool and workplace design. (See tenosynovitis.)

tennis elbow, Sometimes called epicondylitis. An inflammatory and/or degenerative reaction of tissues in the lateral elbow region. In an industrial environment it may follow a single, violent effort or repetitive subsymptomatic strain in supination (q.v.) against resistance (as in screwdriving) or violent extension of the wrist with hand pronated. Can frequently be avoided by assuring that the axis of rotation of a tool or machine control coincides with the orthoaxis (q.v.) of rotation of the forearm.

tenosynovitis, A disease of the tendon sheath, commonly occurring in the fingers, hand, and wrist. Inflammation of the tendon sheaths of the wrist often associated with continual ulnar deviation (q.v.) during rotational movements (e.g., screwdriving) or by other overwork and trauma. In industry, extensor sheath inflammation is more frequent. Work tolerance is reduced because of pain during wrist and finger movement.

thermal environment, The surroundings in which a worker is situated which affect his ability to reject or absorb heat. It includes the local temperature, humidity and air velocity. Presence of radiating surfaces also affect the thermal environment. It is significant in the workplace for its effect on work rate, quality control, accidents, and physiological well-being. (See effective temperature.)

thermal stress, Stress on a subject caused by the effective temperature (q.v.) of his surroundings interacting with internally generated heat. Can cause changes in body temperature, body chemistry and function of heart, lungs, kidneys and other organs. Influences ability to evaluate sensory feedback as well as many work skills.

thermodynamic efficiency, In biomechanics, the ratio of the work performed to the energy consumed by the human body. This is commonly estimated at 20% unless man goes into oxygen debt (q.v.). It is computed by dividing the external work performed (e.g., the number of bricks lifted multiplied by weight and height) by the energy consumption of the human body during the task. Important index for effort rating. (See metabolism.)

thermograph, A device for producing graphic representation of temperature distribution on the body surface. Commonly an infrared sensor connected to a suitable display device, e.g., video projector, with either graded black-white shades or discrete colors to indicate different temperature levels. Thermographs can accurately indicate local hot spots when such spots are not otherwise readily visible. An important tool in locating excessive work strain and in the maintenance of industrial health.

theoretical biomechanics, The analytical study of the human body in relation to its immediate mechanical environment. The principles of classical mechanics and the life sciences are employed in the formation of models which predict the effects of mechanical stress on the body. These models can be used as guides in the design of man-equipment interactions such as the workplace and for predictive hypotheses of performance efficiency. In situations where experimentation with man is not feasible, theoretical biomechanics is used to evaluate and predict potentially destructive task situations, e.g., strong vibrations, highway collisions.

tibia, The shinbone. The larger of the two bones of the leg below the knee. The tibia is the major weight-bearing bone of the leg and transmits forces between the ankle and the knee joint.

tissue rheology, Study of the deformation and flow of live or excised tissues under external stress. Usually evaluated as a function of time. Can be used to evaluate industrial hardware to avoid stress concentration and instantaneous or cumulative trauma.

tonus, Commonly known as muscle tone. The tension inherent in muscle which is nominally at rest. A minimal tonus condition is characteristic of healthy muscle. Flabbiness and poor postural control result from poor tonus. Exercise and regular activity aid in maintaining healthy tonus which contributes to worker efficiency.

torque, Refers to the moment of force (q.v.) exerted around any joint during the performance of a task. A moment, or torque, is the product of the perpendicular component of force exerted and the distance from the point of application to the fulcrum or "pivot point" (the joint). Isometric muscle torque which does not result in motion causes high joint forces with physiologic effects dependent on the duration and magnitude of the torque and the joint configuration. E.g., in lifting an object and holding it elevated, torque is exerted by the paraspinal muscles with resultant heavy loads on the lumbar vertebrae. Formerly measured in foot-pounds, inch-ounces, but now in Newtonmeters (Nm). If a force, F, acts to produce rotation about a center at a distance, d, from the line in which the force acts, the force has a torque, $L = Fd$. (See moment of force.)

torsional moment, See torque or moment.

transverse ligament, There are several transverse ligaments (q.v.) in the body. Usually identified by the local anatomical structure with which they are associated. E.g. flexor retinaculum.

transverse plane, An anatomical reference plane parallel with the ground (i.e., horizontal) when considering a man in anatomical position (q.v.). Also that plane at right angles to the long axis of an isolated organ. (See sagittal plane, coronal plane.)

trapezius muscle, Broad flat muscle of the upper back. Maintains posture and is an important mooring and activating muscle of the shoulder blade. Tasks requiring the worker to hold elbows high will cause fatigue in the trapezius muscle and loss of postural stability. Important muscle in elevating the arms.

trauma, An injury or wound, generally caused by a physical agent. Cuts, bruises and abrasions are examples of trauma, but trauma may be present even though it is not visible, e.g., strained muscle. The causes of trauma must be anticipated in workplace design or tool design. Protective devices and special clothing (work shoes, gloves) are used to avoid trauma.

traumatic arthritis, Inflammation, pain and swelling of a joint caused by the application of excessive forces at the joint. Can be caused by a direct blow or sudden strain and leads to restricted motion. Task design should avoid repeated mild trauma or prolonged strain which can cause traumatic arthritis or exacerbate preexisting arthritic conditions.

traumatic fibrositis, Inflammation in muscle and increase in fibrous tissues in muscle caused by application of excessive force or by disease. Can result from direct blow or repeated mild trauma at the workplace. Weakness and disability are consequences.

tremor, A rapid involuntary alternation of motion of a part of the body. May occur at rest, or only on initiation of movement (intention tremor). Directly caused by alternate rapid contractions of muscle groups and their antagonists (q.v.). Originates from fatigue, disease of nerves or muscles, injury, or response to environment, e.g., shivering from the cold. Presence of tremor disables the worker in precision tasks and increases the risk of accidents.

trigger finger, Also known as snapping finger. A condition of partial obstruction in flexion or extension (q.v.) of a finger. Once past the point of obstruction, movement is eased. May occur without apparent cause. A symptom of tenosynovitis (q.v.), (stenosing tenosynovitis). Caused by thickening of a tendon of localized reduction of the tendon sheath. In the workplace, flexing against strong antagonists and flexing of the distal phalanx (q.v.) without middle phalanx movement is suspected. Tool handles should be designed for trigger operation by the thumb.

trochlea, Bobbin-shaped outcrop of bone at distal end of humerus (q.v.), forming axis shaft of elbow hinge joint, i.e., humero-ulnar joint.

tripodal grasp, One of the five basic grasps of the hand. Also called manipulative grasp. Object is retained by thumb, index finger and middle finger, which forms a tripod with object possibly in contact with palm. Delicate rotational control is available to the worker for fine screwdriver manipulation, etc.

tuberosity, Large rough outgrowth on a bone which serves as a point of attachment for tendons. It may be used as a convenient anatomical landmark (q.v.), particularly with respect to design of seats and equipment controls.

tubular bone, Construction of the long bones of the body, e.g., humerus. Provides maximum strength/weight properties while providing centrally located and protected site for blood vessels, nutrients, and bone marrow. Workplace design should avoid imposition of loads with torsional and transverse components.

ulna, One of the two bones of the forearm. It forms the hinge joint at the elbow and does not rotate about its longitudinal axis. It terminates at the wrist on the same side as the little finger. Task design should not impose thrust loads through the ulna.

ulnar deviation, A position of the hand in which the wrist is bent toward the little finger. Ulnar deviation is a poor working position for the hand and causes nerve and tendon damage. It reduces the useful range of pronation and supination (q.v.) by approximately 50 percent, and work performed in ulnar deviation proceeds at low efficiency. Hand tool design should avoid ulnar deviation.

ulnar nerve, One of three large nerves of the hand. It supplies muscles of fine manipulation (q.v.) located in the hand and provides sensory feedback (q.v.) via the skin of the little finger and part of the ring finger. Because the ulnar nerve enters the hand immediately below the skin, it is susceptible to damage from poorly designed tool handles which apply pressure to the palm on the side of the little finger.

ultrasonics, The use of high frequency vibrations for cleaning, heating tissues, dispersing solid mass, and imaging objects by reflection. (1) Ultrasonic Cleaning — The use of high frequency vibrations in a solvent bath to loosen and disperse surface contaminants. Typical frequencies for this application are 20-50 kHz. Also used in dentistry to remove deposits from tooth structure. (2) Ultrasonic Therapy — The use of high frequency mechanical vibrations to transmit energy into and through body tissue to heat or disperse material. Typical frequency for this application is about one MHz. Ultrasonic phacoemulsification for removal of eye cataracts. Ultrasonic disintegration for pulverization of bladder calculi. (3) Ultrasonic Imaging — The use of high frequency vibrations to obtain a visual image of internal organs and soft tissue, most often using pulse echo techniques. Typical frequencies for this application are 1-15 MHz. Examples are echocardiology and echoencephalography.

ultrasonic transducer, A device to convert electrical energy to mechanical energy, using pieze-electric materials such as quartz, barium titanate or lead metaniobate.

univocal, Having a unique association. A relationship wherein one single specific effect can only be produced by one single specific cause, e.g., an ulnar claw can be produced only by damage to the ulnar nerve (q.v.).

vasoconstriction, Decrease in the cross-sectional area of blood vessels. This may result from contraction of a muscle layer within the walls of the vessels or may be the result of mechanical pressure. Reduction in blood flow results. Vasoconstriction can be controlled by the nervous system and is influenced by certain drugs. Improper design of the manequipment interface (q.v.) may cause vasoconstriction, i.e., poor seat design, improper tool handles.

velocity signature, An analogue recording which shows the velocity history of a body part in motion. Signature size and shape are indices of strength and motor skills of an individual. Repeatability of velocity signatures is indicative of status of health, fatigue and/or training. An important component of the biomechanical profile (q.v.).

vertebra, One of the irregularly shaped bones of the spine (q.v.). Its unique shape allows the longitudinal passage of the spinal cord as well as exit of the spinal nerves (q.v.). Each vertebra provides stabilizing surfaces for adjoining vertebra. (See spinous process, vertebral column.)

vertebral column, Bony structure commonly known as the spine. Provides the primary supporting structure of the trunk of the body, and houses and protects the nerves of the spinal cord. Extends from the head to the coccyx. Consists of 24 bones (vertebrae) separated by shock-absorbing discs. Trauma to the vertebral column (shock, twisting, compression, fracture, or stretching) can cause severe pain and/or disability. Curvature of the vertebral column is an important factor in the design of seats. (See vertebra, spine, spinous process.) Note: Because of the irregularities (especially at the lower end of the spinal column), texts may differ in their count of the number of vertebrae. E.g., *Dorland's Illustrated Medical Dictionary* defines VERTEBRAE as "any of the thirty-three bones of the spinal column, comprising seven cervical, twelve thoracic, five lumbar, five sacral, and four coccygeal vertebrae." *Blakiston* also lists 33. In lay terms, the five regions of the spinal column are: cervical (neck), thoracic (chest), lumbar (lower back), sacral and coccygeal (both in the pelvis). The first three regions have separate and movable vertebrae throughout life, but, in the adult, the sacral and coccygeal vertebrae become fused to form the sacrum and coccyx. This fusion — and, hence, resulting irregularity — in the lower regions give rise to what may seem to be contradictions in texts to lay readers. Again, there may also seem to be confusion at the other (upper) end, depending on whether the uppermost bone is considered part of the skull or spinal column. Thus, it may be noted that the 35th Edition (British) of *Gray's Anatomy* lists

31 vertebrae (8 cervical, 12 thoracic, 5 lumbar, 5 sacral, and 1 coccygeal). Thus, orthopedists are quite definite when they note, for example "slipping of L5 over S1" (referring to the "fifth lumbar" and "first sacral") — probably one of the most common of lower back ailments. Fifth lumbar vertebra (q.v.).

visual acuity, The ability to recognize objects or sets of objects by eye as distinct, specific and separate. Depends on illumination, object size, distance from object to observer, object configuration and brightness of contrast. Expressed quantitatively as the ratio of distance at which the observer recognizes to the distance at which a normal eye recognizes. 20/100 indicates that a person recognizes at 20 feet a sight chart which can be seen by the normal eye at 100 feet. Visual acuity of individuals should be matched to the needs of their task assignments.

visual field defect, A condition of impaired eyesight where the eye cannot see portions of the area in its field of vision. Blind spots of various size and position exist. Visual field defects are sufficiently common to warrant consideration in work placement. Not usually detected in eye examination for visual acuity (q.v.). Should be suspected whenever a person appears to be accident-prone.

work, (1) In mechanics, the product of force and distance through which the force acts. (2) In medicine, a measure of physical or mental effort expended in any task. Physical work results in metabolic activity locally and/or sytemically. In biomechanics, physical work is classified as positive work (q.v.), negative work (q.v.) or isometric work (q.v.). Isotonic work, used in physiology, means under a constant tension.

work measurement, In biomechanics, the application of work physiological methods to the measurement of task severity and/or conjointly with the construction of a biomechanical profile (q.v.) for a task or individual in the evaluation of the compatibility of the available and demanded motions inventories (q.v.).

work metabolism, Physiological energy consumption ascribable to the performance of a specific task in excess of resting metabolism (q.v.). Usually expressed as an index of energy consumed per unit time, e.g., calories per minute. In industry, working metabolism is obtained by analysis of respiratory gas exchange.

work physiology, Application of principles of physiology to ergonomics (q.v.) and industrial engineering in design of tasks and workplace. Concerned with moderation of metabolic cost (q.v.), measurement and prevention of harmful work strain (q.v.). Uses physiological measurements of gas exchange, pulmonary ventilation (q.v.), metabolism (q.v.) and heart rate to measure the effects of workplace-related environmental stress vectors.

work strain, The natural physiological response reaction of the body to the application of work stress (q.v.). The locus of the reaction may often be remote from the point of application of work stress. Work strain is not necessarily traumatic but may appear as trauma when excessive, either directly or cumulatively, and must be considered by the industrial engineer in equipment and task design. Thus, increase of heart rate is nontraumatic work strain resulting from physical exertion, but tenosynovitis (q.v.) may represent pathological work strain resulting from undue work stress on a tendon and/or its sheath.

work stress, Biomechanically, any external force acting on the body during the performance of a task. It always produces work strain (q.v.). Application of work stress to the human body is the inevitable consequence of performance of any task and is, therefore, only synonymous with "stressful work conditions" when excessive. Work stress analysis is an integral part of task design.

work tolerance, Span of time during which a worker can effectively perform a task without rest period at levels of physiological and emotional well-being acceptable to the individual and without pathogenesis. Work tolerance is affected by environmental inputs (q.v.), workplace layout (q.v.), and state of health.

workplace layout, The designed arrangement of a work situation comprising man, materials, equipment and the physical environment of the workplace. Optimally it should be the result of deliberate task design which considers the implications of ecological stress vectors on performance efficiency, occupational safety and health.

wraparound grasp, One of five basic grasps of the hand. The grasped object is held against the palm by the fingers wrapped around it with the thumb opposing the index finger (holding a broomstick). The thumb is of minimal importance in this grasp, which is an innate aptitude of man. It requires almost no learning to apply. A straight wrist is needed for successful application. Flexion of the wrist "breaks" the strength of the grasp and may lead to accidents at the workplace.

References

Astrand, P. and Rodahl, K. *Textbook of Work Physiology.* McGraw-Hill Book Company, 1970.

Barham, J. N. and Wooten, E. P. *Structural Kinesiology.* Macmillan Company, 1973.

Barnes, R. M. *Motion and Time Study.* Sixth Edition. John Wiley & Sons, Inc., 1968.

Basmajian, J. V. *Grant's Method of Anatomy.* Ninth Edition. Williams and Wilkins Company, 1975.

Brunnstrom, S. *Clinical Kinesiology.* Third Edition. F. A. Davis Company, 1972.

Clarke, D. H. *Exercise Physiology.* Prentice-Hall, Inc., 1975.

Dorland, W. A. *Dorland's Illustrated Medical Dictionary.* Twenty-fifth Edition. W. B. Saunders Company, 1974.

Downey, J. A. and Darling, R. C. *Physiological Basis of Rehabilitation Medicine.* W. B. Saunders Company, 1971.

Easton, D. M. *Mechanisms of Body Function.* Second Edition. Prentice-Hall Inc., 1974.

Fung, Y. C., Perrone, N. and Anliker, M. *Biomechanics: Its Foundations and Objectives.* Prentice-Hall Inc., 1972.

Grandjean, E. *Fitting the Task to the Man.* Second Edition. Taylor & Francis Ltd., 1963.

Guyton, A. C. *Textbook of Medical Physiology.* Fifth Edition. W. B. Saunders Company, 1976.

Hunter, D. *The Diseases of Occupation.* Fifth Edition. Little, Brown and Company, 1975.

Jacob, S. W. and Francone, C. A. *Structure and Function in Man.* Third Edition. W. B. Saunders Company, 1974.

Karpovich, P. V. and Sinning, W. E. *Physiology of Muscular Activity.* Seventh Edition. W. B. Saunders Company, 1971.

Katz, B. *Nerve, Muscle, and Synapse.* McGraw-Hill Book Company, 1966.

Maynard, H. B. *Industrial Engineering Handbook.* Third Edition. McGraw-Hill Book Company, 1971.

Mountcastle, V. B. *Medical Physiology.* Thirteenth Edition. C. V. Mosby Company, 1974.

Murrell, K. F. H. *Human Performance in Industry.* Reinhold Publishing Corp., 1965.

Niebel, B. W. *Motion and Time Study.* Sixth Edition. Richard D. Irwin, Inc., 1976.

Rasch, P. J. and Burke, R. K. *Kinesiology and Applied Anatomy.* Fifth Edition. Lea and Febiger, 1974.

Tichauer, E. R. *Biomechanics of Lifting.* Report RD-3130-MPO-69 for Dept. of Health, Education and Welfare. Institute of Rehabilitation Medicine, February 15, 1970.

Tichauer, E. R. *Ergonomics: The State of the Art.* American Industrial Hygiene Association Journal. Vol. 28. March 1967.

Tichauer, E. R. *Gilbreth Revisited.* ASME Publication 66—WA/BHF-7. November 27, 1965.

Tichauer, E. R. *Human Capacity, A Limiting Factor in Design.* Institution of Mechanical Engineers Publication P3764, July 1963.

Tichauer, E. R. *More Bricks, Less Sweat.* ASME Publication. Frank Gilbreth Centennial, 1969.

U.S. Department of Health, Education, and Welfare. *The Industrial Environment – Its Evaluation and Control.* Government Printing Office, 1973.

Cost Engineering
Z94.2

2

This chapter presents a compilation of terms that are widely employed among cost engineers. The definitions represent a consensus among the experts in the field — especially as defined by the American Association of Cost Engineers; with special emphasis on cost control, cost estimating, operating cost, business management and planning.

To avoid duplication, many definitions that the reader may expect to find in this chapter will be found in other sections of this TERMINOLOGY — especially Sections 5 (ENGINEERING ECONOMY) and 8 (APPLIED MATHEMATICS). The reader should consult the overall index.

The Cost Engineering Subcommittee, whose efforts, time, and expertise culminated in this presentation, include the following:

Dr. Carl Wolf, P.E. (Chairman)
Professor, Management Engineering, New Jersey Institute of Technology.

Robert E. Templeton (Vice-Chairman)
Division Manager, Project Cost Services, Pullman Kellogg Company.

Kenneth J. Branch
Manager, Systems Consulting, Dravo-Van Houte.

A. S. Corell
Senior Engineering Associate, Exxon Production Research Company.

Frank G. Eckert
Director, Administration, Lockheed Electronics Company, Inc.

Oliver D. Filley
Filley and Company.

Robert C. Horman
Planning and Development Collaborative International.

Robert S. Straith, P.E.
Formerly, Associate Director, Cost Accounting Standards Board.

George Werner
Principal Estimator, Pullman Kellogg Company.

accounts payable, The value of goods and services purchased for use on which payment has not yet been made.

accounts receivable, The value of goods shipped or services rendered to a customer for which payment has not yet been received.

administrative expense, The overhead cost due to the general direction of the enterprise as a whole as contrasted with the expense of a more specific function. Generally includes top management salaries and the costs of legal, central purchasing, traffic accounting and other staff functions.

allowance, An amount that is included in a cost estimate to cover the cost of expenditures that are expected at the time the estimate is made but cannot be precisely defined, i.e., allowance for site preparation.

battery limit, Comprises one or more geographic boundaries, imaginary or real, enclosing a plant or unit being engineered and/or erected, established for the purpose of providing a means of specifically identifying certain portions of work as being related groups of individual items or associated facilities. Generally refers to the processing area and includes all process equipment, and excludes such other facilities as storage, utilities, administration buildings or auxiliary facilities, unless so specified.

benefit/cost ratio, A comparison of the benefits derived to the costs incurred of a prospective capital investment. Used in the non-profit sector to evaluate feasible alternatives.

bid, An offer to provide or produce a service in accordance with specifications for a specified price within a specified time.

book value (net), (1) Original cost less depreciation accruals. (2) The value of an outstanding share of stock of a corporation at any one time, determined by adding the par (or stated) value of the stock outstanding to the surplus applicable to that class of stock and dividing by the number of shares of that class outstanding.

break-even point, The operating level or financial level at which incomes of a company or facility just cover all fixed and variable costs. Depreciation is normally included.

budget, A plan which defines an anticipated activity or objective in terms useful to its management. Though usually expressed in financial terms it can also be expressed in time or other designation considered appropriate. Serves as a basis for managerial control.

burden, An amount added to the direct cost of a function to represent the indirect cost, i.e., manufacturing burden, payroll burden.

capital budgeting, The task of classifying, evaluating, and ranking proposed capital expenditures for the purpose of comparison and selection, combined with the analysis of the financing requirements.

capital, cost of, The weighted average of (1) the after-tax cost of long term debt, (2) the yield on any outstanding preferred stock, (3) the cost of common equity capital, and (4) the cost of retained earnings.

capital, fixed, The investment in a plant. Represents real estate, equipment, buildings, furniture and fixtures, transportation equipment used directly in the production of a product or service. It includes all the costs incident to getting the facility in place and in operating condition including legal costs, purchased patents, paid-up licenses, and installation and start-up expenses.

capital, working, The funds in addition to fixed capital which a company must contribute to the project (excluding start-up expense) to get the project started and meet subsequent obligations as they come due. It is calculated as the difference between current assets (cash, receivables, and inventories), and current liabilities (accounts, notes, and taxes payable). Characteristically, these funds can be converted readily into cash. Working capital is normally assumed to be recovered at the end of the project without loss.

capitalized costs, The present worth of the cost of renewing and operating a facility perpetually. As a system for comparing proposed investments, it requires assumption of a specific minimum acceptable interest rate. Also used to describe the amount expended for the purchase or manufacture of assets which have an expected utility beyond the current accounting period.

cash flow, The net flow of dollars into or out of the proposed project. The algebraic sum, in any time period, of all cash receipts, expenses, and investments. Also called cash generated.

change notice, A formal notification of an alteration to a contract which will affect its terms and conditions. A means of evaluating, controlling, and possibly obtaining reimbursement for variances in the anticipated cost of the project from an established budget or control estimate.

code of accounts, A system used for classification and categorization of expense and revenue items.

construction cost, A general term used to describe either the total cost associated with the construction of a facility or the costs incurred at the construction site. The sum of all costs, direct and indirect inherent in converting a design plan for material and equipment into a project ready for operation, i.e., sum of field labor, supervision, administration, tools, field office expense and field purchased material costs, engineering, etc.

contingencies, Specific provision for unforeseeable elements of cost within the defined project scope; particularly important where previous experience relating estimates and actual costs has shown that statistically, unforeseeable events which will increase costs are likely to occur.

contracts, Legal agreements between two or more parties, some types of which are enumerated below:

1. In Cost Plus contracts the Contractor agrees to furnish to the client services and material at actual cost, plus an agreed-upon fee for his services. This type of contract is employed most often when the scope of services to be provided is not well defined.

   a) *Cost Plus Percentage Burden* and Fee,* The client will pay all costs as defined in the terms of the contract, plus "burden and fee" at a specified percent of the labor costs which he is paying for directly. This type of contract generally is used for engineering services.

   *In contracts with some Governmental agencies, these items are included in indirect cost.

   b) *Cost Plus Fixed Fee,* The client pays costs as defined in the contract document. Burden on Reimbursable Technical Labor cost is considered in this case as part of cost. In addition to the costs and Burden, the client also pays a fixed amount as the contractor's "fee."

c) *Cost Plus Fixed Sum,* The client will pay costs defined by contract plus a fixed sum which will cover "non-reimbursable" costs and a fee. This type of contract is used when the client wishes to have the contractor assume some of the risk for items which would be reimbursable under a Cost Plus Fixed Fee type of contract.

d) *Cost Plus Percentage Fee,* The client pays all costs, plus a percentage for the use of the contractor's organization.

e) *Cost Plus Incentive Fee,* The client pays all costs, plus a percentage for the use of the contractor's organization. In addition, the contract defines an additional fee for performance which exceeds that required by the contract and a penalty should the latter not be met.

2. Fixed Price types of contract are ones wherein a Contractor agrees to furnish services and material at a specified price, possibly with a mutually agreed-upon escalation clause. This type of contract is most often employed when the scope of services to be provided is well defined.

a) *Lump Sum,* Contractor agrees to perform all services as specified by the contract for a fixed amount. A variation of this type may include a Turn-Key arrangement where the contractor guarantees quality, quantity and yield on a process plant or other installation.

b) *Unit Price,* Contractor will be paid at an agreed-upon unit rate for services performed. For example, technical manhours will be paid for at the unit price agreed upon.

c) *Guaranteed-Maximum (Target Price),* A Contractor agrees to perform all services as defined in the contract document guaranteeing that the total cost to the client will not exceed a stipulated maximum figure. Quite often, these types of contracts will contain special share-of-the-saving arrangements to provide incentive to the Contractor to minimize costs below the stipulated maximum.

d) *Bonus-Penalty,* A special contractual arrangement usually between a Client and a Contractor wherein the Contractor is guaranteed a bonus, usually a fixed sum for each day the project is completed ahead of a specified schedule and/or below a specified cost, and agrees to pay a similar penalty for each day of completion after the schedule date or over a specified cost up to a specified maximum either way. The penalty is sometimes referred to as liquidated damages.

cost accounting, A subsidiary accounting system designed to compile detailed data on the expenditures required to produce a product or service.

cost capacity factor, An estimating technique whereby two quantities are compared, one with a known cost in order to evaluate the unknown cost. An exponential power is often used to extend the capacity ratio.

cost center, A designated function or organizational unit for which all costs that are incurred by and attributable to it are compiled.

cost control, The application of procedures to follow the progress of design and construction projects or manufacturing operations in order to minimize cost with the objective of increasing profitability and assuring efficient operations. There are three essential elements of control. The first is to establish the optimum condition, the second is to measure variation from the optimum, and the third is to take corrective action in order to minimize this variation. The application of these procedures attempts to limit costs to those authorized for capital projects or cost standards, focuses control efforts where they will be most effective, and achieves maximum control at minimum operating cost.

cost, controllable, That classification of costs manageable by individuals responsible for authorizing these expenditures. Used for cost control.

cost engineering, Engineering discipline in which judgement and experience are utilized in the application of scientific principles and techniques to problems of cost estimation, cost control, business planning, and management science. A certified cost engineer is one who has met the certification requirements of the American Association of Cost Engineers.

cost function, An expression defining the relationship between total cost and output.

cost index, A number which relates the cost of an item at a specific time to the corresponding cost at some arbitrarily specified time in the past.

cost removal, The costs which must be incurred to remove a capital asset from its location.

cost trending, An ongoing projection of project cost due to variations in cost, scope, design evolution, or market condition from a predetermined budget. Used to evaluate fiscal impact of such changes on a project.

cost variance, The difference between the pre-established cost, budget or standard cost and the actual cost incurred.

depletion, The allocation of the value of the quantity of natural resources extracted from a deposit (applying to resources not replaceable by human beings). A non-cash expense.

depreciation, The allocation of the cost of fixed capital assets less salvage (if any), over the estimated useful life of the unit in a systematic and rational manner.

development costs, Those costs specific to a project, either capital or expense items, which occur prior to commercial sales and which are found necessary to evolve the potentialities of that project for consideration and eventual promotion. Major cost areas include process, product, and market research and development.

direct costs, Any costs which are directly identifiable to a particular final cost objective.

discounted cash flow, A technique for converting future cash flow to a present value. (See present value.)

escalation, A variation in the cost of equipment, bulk material, or labor due to price level changes.

equivalent annual cost, The conversion by an interest rate and present worth technique of all capital and operating costs to a series of equal annual costs. As a system for comparing proposal investments, it requires assumption of a specific minimum acceptable interest rate.

estimate, A summation of the anticipated costs of all the elements of a project or effort as defined by an agreed upon scope. Generally, the accuracy of an estimate is dependent on the extent of information available at the time the estimate is made, the effort expended on it, and is contingent on the purpose for which the estimate is made. Ranked below are typical types of estimates in order of increasing detail and accuracy.

1. *Order of Magnitude,* An estimate made without detailed engineering data. Such an estimate could be made from cost capacity curves, use of scale up or down factors, and an appropriate ratio estimate.
2. *Conceptual Estimate,* Often used in the same sense as the order of magnitude estimate.
3. *Preliminary or Factored Estimate,* An approximation made with little engineering data. It may be prepared from cost capacity data, scale factors, or capacity ratios normally made from study drawings, preliminary plans, preliminary sizing, etc.
4. *Engineer's Estimate (Owner's Estimate),* A very close approximation made for bid evaluation purposes on a complete and approved bid package. Sometimes used as the basis for capital appropriations.
5. *Definitive Estimate (Budget Estimate),* An estimate prepared from very defined data. The engineering data includes as a minimum fairly complete plot plans and elevations, piping and instrument diagrams, on line electrical diagrams, equipment data sheets and quotations, structural sketches, soil data and sketches of major foundations, building sketches, and a complete set of specifications. It can be used as a project control document.
6. *Control Estimate,* A detailed estimate prepared in order that actual cost returns can be compared to the control estimate for cost control purposes. The Definitive Estimate often serves this function.

fee, The charge for the use of the contractor's organization for the period and to the extent specified in the contract.

field costs, All costs normally incurred at the job site.

field overhead costs (field distributable costs), All those costs, other than direct labor and material such as field administration, direct supervision, capital tools, etc., involved in producing a finished project.

fixed costs, Those costs independent of short-term variations in output of the system under consideration. Includes such costs as indirect labor, maintenance, technical service and laboratory expense, taxes and insurance, plant overhead, and administrative, selling and research expense.

fringe benefits, Employee welfare benefits; expenses of employment other than wages and salaries, such as holiday, sick leave, supplemental unemployment benefits, insurance, etc.

general and administrative expenses, Any management, financial or other expense which is incurred by or allocated to a segment of an organization, and which is for the general management and administration of the organization as a whole.

home office costs, Those necessary services, general and administrative costs involved in the conduct of everyday business which can be directly assigned to projects, processes, or end products, i.e., Engineering, Procurement, Expediting, Inspection, Estimating, Reproduction, Telephone, Telegraph, etc.

incremental costs (revenues), The cost (revenue) difference between two alternate courses of action. These should be based on changes which can actually be foreseen as occuring rather than having accounting conventions to approximate overheads, etc. Costs refer both to investment and annual operating costs.

indirect costs, In a project, all costs which cannot be specifically identified with direct accounts and are required for the completion of the installation includes material and labor which do not become a final part of the installation but which are required for the orderly completion of the installation. May include starting costs, contractor's fees, insurance, taxes, etc. In manfacturing, those costs not directly assignable to the end product or process and which may include plant overhead, material handling and maintenance charges, etc.

intangible capital asset, An asset which has no physical substance, but has a value which is generally the present worth of its future estimated earnings.

investment, Used interchangeably with capital required to expand an existing facility or initiate a new venture.

labor cost, The cost of time, including the base hourly pay, fringe benefits, and supplementary compensation paid to workmen up to but not including the level of general foreman.

leaseback, A method of financing capital equipment or facilities where in a simultaneous transaction an asset is sold by the user to a financial institution which executes a lease for the use of the asset by the seller.

leverage (trading on equity), The use of borrowed funds or preferred stock in the intent of employing these "senior" funds at a rate of return higher than their cost, in order to increase the return on the investment of the residual owners.

life cycle cost, The original in place cost of an asset plus the anticipated operation and maintenance costs attributed to it during its useful life.

maintenance costs, The expense, both for labor and materials, required to keep equipment or other installations in suitably operable condition. Maintenance does not usually include those items which cannot be expended within the year purchased. These are considered fixed capital.

manufacturing costs, The total of operating costs chargeable to production of a given product, usually expressed in cents or dollars per unit of production. Includes direct expenses such as ingredients, labor, materials, and also such fixed or allocated charges as maintenance, depreciation, insurance, and property taxes. Overheads within the plant are usually considered within manufacturing costs. Sales, administrative and research overheads, and freight are usually not included.

MAPI method, A system of analysis based on the discounted cash flow procedure. By formula it compares making an investment now with going on without investing for one more year. (MAPI = Machinery and Allied Products Institute.)

marginal cost, The change in total cost caused by varying the output by one unit.

market factors, Those economic forces which may operate to cause price changes between anticipated and actual price. These can be expressed as a numerical index which relates current market level to a base or standard level at a previous calendar time.

Monte Carlo method, A simulation method used to deal with the problem of uncertainty in conformance with probabilities involved through the use of random numbers. Used in contingency analysis to determine the probabilistic spread of an estimate in order to evaluate the risk involved.

offsites, General facilities outside battery limits such as file storage, auxiliary systems, utilities, and administrative buildings. Also facilities geographically remote to central facilities but which are considered operationally and financially a part of the central facilities.

operating costs, Charges for labor, material and contracted services which can be assigned, or allocated, to a specific manufacturing operation.

opportunity cost, The profits from alternative ventures that are foregone by using limited facilities for a particular purpose.

optimum plant size, The plant capacity which represents the best balance between the economics of size and the cost of carrying excess capacity during the initial years of sales.

overhead, A cost or expense inherent in the performing of an operation, i.e., engineering, construction, operating, or manufacturing which cannot be charged to or identified with a specific part of the work, product, or asset and, therefore, must be allocated on some arbitrary basis believed to be equitable, or handled as a business expense, independent of the volume of production.

parametric estimating, An estimating technique which is based on the definition and projection of the major parameters determining the estimate, i.e., cost, time, etc. These parameters and their behavior are developed from historical data, sometimes by application of econometric analysis. Care must be taken that should conditions or technology have changed from the time that these parameters were defined, that these be updated to reflect these or anticipated changes. Used to obtain preliminary estimates and establish value of cost trends.

payout time (payback), The time required to recover the original fixed investment from profit and depreciation. Usually, but not always, calculated after taxes and based on capacity operations. Most recent practice is to base payout time on an actual sales projection. Also called pay-off or payback period. It is not satisfactory for comparing projects economically.

present worth (present value), The discounted value of a series of cash flows at any arbitrary point in time. Also, the system of comparing proposed investments which involves discounting at a known interest rate (representing a cost of capital or a minimum acceptable rate of return) in order to choose the alternative having the highest percent value per unit of investment.

productivity, A relative measure of work output calculated as the ratio of output to input or as the quantity of work produced compared to a norm based on measurement, experience, or index. Often used as a relative measure of labor efficiency which is indicated by a comparison of the actual work produced to the expected work output.

productive capacity, A measure of an organization's ability to produce within a given time frame.

profit-gross, The figure determined by subtracting the cost of goods sold from sales revenue.

profit-net, Residue after subtracting all expenses, including interest and income taxes, from revenues generated.

profitability, A measure of the excess of income over expenditure during a given period of time.

profitability index (PI), A measure used to evaluate and compare prospective capital investments. It reflects in a single number both the dollar and time values of all money involved in a project. The PI of a project is the ratio of the present value of its estimated cash inflow compared to the present value of its estimated cash outflow. For a project to be acceptable its PI must be equal or greater than one.

project life (economic life), Total years of operation for any facility. Sometimes, but not necessarily, equal to depreciable life.

project scope, The documented definition of a facility expressed in quality and quantity. Expresses the institutional purpose and capacity of the project.

proposal, Any offer or other submission used as a basis for pricing a contract, contract modification, or termination settlement of for securing payments thereunder.

rate of return on investment, A ratio relating profit or cash flow incomes to investments. Several methods of calculating this ratio are in use.

replacement, A facility proposed to take the place of an existing facility, without increasing its capacity, caused either by obsolescence or physical deterioration.

replacement cost, The lowest amount that would have to be paid in the normal course of business to obtain a new asset of equivalent operating or productive capacity. Replacement cost (new) is the total estimated current cost of replacing total productive capacity at the end of the year while depreciated replacement cost is the replacement cost (new) adjusted for the already expired service potential of such assets.

reproduction cost, The amount required to exactly reproduce the existing facilities.

research expense, Those continuing expenses required to provide and maintain the facilities to develop new products and improve present products.

risk, The degree of dispersion or variability around the expected or "best" value which is estimated to exist for the economic variable in question; i.e., a quantitative measure of the upper and lower limits which are considered reasonable for the factor being estimated.

risk analysis, A procedure used to assess the amount of exposure a particular estimate is subject to by defining the accuracy of the estimate in probabilistic terms.

sales profile, The growth or decline of historical or forecast sales volume, by years.

sales price, The revenue received for a unit of product. Gross sales price is the total amount paid. Net sales are gross sales less returns, quantity discounts, and allowances.

schedule, A logical plan showing the sequence and duration of all activities which must be accomplished to achieve some desired end by a specific date.

selling expense, The total expense involved in marketing the products in question. This normally includes direct selling costs, advertising, and customer service.

sensitivity analysis, An analysis undertaken to determine how the rate of return varies when key factors which determine this rate change; e.g., revenue, cost capacity. Used to factor in uncertainty in the analysis of a prospective capital investment.

shutdown point, The operating level at which it becomes less expensive to close the plant and pay remaining fixed expenses out-of-pocket, rather than continue operation.

site preparation, Preparation of a site for construction activities, grading, landscaping, installation of roads and siding on an area of ground upon which anything previously located had been cleared so as to make the area free of obstructions, entanglements or possible collisions with the positioning or placing of anything new or planned.

standard cost, A benchmark cost which is based on pre-established measures that are expected to remain unchanged over an extended period of time. Should be periodically reviewed.

start-up costs, Presale operating costs to bring the plant on stream. Includes employee training equipment tests, process adjustments, as well as the usual manufacturing costs. It is generally assumed that no revenue is obtained from products made during this period.

sunk cost, Money which has been spent and cannot be recovered.

taxes payable, Tax accruals due within a year.

turnover ratio (capital), The ratio of annual sales to investment. Inclusion of working capital is preferrable, but not always done. Considered by some to be a reasonable basis for an estimate of facilities cost for new products similar to existing products. Ranges around 1.0 for many chemical plants. The product of turnover ratio and profit margin on sales gives a return-on-investment measure.

uncertainty, Unknown future events which cannot be predicted quantitatively within useful limits; i.e., accidents which destroy invested facilities, a major strike, a competitor's innovation which makes the new product obsolete, etc.

unallowable cost, Any cost which under the provisions of any pertinent law, regulation or contract cannot be included in prices, cost, reimbursement or settlement under a contract to which it is allocable.

unit cost, $/unit of production. Usually total cost divided by units of production, but a major cost divided by units of production is frequently referred to as a unit cost; i.e., the total unit cost is frequently sub-divided into the unit costs for labor, chemicals, etc. Sometimes expressed in man-hours/unit of production.

unitizing plant capital cost, Process of allocating all applicable cost to a production facility.

value engineering, The evaluation of a project or product for the purpose of reducing cost or increasing utility without increasing cost.

variable costs, Raw materials costs, by-product credits, and those processing costs which vary with plant output (such as utilities, catalysts, chemicals, packaging, and labor for batch operations).

venture capital, Capital invested in technology or markets new to the particular organization and which are considered to involve a relatively high degree of risk and a superior potential payoff.

Data Processing and Systems Design Z94.3

3

In the early years of digital computer proliferation, there was a rapidly increasing need for many technical and managerial personnel to learn to speak the new language of data processing. This need led to the publication of a number of different glossaries of terms. Some of these glossaries were pocket-size lists of the most common data processing terms. Others were voluminous editions of every conceivable word associated with computers. Some were published by computer manufacturers. Others were published outside the industry.

Now, with computer usage well-established, there is less of a need to publish volume after volume of duplicate definitions. Many of the terms are no longer new (having been used for a few decades by now); and, many people already know most of the terminology they need.

Hence, what is needed is not *another* data processing glossary, but, rather, a *widely distributed standard* glossary.

The present Subcommittee has examined both the 1970 version and the forthcoming revised version of the ANSI X3.12 Vocabulary for Information Processing. Since the 1972 version of the Z94.3 Terminology is mostly a subset of the revised version of the X3.12 manual, and since the X3.12 manual (even in its revised form) is not physically much larger than the Z94.3-1972 edition, it was the Subcommittee's recommendation that the X3.12 manual become the widely distributed standard glossary mentioned above.

However, the Subcommittee felt that there were a number of data processing terms omitted from the revised X3.12 glossary which were important to Industrial Engineers. These terms are contained in this "revised" and dramatically shortened Z94.3-1983.

It is our hope that future revisions of Z94.3 will continue this practice of supplementing the X3.12 manual with terms appropriate to Industrial Engineers which are not included in X3.12.

It is the Subcommittee's strong belief that a proliferation of similar glossaries in any field is a *disservice* to that field. This is the reason we have chosen to take our current approach towards the Z94.3 revision.

In addition to the Z94.3-1983 and X3.12 glossaries, the following volume is recommended to all Industrial Engineers intending to get *deeply* involved in the very technical aspects of the data processing field:

Data Processing Glossary
IBM Corporation
Publication number: GC20-1699-4

The Subcommittee wishes to thank Mr. Martin Weik, Chairman of the X3.5 Subcommittee, for providing us with a pre-publication copy of the revised X3.12 standard.

Members of the Subcommittee were:

Dr. Victor B. Godin, P.E., Chairman
Associate Professor of Business Administration
Northeastern University

Dr. Lewis H. Geyer
Professor of Industrial Engineering
Northeastern University

Dr. Ronald F. Perry
Associate Professor of Industrial Engineering
Northeastern University

acoustic coupler, A modem which converts digital data into signals which can be transmitted via a conventional telephone hand set. At the other end of the line, another modem converts the data back to digital form.

acronym, A word formed by combining the initial letter or letters of a name or phrase, for example, FORTRAN for FORmula TRANslating system.

APL, A Programming Language, A problem solving language designed for use at remote terminals. It offers special capabilities for handling arrays and for performing mathematical functions.

application, The problem or system to which some processing equipment or technique is applied.

application package, A set of computer routines designed for a designated application. Usually, the routines are written in a general way and have to be modified to meet each user's specific needs.

applications programmer, Writes and tests programs for applications, usually as deliniated by the systems analyst.

audio response unit, A device that accepts digitally-coded input and converts it to voice messages which can be transmitted over telephone lines.

backup, Pertaining to equipment or procedures available for use in the event of failure or overloading of the visual equipment or procedures.

BASIC, Beginner's All-purpose Symbolic Instruction Code, A programming language commonly used by time-sharing terminals.

bpi, (1) Bits per inch. (2) Bytes per inch.

bps, bits per second, In serial transmission, the instantaneous bit speed with which a device or channel transmits characters.

batch total, A sum of a set of items used to check the accuracy of operations on a particular batch of records.

baudot code, A 5-bit code used in telegraphy for more than a hundred years and still used in data communications and punched tape.

brush, An electrical conductor to sense the presence of holes in a punched card.

CAI, Computer Assisted Instruction.

computer assisted instruction, A data processing application in which a computing system is used to assist in the instruction of students. The application usually involves a dialogue between the student and a computer program which informs the student of mistakes as they are made. Abbreviated CAI.

configuration, The group of machines, devices, and programs that make up a data processing system.

control card, A punched card containing input data required for the application of a general routine; e.g., one of a group of cards directing the operating system to load and execute a particular program.

cylinder, The group of locations that the read/write heads of a magnetic disk storage device can access without any physical movement.

data base administrator, Administers, controls, and safeguards the organization's data base in centralized data processing systems.

Data-phone, The Bell System's trademark for its data communication which operates on the public dial telephone network.

data preparation personnel, Convert data or program material into machine readable form from formatted alphanumeric text using keypunch, keytape, keydisc, or similar devices.

data set, (1) A collection of data items in a specified arrangement. (2) A device providing the interface between a communications link and a data processing system; e.g., a modem.

end of file mark, A special character or symbol denoting the end of a file. Abbreviated EOF.

EOF, End Of File mark.

facilities management, The management of the data processing facilities of a firm by an outside service organization.

FD or FDX, Full Duplex.

feasibility study, The initial formal investigation of an idea or concept culminating in a report containing (1) a user-needs analysis, (2) descriptions of apparently feasible system(s) which appear to meet the requirements, (3) an estimate of the development, implementation, and operating costs, and (4) a recommendation for action.

frequency modulation, In data communications the process of varying the frequency of a carrier signal to transmit binary data. Abbreviated FM.

GIGO, A contraction of "Garbage In, Garbage Out" which emphasizes that the output of a data processing system cannot be of a higher quality than the input data or the instructions.

hard copy, Documents containing data in a form suitable for permanent retention; e.g., printed reports.

HD or HDX, Half Duplex.

implementation, The steps involved in installing and starting successful operation of a system.

information systems planner, Plans the future course of information system development and services in liaison with management and present and potential user personnel.

input output control system, A standard set of routines which initiate and control the input/output process of a computer system. Abbreviated IOCS.

integrated circuit, An electronic circuit whose component parts are fabricated and assembled in a single integrated process. The circuit cannot be disassembled without destroying it.

IOCS, Input Output Control System.

iterative, A sequence of operations which is executed repeatedly until some condition is satisfied.

keyboard, The whole arrangement of keys on a card punch, teletype, or other data communication or origination device, consisting of special symbols, numerals, and letters.

large-scale integration, Design concept which embodies

miniaturized circuits containing perhaps 1000 circuits on a single silicon chip (semiconductor). Abbreviated LSI.

librarian, Maintains system and program documentation. Maintains system for safeguarding, issuing and re-storing data files on tape or disc.

listing, A list of instructions or statements usually produced on a printer.

logon or login, The procedure by which a user begins a terminal session.

logoff or logout, The procedure by which a user ends a terminal session.

LSI, Large-Scale Integration.

maintenance programmer, Makes and tests necessary changes to existing applications programs.

microcomputer, A microprocessor used in conjunction with memory components and input/output devices. (See microprocessor.)

microprocessor, One or more large-scale integrated circuits (LSI chips) that perform the basic functions of a computer processing or control unit (i.e., retrieving instructions from memory and executing them). (See microcomputer.)

microsecond, One millionth of a second. Abbreviated μsec or μs.

minicomputer, A small digital computer capable of performing the same functions as larger digital computers. Lower price and smaller capacity are usually the distinguishing features.

MOS, Metal Oxide Semiconductor, A silicon chip containing many miniature circuits used in computer control components and main memory.

nanosecond, One billionth of a second. Abbreviated nsec or ns.

picosecond, One thousandth of a nanosecond. Abbreviated psec or ps.

position description, See the following customary positions: (1) systems analyst, (2) programmer, (3) operator, (4) data preparation personnel, (5) information systems planner, (6) systems programmer, (7) applications programmer, (8) maintenance programmer, (9) data base administrator, (10) librarian.

preliminary design, See feasibility study.

primary storage, Same as main storage.

program compatibility, The characteristic that enables programs written for one computer system to be executed on another with identical results.

record mark, A special character used in some computers to limit the number of characters in a data transfer operation or to separate blocked records on tape.

RPQ, Request for Price Quotation.

rewind, To return a magnetic tape, punched tape or film to its beginning.

secondary storage, Same as auxiliary storage.

semiconductor, A solid with electrical conductivity between that of metals and insulators. Examples of semiconductor elements are transistors and crystal diodes.

source document, A document containing data which is to be extracted for input to a data processing system.

subprogram, Part of a larger program that can be converted to machine language independently.

systems analyst, Works with users to define information system performance requirements and specifications. Designs and develops processing system and user procedures to meet requirements and specifications.

systems programmer, Writes and tests software for operating systems and data base management systems.

teleprocessing, A trademark of IBM's line of data communications equipment.

time slice, (1) An interval of time on the central processing unit allocated for use in performing a task. Once the interval has expired, CPU time is allocated to another task; thus a task cannot monopolize CPU time beyond a fixed limit. (2) In systems with time sharing, a segment of time allocated to a terminal job.

unit record, A card containing one complete record; a punched card.

update, To incorporate into a master file the transactions recorded on a detail file.

Distribution and Marketing
Z94.4

4

Col. Lyndall Urwick, in THE GOLDEN BOOK OF MANAGEMENT, which he edited for CIOS (Comité International de l' Organizaion Scientifique), credits Dexter Kimball (1865-1952), mechanical engineer and first Dean of Cornell's Unified School of Engineering, with having instituted the first course in works administration in 1904. It was not until later that the newly established Harvard School of Business Administration followed his example in presenting a course based on the principles of management, with proper credit to Taylor's work in shop management. It was Dean Kimball, too, who felt that the field of distribution deserved individual attention, and ASME eventually had a distinct division for it. In fact, distribution forms an integral part of all of the ASME TEN YEAR MANAGEMENT REPORTS. This was especially true during the Great Depression when it was felt that a major problem was improper distribution.

In further development, it did not take long for industrial engineers to realize that designing and producing a product was only part of the problem. Why was a product needed in the first place? What features must it display? And, after it was produced, how was it to get into the hands of the consumer? As one industrial engineer expressed it, "Marketing and Distribution are concerned with everything that comes before a product is designed and everything that comes after it is produced."

Hence, this section is devoted to these aspects. Moreover, much progress has been made since the 1972 edition of Z94.4: DISTRIBUTION AND MARKETING. These fields have matured sufficiently so that many techniques of statistics and operations research, then only suggested, are now in common use. Although many of these terms are properly defined in this edition — as well as in the 1972 edition — they are listed here and cross-referenced with other sections whenever possible and whenever the proper expertise is needed. (Many of these cross-references have been suggested by Dr. D. R. Herzog, Governors State University, and Mr. George Hawley, who was Chairman of the Z94.8 Subcommittee on APPLIED MATHEMATICS.) Finally it must be noted that many of the terms contained in this present section may not seem to be specifically related to industrial engineering, but they are essential for a working knowledge for an IE working in these and related fields.

Many terms have been added in this section to the basic Z94.4: DISTRIBUTION AND MARKETING which was published in 1972. For example, many techniques of operations research and statistics have, since then, matured sufficiently to become part of the daily vocabulary of persons — especially industrial engineers — working in the field. Other definitions, having become somewhat dated, have been reinterpreted. In addition, a number of modifications have also been made.

Among those who contributed their time, effort, and expertise are the following:

Richard A. Forberg (Co-chairman)
Manager, Management Systems Division, The Procter & Gamble Company (Retired)

George D. Montillon (Co-chairman)
Manager, Design and Development Department, Management Systems Division, The Procter & Gamble Company (Retired)

Alan M. Gast
Manager, Technical Staff Department, Management Systems Division,The Procter & Gamble Company

Donald E. Stiling
Director, Materials Planning, International Paper Company

accounting prices, See Applied Mathematics.

activity analysis problem, See Applied Mathematics.

advertising evaluation, A group of techniques designed to reveal the relative effectiveness of various advertising strategies and tactics in the light of competitive activity. (See marketing research.)

advertising research, See marketing research.

aggregate inventory, See Production Planning and Control.

analysis of variance, See Applied Mathematics.

anticipation, Stock built up for buffer production to meet, e.g. seasonal fluctuations in sales or a planned sales campaign, or to carry sales over plant vacation or maintenance shutdown. (See Production Planning and Control.)

assembly line balancing problem, See Applied Mathematics.

assignment problem (personnel assignment problem), See Applied Mathematics.

available stock, Amount of finished product in inventory to cover an order which has been (or may be) placed. (See Production Planning and Control.)

average inventory, In a sample inventory system this is the sum of half the lot sizes plus reserve stock; or half the maximum number of units in stock. (See Production Planning and Control.)

backhaul, Use of a common carrier for a return load to source after completion of a shipment.

backorder, See Production Planning and Control.

base inventory level, The normal aggregate inventory level made up of the aggregate lot size inventory plus the aggregate safety stock inventory but not taking into account the anticipation inventory that will result from the production plan. The base inventory should be made known before the production plan is made. Essentially, it is that inventory level necessary to minimize the probability of running out of stock — i.e., buffer stock (q.v.). (See Production Planning and Control.)

blind tests, Consumer preference tests in which care is exercised to avoid identification of the products or brands involved.

brand management, Product management as applied to a specific class of goods identified by name as the product of a single firm or manufacturer. (See product management.)

brand recall, Extent to which consumers can remember and state the brand name used in a previously seen or heard advertisement. (Contrast with product recall.)

brand switching, Sequential choices by consumers of different brands having the same or similar characteristics and functions.

breakeven point, See Cost Engineering.

buffer stock, That quantity of an item of inventory held in stock for absorbing expected variations in usage between the time reorder action is initiated and the first part of the new order is received in stock. (See Production Planning and Control.)

capacity planning, See Production Planning and Control.

capacitated transportation problem, A bounded transportation problem. (See Applied Mathematics.)

capital, working, See Cost Engineering.

capital, sustaining, The fixed capital requirements to (1) maintain the competitive position of a project throughout its life by improving product quality, related services, safety, or economy or (2) replace facilities which wear out before the end of project life. (See Cost Engineering.)

cash flow, See Cost Engineering.

centralized dispatching, See Production Planning and Control.

code dating, Identifying attributes of a product — especially date of manufacture, lot number, and recommended usage date — by a system of symbols (letters, numbers, or words).

commodity code, See Facility Planning.

commodity rate, See Facility Planning.

conjoint measurement (or tradeoff analysis), Methods for measuring the relative importance of various combinations of product attributes in terms of overall consumer satisfaction.

consumer research, See marketing research.

cost function, See Applied Mathematics.

critical path, Sequence of jobs or activities in a network analysis (q.v.) project such that the total duration equals the sum of the durations of the individual jobs in the sequence. There is no time leeway or slack (float) in activity along critical path (i.e., if the time to complete one or more jobs in the critical path increases, the total production time increases). (See Production Planning and Control.)

critical path scheduling or Critical Path Method (CPM), A network planning technique used for planning and controlling elements in a project. By showing each of these elements and associated lead time, the "critical path" can be determined. The critical path identifies those elements that actually control the lead time of the project. CPM uses vector diagrams. It is easily run on a computer. (See Production Planning and Control.)

customer service ratio, See Production Planning and Control.

cycle stock, One of the two main components of any item of inventory, the cycle stock is the most active part, i.e., that which depletes gradually and is replenished cyclically when orders are received. Another part of the item inventory is the safety stock which is a cushion of protection against uncertainty in the demand or in the replenishment lead time. The stock level necessary to prevent shortage during ordering, producing, and shipping of replacement work. (See Production Planning and Control.)

decision function, See Applied Mathematics.

decomposition principle, See Applied Mathematics.

demographics, Periodic measures relating to people in specific geographical areas — incuding income, living conditions, family size occupations, ethnic backgrounds, and educational levels (e.g., census tract data).

direct cost, See Cost Engineering.

direct planning, "Day-to-day" production planning and control, involving "how much" and "when" to make the order quantity calculated on a discount that the vendor will give if a fairly large minimum quantity is ordered. (See Production Planning and Control.)

discriminatory analysis, Refers to statistical methods used in classification problems. Given that an individual may have emanated from one of K populations, the major problem is to allocate it to the correct population with minimum error, usually based on multiple measurements on the individual and a prior set of similar measurements on individuals whose origin is known. (See Applied Mathematics.)

distribution, The broad range of activities concerned with efficient movement of finished products from the end of the production line to the consumer; in some cases it may include the movement of raw materials from the source of supply to the beginning of the production line. These activities include freight transportation, warehousing, material handling, protective packaging, inventory control, plant and warehouse site selection, order processing, market and sales forecasting, customer service, and attendent management information systems; in some cases it may include buying activities.

distribution center, Intermediate warehouse(s) where products from different sources are assembled for shipment to specific customer locations.

distribution cost analysis, See marketing cost analysis.

distribution function, The distribution function, F(x), of a variate X is the total frequency of members with variate values less or equal to X. (See Applied Mathematics.)

dynamic programming, See Applied Mathematics.

Economic Order Quantity (EOQ), See Production Planning and Control.

exponential smoothing, See Production Planning and Control.

factor analysis, See Applied Mathematics.

FIFO, First-In, First-Out. An inventory valuation method, in which costs of material are transferred in chronological order. (See LIFO, also Production Planning and Control.)

fishy-back, See Facility Planning.

fixed charge problem, See Applied Mathematics.

fractional factorial, An experimental design where only a fraction of all treatment combinations are used. These designs are set up to reduce the number of experimental units and yet estimate certain effects of interest. (See Applied Mathematics.)

freight classification, See commodity code, Facility Planning.

freight forwarder, See Facility Planning.

frequency distribution, A specification of the way in which frequencies of members of a population are distributed according to the values of the variates which they exhibit. (See Applied Mathematics.)

incremental cost, See Cost Engineering.

input-output coefficient, See Applied Mathematics.

integer linear programming, See Applied Mathematics.

interindustry (input-output) anaylsis, See Applied Mathematics.

interline, See Facility Planning.

inventory, See Production Planning and Control.

inventory control, See Production Planning and Control.

inventory management, See Production Planning and Control.

LIFO, Last-In, First-Out. An inventory valuation method, in which costs are transferred in reverse chronological order. (See FIFO, also Production Planning and Control.)

linear programming, See Applied Mathematics.

machine-assignment problem, See Applied Mathematics.

market, (1) The aggregate of forces or conditions within which buyers and sellers make decisions that result in the transfer of goods and services. (2) The aggregate demand of the potential buyers of a product or service under given conditions (such as price, availability, buyer awareness). In either case, the definition usually implies a specific geographical area. (See marketing.)

market analysis, See marketing research.

market research, See marketing research.

market share (or sales share), The ratio of the sales of a specific product or service to the total sales of a group of products or services which perform the same or similar functions. Various units may be appropriate for measuring sales in different applications; physical units and monetary units are most common.

market testing, The marketing for test purposes of a new or improved product and/or a new or improved advertising or sales promotion plan. Typically, market testing is done within a limited and self-contained territory. Tools of marketing research are used in interpreting the test. (See marketing research.)

marketing, The broad range of activities concerned primarily with the determination of consumer or user demands or desires, both existing and potential; the satisfaction of these demands or desires through innovation or modification; and the building of buyer awareness of product or service availability through sales and advertising efforts. (See market.)

marketing cost analysis, The study and evaluation of different marketing operations in terms of customers, marketing units, commodities, territories, or marketing activities and the determination of the relative cost or profitability of these factors. Typical tools include cost accounting and operations research techniques.

marketing information systems, Methods, either manual or computerized, for systematic gathering, editing, storage, and retrieval of data relevant to the marketing of goods and services. Such systems may include methods for summarizing and analyzing the data for either short-range or long-range purposes.

marketing model, Any representation which aids understanding of one or more characteristics of a market. Models take many forms, varying from concrete versions, such as physical models of equipment and facilities, to abstract versions, such as mathematical or statistical models of share interchanges among competing brands.

marketing planning, The work of setting up objectives for marketing activity and of determining and scheduling the detailed steps necessary to achieve such objectives, including priorities where necessary.

marketing policy, A course of action established to obtain consistency of marketing decisions and operations under recurring and essentially similar circumstances.

marketing research, The systematic gathering, recording, and analyzing of data about problems relating to the marketing of goods and services. Such research may be undertaken by impartial agencies or by business firms, or their agents. Marketing research is an inclusive term which includes various subsidiary types:

(1) market analysis, of which product potential is a type, which is the study of the size, location, nature, and characteristics of markets.

(2) sales analysis (or research), which is the systematic study and comparison of sales (or consumption) data.

(3) consumer research, of which motivation research is a type, which is concerned chiefly with the discovery and analysis of consumer attitudes, reactions, and preferences.

(4) advertising research, of which advertising evaluation is a type, which is aimed toward improving the use of nonpersonal aids to marketing, primarily the mass media.

The techniques of operation research and statistics are often useful in marketing research.

markoff process, A stochastic process such that the conditional probability distributions for the state at any future instant is unaffected by any additional knowledge of the past history of the system. (See Applied Mathematics.)

materials control, The function of maintaining a constantly available supply of raw materials, purchased parts and supplies needed for production — including requisitioning materials in economic quantities at their proper time; also includes receipts, storages, protection, etc. (See materials management, also Production Planning and Control.)

materials handling, An aspect of materials management relating specifically to the physical transfer of materials.

materials management, See Production Planning and Control.

maximum inventory, Sum of economic lot size plus twice the reserve stock. Especially of interest in allocating stockroom size. (See Production Planning and Control.)

Monte Carlo method, See Applied Mathematics.

Monte Carlo technique, Any procedure involving random sampling techniques to obtain a probabilistic approximation to the solution of a mathematical or physical problem. A procedure by which can be obtained approximate evaluations of mathematical expressions which are built up of one or more probability functions. (See Production Planning and Control.)

motivation research, A group of techniques which may be used to discover consumer oriented factors influencing purchase behavior. (See marketing research.)

moving average, See Applied Mathematics.

multiple regression, See Applied Mathematics.

multivariate analysis, See Applied Mathematics.

network analysis, Technique used in planning a project consisting of a sequence of activities and their interrelationship within a network of activities making up a project. (See critical path; also Production Planning and Control.)

network planning, A broad generic term for techniques used to plan complex projects. Two of the most popular techniques are PERT and CPM. (See Production Planning and Control.)

nonlinear programming, See Applied Mathematics.

non-metric scaling, Methods for constructing a spatial configuration of objects whose rank order of distances best reproduces the rank order of dissimilarities among the objects from original data.

non-parametric statistics, Methods of statistical inference that do not depend upon assumptions about an underlying probability distribution.

normal distribution, Continuous frequency distribution of infinite range represented by the equation

$$dF = \frac{}{\sigma\sqrt{2\pi}} e^{\left[-\frac{1}{2}\left(\frac{x-m}{\sigma}\right)^2\right]} dx,$$

$$-\infty \leqslant x \leqslant \infty, -\infty < m < \infty, 0 < \sigma < \infty$$

where m is the mean and σ the standard deviation. (See Applied Mathematics.)

operations research, See Applied Mathematics.

order bill of lading, See Facility Planning.

order picking, Selection and gathering of items which a customer is buying in a single lot at one time.

organization inventories, See Production Planning and Control.

outside warehouse, A warehouse separated physically from an associated factory by land owned by others.

over, short & damaged (O, S & D), A recapitulation of items requiring reconciliation with a supplier, a common carrier, or a customer, because goods received do not correspond to a bill of lading regarding quantity or quality.

parametric programming, See Applied Mathematics.

pareto curve, See Applied Mathematics.

payout time, See Cost Engineering.

perceptual mapping, Methods for measuring consumer perceptions of a product relative to competitive products based on a given set of product attributes.

piggy-back, See VOFC; see also Facility Planning.

plant site selection, Identification, measurement, and comparison of total costs for providing sets of goods or services to all buyers from existing or contemplated sources. Objectives of such analyses are to optimize total costs while maintaining appropriate customer service. Short term analyses are used to influence operating decisions re scheduling and distribution; long term analyses are used to influence capital investment decisions re adding capacity within existing plants or establishment of new plants. (See distribution.)

pool shipment, Goods ordered for two or more customers and/or destinations loaded into a single common carrier unit (as truck or rail car).

prebuilt display, A unit load which includes attention-attracting display material, designed for direct placement on the floor of a store.

preference tests, Consumer tests of competing products or brands to determine preferences regarding overall attributes.

present value, See Cost Engineering.

probability distribution, A distribution giving the probability of a value x as a function of x; or more generally, the probability of joint occurrence of a set of variables x_1 $x_2, \ldots, x_p$ as a function of those quantities. (See Applied Mathematics.)

product management, The planning, direction and control of all phases of the life cycles of products, including the creation or discovery of ideas for new products, the screening of such ideas, the coordination of the work of research and physical development of products, their packaging and branding, their introduction on the market, their market development, their modification, the discovery of new uses for them, their repair and servicing, and their deletion.

product manager, The executive who has the responsibility for the major portion of the activities listed under product management. The dimensions of this position vary widely among companies, ranging from sales promotion alone to the full scope of product management activity.

product potential, A class of studies designed to express the potential market for a product in such terms as the personal characteristics of individual consumers or the demographic characteristics of market areas. (See marketing research.)

product recall, Return of product to manufacturer for reprocessing or destruction because of failure to meet specifications. (Contrast with brand recall.)

product roll-out, Process of distributing a new product, or variation thereof, territory-by-territory, until entire marketing area has been covered.

production control, See Production Planning and Control.

profit objective (or profit forecast), A goal set by product management for profit to be achieved on a specified item of merchandise or on an entire line, based upon a sales forecast and a total cost analysis.

profitability index, See Cost Engineering.

project management, This includes the areas of project evaluation and scheduling plus project coordination and control. Useful techniques in project management include the concepts of cash flow and present worth, decision tree analysis, critical resource analysis and critical path scheduling.

promotion evaluation, A group of techniques designed to reveal the relative effectiveness of various promotion strategies and tactics which may include consideration of competitive activity. (See marketing research, sales analysis.)

quality assurance, See Applied Mathematics.

queueing theory, See Applied Mathematics.

random process, In general the term is synonymous with the more usual and preferable "stochastic" process. It is often used to denote a process where movement from one state to the next is denoted by a variate independent of the initial and final states. (See Applied Mathematics.)

random walk, See Applied Mathematics.

rate of return on investment, See Cost Engineering.

reorder point, See Production Planning and Control.

reorder quantity, See Production Planning and Control.

safety inventory, Quantity of an inventory item carried in stores or stock as a hedge against stockout resulting from above average or unexpected demand during procurement lead time. Other terms: MINIMUM OR PROTECTIVE INVENTORY, BUFFER INVENTORY, CUSHION, or RESERVE. (See Production Planning and Control.)

safety stock, See Production Planning and Control.

sales analysis (or research), Systematic study and comparison of sales or consumption data along the lines of market areas, organizational units, products or product groups, customers or customer groups, or such other units as may be useful. Typical analyses would include: (1) promotion evaluation, (2) quota assignment, (3) territory assignment. (See marketing research.)

sales forecast, An estimate of sales, in dollars or physical units, for a specified future period under a proposed marketing plan or program and under an assumed set of economic and other forces outside the unit for which the forecast is made. The forecast may be for a specified item of merchandise or for an entire line. (See profit objective.)

sales promotion, Those marketing activities, other than personal selling, advertising, and publicity, that stimulate consumer purchasing and dealer effectiveness, such as display, shows and exhibitions, demonstrations, and various selling efforts such as price-off, couponing, dealer allowances, and special offers.

sales quota assignment, A projected volume of sales assigned to a sales unit for use in managing marketing efforts. (See sales analysis.)

seasonal, Daily, weekly, or monthly sales data showing a more or less repetitive pattern. (See Production Planning and Control.)

selling expense, See Cost Engineering.

sensitivity analysis, See Applied Mathematics.

separable programming, See Applied Mathematics.

served available market, The portion of the total available market in which an operating unit can compete with its product lines.

sequential analysis, More correctly sequential probability analysis, a process by which statistical data are analyzed continuously as the sample accumulates. (See Applied Mathematics.)

shadow prices, See Applied Mathematics.

shipment planning, See direct planning.

shortest route problem, See Applied Mathematics.

simulation, See Applied Mathematics.

standard deviation, See Applied Mathematics.

Standard International Trade Classification (SITC), See Facility Planning.

statistical quality control, See Applied Mathematics.

stockout, See Production Planning and Control.

straight bill of lading, See Facility Planning.

sub-optimization, An operations research term describing a problem solution that is best from a narrow (but not overall) point of view. (See Production Planning and Control.)

territory assignment, A geographical area assigned to a sales unit for use in managing marketing efforts or to a producing unit for use in managing distribution efforts. (See sales analysis.)

time series, See Applied Mathematics.

total available market, The non-captive portion of the total market.

total market, The total consumption of a given product or service, including both captive and non-captive markets for products within the scope of an operating unit.

transportation problem, See Applied Mathematics.

transshipment problem, See Applied Mathematics.

traveling salesman problem, See Applied Mathematics.

unit load, An assembly of many units of materials and/or products into a compact group which can be transported or stored in a single move. Stability can be provided by stacking in interlocking patterns, by plastic sheet wrapping, by strapping, by using trays, etc. Movement can be made by forklift truck, clamp truck, hand truck, etc. A horizontal platform device such as a pallet, is frequently used as a base.

Universal Product Code (UPC), The 10 digit, all-numeric, code for representing grocery products. The first five digits uniquely identify the grocery manufacturer and the last five assigned by the manufacturer to identify his products.

value added by distribution, That portion of the value of a product or service to the consumer or user which results from distribution activities. This value includes such components as time utility and place utility.

value added by marketing, That portion of the value of a product or service to the consumer or user which results from marketing activities. This value includes such components as price reduction through economies of scale and buyer awareness of more desirable innovations in products or services.

Van-On-Freight-Car (VOFC), Transportation of truck trailers on railroad flatcars between major points, with pickup and delivery by truck.

warehouse site selection, Same as plant site selection, except that the number of alternatives to be considered is greater because of potential intermediate storage between sources and buyers. (See distribution.)

waybill, See Facility Planning.

Engineering Economy

Z94.5

5

This presentation consists of two parts:

Part A: SYMBOLOGY — STANDARD NOTATION FOR ENGINEERING ECONOMY — PARAMETERS AND INTEREST FACTORS

Part B: TERMINOLOGY — GLOSSARY OF TECHNICAL TERMS USED IN ENGINEERING ECONOMY

Both aspects were developed under the full cooperation of the Engineering Economy Division of the American Society for Engineering Education.

The SYMBOLOGY and TERMINOLOGY were compiled by using, as a basis, terms from current texts and brochures. Attempting to discover the most *common* definitions of these terms, the list (together with the definitions found) was distributed to some five hundred members of the DIVISION — with the request that each individual either choose that definition he/she considered most appropriate or supply his/her own. There was an approximate return of about twenty percent. The choices were tabulated, and the most widely chosen definition selected. In an overwhelming number of cases, there was a single outstanding definition for each term, and, consequently, no question as to preference. In a few cases votes were almost evenly divided between two definitions: in such cases, *both* definitions appear.

Among the texts examined were:

American Telephone and Telegraph: Engineering Economy, 3rd ed., 1977
Canada: Intermediate Economic Analysis for Management and Engineering, 1972
DeGarmo and Canada: Engineering Economy, 5th ed., 1973
Fabrycky and Thuesen: Economic Decision Analysis, 1974
Grant, Ireson, and Leavenworth: Principles of Engineering Economy, 6th ed., 1976
Morris: Engineering Economic Analysis, 1976
Newnan: Engineering Economic Analysis, rev. ed., 1977
Riggs: Engineering Economics, 1977
Roscoe: Project Economy, 1963
Rose: Engineering Investment Decisions, 1976
Smith: Engineering Economy, 3rd ed., 1980
Stermole: Economic Evaluation and Investment Decision Methods, 1974
Tarquin and Blank: Engineering Economy, 1976
Taylor: Managerial and Engineering Economy, 2nd ed., 1975
Thuesen, Fabrycky, and Thuesen: Engineering Economy. 5th ed., 1977
White, Agee, and Case: Principles of Economic Analysis, 1977
Winfrey: Economic Analysis for Highways, 1969

Finally, it is fitting and proper to acknowledge the debt to the following for their efforts and expertise:

Z94 SUBCOMMITTEE ON ENGINEERING ECONOMY

Co-CHAIRMEN

Gerald A. Fleischer
University of Southern California
John Mihalasky
New Jersey Institute of Technology

MEMBERS

Daniel L. Babcock
University of Missouri (Rolla)
John R. Canada
North Carolina State University
Gerald W. Smith
Iowa State University
Gerald J. Thuesen
Georgia Institute of Technology

Also:

COMMITTEE MEMBERS from the ASEE COMMITTEE ON ENGINEERING ECONOMY NOTATION STANDARDIZATION

Donald G. Newnan (Chairman)
San Jose State University
James R. Buck
Purdue University

John R. Canada
North Carolina State University
Walter J. Fabrycky
Virginia Polytechnic Institute and State University
Richard S. Leavenworth
University of Florida
James L. Riggs
Oregon State University
Gerald W. Smith
Iowa State University
Ralph O. Swalm
Syracuse University

The standards suggested herein were, after considerable deliberation, endorsed by the Engineering Economy Division of the American Society for Engineering Education. They are suggested for use in all future books and articles requiring symbols common to engineering economy. Cooperation by authors and publishers will result in the improved communication benefits of standardization which have been so abundantly demonstrated in many areas of endeavor.

Part (A) SYMBOLOGY MANUAL OF STANDARD NOTATION FOR ENGINEERING ECONOMY PARAMETERS AND INTEREST FACTORS

Table 1 lists suggested standard definitions and symbols for common parameters.

Table 2 lists the suggested functional format for designating compound interest factors.

Table 3 contains diagrams, algebraic forms, and simple explanation of uses for the compound interest factors, and is merely an explanatory supplement to Table 2. The form of the diagrams is not given as suggested standard.

TABLE 1
DEFINITIONS AND SYMBOLS USED FOR PARAMETERS
(Suggested Standards)

No.	Definition of Parameter	Symbol
1.	Effective interest rate per interest period (discount rate).	i
2.	Nominal interest rate per year	r
3.	Number of compounding periods	n
4.	Number of compounding periods per year	m
5.	Present sum of money. The letter "P" implies present (or equivalent present value).	P
6.	Future sum of money. The letter "F" implies future (or equivalent future value).	F
7.	End-of-period cash flows (or equivalent end-of-period values) in a uniform series continuing for a specified number of periods. The letter "A" implies annual or annuity.	A
8.	Uniform period-by-period increase or decrease in cash flows (or equivalent values); the arithmetic gradient	G
9.	Amount of money (or equivalent value) flowing continuously and uniformly during a given period	$\overline{P}$ or $\overline{F}$
10.	Amount of money (or equivalent value) flowing continuously and uniformly during each and every period continuing for a specific number of periods.	$\overline{A}$

TABLE 2
FUNCTIONAL FORMS OF COMPOUND INTEREST FACTORS

Ref. No.	Name of Factor	Functional Format
Group I. All cash flows discrete: end-of-period compounding		
1.	Compound Amount Factor (Single Payment)	(F/P, i%, n)
2.	Present Worth Factor (Single Payment)	(P/F, i%, n)
3.	Sinking Fund Factor	(A/F, i%, n)
4.	Capital Recovery Factor	(A/P, i%, n)
5.	Compound Amount Factor (Uniform Series)	(F/A, i%, n)
6.	Present Worth Factor (Uniform Series)	(P/A, i%, n)
7.	Arithmetic Gradient Conversion Factor (to Uniform Series)	(A/G, i%, n)
8.	Arithmetic Gradient Conversion Factor (to Present Value)	(P/G, i%, n)
Group II. All cash flows discrete: continuous compounding		
9.	Continuous Compounding Compound Amount Factor (Single Payment)	(F/P, r%, n)
10.	Continuous Compounding Present Worth Factor (Single Payment)	(P/F, r%, n)
11.	Continuous Compounding Sinking Fund Factor	(A/F, r%, n)
12.	Continuous Compounding Capital Recovery Factor	(A/P, r%, n)
13.	Continuous Compounding Compound Amount Factor (Uniform Series)	(F/A, r%, n)
14.	Continuous Compounding Present Worth Factor (Uniform Series)	(P/A, r%, n)
Group III. Continuous, uniform cash flows: continuous compounding (payments during one period only)		
15.	Continuous Compounding Present Worth Factor (Single, Continuous Payment)	(P/$\overline{F}$, i%, n)
16.	Continuous Compounding Compound Amount Factor (Single, Continuous Payment)	(F/$\overline{P}$, i%, n)
Group IV. Continuous, uniform cash flows: continuous compounding (payments during a continuous series of periods)		
17.	Continuous Compounding Sinking Fund Factor (Continuous, Uniform Payments)	($\overline{A}$/F, i%, n)

18.	Continuous Compounding Capital Recovery Factor (Continuous, Uniform Payments)	$(\overline{A}/P, i\%, n)$
19.	Continuous Compounding Compound Amount Factor (Continuous, Uniform Payments)	$(F/\overline{A}, i\%, n)$
20.	Continuous Compounding Present Worth Factor (Continuous, Uniform Payments)	$(P/\overline{A}, i\%, n)$

TABLE 3

DIAGRAMS, ALGEBRAIC FORMS, AND USES FOR COMPOUND INTEREST FACTORS

(Explanatory Supplement to Table 2)

Group I. All cash flows discrete: end-of-period compounding

Cash Flow diagram for factors 1 through 6 (and 9 through 14):

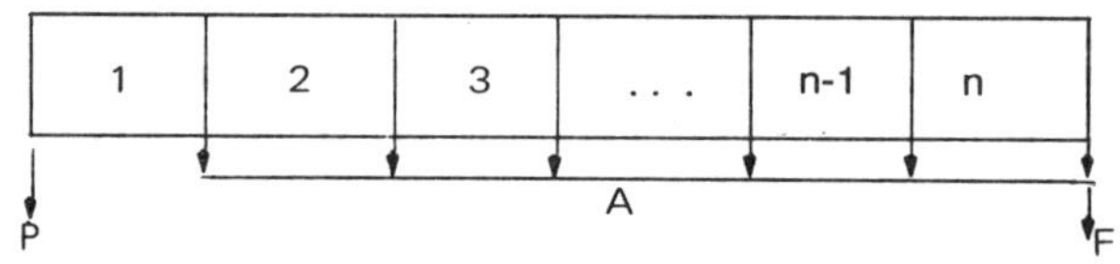

No.	Name of Factor	Algebraic Form	Use when:
1.	Compound Amount Factor (Single Payment)	$(1 + i)^n$	Given P, to find F
2.	Present Worth Factor (Single Payment)	$(1 + i)^{-n}$	Given F, to find P
3.	Sinking Fund Factor	$\frac{i}{(1 + i)^n - 1}$	Given F, to find A
4.	Capital Recovery Factor	$\frac{i(1 + i)^n}{(1 + i)^n - 1}$	Given P, to find A
5.	Compound Amount Factor (Uniform Series)	$\frac{(1 + i)^n - 1}{i}$	Given A, to find F
6.	Present Worth Factor (Uniform Series)	$\frac{(1 + i)^n - 1}{i(1 + i)^n}$	Given A, to find P

Cash flow diagram for factors 7 & 8

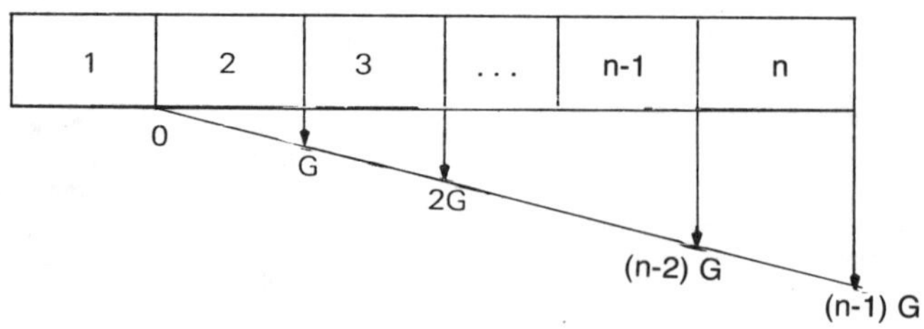

No.	Name of Factor	Algebraic Form	Use when:
7.	Arithmetic Gradient Conversion Factor (to uniform series)	$\left[\frac{1}{i} - \frac{n}{(1 + i)^n - 1}\right]$	Given G, to find A
8.	Arithmetic Gradient Conversion Factor (to present value)	$\left[\frac{1}{i} - \frac{n}{(1 + i)^n - 1}\right]\left[\frac{(1 + i)^n - 1}{i(1 + i)^n}\right]$	Given G, to find P

Group II. All cash flows discrete: continuous compounding

9.	Continuous Compounding Compound Amount Factor (Single Payment)	e^{rn}	Given P, to find F
10.	Continuous Compounding Present Worth Factor (Single Payment)	e^{-rn}	Given F, to find P
11.	Continuous Compounding Sinking Fund Factor	$\dfrac{e^r-1}{e^{rn}-1}$	Given F, to find A
12.	Continuous Compounding Capital Recovery Factor	$\dfrac{e^{rn}(e^r-1)}{e^{rn}-1}$	Given P, to find A
13.	Continuous Compounding Compound Amount Factor (Uniform Series)	$\dfrac{e^{rn}-1}{e^r-1}$	Given A, to find F
14.	Continuous Compounding Present Worth Factor (Uniform Series)	$\dfrac{e^{rn}-1}{e^{rn}(e^r-1)}$	Given A, to find P

Group III. Continuous, uniform cash flows: continuous compounding
(Payments during one period only)

Cash flow diagram for factors 15 and 16

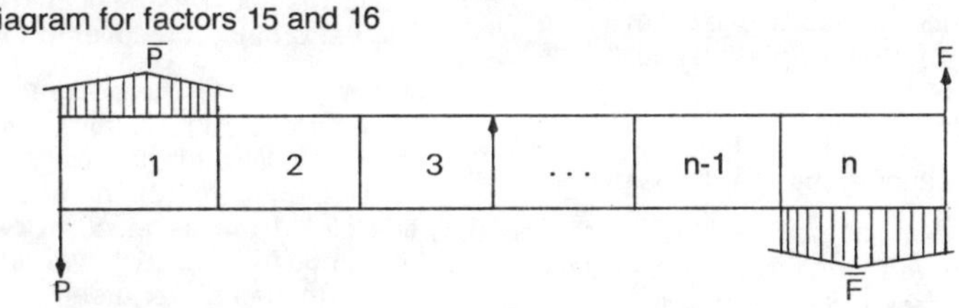

15.	Continuous Compounding Present Worth Factor (Single Continuous Payment)	$\dfrac{e^r-1}{re^{rn}} = \dfrac{i(1+i)^{-n}}{\ln(1+i)}$	Given F, to find P
16.	Continuous Compounding Compound Amount Factor (Single Continuous Payment)	$\dfrac{e^{rn}(e^r-1)}{re^r} = \dfrac{i(1+i)^{n-1}}{\ln(1+i)}$	Given P, to find F

Group IV. Continuous, uniform cash flows
(Payments during a continuous series of periods)

Cash flow diagram for factors 17 through 20

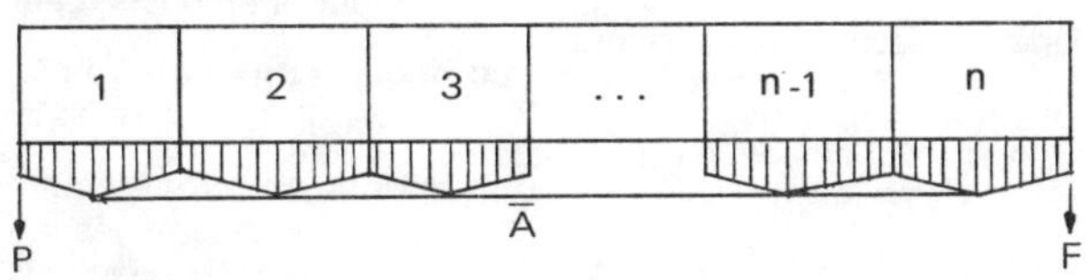

17.	Continuous Compounding Sinking Fund Factor (Continuous, Uniform Payments)	$\dfrac{r}{e^{rn}-1} = \dfrac{\ln(1+i)}{(1+i)^n-1}$	Given F, to find A
18.	Continuous Compounding Capital Recovery Factor (Continuous, Uniform Payments)	$\dfrac{re^{rn}}{e^{rn}-1} = \dfrac{(1+i)^n \ln(1+i)}{(1+i)^n-1}$	Given P, to find A
19.	Continuous Compounding Compound Amount Factor (Continuous, Uniform Payments)	$\dfrac{e^{rn}-1}{r} = \dfrac{(1+i)^n-1}{\ln(1+i)}$	Given A, to find F
20.	Continuous Compounding Present Worth Factor (Continuous, Uniform Payments)	$\dfrac{e^{rn}-1}{re^{rn}} = \dfrac{(1+i)^n-1}{(1+i)^n \ln(1+i)}$	Given A, to find P

Part (B) TERMINOLOGY — GLOSSARY OF TECHNICAL TERMS USED IN ENGINEERING ECONOMY

accelerated depreciation, A faster than historical rate of depreciation.

accounting life, The period of time over which asset cost will be allocated by the accountant.

amortization, (1) a) As applied to a capitalized asset, the distribution of the initial cost by periodic charges to operations as in depreciation. Most properly applies to assets with indefinite life. b) The reduction of a debt by either periodic or irregular payments. (2) A plan to pay off a financial obligation according to some prearranged program.

annual equivalent, In time value of money, one of a sequence of equal end-of-year payments which would have the same financial effect when interest is considered as another payment or sequence of payments which are not necessarily equal in amount or equally spaced in time.

annuity, (1) An amount of money payable to a beneficiary at regular intervals for a prescribed period of time out of a fund reserved for that purpose. (2) A series of equal payments occurring at equal periods of time.

annuity factor, The function of interest rate and time that determines the amount of periodic annuity that may be paid out of a given fund.

annuity fund, A fund that is reserved for payment of annuities. The present worth of funds required to support future annuity payments.

annuity fund factor, The function of interest rate and time that determines the present worth of funds required to support a specified program of annuity payments.

apportion, In accounting or budgeting, to assign a cost responsibility to a specific individual, organization unit, product, project, or order.

average-interest method, A method of computing required return on investment based on the average book value of the asset during its life or during a specified study period.

bayesian statistics, The translation of subjective forecasts into probability curves.

benefits, Consequences of an investment either measured in, or converted to, economic terms.

benefit-cost analysis, An analysis in which all consequences of the investment are measured in, or converted to, economic terms.

benefit-cost ratio, A measure, expressed as the ratio of benefits to costs.

benefit-cost ratio method, An analytical method in which benefit — cost ratios are computed to evaluate the desirability of alternatives.

book value, Original cost of an asset less the accumulated depreciation.

break-even chart, A graphic representation of the relation between total income and total costs for various levels of production and sales indicating areas of profit and loss.

break-even point, The rates of operations, output, or sales at which income will just cover expenses.

capacity factor, The ratio of output to maximum capacity.

capital, (1) The financial resources involved in establishing and sustaining an enterprise or project. (2) A term describing wealth which may be utilized to economic advantage. The form that this wealth takes may be as cash, land, equipment, patents, raw materials, finished products, etc. (See investment and working capital.)

capital budgeting, The allocation of funds to projects.

capital recovery, (1) Charging periodically to operations amounts that will ultimately equal the amount of capital expenditure. (2) The replacement of the original cost of an asset plus interest. (3) The process of regaining the new investment in a project by means of revenue in excess of the costs from the project. (Usually implies amortization of principal plus interest on the diminishing unrecovered balance.) (See amortization, depletion, and depreciation.)

capital recovery factor, A factor used to calculate the sum of money required at the end of each of a series of periods to regain the net investment of a project plus the compounded interest on the unrecovered balance.

capitalized cost, (1) The present worth of a uniform series of periodic costs that continue for an indefinitely long time (hypothetically infinite). Not to be confused with a capitalized expenditure. (2) The value at the purchase date of the first life of the asset of all expenditures to be made in reference to this asset over an infinite period of time. This cost can also be regarded as the sum of capital which, if invested in a fund earning a stipulated interest rate, will be sufficient to provide for all payments required to maintain the asset in perpetual service.

cash flow, (1) The real dollars passing into and out of the treasury of a financial venture. (2) The flowback of profit plus depreciation from a given project.

challenger, The proposed property or equipment which is being considered as a replacement for the presently owned property or equipment (defender).

common costs, Costs which cannot be identified with a given output of products, operations, or services.

compound amount, The future worth of a sum invested (or loaned) at compound interest.

compound amount factor, (1) The function of interest rate and time that determines the compound amount from a stated initial sum. (2) A factor which when multiplied by the single sum or uniform series of payments will give the future worth at compound interest of such single sum or series.

compound interest, (1) The type of interest that is periodically added to the amount of investment (or loan) so that subsequent interest is based on the cumulative amount. (2) The interest charges under the condition that interest is charged on any previous interest earned in any time period, as well as on the principal.

compounding, continuous, (1) A compound interest situation in which the compounding period is zero and the number of periods infinitely great. A mathematical concept that is practical for dealing with frequent compounding and small interest rates. (2) A mathematical procedure for evaluating compound amount factors based on a continuous interest function rather than discrete interest periods.

compounding period, The time interval between dates at which interest is paid and added to the amount of an investment or loan. Designates frequency of compounding.

contingent alternative, One that requires that some other alternative be accepted first, before it can be accepted.

cost-benefit analysis, See benefit-cost analysis.

cost effectiveness analysis, An analysis in which all prospective consequences of a proposed investment, economic as well as non-economic, are included.

cost of capital, The cost to borrow capital, usually expressed as a percentage.

costs, Consequences of an investment either measured in, or converted to, economic terms.

cutoff rate of return, The rate of return after taxes, that is determined by management, based on the supply and demand for funds, that will be used as a criterion for approving projects or investments. The cutoff rate of return may or may not be equal to the minimum attractive rate of return.

decision tree, A graphical representation of the anatomy of a decision, showing the interplay between a present decision, chance events, possible future decisions, and their results or payoffs.

decisions under certainty, Decisions that assume complete information and no uncertainty connected with the analysis of the decisions.

decisions under risk, Decisions in which the analyst elects to consider several possible futures, the probabilities of which can be estimated.

decisions under uncertainty, Decisions which the analyst elects to consider several possible futures, the probabilities of which *cannot* be estimated.

declining balance depreciation, Also known as percent on diminishing value. A method of computing depreciation in which the annual charge is a fixed percentage of the depreciated book value at the beginning of the year to which the depreciation applies.

defender, The presently owned property or equipment being considered for replacement.

deflation, A fall in the general price level, associated with a contraction in the supply of money and credit.

demand factor, (1) The ratio of the maximum instantaneous production rate to the production rate for which the equipment was designed. (2) The ratio between the maximum power demand and the total connected load of the system.

depletion, (1) A form of capital recovery applicable to extractive property (e.g., mines). Can be on a unit-of-output basis the same as straight-line depreciation related to original or current appraisal of extent and value of deposit. (Known as cost depletion.) Can also be a percentage of income received from extractions. (Known as percentage depletion.) (2) A lessening ofthe value of an asset due to a decrease in the quantity available. It is similar to depreciation except that it refers to such natural resources as coal, oil, and timber in forests.

depletion allowance, See depletion.

depreciation, (1) a). Decline in value of a capitalized asset; b) A form of capital recovery applicable to a property with two or more years' life span, in which an appropriate portion of the asset's value is periodically charged to current operations. (2) The loss of value because of obsolescence or due to attrition. In accounting, depreciation is the allocation of this loss of value according to some plan.

depreciation allowance, See depreciation.

development cost, The sum of all the costs incurred by an inventor or sponsor of a project up to the time that the project is accepted by those who will promote it.

direct cost, The traceable cost that can be segregated and allocated against specific products, operations, or services.

discounted cash flow, (1) The present worth of a sequence in time of sums of money when the sequence is considered as a flow of cash into and/or out of an economic unit. (2) An investment analysis which compares the present worth of projected receipts and disbursements occurring at designated future times in order to estimate the rate of return from the investment or project.

earning value, The present worth of an income producer's probable future net earnings, as prognosticated on the basis of recent and present expense and earnings and the business outlook.

economic life, The period of time, extending from the date of installation to the date of retirement from the intended service, over which a prudent owner expects to retain the property, so as to obtain a minimum cost.

economic return, The profit derived from a project or business enterprise without consideration of obligations to financial contributors and claims of others based on profit.

economy, The cost or profit situation regarding a practical enterprise or project, as in economy study, engineering economy, project economy.

effective interest, See interest rate, effective.

effectiveness, Consequences of an investment not measured in economic terms, e.g., reliability, maintainability, and safety.

endowment, A fund established for the support of some project or succession of donations or financial obligations.

endowment method, As applied to economy study, a comparison of alternatives based on the present worth of the anticipated financial events.

engineering economy, (1) The application of engineering or mathematical analysis and synthesis to economic decisions. (2) A body of knowledge and techniques concerned with the evaluation of the worth of commodities and services relative to their cost. (3) The economic analysis of engineering alternatives.

estimate, The magnitude as closely as it can be determined by the exercise of sound judgement based on approximate computations and is not to be confused with offhand approximations that are little better than outright guesses.

exact method, A method of computing required return combined with depreciation, based on the assumption that the present worth of these plus that of salvage value during the prescribed study period is equivalent to the first cost.

expected return, The profit anticipated from a venture.

expected yield, The ratio expected return/investment, usually expressed as a percentage on an annual basis.

fair rate of return, The rate of return which a public utility is entitled to earn in order to pay interest, dividends, and purchase new plant and equipment. The rate, or percentage, is usually determined by state or federal regulatory bodies.

first cost, The initial cost of a capitalized property, including transportation, installation, preparation for service, and other related initial expenditures.

fixed cost, A cost that tends to be unaffected by changes in the number of units produced.

future worth, (1) The equivalent value at a designated future date based on time value of money. (2) The monetary sum, at a given future time, which is equivalent to one or more sums at given earlier times when interest is compounded at a given rate.

going-concern value, The difference between the value of a property as it stands possessed of its going elements and the value of the property alone as it would stand at completion of construction as a bare or inert assembly of physical parts.

good-will value, That element of value which inheres in the fixed and favorable consideration of customers arising from an established well-known and well-conducted business.

increment cost, The additional cost that will be incurred as the result of increasing the output one more unit. Conversely, it can be defined as the cost that will not be incurred if the output is reduced one unit. More technically, it is the variation in output resulting from a unit change in input. It is known as the marginal cost.

independent alternative, One such that the acceptance of it has no influence on the acceptance of other alternatives in that set.

indirect cost, The traceable or common cost which is not allocated against specific products, operations, or services, but is allocated against "all" products, operations, or services by a predetermined formula.

inflation, A persistent rise in the general price level, usually resulting in the decline of purchasing power.

in-place value, A value of a physical property — market value plus costs of transportation to site and installation.

intangibles, (1) In economy studies, conditions or economy factors that cannot be readily evaluated in quantitative terms as in money. (2) In accounting, the assets that cannot be reliably evaluated (e.g., goodwill.)

interest, (1) a) Financial share in a project or enterprise. b) Periodic compensation for the lending of money. c) In economy study, synonymous with required return, expected profit, or charge for the use of capital. (2) The cost for the use ofcapital. Sometimes referred to as the time value of money. (See time value of money.)

interest rate, The ratio of the interest payment to the principal for a given unit of time, usually expressed as a percentage of the principal.

interest rate, effective, The true value of interest rate computed by equations for compound interest rate for a one year period.

interest rate, nominal, The customary type of interest rate designation on an annual basis without consideration of compounding periods. The usual basis for computing periodic interestpayments.

internal rate of return, The rate of return which takes into account only what a project or investment earns, and disregards earnings on cash flows reinvested in other projects.

investment, (1) As applied to an enterprise as a whole, the cost (or present value) of all the properties and funds necessary to establish and maintain the enterprise as a going concern. The capital tied up in the enterprise or project. (2) Any expenditure which has substantial and enduring value (at least two years' anticipated life) and which is therefore capitalized.

investor's method, See discounted cash flow.

irreducibles, (1) A term that may be used for the class of intangible conditions or economy factors that can only be *qualitatively* appraised (e.g., ethical considerations. (2) Matters that cannot readily be reduced to estimated money receipts and disbursements.

leaseback, A business arrangement whereby the owner of land, buildings, and other equipment sells such assets, and simultaneously leases them back under a long term lease.

life, (1) Economic: that period of time after which a machine or facility should be discarded or replaced because of its excessive costs or reduced profitability. The economic impairment may be absolute or relative. (2) Physical: that period of time after which a machine or facility can no longer be repaired in order to perform its design function properly. (3) Service: the period of time that a machine or facility will satisfactorily perform its function without major overhaul.

load factor, (1) A ratio that applies to physical plant or equipment: average load/maximum demand, usually expressed as a percentage. Equivalent to percent of capacity operation if facilities just accommodate the maximum demand. (2) Is defined as the ratio of average load to maximum load.

MAPI method, (1) A procedure for replacement analysis sponsored by the Machinery and Allied Products Institute. (2) A method of capital investment analysis which has been formulated by the Machinery and Allied Products Institute. This method uses a fixed format and provides charts and graphs to facilitate calculations. A prominent feature of this method is that it explicity includes obsolescence.

marginal cost, (1) The cost of one additional unit of production, activity, or service. (2) The rate of change of cost with production or output.

marginal analysis, An economic concept concerned with those elements of costs and revenue which are associated directly with a specific course of action, normally using available current costs and revenue as a base and usually independent of traditional accounting allocation procedures.

Matheson formula, A title for the formula used for declining balance depreciation. (See declining balance depreciation.)

maximax criterion, In decision making under uncertainty, the highest value of the range for the value of the highest return anticipated for an alternative.

maxim criterion, In decision making under uncertainty, the maximum of any combination of the maximin and the maximax returns.

maximin criterion, In decision making under uncertainty, the highest value of the range for the value of the lowest return anticipated for an alternative.

minimin criterion, In decision making under uncertainty, the minimum of any combination of the minimax and the minimin returns.

minimax criterion, In decision making under uncertainty, the lowest value of the range for the value of the highest cost anticipated for an alternative.

minimin criterion, In decision making under uncertainty, the lowest value of the range for the value of the lowest cost anticipated for an alternative.

minimax regret criterion, In decision making under uncertainty, the least potential profit regret.

minimum attractive rate of return, The effective annual rate of return on investment, after taxes, which must meets the investor's threshold of acceptability.

minimum cost life, See economic life.

multiple straight-line depreciation method, A method of depreciation accounting in which two or more straight-line rates are used. This method permits a predetermined portion of the asset to be written off in a fixed number of years. One common practice is to employ a straight-line rate which will write off 3/4 of the cost in the first half of the anticipated service life; with a second straight-line rate to write off the remaining 1/4 in the remaining half-life.

mutually exclusive alternative, One such that the selection of it rules out the selection of all other alternatives in the set. Mutually exclusive alternatives tend to be substitutes for each other.

nominal interest, See interest rate, nominal.

obsolescence, (1) The condition of being out-of-date. A loss of value occasioned by new developments which place the older property at a competitive disadvantage. A factor in depreciation. (2) A decrease in the value of an asset brought about by the development of new and more economical methods, processes, and/or machinery. (3) The loss of usefulness or worth of a product or facility as a result of the appearance of better and/or more economical products, methods, or facilities.

opportunity cost, The cost of not being able to invest in an alternative, due to limited resources being applied to another "approved" alternative, and thus not being available for investment in other income-producing alternatives.

payback period, (1) Regarding an investment, the number of years (or months) required for the related profit or savings in operating cost to equal the amount of said investment. (2) The period of time at which a machine, facility, or other investment has produced sufficient net revenue to recover its investment costs.

payoff period, See payback period.

payoff table, A tabular presentation of the results of complex decision questions involving many alternatives and events.

payout period, See payback period.

perpetual endowment, An endowment with hypothetically infinite life. (See capitalized cost and endowment.)

planning horizon, The largest single dollar amount that a decision maker would recommend to be spent.

present worth, (1) The equivalent value at the present, based on time value of money. (2) a) The monetary sum which is equivalent to a future sum(s) when interest is compounded at a given rate. (3) The discounted value of future sums.

present worth factor, (1) A mathematical expression also known as the present value of an annuity of one. (2) One of a set of mathematical formulas used to facilitate calculation of present worth in economic analyses involving compound interest.

profitability index, The rate of return in an economy study or investment decision when calculated by the discounted cash flow method or the investor's method. (See discounted cash flow method; investor's method.)

promotion cost, The sum of all the expenses found to be necessary to arrange for the financing and organizing of the business unit which will build and operate the project.

rate of return, (1) The interest rate at which the present worth of the cash flows on a project is zero. (2) The interest rate earned by an investment.

replacement policy, A set of decision rules (usually optimal) for the replacement of facilities that wear out, deteriorate, or fail over a period of time. Replacement models are generally concerned with weighing the increasing operating costs (and possibly decreasing revenues) associated with aging equipment against the net proceeds from alternative equipment.

replacement study, An economic analysis involving the comparison of an existing facility and a facility proposed to supplant the existing facility.

required return, The *minimum* return or profit necessary to justify an investment. Often termed interest, expected return or profit, or charge for the use of capital. It is the minimum acceptable percentage, no more and no less.

required yield, The ratio of required return over amount of investment, usually expressed as a percentage on an annual basis.

retirement of debt, The termination of a debt obligation by appropriate settlement with lender — understood to be in full amount unless partial settlement is specified.

risk, (1) The probability of not achieving a targeted return. (2) The degree of expected variation in the expected return.

salvage value, (1) The cost recovered or which could be recovered from a used property when removed, sold, or scrapped. A factor in appraisal of property value and in computing depreciation. (2) The market value of a machine or facility at any point in time. Normally, an estimate of an asset's net market value at the end of its estimated life.

sensitivity, The relative magnitude of the change in one or more elements of an engineering economy problem that will reverse a decision among alternatives.

service life, See economic life.

simple interest, (1) Interest that is not compounded — is not added to the income-producing investment or loan. (2) The interest charges under the condition that interest in any time period is only charged on the principal.

sinking fund, (1) A fund accumulated by periodic deposits and reserved exclusively for a specific purpose, such as retirement of a debt or replacement of a property. (2) A fund created by making periodic deposits (usually equal) at compound interest in order to accumulate a given sum at a given future time for some specific purpose.

sinking fund deposit factor, The function of interest rate and time that determines the periodic deposit required to accumulate a specified future amount. Reciprocal of sinking fund factor. (See sinking fund factor.)

sinking fund depreciation, (1) A method of computing depreciation in which the periodic amount is presumed to be deposited in a sinking fund that earns interest at a specified rate. Sinking fund may be real but is usually hypothetical. (2) A method of depreciation where a fixed sum of money is regularly deposited at compound interest in a real or imaginary fund in order to accumulate an amount equal to the total depreciation of an asset at the end of the asset's estimated life. The sinking fund depreciation in any year equals the sinking fund deposit plus interest on the sinking fund balance.

sinking fund factor, The function of interest rate and time that determines the cumulative amount of a sinking fund resulting from specified periodic deposits. Future worth per unit of uniform periodic amounts.

straight-line depreciation, Method of depreciation whereby the amount to be recovered (written off) is spread uniformly over the estimated life of the asset in terms of time periods or units of output. May be designated percent of initial value.

study period, In economy study, the length of time that is presumed to be covered in the schedule of events and appraisal of results. Often the anticipated life of the project under consideration, but a shorter time may be more appropriate for decision making.

sum-of-the-years-digits method, A method of computing depreciation in which the amount for any year is based on the ratio: (years of remaining life) / (1 + 2 + 3 . . . + n), n being the total anticipated life.

sunk cost, (a) A cost, already paid, that is not relevant to the decision concerning the future that is being made. Capital already invested that for some reason cannot be retrieved. (b) A past cost which has no relevance with respect to future receipts and disbursements of a facility undergoing an engineering economy study. This concept implies that since a past outlay is the same regardless of the alternative selected, it should not influence the choice between alternatives.

tangibles, Things that can be *quantitatively* measured or valued, such as items of cost and physical assets.

time value of money, (1) The cumulative effect of elapsed time on the money value of an event, based on the earning power of equivalent invested funds. (See future worth and present worth.) (2) The expected interest rate that capital should or will earn.

traceable costs, Costs which can be identified with a given product, operation, or service.

unallocated balance, The undepreciated portion of the first cost of an asset at any given time.

uniform gradient series, A pattern of receipts or disbursements for a series, that has a uniform linear (arithmetic) or geometric increase in each time period.

unrecovered balance, See unallocated balance.

utility, A preference among various choices available in risk situations, based on the decision making environment, the alternatives being considered, and the decision makers, personal tastes.

utility function, The relationship between utility, measured in "utiles", and a monetary measurement such as dollars.

valuation or appraisal, The art of estimating the fair-exchange value of specific properties.

value, The worth to an individual of a product or service.

variable cost, A cost that tends to fluctuate according to the changes in the number of units produced.

working capital, (1) That portion of investment represented by current assets (assets that are not capitalized) less the current liabilities. The capital necessary to sustain operations. (2) Those funds that are required to make the enterprise or project a going concern.

yield, The ratio of return or profit over the associated investment, expressed as a percentage or decimal usually on an annual basis.

Facility Planning
Z94.6

6

The need for compiling a standard terminology continues to be increasingly apparent to engineers and lay public alike. Many terms peculiar to facility planning have definitions that have evolved through common usage. However, many terms may have a meaning other than that given in the commonly used popular dictionaries.

As in the 1972 edition (Z94.6), which served as the basis for the present revision, intermodal transportation continues to be an important aspect.

This compilation of facility planning terms is by no means exhaustive, and sources for additional terms may be found in the references appended to this chapter.

For some time, after 1972, this subcommittee was headed by Mr. William W. Olsen, Product Assurance Manager, Industrial Truck Division, Eaton Corporation, whose aid continued throughout this latest endeavor. Acknowledgment is also due to the subcommittee members, past and present, who devoted their time, effort, and expertise to this FACILITY PLANNING TERMINOLOGY: Messrs T. E. Allegri (Materials Handling Manager, Caterpiller Tractor), R. A. Bell (IE, Consumer Protection, Environmental Health Service), D. Bowman (Rohr Corporation), D. Carroll (Materials Handling Specialist, Engineering, Eastman Kodak), F. G. Melham (Director, Engineering and Storage, Office of Supply Distribution, General Services Administration), C. A. Monis (Manager, Resource Application, American Airlines), J. M. Murray (President, CERCO Corporation), I. Ross (Manager, Proposal and Contract Engineering, Mechanical Handling Systems, Inc.), M. J. Rowan (Editor and Assistant Publisher, MODERN MATERIALS HANDLING), R. W. Saigh (Manager, Trambeam and Pressuregrip, Whiting Corporation), and S. L. Werner (Werner-Dyer & Associates), and Dr. P. T. Eaton (School of Industrial Engineering, Georgia Institute of Technology, the late Professor Ruddell Reed, Jr. (School of Industrial Engineering, Purdue), and Edward R. Sirotak, Chairman (Chief Engineer, Towmotor Corporation).

acceleration resistance, Inertial reaction opposing increase in speed of a vehicle.

accessory, Any supplemental device for an INDUSTRIAL TRUCK that cannot be classified as an ATTACHMENT.

accident, Any unplanned occurrence which may result in damage to property or equipment, injury or illness to persons, and/or adversely affects an activity or function. (See Human Factors.)

accident prevention, The application of basic scientific and technical principles including education and training for the detection, analysis, and minimization of hazards with the objective of avoiding accidents. (See Safety.)

accordion roller conveyor, A ROLLER CONVEYOR with a flexible latticed frame which permits variation in length. (See roller conveyor.)

accumulator conveyor, Any CONVEYOR designed to permit accumulation of packages, objects, or carriers. May be roller, live roller, roller slat, belt, power-and-free, or tow conveyors. (See minimum pressure accumulating conveyor.)

airbill, A nonnegotiable shipping document used by U.S. airlines for domestic AIR FREIGHT. It contains shipping instructions, a description of the commodity and applicable transportation charges. May be used as through documents for coordinating other transportation.

air cargo, In the United States, goods transported by AIR MAIL, AIR FREIGHT, or AIR EXPRESS. In other countries, AIR FREIGHT only.

aircraft accident, Any accident that occurs between the time the aircraft engine(s) is started with the intent of flight until the aircraft comes to rest with engine(s) stopped upon completion of flight.

air express, A nationwide priority AIR CARGO service providing door-to-door delivery under the Air Express Tariff for packages usually weighing between five pounds and 50 pounds.

air-float chain conveyor, A chain and roller conveyor in which the chain is supported by an inflatable air hose to control and minimize line pressure.

air freight, AIR CARGO service for the transport of goods not classified as AIR MAIL or AIR EXPRESS.

air freight forwarder, An agent who collects, ships and distributes cargo under his own tariff usually consolidating the cargo into a larger unit that is tendered to the airline. Clearing of cargo through Federal Customs may be handled by the agent.

air mail, Mail service having priority over all other AIR CARGO.

air parcel post, AIR MAIL weighing more than eight ounces and not exceeding 70 pounds and 100 inches in combined length and girth. In common practice, AIR PARCEL POST is most widely used for packages of five pounds or less.

air waybill, A nonnegotiable shipping document used for international AIR FREIGHT.

aisle, That SPACE used to accommodate the movement of men, material, and/or equipment.

alternate capacity, A capacity, other than BASIC CAPACITY, that adjusts for variations such as LOAD CENTER, FORK HEIGHT, and application of an ATTACHMENT.

approved industrial truck, A truck that is listed or approved for fire safety purposes for the intended use by a nationally recognized testing laboratory using nationally recognized testing standards.

apron, A series of pans or plates pivotally attached to one another to form the conveying medium of a CONVEYOR.

apron conveyor, A conveyor in which a series of pans or plates attached to a chain or pivotally attached to one another form the moving bed.

area, A functional subdivision of a facility, both exterior and interior.

areaway, An open subsurface SPACE, exterior and adjacent to the building, used to admit light or air, or as a means of access.

arm conveyor, A conveyor consisting of an endless belt, one or more chains, to which are attached projecting arms, or shelves, for handling packages or objects in vertical or inclined path.

arm to column connector, A bracket affixed to the arm and is used to attach a complete arm to a COLUMN.

ASME type LP gas cylinder, A fuel container for LIQUEFIED PETROLEUM GAS made and inspected under the ASME Boiler and Pressure Vessel Code, Section VIII, for Unfired Pressure Vessels. Where used for storage of an INDUSTRIAL TRUCK's fuel supply, it may be permanently or removably affixed to the vehicle.

assembly service (air transport), A service under which an airline consolidates shipments from many shippers and transports them as one shipment to one receiver.

attic, The SPACE between the ceiling framing of the top STORY and the roof framing.

automatic coupler, A trailer connection that may be connected without manual actuation.

bag conveyor, double helical, Closely spaced parallel tubes with right and left hand helical threads rotating in opposite directions on which bags or other objects are carried while being conveyed.

back cross brace, Diagonal members forming a rigid cross between the back post ties on the uprights at the end of the drive-in rack bay.

back diagonal brace, A diagonal member connecting back post ties at the end of the drive-in rack bay.

back post ties, Horizontal ties across the last posts at the end of the drive-in rack bay.

balcony, An elevated PLATFORM projecting from an exterior or interior wall, supported by columns or the wall, and enclosed by a railing.

base (cantilever rack), The bottom member of a rack in contact with the floor and is permanently affixed to the COLUMN. May also be affixed to the floor and used as a load bearing arm. In free standing racks, the base usually extends beyond the length of the arms. Purpose of the base is to prevent the rack from tipping over.

base (portable metal stacking rack), Consists of structural cold-formed or tubed members joined together into a square or rectangular load carrying perimeter to which posts or sockets are attached, when required. The four basic styles are as follows:

a) Base No. 1 — base is constructed with two center structural members or intermediate framing inside the basic rack for allowing 4-way entry of light truck forks. b) Base No. 2 — base is constructed with six structural members or intermediate framing for allowing 4-way entry for racks that are wider than the length of the lift truck forks. c) Base No. 3 — base is constructed with X-type structural members or intermediate frame for 4-way entry of lift truck forks. d) Base No. 4 — base is constructed *without* corner sockets, but with open style intermediate bracing for load only. This style for transportation by crane with special tongs or slings.

base guide rail, Member attached to the outside edge of the base and is used to align a row or pair of complete frames. Used also as a truck guide.

basement, Any story having at least 1/2 of its volume below the adjacent finished grade.

batch process, An industrial manufacturing method in which one or several units are produced at a time, in contrast to CONTINUOUS PROCESS (q.v.).

battery (bat.) — AIEE (60.06.010), A battery consists of two or more cells electrically connected for producing electric energy. (Common usage permits this designation to be applied also to a single cell used independently. In the following definitions, unless otherwise specified, the term battery will be used in this dual sense.)

battery-electric truck, An ELECTRIC TRUCK in which the power source is a storage BATTERY.

bay, The space defined by a pair of UPRIGHT FRAMES, the OVERALL DEPTH, and the clear height between shelves.

bay, A cubic SPACE with limits normally defined by functional or physical constraints; as example:

structural bay, The SPACE defined by four columns.

storage bay, The SPACE defined by the size of a block of material stored within it.

truck bay, The SPACE defined by the area required to park a truck.

service bay, The SPACE defined by the area required to park and service a vehicle.

beam clamp, A type of suspension fitting used to support tracks from overhead structure.

beams, A horizontal load-carrying member on each end of which is a connector used in tying together two upright frames.

bearing brace, The member of an UPRIGHT FRAME with full length floor contact.

bearing plate, A metal plate affixed to the bottom of a POST for load distribution, permitting lagging or bolting to the floor.

belt conveyor, An endless fabric, rubber, plastic, leather, or metal belt operating over suitable drive, tail end and bend terminals and over belt idlers or slider bed for handling bulk materials, packages, or objects placed directly upon the belt.

belt conveyor, closed, Moving, endless, flexible belt, or belts, which may be formed into a tubular shape by joining of edges, and which are opened while in motion to receive load, closed to convey or elevate, and opened to discharge.

belt conveyor, flat, A BELT CONVEYOR in which the carrying run of the conveyor belt is supported by flat belt idlers or by a flat surface.

belt conveyor, magnetic, An inclined BELT CONVEYOR operating over a SLIDER BED containing permanent magnets for handling ferrous metal parts.

belt conveyor, multiple cord, A BELT CONVEYOR composed of two or more spaced strands of Vee, double Vee, or round belts.

belt conveyor, multiple ribbon, A BELT CONVEYOR having a conveying surface of two or more spaced strands of narrow conveyor belts.

belt conveyor, portable, A PORTABLE CONVEYOR in which a belt is used as the conveying medium.

belt conveyor, troughed, A BELT CONVEYOR with the conveyor belt edges elevated on the carrying run to form a trough by conforming to the shape of the troughed carrying idlers or other supporting surface.

belt feeder, A BELT CONVEYOR adapted for feeder service. (See feeder; conveyor type feeder.)

bill of lading, A receipt given by the carrier or his agent for goods received for shipment. (See straight bill of lading and order bill of lading.)

bin storage, An area of a plant or facility devoted entirely to loose item placement.

blending conveyor, See mixing conveyor, paddle type.

block, A storage term denoting a rectangular area within a facility or building in which items-objects are densely located, and in which only the exterior faces of the area may be conveniently reached.

block, leg or column, DECK SPACERS used to make FULL FOUR-WAY or ALL-WAY ENTRY PALLETS.

block stacking, The action of putting objects into a block pattern in a storage area, usually more than one tier in height.

blocked space (air transport), A program of reduced rates and reserved space granted a shipper who assures a minimum of freight (15,000 lbs. a week, for example) between the same two cities for a given period (90 days, for example).

boom, A cantilever member, with a load-carrying hook at the free end.

boom conveyor, Any type of conveyor mounted on a boom.

booster conveyor, Any POWER CONVEYOR used to regain elevation lost in gravity ROLLER or WHEEL CONVEYOR lines.

bottom discharge bucket conveyor, A CONVEYOR for carrying bulk materials in a horizontal path consisting of an endless chain to which roller-supported, cam-operated conveyor buckets are attached continuously.

brace, Stiffening member(s) connecting two POSTS in an upright frame, i.e., diagonal, cross, or horizontal.

bridge rail, The rail supported by the bridge girders on which the TROLLEY travels.

bucket conveyor, Any CONVEYOR in which the material is carried in a series of buckets. (See bucket elevator; gravity discharge bucket conveyor; pivoted bucket conveyor.)

bucket elevator, A CONVEYOR for carrying bulk materials in a vertical or inclined path, consisting of endless belt, chain or chains, to which elevator buckets are attached, the necessary head and boot terminal machinery and supporting frame or casing. (See internal discharge bucket elevator; positive discharge bucket elevator.)

bucket elevator, gravity discharge, See gravity discharge bucket conveyor.

bucket elevator, pivoted, See pivoted bucket conveyor.

bucket loader, A form of portable, self-feeding, inclined BUCKET ELEVATOR for loading bulk materials into cars, trucks, or other conveyors. (See bucket elevator; portable conveyor.)

building, A roofed structure for the shelter, support, or enclosure of persons or property.

bulk storage, The area within a facility or warehouse which is devoted to the placement of large, greater than loose-issue quantities of items, or in which each single item is too large to be placed in a bin storage location.

cab, An equipment-mounted enclosure for operators from which motion of equipment may be controlled.

cab controlled, Equipment controlled from an operator's CAB.

cable connector, A length of insulated cable terminating at each end in a casting or a lug; and used to connect one cell to another. CABLE CONNECTORS differ from TERMINAL CONNECTORS in being used between cells rather than as a connection between TERMINAL and external circuits.

cable-screw conveyor, A one-way or CLOSED-CIRCUIT CONVEYOR of which the propelling medium is a flexible torque-transmitting cable of which helical (screw) threads are an integral part. Loads or load carriers engage the thread and advance a distance equal to one pitch for each revolution of the cable screw.

cableway, A cable, or rope-supported system in which the material handling carriers are not detached from the operating mechanism and the travel is wholly within the span.

cableway, slack line, A cable or rope-supported system in which the supporting cable is adjusted in length to provide the lifting function of the unit.

caged storage, The area within a facility or warehouse which has been set aside for pilferable items and is afforded a measure of security by means of a partition or other enclosure, usually wire screening.

camber (girder), The slight vertical curve given to girders to compensate partially for deflection due to load.

canopy, A roofed projection from the wall of a building, either exterior or interior, without enclosures below its edges, and providing protection from above.

cantilever rack, A rack consisting of arms cantilevered from COLUMN(S).

cantilever truck, A self-loading COUNTERBALANCED or non-counterbalanced truck, equipped with cantilever load-engaging means such as FORKS.

capacity, The maximum load in pounds, or the maximum load in pounds at a given LOAD CENTER, that a truck can safely transport and/or stack to a specified height.

captive container, A container whose use cycle remains within a single enterprise (private, corporate, or government).

cargo, The loading of an aircraft, car, ship or truck.

cargo container capacity, The inside usable cubic volume of the container.

cargo container displacement, The outside cubic volume of the container. (Cargo Container Length × Cargo Container Width × Cargo Container Height.)

cargo container height, The vertical overall outside dimension of the container.

cargo container length, The horizontal overall outside dimension usually parallel to the longitudinal axis of carrier equipment.

cargo container width, The horizontal overall outside dimension at right angles to the container length.

carriage, A support structure for FORKS or ATTACHMENTS, generally roller mounted, traveling vertically within the MAST of a CANTILEVER TRUCK or STACKER CRANE.

carrier, A person or firm engaged in the business of transporting goods.

carrier, An assembly usually with four or more wheels which will run on monorails or underhung crane girders, and which will support a load or a HOIST and load.

carrier head, A two-wheeled assembly used with load bars to form a CARRIER or an END TRUCK.

carrousel conveyor, A continuous platform or series of spaced platforms which move in a circular horizontal path. Note: The term "carrousel" has been applied to other forms of conveyors such as CAR TYPE and PALLET TYPE.

car type conveyor, A series of cars attached to and propelled by an endless chain or other linkage running on the horizontal or a slight incline.

catwalk, An elevated service platform or walkway constructed to permit access to equipment, controls or other devices not frequently used. The catwalk generally is designed to support only itself and the weight of men required for access to the aforementioned items.

ceiling tie, A member tying the rack to the overhead building structure.

center control truck, A POWERED INDUSTRIAL TRUCK in which the driver's position is located near the longitudinal center of the truck.

chain elevator, See bucket elevator.

chain conveyor, Any CONVEYOR in which one or more chains act as the conveying element. A British term for trolley conveyor. (See drag, rolling, and sliding chain conveyors.)

chain feeder, See conveyor type feeder.

charged and dry, Charged and dry is the condition of a storage battery when it is assembled with charged plates, dry separators, and with no ELECTROLYTE. Charged and dry batteries can be used as soon as they are filled but often are given an initial charge after filling and before use.

charged and wet, Charged and wet is the condition of a storage BATTERY when it is filled with ELECTROLYTE and is fully charged.

charging plug, The male half of a quick CONNECTOR, housing both positive and negative leads. When used on a BATTERY the connection is readily made to the load or charging circuit, each of which is equipped with a CHARGING RECEPTACLE.

charging receptacle, The female half of a quick CONNECTOR housing both positive and negative leads. When mounted on a BATTERY it is only necessary to insert a CHARGING PLUG to connect either the load or the charging circuit.

city terminal service (air transport), Shipments accepted or delivered by airlines at the terminals of cartage agents or other designated in-town terminals, at lower rates than those charged for the door-to-door airport pick-up and delivery service.

clamp, An ATTACHMENT which uses a clamping action to engage the load from opposite sides. (See clamp arms.)

clamp arms, Auxiliary members that bolt or are welded to clamp arm supports. Various types are available for CLAMP(S) to suite them for specific loads, such as bag, bale, drum, carton, paper roll, tire, etc.

closed circuit conveyor, An arrangement of a CONVEYOR or conveyors capable of moving material through all portions of a circuit, and returning the undistributed portion to the starting point.

closet, A small room designed to provide limited storage.

collapsible container, A container, the major components of which can be disassembled and later reassembled for use.

collector, shoe, The portion of a COLLECTOR which makes contact by sliding on the CONDUCTOR bar.

collector, wheel, The portion of a COLLECTOR which makes contact by rolling on the CONDUCTOR.

collectors, Electrical contacting devices providing a path for current flow from stationary CONDUCTORS to moving equipment.

column, A single vertical member of the CANTILEVER RACK that may have a means for adjusting arms on preset centers.

commodity code, A system for identifying a given commodity by a number.

commodity rate, A charge applicable on a specific commodity between certain specified points.

common carrier, A CARRIER engaged in the business of transporting goods and/or persons for compensation and without discrimination.

compartment (battery), The chamber on a truck in which a storage BATTERY is housed and mounted.

conductors, Bar or wire used to transmit an electrical current.

conductors, enclosed, CONDUCTORS enclosed in a nonconducting material to minimize accidental contact with the CONDUCTOR.

connecting carrier, A CARRIER maintaining a direct physical connection with one or more carriers.

connector (battery), An item for quick connection composed either of male and female halves or two halves that differ only in method of mounting. Generally the two halves are known as CHARGING PLUG and CHARGING RECEPTACLE.

connector lock, A safety device used to lock the RAIL SUPPORT CONNECTOR to the POST.

container corner fittings, The fittings located at the corners of the container which provide means for handling, supporting and/or securing the container.

container load, A shipment sufficient in size to "fill" a container, either by cubic measurement or weight, depending upon governing tariff to meet the provided minimums.

continuous process, An operation wherein raw materials enter into the process in an unbroken stream and proceed to a final form, in contrast to batch processing (q.v.).

contract carrier, A person or firm, not a COMMON CARRIER, that transports goods and/or persons for compensation under private or special contract agreements.

control bars, CONDUCTOR bars that carry the control current.

control, single speed, A drive control system providing one speed operation in either direction.

control, variable speed, A drive control system providing more than single-speed operation in either direction.

control voltage, The voltage impressed on the control device(s).

controlled velocity roller conveyor, A ROLLER CONVEYOR having means to control the velocity of objects being conveyed. (See roller conveyor.)

controllers, Electromechanical devices by means of which an operator controls the speed and direction of motor driven equipment.

conveyor, A horizontal, inclined or vertical device for moving or transporting bulk materials, packages or objects in a path predetermined by the design of the device and having points of loading and discharge fixed, or selective; included are skip hoists and vertical reciprocating and inclined reciprocating conveyors: typical exceptions are those devices known as industrial trucks, tractors and trailers, tiering machines (truck type), cranes, hoists, monorail cranes, power and hand shovels, power scoops, bucket drag lines, platform elevators designed to carry passengers or the elevator operator, and highway or rail vehicles.

conveyor-elevator, A CONVEYOR which follows a path, part of which is substantially horizontal or on a slope less than the angle of slide of the material and part of which is substantially vertical or on a slope steeper than the angle of slide.

conveyor type feeder, Any CONVEYOR such as apron, belt, chain, flight, pan, oscillating, screw, or vibrating, adapted for feeder service.

counterbalanced front sideloader truck, A self-loading high lift counterbalanced truck (equipped with a fixed or tiltable elevating mechanism) capable of transporting and tiering a load in both the counterbalanced forward position and any location up to and including 90° from the longitudinal centerline of the truck, which possessing the capability of traversing the load laterally.

counterbalanced truck, A truck equipped with load engaging means wherein all the load during normal transporting is external to the polygon formed by the wheel contacts.

coupler height, The vertical distance from floor to the effective horizontal center line of coupler.

crane, A machine for lifting, and/or moving a load with the hoisting mechanism an integral part of the machine. CRANES may be traveling, portable, or fixed type. (See traveling, portable, or fixed cranes.)

crane bar (portable metal stacking rack), See tiering guide.

crane bridge (top-running), The girders, trucks, END TIES, FOOTWALK (if any), and drive mechanism which carries the trolley.

crane drive, individual, Separate drives mounted on individual END TRUCKS.

crawler crane, A PORTABLE CRANE mounted on a base equipped with crawler treads.

cross aisle, The SPACE used to accommodate secondary movement of men, material, and/or equipment, and intersecting another aisle.

cross bar conveyor, A CONVEYOR having endless chains supporting spaced cross members from which materials are hung or festooned while being processed.

curved belt conveyor, A unit load BELT CONVEYOR usually operating horizontally through 90° or 180° turns.

cushion tire, A SOLID TIRE having a high rounded profile in cross section.

dangling leads, A reference adopted to designate that the cable ends of the TERMINAL CONNECTORS hang free from the edge of the BATTERY being connected to an unmounted CONNECTOR or CHARGING PLUG.

dead loads, The loads on a structure which remain in a fixed position relative to the structure.

deck, The horizontal load-carrying or load-bearing surface of a PALLET. TOP DECK — Load-carrying surface. BOTTOM DECK — Load-bearing surface.

deck (pallet storage rack), A solid top or platform supported by BEAMS.

deck (portable metal stacking rack), The platform attached to the BASE.

deck opening, Any void in deck caused by spacing of surface elements.

deck spacer, A structural member which supports the top DECK and/or separates the top and bottom DECKS.

deckmember, A surface element used in the construction of a PALLET DECK.

deckmember chamfer, A beveled edge on the side of the bottom DECKMEMBER for the purpose of easing the entry and exit of pallet truck load wheels.

declared value (air transport), The value of the shipment declared by the shipper. In AIR FREIGHT, declared value on most shipments has historically been 50 cents a pound or 50 dollars, whichever is greater, assumed unless the shipper declares a higher value.

deferred rate (air transport), Rate for DEFERRED SERVICE in return for the greater flexibility given the airlines in determining when to move such freight. The rates are usually lower than general commodity rates.

deferred service (air transport), A service under which delivery is deferred for a specified number of days (to the third or fourth day, for example).

delivery table, (1) a CONVEYOR which transports material from the discharge of a machine; (2) a table to which a chute discharges.

demountable box, A steel case with a cover in which a BATTERY is assembled for use on a truck that does not have its own integral BATTERY COMPARTMENT.

demountable cargo container, A reusable weatherproof cargo container for shipping or storing materials as a unit that can be secured to carrier equipment.

demountable cargo container — standard Group I, A DEMOUNTABLE CARGO CONTAINER of rigid construction, rectangular configuration, with end doors, in nominal lengths of 10, 20, 30 and 40 feet and width and height of 8 feet. Containers can be stacked and are normally positioned longitudinally on carrier equipment.

demountable cargo container — standard Group II, A DEMOUNTABLE CARGO CONTAINER of rigid construction, rectangular configuration, with end doors, in nominal lengths of 6 2/3 feet and 5 feet, with a width of 8 feet and optional heights of 8 feet or 6 feet 10 inches, capable of being stacked, supported and mounted on 6-inch side runners for either platform or forklift handling. Containers are normally positioned transversely on carrier equipment.

demurrage, A penalty charge assessed by carriers for detention of cars, ships or trucks by shippers or receivers of goods beyond a specified free time.

derrick, A FIXED CRANE consisting of a mast or equivalent member held at the head by guys or braces, with or without a boom, for use with a hoisting mechanism.

diesel-electric truck, An ELECTRIC TRUCK in which the power source is a diesel engine-driven generator.

dimensional weight, The density, or weight per cubic foot of a shipment.

dimensional weight rule, A practice applicable to low density shipments under which the transportation charges are based on a dimensional weight rather than upon actual weight.

discharged and wet, Is the condition of a storage BATTERY when it is filled with ELECTROLYTE and is discharged. An INITIAL CHARGE is required before it is ready for use.

disconnect device, main line, A device used to open the electric circuit.

distribution service (air transport), A service under which an airline accepts and transports one shipment from one shipper and separates the shipment into a number of parts at destination and distributes them to many receivers.

dock, A horizontal PLATFORM designed to place the floor of a building level with the bed of a vehicle.

dock leveler, An adjustable rectangular platform of approximately highway truck width built into the dock edge used to compensate for the difference between varying truck bed heights and delivery platforms of a facility.

dockboard, A portable or fixed device for spanning the gap or compensating for difference in level between loading platforms and carriers.

double arm, Horizontal load member affixed to the COLUMN for loading on either side of the column.

double leg bucket elevator, A BUCKET ELEVATOR having the carrying and return runs enclosed in separate casings between the head and boot. (See bucket elevator.)

double leg en masse conveyor, An EN MASSE CONVEYOR or elevator in which the carrying and return runs are operated in separated parallel and adjacent casings.

drag chain conveyor, A CONVEYOR having one or more endless chains which drag bulk materials through a trough. (See sliding chain conveyor.)

drag conveyor, portable, See portable drag conveyor.

drawbar pull, maximum, The maximum towing force in pounds at a specified coupler height that a vehicle will develop on a level surface based upon a prescribed coefficient of friction existing between the driving wheel(s) and the supporting surface when the load is being moved at a uniform rate of not less than 1 percent of the top level travel speed.

drawbar pull, rated normal, The greatest sustained towing force in pounds at a specified coupler height that a vehicle will develop on a level surface and within a given duty cycle without exceeding the allowable continuous temperature rating for the components.

drive girder, The girder on which the bridge drive machinery is mounted.

drive-in rack, A rack consisting of UPRIGHT FRAMES, RAILS, and TIES permitting a vehicle to enter the structure from one side only to pickup up or deposit pallets on continuous rails. The structure dictates first-in, last-out storage.

drive-thru rack, A rack consisting of UPRIGHT FRAMES, RAILS, and TIES permitting a vehicle to enter the structure from opposite sides to pick up or deposit pallets on continuous rails. The structure permits first-in, first-out storage.

driving head, A motor-driven carrier head, the load-bearing wheels of which are driven.

drop section, A section of track which can be lowered out of alignment with a stationary track.

drum grooves, Helical grooves cut in the periphery of the drum to guide the hoisting rope.

drum handler, An ATTACHMENT to engage a drum(s). Drum handlers are available which engage the drum(s) in various ways and positions. Certain models may also manipulate drum(s).

drum support, A unit composed of a pair of front to back members placed between a pair of BEAMS used to support cylindrical objects.

dummy (battery), Generally of wood, is employed in LEAD BATTERIES to fill the space left in a TRAY when the specified number of cells is less than the TRAY'S capacity.

dumper, An ATTACHMENT for dumping containers. Various types are available to suit container and operational requirements.

dunnage, Material used to protect or support cargo in aircraft, car, ship, or truck.

dynamic lowering, A method of control by which the hoist motor is so connected in the lowering direction, that when it is overhauled by the load, it acts as a generator and forces current either through the resistors or back into the line, thus acting as a brake or speed limit.

economy service (air transport), See deferred service.

eddy current clutch, A magnetic torque transmitting component in which the slip is varied by the voltage impressed in its electromagnetic coils, and no mechanical friction contact takes place between the driving and driven member.

edgemember, A DECK MEMBER assembled at right angles to and at the extreme ends of STRINGERS or STRINGER-MEMBERS.

el conveyor, A trough type ROLLER or WHEEL CONVEYOR consisting of two parallel rows of rollers or wheels set at a 90° included angle, with one row providing a sloped carrying surface and the other acting as guard. (See roller conveyor; wheel conveyor.)

electric tractor, An INDUSTRIAL TRACTOR in which the principal energy is transmitted from power source to motor(s) in the form of electricity and can be either BATTERY, GAS-ELECTRIC, DIESEL-ELECTRIC or TETHERED-ELECTRIC.

electric truck, A truck in which principal energy is transmitted from power source to motor(s) in the form of electricity.

electrically baffled, Wired in such a way that electric current is cut off from motor-driven equipment approaching units which are not properly set for passage of the equipment.

elevating conveyor, Any CONVEYOR used to discharge material at a point higher than that at which it was received. Term is specifically applied to certain underground mine conveyors.

empty pallet stacker, A device for receiving empty pallets and automatically stacking them for re-use or shipment.

en masse conveyor, A CONVEYOR comprised of a series of skeleton or solid flights on an endless chain or other linkage which operates in horizontal, inclined, or vertical paths within a closely fitted casing for the carrying run. Bulk material is conveyed and elevated in a substantially continuous stream with a full cross section of the casing.

enclosed track trolley conveyor, A TROLLEY CONVEYOR

where the propelling medium such as chain, cable or other linkage and the trolleys or load-carrying wheels are supported and completely enclosed by a tubular type track. (See trolley conveyor.)

end control truck, A POWERED INDUSTRIAL TRUCK in which the driver's position is located at the end opposite the load end of the vehicle.

end frames (portable metal stacking rack), A completely removable unit making up one side of the rack. Consists of two posts with appropriate bracing separating the posts so as to nest with two corners of the base rack.

end stop, A device located at the end of a track or CRANE BRIDGE to prevent the CARRIER or CRANE from leaving the track or CRANE BRIDGE.

end tie, A structural member other than the END TRUCK which connects the ends of the girders to maintain the squareness of the bridge.

end trucks, The assemblies consisting of truck frame and wheels which support the crane girders.

environmental hazard, Any condition which may in case of excessive exposure prove detrimental to the health of exposed persons.

escalator, A PASSENGER CONVEYOR in which the passenger-carrying surfaces form stairs where travel is on a slope.

expendable container, A container intended to be discarded after one cycle of use.

extendable conveyor, (1) For bulk materials CONVEYOR is usually of troughed design and may be lengthened or shortened while in operation. Commonly used in underground mine conveyor work. (2) For packaged materials, objects, or units, CONVEYOR may be one of several types including roller, wheel, or belt conveyors. Construction is such that the CONVEYOR may be lengthened or shortened within limits to suit operating needs.

facility, A combination of all of the physical elements which are required to manufacture a product, or to provide a service resulting from planning and the scientific development of a system.

FCL, See full container load.

factory mutual listing (FM), A list of truck models approved as complying with the requirements of and used by the mutual group of insurance companies to determine the insurance premium rate for various areas of operation.

fastener, Any device or medium used to hold together the components of an assembled PALLET.

feeder, A CONVEYOR or other mechanism adapted to control the rate of delivery of bulk material, packages or objects or a device that controls, separates, or assembles objects.

feeder and catchers table, A pair of reversible conveyors, entry and exit, which provide for repeat feeding of metal being processed through a rolling mill.

fishy-back, The transportation of truck trailers or containers on vessels.

fire accident, Any instance of unintentional ignition, uncontrolled burning, and/or explosion of combustible materials.

fire prevention, The application of basic scientific and technical principles including education and training for the detection, analysis, and minimization of hazards with the objective of avoiding the inception of fire.

first floor, The floor of a story which is closest to grade.

fire safety, See fire prevention.

fixed cranes, Cranes which are nonmobile. (See derrick crane and jib crane.)

fixed platform truck, An INDUSTRIAL TRUCK equipped with a load platform and not capable of self-loading.

fixture conveyor, Usually a SLAT or APRON CONVEYOR on which are mounted "pedestals" or fixtures for mounting loads such as engine blocks, etc.

flat top conveyor, Special SLAT CONVEYORS which form a continuous top and are supported by chain rollers.

flat wire conveyor belt, A belt composed of flat metal strips formed in a series of lateral, rectangular loops held together by lateral wires passing through holes in the flat strips to provide articulation similar to a conveyor chain and arranged to be driven by sprockets.

flight, (1) Plain or shaped plates suitably made for attachment to the propelling medium of a FLIGHT CONVEYOR. (2) A term applied to any CONVEYOR in a tandem series.

flight conveyor, A CONVEYOR comprised of one or more endless propelling media, such as chain, to which FLIGHTS are attached, and a trough through which material is pushed by the FLIGHTS.

floor, The walking level of a story.

floor controlled units, Motor propelled units which are controlled by an operator on the floor by means of a control suspended from the overhead equipment.

floor conveyor, Any of several types of CONVEYORS using chain, cable, or other linkage mounted near or flush with the floor for the purpose of assembling, or finishing built-up products and subassemblies. (See car type conveyor; pallet type conveyor; slat conveyor.)

flow analysis, The detailed examination of the progressive travel, either of men or material, from place to place and/or from operation to operation. The examination consists of questioning the reason for the existence of every aspect of the operation, or the travel; and, also, the examination seeks reasons for determining how the progressive travel, or the operation, may be changed or modified to achieve the utmost economies in both time and material, all other things being equal.

flow diagram, A representation of the location of activities or operations and the flow of materials between activities on a pictorial layout of a process. Usually accompanied by a flow process chart.

flow line, See line of flow.

flow path, The space occupied by material as it moves through the manufacturing or clerical process.

flow process chart, A graphic, symbolic representation of the work performed, or to be performed, on a product as it passes through some or all of the stages of a process.

Typically, the information included in this type of chart concerns the quantity, distance moved, type of work done (by symbol with explanation), and equipment used. A composite of the process charts of individual components, plan assembly and related operations. Work times may also be included. Process symbols generally used are:

Operation: A subdivision of a process that changes or modifies a part, material, or product, and is done essentially at one workplace location.

Move: Change in location of a part, material, or product from one workplace to another.

Inspection: Comparison of observed quality or quantity of product with a quality or quantity standard.

Storage: Keeping a product, material, or part protected against unauthorized removal.

Delay: An event which occurs when conditions (except those which intentionally change the physical or chemical characteristics of the part, material, or product) do not permit or require immediate performance of the next planned step.

The above symbols are ASME Standard Symbols, 101.

fluid coupling, A hydrodynamic drive which transmits power without ability to change torque. (Torque ratio is unity for all speed ratios.)

flume conveyor, A channel through which a liquid flows and conveys materials.

folding container, A container having hinged components that can be folded with all parts and/or hardware attached.

footwalk, The walkway with handrail, attached to the bridge or trolley for access purposes.

fork adapter, Quickly detachable attachment that mounts on a FORK to adapt the truck to handle specific loads.

fork adjuster, A manual or power actuated ACCESSORY to facilitate lateral positioning of FORKS.

fork height, The vertical distance from the floor to the load-carrying surface adjacent to the heel of the FORKS with MAST VERTICAL, and in the case of REACH TRUCKS, with the FORKS extended.

forklift opening, See fork pockets.

forklift truck, A high lift self-loading truck, equipped with load carriage and forks for transporting and tiering loads.

fork opening, An opening in a PALLET to allow insertion and withdrawal of a lifting device.

fork pockets, Openings in container base for inserting the tines of a fork-lifting device.

fork, safety, A mechanical device which prevents passage of CARRIERS when the elements of interlocking equipment are not interlocked.

fork spacing, (a) MAXIMUM INSIDE FORK SPACING — the nominal distance between the inside edges of adjustable forks at their greatest separated position. (b) MAXIMUM OUTSIDE FORK SPACING — the nominal distance between the outside edges of adjustable forks at their greatest separated position. (c) MINIMUM INSIDE FORK SPACING — the nominal distance between the inside edges of adjustable forks at their least separated position. (d) MINIMUM OUTSIDE FORK SPACING — the nominal distance between the outside edges of adjustable forks at their least separated position.

forks, Horizontal tine-like projections, normally suspended from the CARRIAGE, for engaging and supporting loads.

four-way container, A container whose configuration permits retrieving or discharging from adjacent right angle directions in the same horizontal plane.

frame (cantilever rack), One UPRIGHT complete with arms.

free lift, The attainable LIFT of the CARRIAGE from the extreme lowered position before the stated OVERALL LOWERED HEIGHT of the MAST is exceeded by any standard part of the FORKS, MAST or CARRIAGE assemblies.

free standing rack, Is a structure composed of one or more racks without external bracing, but which may be lagged to the floor.

freight, A method or service for transporting CARGO by air, land, or water. CARGO carried by transportation lines. The cost for such transportation.

freight forwarder, A person or firm engaged in the business of collecting, consolidating, shipping, and distributing goods acting on behalf of the shipper and attending to the necessary details of shipping, insuring, and documenting of goods. Services may also include the preparation of consular invoices and the clearing of goods through customs.

front to back members, Formed structural shapes placed between two BEAMS to form a single shelf unit. Among common types are welded, bolted, or drop-in-place.

WELDED, members which are welded at both ends to a pair of BEAMS.

BOLTED, members which are bolted at both ends to a pair of BEAMS.

DROP-IN-PLACE, members which are placed between a pair of BEAMS.

full container load — FCL, Where the load carried in a container equals one of the two operating maxima — in weight or volume.

gantry crane, A TRAVELING CRANE similar to an overhead crane, except that the bridge for carrying the hoisting mechanism is rigidly supported on two or more legs running on fixed rails or other runway.

gas-electric truck, An ELECTRIC TRUCK in which the power source is a gasoline or LP gas engine-driven generator.

gateway city (air transport), That city at which the aircraft first lands upon arrival in the USA and/or the last city from which the aircraft takes off when leaving the USA.

gauge of double girder cranes, the center-to-center distance between two CRANE GIRDERS.

gauger table, The combination of a CONVEYOR and any mechanism to stop and gauge the cutting length at a shear in a metal processing line.

general commodity rate (air transport), Published AIR FREIGHT rates for all commodities except those for which specific rates have been filed.

girder, crane, Horizontal beam of the crane bridge which supports the TROLLEY and is supported by the END TRUCKS.

grade clearance, The maximum grade change that the vehicle will clear empty and LOADED without abnormal mechanical contact between the floor and the vehicle.

grade resistance, The component of gravitational force affecting movement up an inclined plane and acting at center of gravity of a vehicle in a direction parallel to the surface on which the wheels are supported. It is numerically equal to the product of gross vehicle weight and the sine of the angle of incline.

gravity discharge bucket conveyor, A CONVEYOR using gravity discharge buckets attached between two endless chains and which operate in suitable troughs and casings in horizontal, inclined and vertical paths over suitable drive, corner and take-up terminals.

gravity hopper, A surge bin attached to the tail end of a PORTABLE CONVEYOR which utilizes the CONVEYOR BELT as a live bottom.

guided rail (drive-in, drive-thru racks), A combination of rail and rail guide.

hand, (1) The right hand or left hand of a CONVEYOR is determined by facing the direction in which the material is flowing. In the case of a reversible conveyor, the hand is determined when the material is flowing toward the drive end. (2) In a SCREW CONVEYOR a right-hand screw when rotated clockwise will move material toward the observer; a left-hand screw when rotated clockwise will move material away from the observer.

hand truck, A general group of nonpowered INDUSTRIAL TRUCKS for operation by a pedestrian.

hanger rods, Steel rods which, together with other suspension fittings, are used to suspend equipment from the supporting structure.

hazardous material or operation, Any material or operation which has the potential for producing ENVIRONMENTAL HAZARDS, PERSONAL INJURY and/or PROPERTY DAMAGE.

heel (cantilever rack), The back part of the base of a single sided UPRIGHT.

height above deck (portable metal stacking rack), The clear vertical opening from the deck surface to the underside of the target.

high lift truck, A self-loading truck equipped with an elevating mechanism designed to permit tiering.

high lift platform truck, A self-loading truck equipped with a load platform intended primarily for transporting and tiering loaded skid platforms.

hinged feeder, One which vertically reciprocates one end of an adjoining hinged horizontal conveyor to provide synchronization with the movement of a vertical conveyor during the period required to permit transfer of objects from the vertical to the horizontal conveyor; or from the horizontal to the vertical conveyor.

hoist, Mechanism for lifting and lowering loads.

honeycombing, The practice of removing merchandise in pallet-load quantities where the space is not exhausted in an orderly fashion. This results in inefficiencies due to the fact that the received merchandise may not be efficiently stored in the space which is created by the honeycombing.

hook approach, The minimum horizontal distance between the center of the runway rail and the hook.

horizontal bay clearance, The internal dimension between upright POSTS parallel to the aisle.

horizontal rail clearance, The dimension between the two nearest edges of the rails.

hugger belt conveyor, Two BELT CONVEYORS whose conveying surfaces combine to convey loads up steep inclines or vertically.

hydraulic conveyor, A CONVEYOR in which water jets form the conveying medium for bulk materials through pipes and troughs.

hydrostatic transmission, A transmission using fluid placed under pressure by a pump to drive a hydraulic motor.

ICC type LP gas cylinder, A fuel container for LIQUEFIED PETROLEUM GAS made and inspected under Interstate Commerce Commission regulations. When used for storage of an industrial truck's fuel supply, it may be permanently or removably affixed to the vehicle.

ICHCA, International Cargo Handling Coordination Association.

idler sheave, A sheave used to equalize tension in opposite parts of a rope.

impact factor, An allowance for additional load resulting from the dynamic effect of the live load.

in bond cargo, Cargo that moves under the carrier's bond from the GATEWAY CITY until released by customs at an inland point.

inclined reciprocating conveyor, A reciprocating power- or gravity-actuated unit which receives only inanimate objects on a track, roller conveyor or other form of carrying surface not designed to carry passengers or the operator. These units operate on inclines of less than 90° to the horizontal and usually at less than 45°. This type of conveyor is never designed to carry an operator or passengers and no person may be permitted to ride on it. No controls for movement of the carrier may be located on the carrier nor within reach of any person who might be standing on the carrier. For carriers operating on steeper than stairway slopes further safeguards are necessary and the construction should follow the provisions given for VERTICAL RECIPROCATING CONVEYORS. (See vertical reciprocating conveyor.)

indexing, Controlled spacing or feeding.

industrial tow tractor, A powered industrial truck designed primarily to draw one or more non-powered trucks, trailers or other mobile loads.

industrial truck, A wheeled vehicle, primarily for the movement of objects or materials, and usually associated with manufacturing, processing, or warehousing, but not including vehicles intended primarily for earth-moving or over-the-road hauling.

interline, The movement of a shipment via two or more carriers.

interlocking mechanism, A mechanical device to lock together the adjacent ends of two CRANES or a CRANE and SPUR TRACK to permit the transfer of CARRIERS from one CRANE or track to the other.

interlocks, electrical, Electrical devices in the starter circuit which prevent a short circuit when opposite controls are operated at the same time.

interlocks, mechanical, Mechanical devices which prevent operation of opposite controls at the same time.

internal combustion engine truck, A truck in which the power source is a gasoline, LP gas or diesel engine.

internal combustion tractor, An INDUSTRIAL TRACTOR in which the power source is a gasoline, LP gas, or Diesel engine and the power is transmitted to the driving wheels through a mechanical, hydrodynamic, or HYDROSTATIC TRANSMISSION.

internal discharge bucket elevator, A BUCKET ELEVATOR having continuous buckets abutting, hinged, or overlapped and designed for loading and discharging along the inner boundary of the closed path of the buckets. (See bucket elevator.)

internal ribbon conveyor, A trunnion-supported revolving cylinder the inner surface of which is fitted with continuous or interrupted ribbon fighting.

inventory, The merchandise which is maintained in the facility.

jib crane, A FIXED CRANE with a cantilevered bridge supported from a stationary vertical support.

job shop, A manufacturing enterprise devoted to producing parts or products in small quantities.

joint rate, A single through rate on cargo moving via two or more carriers.

knockdown container, See collapsible container.

laboratory, The space and resources assigned to perform analysis, research, testing or experimental activities.

lanyard, A rope suitable for supporting one person. One end is fastened to a safety belt or harness and the other end is secured to a substantial object or a lifeline.

layout, A graphical representation to scale of a facility, usually in two dimensions, but sometimes in three.

lead storage battery — AIEE (60.08.011), A lead storage battery is a storage battery the electrodes of which are made of lead and the electrolyte consisting of a solution of sulphuric acid. (These cells have a nominal voltage of 2.0.)

level, Pertains to the storage height of a row of arms, or base or shelf, on the same horizontal plane. For example: a third level would indicate the third load or storage plane from the floor.

lifeline, A rope, suitable for supporting one person, to which a lanyard or safety belt (or harness) is attached.

lift, The vertical travel of the lifting device.

lifting magnet, An electromagnetic device for attracting and holding ferrous materials for hoisting.

lift-out links or track openers, Sections of monorail or runway track arranged to lift or swing out of the line of trackto permit door operation.

lift section, A section of track which can be raised out of alignment with a stationary track.

lift speed, The vertical velocity of the lifting mechanism.

lift truck, See forklift truck.

limit switch, A device actuated by the motion of another object to alter an electric circuit.

line of flow, The route of travel for man and/or material during a process or procedure.

liquefied petroleum gas (LP gas) (NBFU #58), A fuel which is composed predominantly of any of the following hydrocarbons, or mixtures of them: propane, propylene, butanes (normal butane or isobutane), and butylenes.

live load, The load being moved not including the weight of equipment being used to effect the movement.

load axle, The vehicle axle nearest the load.

load backrest extension, A device extending vertically from the fork carriage frame.

load bar, A load-carrying member between carrier heads or carriers.

load block, The assembly of hook, swivel, bearing, sheaves, pins and frame suspended by the hoisting ropes.

load carrying flange, The lower flange of the track on which the load bearing wheels roll.

load center, The horizontal longitudinal distance from the intersection of the horizontal load-carrying surfaces and vertical load-engaging face of the FORKS (or equivalent load positioning structure) to the center of gravity of the load.

load moment, In COUNTERBALANCED TRUCKS, REACH TRUCKS, and SIDELOADERS, the nominal moment produced by the load tending to overturn the truck. Mathematically it is represented by the following formula:

$$\text{Load Moment} = \text{Load} \times \left(\frac{\text{Load Center}}{\text{Load Moment Constant}}\right)$$

load moment constant, In a COUNTERBALANCED TRUCK, the longitudinal horizontal dimension from the overturning axis to the vertical load-engaging face with MAST VERTICAL.

loaded, When not otherwise specified, "loaded" is understood to mean the condition when the truck is handling the equivalent of a symmetrical cubic load.

loading platform, An area of a warehouse or other facility where merchandise is received or shipped. The height of the structure is usually designed to conform to truck bed height on the truck side of the facility, and to rail car bed height on the rail car facility.

locomotive crane, A PORTABLE CRANE mounted on a base or car equipped for travel on a railroad track which may be self-propelled or propelled by an outside source.

loft, A large room or space above the first floor, generally enclosed only by the exterior walls of the building. A loft can be either the top story of any type of building or all of the upper stories of a multistoried industrial building, and generally is intended for division into one large space only.

looping table, A roller, wheel, or other suitable type of conveyor located between two metal strip processing units over which the strip can loop when the forward machine slows down.

loss control, See accident prevention.

loss prevention, See accident prevention.

low-lift pallet truck, A self-loading industrial truck equipped with wheeled forks of dimensions to go between the top and bottom boards of a double-faced pallet, and having wheels capable of lowering into spaces between the bottom boards so as to raise the pallet off the floor for transporting.

low-lift platform truck, A self-loading industrial truck equipped with a load platform intended primarily for transporting loaded skid platforms.

low-lift truck, A self-loading truck equipped with an elevating mechanism designed to raise the load only sufficiently to permit horizontal movement. (Popular types are low lift platform truck and pallet truck.)

low solid tire, A solid tire having a low flat profile in cross section.

lowering conveyor, Any type of vertical conveyor for lowering objects at a controlled speed. (See arm conveyor; suspended tray conveyor; inclined reciprocating conveyor; vertical reciprocating conveyor.)

magazine, A type of warehouse usually covered with earth or depressed, where ammunition or other hazardous material is stored.

magnetic controller, A device or system of devices having all basic functions operated by electromagnets.

main aisle, The space used to accommodate the primary movement of men, material, and/or equipment, and usually wide enough to permit passage in opposite directions.

main floor, The floor of a story in which the principal entrance of a building is located.

manifest, A document giving the description of aircraft, car, ship, or truck cargo.

manlift, A PASSENGER CONVEYOR consisting of a vertical, endless belt with projecting steps and handles on its outer surface for transporting passengers between different elevations.

marine accident, Any grounding, fire, explosion, or other accident involving a vessel while afloat.

marine leg, A self-feeding BUCKET ELEVATOR with a means for lowering into the hold of a vessel. (See bucket elevator.)

mast, The support member providing the guideways permitting vertical movement of the CARRIAGE. It is usually constructed in the form of channels or similar sections providing the supporting pathway for the CARRIAGE.

mast vertical (fork truck with tilting mast), Mast is considered vertical at any given FORK HEIGHT when adjusted so that the intersection of load-carrying surfaces and forward faces of forks or their equivalent, is the same horizontal distance from the LOAD AXLE as it is when at the height of the mast pivot.

master switch (crane), A manually operated device which dominates the operation of contactors and auxiliary devices of an electric circuit.

matte feeder, An extremely heavy duty type of apron feeder consisting of thick steel flights attached to a solid mat of chain links supported, in turn, by closely spaced rollers.

maximum fork height, The fork height attainable in fully raised position when loaded.

maximum lift, The lift from the extreme lowered position of the lifting device to the fully elevated position when loaded.

measurement cargo, A CARGO on which transportation charges are based on measurement of dimensions.

mechanical load brake, An automatic type of brake used for controlling loads in the lowering direction when simple reversing control is used. The brake remains released as long as the motor is driving in the down direction but automatically sets to provide braking action when the load begins to overhaul the motor.

mechanical shift transmission, A transmission requiring manual actuation of friction clutch, reversing gears, and speed gears, when used.

memorandum tariff, Nonofficial documents which contain rule and rate information extracted from official tariffs and published by carriers.

merging feeder, One which consolidates the movement of objects from two or more lanes into a single lane.

merry-go-round conveyor, See carrousel conveyor.

metering conveyor, In package handling, a POWER CONVEYOR whose speed and motion are controlled to release units at a predetermined rate.

mezzanine, A platform within a story between floor of the story, and the floor or roof next above it; elevated to a height which permits movement of men both below, and on the mezzanine; and generally having an area of not more than 1/3 the area of the floor immediately below.

minimum pressure accumulating conveyor, A conveyor designed to minimize buildup of pressure between adjacent packages or cartons.

minimum underclearance, The vertical dimension from the lowest point on the LOADED or empty vehicle, excluding wheel assemblies, to the vehicle supporting plane.

mixing conveyor, paddle type, A conveyor consisting of one or more parallel PADDLE CONVEYOR SCREWS.

mixing conveyor, screw type, A SCREW CONVEYOR consisting of one or more conveyor screws, ribbon flight, cut-flight, or cut-and-folded flight conveyor screws with or without auxiliary paddles.

motor brake, A mechanical device which applies braking action to stop the rotation of the armature.

motorized hand/rider truck, A dual purpose truck that is designed to be controlled by a walking or a riding operator.

motorized hand truck, A truck that is designed to be controlled by a walking operator.

motor vehicle accident, An accident involving any powered vehicle designed primarily for on-the-road operations. This excludes equipment designed primarily for off-the-road operations sucsh as material handling equipment, amphibious vehicles when afloat, tractors, etc.

multiple flight conveyor screw, A conveyor screw having two or more conveyor screw flights mounted at equal intervals around the pipe or shaft. The axial distance between adjacent flights is equal to the pitch of the conveyor screw flight divided by the number of flights.

multiple screw feeder, A series of conveyor screws installed side by side, usually in a flat bottom bin.

multiple strand conveyor, (1) Any conveyor which employs two or more spaced strands of chain, belts, or cords as the load-supporting medium. (2) Any conveyor in which two or more strands are used as the propelling medium connecting pans, pallets, etc.

narrow aisle truck, A self-loading truck primarily intended for right angle stacking in aisles narrower than those normally required by counterbalanced trucks of the same capacity.

natural frequency vibrating conveyor, A VIBRATING CONVEYOR in which the rate of free vibration of the trough on its resilient supports is approximately the same as the rate of vibration induced by the driving mechanism.

negative terminal (NEG or N or −): AIEE (60.06.025), The negative TERMINAL of a BATTERY is the terminal toward which positive electric charge flows in the external circuit from the POSITIVE TERMINAL.

nesting target (portable metal stacking rack), A formed metal piece (female form) affixed to the top of the post designed to retain the plug on the leg.

NEMA, National Electrical Manufacturers Association.

nickel-iron storage battery, A storage battery in which the positive active material consists of oxides of nickel, the negative active material consists of a mixture of iron and iron oxides and the alkaline electrolyte is usually an aqueous solution of potassium hydroxide. (The cells have a nominal voltage of 1.2.)

nonrigid tie, A member connecting two single racks maintaining a constant space, but does not add to the rigidity of the assembly.

nontelescoping mast, A MAST in which the support member or members providing the guideways for vertical movement of the fork carriage do not move vertically with respect to the truck.

normal and proper usage, Operation of the equipment with a program of regular maintenance in accordance with generally accepted practices and within the rated capacity and service classification for which it was specified and designed.

notched stringer, A STRINGER which has openings cut out for insertion of pallet lifting devices.

office, A room in which the administrative functions of a facility are conducted or a service is supplied; generally having a high lighting level, low noise level, and finished floors, ceilings and walls.

optional feature, Any substitution for the standard ATTACHMENTS, ACCESSORIES or components or any deviation from standard operational methods.

order bill of lading, A negotiable document by which a carrier acknowledges receipt of freight and contracts for its movement. The surrender of the original order bill of lading, properly endorsed is required by transportation lines upon delivery of the freight, in accordance with the terms of the BILL OF LADING.

order picker truck, high lift, A high-lift truck controllable by the operator stationed on a platform movable with the load-engaging means and intended for (manual) stock selection. The truck may be capable of self-loading and/or tiering.

order picking and stacking truck, A self-loading industrial truck, primarily intended for right-angle stacking and ordering picking, equipped with an operator control platform that moves vertically on the mast with the load-engaging means.

order picking truck, A manually loaded industrial truck equipped with both a load platform, and an operator control platform movable as a unit on a mast.

oscillating conveyor, A VIBRATING CONVEYOR having a relatively low frequency and large amplitude of motion. (See vibrating conveyor.)

oscillating feeder, See conveyor type feeder.

outrigger, a stabilizing structural member extending beyond the main body of the vehicle.

overall depth, The horizontal front-to-back dimension of a rack.

overall height, The vertical dimension from the floor to the topmost point of the UPRIGHT FRAME.

overall lowered height (OALH), The maximum vertical dimension from the vehicle supporting plane (the ground) to the extreme top point of the mast with the fork carriage in the fully lowered position and unloaded. (Not to be confused with OVERALL VEHICLE HEIGHT.)

overall vehicle height, The minimum height of the vehicle measured from the supporting plane.

over-and-under conveyor, Two endless chains or other linkage between which carriers are mounted and controlled so that the carriers remain in an upright and horizontal position throughout the complete cycle of the conveyor.

overhead conveyor, See trolley conveyor; power-and-free conveyor.

overhead crane, A TRAVELING CRANE with a movable bridge running on the top surface of rails of an overhead fixed runway structure and carrying a movable or fixed hoisting mechanism.

overhead trolley conveyor, See trolley conveyor.

paddle conveyor, See mixing conveyor, paddle type.

paddle conveyor screw, A conveyor screw in which paddles are pitched and positioned to form the material propelling means.

paddle mixer, See mixing conveyor, paddle type.

paddle washer, A conveyor consisting of one or two parallel inclined paddle conveyor screws in a conveyor trough having a receiving tank and an overflow weir at the lower end and a discharge opening at the upper end.

pallet, A horizontal platform device used as a base for assembling, storing and handling materials and/or products in a unit load, usually consisting of 3 stages, with upper and lower dockboards.

pallet, all-way entry, A pallet whose configuration permits retrieving or discharging from any direction in the same horizontal plane.

pallet, captive, A pallet whose use cycle remains within a single enterprise (private, corporate, or government).

pallet dispenser, A device for feeding empty pallets as required by a pallet loader or other machine or operation.

pallet, double deck, A pallet having top and bottom decks.

pallet, double entry, See two-way pallet.

pallet, double faced, See double deck pallet.

pallet, double wing, A PALLET whose top and bottom DECKS protrude beyond the outer edgesof the DECK SPACERS.

pallet, expendable, A pallet intended to be discarded after one cycle of use.

pallet, flush, A pallet whose DECKS do not protrude beyond the DECK SPACERS.

pallet, full four-way entry, Full four-way pallets permit four-way insertion and withdrawal by forks of INDUSTRIAL TRUCKS or by load wheeled forks of LOW-LIFT PALLET TRUCK.

pallet height, The total vertical dimension between the outer surfaces of the top and bottom decks.

pallet length, The first dimension stated in designating a pallet size. Based on common practice among pallet manufacturers and users, the first dimension is defined as follows:

ONE-WAY, TWO-WAY, AND PARTIAL FOUR-WAY ENTRY PALLETS — by the length of the STRINGER.

FOUR-WAY ENTRY AND ALL-WAY ENTRY PALLETS HAVING TWO OR MORE DECK MEMBERS AT THE TOP DECK — by the length of the STRINGER.

FOUR-WAY ENTRY AND ALL-WAY ENTRY PALLETS HAVING A SINGLE DECK MEMBER AND A SOLID TOP DECK.

SINGLE DECK OR DOUBLE DECK REVERSIBLE — longest dimension of the pallet.

DOUBLE DECK NONREVERSIBLE — the horizontal dimension at right angles to the length of bottom DECK MEMBER.

pallet load detierer, A device for separating tiered pallet loads.

pallet load tierer, A device for stacking full pallet loads for storage.

pallet loader, An automatic or semiautomatic machine, consisting of synchronized conveyors and mechanisms to receive objects from a conveyor(s) and place them onto pallets according to a prearranged pattern. (See palletizer.)

pallet, noncaptive, A pallet whose use cycle extends through one or more enterprises (private, corporate, or government) and may include a common carrier service.

pallet, nonreversible, A pallet having only one load-carrying surface.

pallet, partial four-way entry, Partial four-way pallets permit four-way insertion and withdrawal by forklift trucks and restrict the load wheeled forks of low-lift pallet trucks to two-way entry.

pallet, reusable, A pallet intended for multiple cycles of use.

pallet, reversible, A pallet having top and bottom decks capable of carrying a load.

pallet, single deck, A pallet having only one deck.

pallet, single faced, See single deck pallet.

pallet, single wing, A pallet whose top deck only protrudes beyond the outer edges of the DECK SPACERS.

pallet size, The horizontal dimensional length and width of the top deck of a pallet.

pallet stacker, See empty pallet stacker.

pallet, stacking, A pallet having a superstructure of vertical members (fixed, removable or collapsible), to support a superimposed load.

pallet storage rack, A structure composed of two or more upright frames, beams, and connectors, for the purpose of supporting palletized materials in storage. Among the common methods of assembly are welded, bolted, or clipped.

pallet storage shelf, At least two parallel BEAMS in the same horizontal plane with connectors in place between upright frames for supporting materials in storage.

pallet, take-it-or-leave-it, A pallet configuration that permits two choices of handling: (1) retrieving the pallet and unit load together, or (2) retrieving the unit load and leaving the PALLET.

pallet-to-post clearance, The dimension between the outside edge of the pallet to the inside edge of the POST on either side of the pallet.

pallet truck, A self-loading, low lift truck equipped with wheeled forks of dimensions to go between the top and bottom boards of a double faced pallet and having wheels capable of lowering into spaces between the bottom boards so as to raise the pallet off the floor for transporting.

pallet truck openings, Openings in the bottom deck of a pallet to allow the load wheels of a low-lift pallet truck to bear on the floor.

pallet, two-way entry, A pallet whose configuration permits insertion and withdrawal of a lifting device from opposite directions along the same horizontal axis.

pallet type conveyor, A series of flat or shaped wheelless carriers propelled by and attached to one or more endless chains or other linkage.

pallet unloader, An automatic machine consisting of synchronized conveyors and mechanisms to disassemble a pallet load and discharge the objects singly.

pallet width, The horizontal dimension at right angles to the PALLET LENGTH.

pallet, wing, A pallet whose deck(s) boards protrude beyond the outer edges of the deck spacers.

palletless loader, An automatic machine consisting of synchronized conveyors and mechanisms to receive objects and arrange them automatically into a unit in accordance with a predetermined pattern.

pan conveyor, A conveyor comprised of one or more endless chains or other linkage to which overlapping or interlocking pans are attached to form a series of shallow, open-topped containers. Some are called APRON CONVEYORS.

pan feeder, See conveyor type feeder.

parcel post air freight, Consolidation of a number of parcel post packages (with destination postage affixed by the shipper) for shipment as AIR FREIGHT to the postmaster at another city for subsequent delivery within local postal zones or beyond.

parking brake, A device to prevent the movement of a stationary vehicle.

passenger conveyor, A conveyor for transporting people who enter and leave the conveyor by walking or stepping and who either stand, sit or walk on the conveyor while being transported.

penthouse, An enclosed floor above the main roof of the building, generally substantially smaller than the area of the main floor of the building, and generally used to house either building or process equipment or in some cases, small habitable rooms, for specialized functions.

percent (%) grade, The measure of the rate of ascent of an inclined plane numerically equal to the vertical rise divided by the horizontal length, multiplied by 100.

personal injury, Any bodily harm sustained by a person as the result of an accident.

personal protective equipment, Any device utilized by the individual to prevent personal injury.

pick-up and delivery service (air transport), An optional service for the surface transport of shipments from shipper's door to originating air terminal and from the air terminal of destination to receiver's door. Pick-up service is provided upon shipper's request. Delivery service is provided automatically by the air carrier at an additional charge unless the shipper requests otherwise. PU&D is normally performed within a 25-mile radius of the airport. The latter is commonly referred to as the terminal area. (For service beyond the terminal area see truck/air/truck.)

piggy-back, The transportation of truck trailers and containers on railroad cars.

pilot plant, A facility devoted to the production of initial lots, or to the continuous production of small quantities of a product for the purpose of experimenting with its design, composition, production methods, and/or equipment used.

pin type slat conveyor, Two or more endless chains to which crossbars are attached at spaced intervals, each having affixed to it a series of pointed rods extending in a vertical plane on which work is carried. Used principally in spraying or washing operations where the least amount of area of the product is contacted.

pins, Fasteners such as rivets, belts, lags or other material used in assembling rack frames.

pit, A depression below the adjacent floor or grade level, generally open above.

pivoted bucket conveyor, A conveyor using pivoted buckets attached between two endless chains which operate in suitable guides or casing in horizontal, vertical, or inclined, or a combination of these paths over drive, corner, and take-up terminals. The buckets remain in the carrying position until they are tipped or inverted to discharge.

plain chain conveyor, See sliding chain conveyor.

planograph, A storage plan usually in two dimensions, on paper, which indicates the storage location of merchandise within a warehousing facility. Its primary function is to achieve the most effective utilization of space within the warehouse complex.

plant location analysis, The compilation and evaluation of data necessary to render a decision concerning site selection.

platform, An area that is raised in relation to one or more adjacent AREAS.

plug (portable metal stacking rack), A formed, cast, or forged metal piece (male form) affixed to the bottom of each leg designed to fit the nesting target on the post.

pneumatic chute or spout, A chute or spout in which air is introduced through the bottom to facilitate movement of bulk material down a slight decline.

pneumatic conveyor, An arrangement of tubes or ducts through which bulk material or objects are conveyed in a pressure and/or vacuum system.

pocket belt conveyor, See pocket conveyor.

pocket conveyor, A continuous series of pockets formed of a flexible material festooned between cross-rods carried by two endless chains or other linkage which operate in horizontal, vertical, or inclined paths.

polarity, The polarity of a battery is an electrical condition determining the direction in which current tends to flow on discharge. By common usage the discharge current is said to flow from the positive electrode through the external circuit.

port of entry, A harbor or airport, designated by the government, at which goods or persons are admitted into the country.

portable conveyor, Any type of transportable conveyor, usually having supports which provide mobility.

portable cranes, CRANES having a revolving superstructure with power plant, operating machinery and boom mounted on a fully mobile carriage not confined to a fixed path. (See crawler, locomotive, or truck crane.)

portable drag conveyor, A portable conveyor upon which endless drag chains are used as the conveying medium. Sometimes applied to a portable FLIGHT CONVEYOR. (See drag chain conveyor.)

portable metal stacking racks, A structure constructed of welded tubing or structural shapes or a combination of both. The rack bed is rectangular with or without short legs on each corner which allow truck forks to penetrate under the bed. Rack legs can be fixed or removable. Some have one or two removable side frames rather than single legs. Decking, when required, can be of wood or metal. When legs or frames are in place, one rack rests on top of another by means of interlocking male and female socket arrangements or tiering guides. The racks can be customized to hold anything from small stampings (by adding wire mesh sides and bottoms) to diesel engines.

port-of-origin cargo clearance, Customs formalities can be handled at certain originating cities in the USA rather than delaying such procedures until the export cargo reaches a gateway city.

positive discharge bucket elevator, A spaced bucket type of elevator in which the buckets are maintained over the discharge chute for a sufficient time to permit free gravity discharge of bulk materials. (See bucket elevator.)

positive terminal (POS or P or +), The positive terminal of a battery is the terminal from which the positive electric charge flows through the external circuit to the negative terminal when the cell discharges.

post, A single upright member, a part of an upright frame that may have provision for vertical adjustment of rails and beams.

power-and-free conveyor, A conveying system wherein the load is carried on a trolley or trolleys which are conveyor propelled through part of the system and may be gravity or manually propelled through another part. This arrangement provides a means of switching the free trolleys into and out of adjacent lines. The spur or subsidiary lines may or may not be powered.

power bars, CONDUCTOR bars which carry motor operating current.

power conveyor, Any conveyor which requires power to move its load.

power curve, A power driven conveyor for moving unit loads around a horizontal curve.

power driven roller conveyor, See roller conveyor, live.

power shift transmission, A transmission wherein the actual drive shifting of the reversing gear is accomplished by power application triggered by a pilot device such as a fingertip lever control or push button. Usually it is used in conjunction with a torque converter, and hydraulically applied clutches are used to accomplish the gear shifts. Speed changing gears may also be power shifted.

powered industrial truck, A mobile, power-driven truck used to carry, push, pull, lift, stack or tier material.

process chart, Tabular representation, using standarized symbols along with explanations depicting the route of either man or material from place to place and/or from operation to operation, in a manufacturing, clerical, or processing sequence of events.

property damage, The total cost to restore the property to the condition that existed prior to the accident.

push button station (crane), A means by which an operator at a fixed or moving location may actuate push buttons to control powered equipment.

pusher bar conveyor, Two endless chains cross-connected at intervals by bars or rotatable pushers which propel the load along the bed or trough of the conveyor.

pusher chain conveyor, One or more endless chains with attachments which propel or retard the movement of packages, objects, trucks, dollies, or cars along stationary wood, metal, or roller beds, troughs, rails, or tracks.

radio controlled, Motor-propelled units which are controlled by an operator by means of a remote radio transmitter.

rail, A horizontal load-carrying member running the full depth of the rack. Within each bay, two rails are used at each level.

rail and support height, The combined height in inches of the guide rail and support bracket in a drive-thru rack.

rail guide, A device attached to the rail for alignment of pallets.

rail projection, The horizontal dimension from the POST to the outside edge of the rail assembly.

rail span, The length of rail between the centerline of the uprights.

rail support, A member between the rail and the upright post.

rail support connector, A device connecting the rail support to the post.

rail tie, A member used to connect a pair of rails or near the end of the rail assembly.

ram, An attachment comprising a cantilever member, extending horizontally forward from a lift truck carriage, which is used to handle hollow core, cylindrical or similar loads, such as rugs, coils, pipe, etc.

ramp, A sloping aisle or roadway connecting two different levels, generally not exceeding 45 degrees.

rated load (crane), The maximum load which the equipment is designed to handle safely.

reach truck, A self-loading truck, generally high lift, having load-engaging means mounted so it can be extended forwardly under control to permit a load to be picked up and deposited in the extended position and transported in the retracted position.

receiving, The function of accepting, recording, and reporting merchandise into a facility.

receiving area, The space reserved for the RECEIVING function.

reciprocating beam conveyor, One or more parallel reciprocating beams with tilting dogs or pushers arranged to progressively advance objects.

removable posts (portable metal stacking racks), A load-carrying column that is inserted at each corner of the BASE by a simple mechanical means. A NESTING TARGET or PLUG is permanently affixed to the top of the column.

resistors, ballast (crane or carrier), Electrical resistance added in motor circuit to provide reduced starting current for gradual acceleration.

restricted articles (air transport), Certain articles (explosives, for example) which are either excluded from air cargo entirely or subject to stringent requirements on volume and packaging. A restricted articles tariff, giving full details on such articles, is published by Air Tariff Publishers, Inc. for the airlines.

retarder, Any device used to slow the rate of travel of bulk material or objects on a conveyor.

retarding conveyor, Any conveyor used to retard the rate of movement of bulk materials, packages or objects where the slope is such that the conveyed material tends to propel the conveying medium.

reusable container, A container intended for multiple cycles of use.

ribbon belt conveyor, See belt conveyor, multiple ribbon.

ribbon conveyor, See ribbon flight screw conveyor.

ribbon conveyor, internal, See internal ribbon conveyor.

ribbon flight, A conveyor screw flight proportioned to provide a space between it and the internal supporting pipe or shaft when mounted thereon.

ribbon flight conveyor screw, A conveyor screw in which the conveyor screw flight is of the ribbon flight type.

ribbon flight screw conveyor, A screw conveyor having a ribbon flight screw. (See screw conveyor.)

rider truck, A truck that is designed to be controlled by a riding operator.

right angle stack aisle width, The aisle width necessary to place in location the maximum dimension load perpendicular to the aisle with a given vehicle.

rigid back-to-back tie, A member connecting back-to-back racks, maintaining a constant space and adding rigidity to the complete assembly.

rigid container, A container with components permanently assembled.

rigid post (portable metal stacking rack), A load-carrying column permanently attached to the corner of the BASE. A NESTING TARGET or PLUG is permanently affixed to the top of the column.

roller conveyor, A series of rollers supported in a frame over which objects are advanced manually, by gravity, or by power.

roller conveyor, herringbone, A roller conveyor consisting of two parallel series of rollers having one or both series skewed.

roller conveyor, hydrostatic, A section of roller conveyor having rollers suitably weighted with liquid to control the velocity of objects being conveyed.

roller conveyor, live, A series of rollers over which objects are moved by the application of power to all or some of the rollers. The transmitting medium is usually belting or chain.

roller conveyor, shock absorbing, See roller conveyor, spring mounted.

roller conveyor, skewed, A roller conveyor having a series of rollers skewed to direct objects laterally while being conveyed.

roller conveyor, speed-up, A powered roller conveyor operating at a higher speed than its feeder to create space between unit loads.

roller conveyor, spool type, A roller conveyor in which the rollers are of conical or tapered shape with diameter at ends of roller larger than at the center.

roller conveyor, spring mounted, (1) A ROLLER CONVEYOR where the ends of each roller are supported on springs. (2) A section of roller conveyor supported on springs.

roller conveyor, troughed, A roller conveyor having two rows of rollers set at an angle to form a trough over which objects are conveyed.

roller spiral, An assembly of curved sections of roller conveyor arranged helically and over which objects are lowered by gravity.

roller table, A table with a surface consisting of a series of rollers.

rolling chain conveyor, A conveyor consisting of one or more endless roller chains on which packages or objects are carried on the chain rollers. The speed of transportation is double that of the chain speed.

rolling resistance, Force, resisting TRACTIVE EFFORT, caused by friction losses. It is preferably expressed in pounds, or as pounds per thousand pounds of vehicle weight, or as a percent of gross vehicle weight.

rollway skid, A roller conveyor with brakes to prevent the rollers from turning except in one direction.

room, An enclosed space within a building.

rope and button conveyor, A series of buttons or FLIGHTS attached to an endless wire rope or cable for the purpose of conveying or retarding the movement of bulk material or objects along a stationary trough.

rotary switch, A switch with a movable inner frame containing straight and/or curved sections of track. The inner frame can be rotated to align these sections of track with other tracks for routing carriers from one track to another.

rotary table feeder, A rotating horizontal, circular table to which material flows from a round bin or hopper opening and from which it is discharged by a plow.

rotary vane feeder, A rotor of cylindrical outline with radial, spaced plates or vanes rotating on a horizontal axis for controlling the flow of bulk materials.

rotator, A power-actuated attachment which provides rotary motion about an axis that is usually perpendicular to the face of the carriage.

rough terrain truck, A wheeled type truck designed primarily as a fork truck, but which may be equipped with attachments. This truck is intended for operation on unimproved natural terrain as well as the disturbed terrain of construction sites.

row, A continuous series of racks.

runners, Integral rails or skids on which demountable containers are supported.

running sheave, A sheave which rotates as hook is raised or lowered.

runway, The rails, beams, brackets and framework on which the CRANE operates.

runway conductors, The conductors mounted on or parallel to the runway which carry current to the crane.

safety, See accident prevention. (See also Safety.)

safety belt, A device, usually worn around the waist, which, by reason of its attachment to a lanyard and lifeline or a structure, will prevent a worker from falling.

safety management, See accident prevention.

safety training, Development of an accident prevention attitude by providing educational opportunities concerning the causes capable of producing ENVIRONMENTAL HAZARDS, PERSONAL INJURY, and/or PROPERTY DAMAGE.

safety work surface, A surface intended to reduce the possibility of a foot slippage.

scoop, An attachment to handle loose or bulk material by a shoveling action. Can be manually or power manipulated and dumped.

screw conveyor, A conveyor screw revolving in a suitably shaped stationary trough or casing fitted with hangers, trough ends and other auxiliary accessories.

screw conveyor, rotating casing, A SCREW CONVEYOR in which a tubular casing rotates.

screw conveyor, vertical, A SCREW CONVEYOR that conveys in a substantially vertical path.

screw, cut-and-folded flight, A conveyor screw with a section or sections of each pitch cut and folded back.

screw, cut flight, A conveyor screw with a section or sections notched from each pitch.

screw, double flight, A conveyor screw having two conveyor screw flights mounted 180° apart on the pipe or shaft. The axial distance between adjacent flights is equal to one-half the pitch of the conveyor screw flight.

section, A division physically arranged within a facility, usually a warehouse, which is arbitrarily assigned to a particular item, product, or group of items, and may be delineated by the structural features of the facility such as columns, partitions, firewalls, etc.

self-feeding portable conveyor, Any power-propelled conveyor designed to advance into a pile of bulk material, thereby automatically feeding itself.

self-feeding conveyor, Any conveyor so arranged as to feed itself automatically without the necessity of using a separate feeder.

self loading, The capability of a POWERED INDUSTRIAL TRUCK to pick up, carry, and deposit its load without the aid of external handling means.

semi-gantry crane, A TRAVELING CRANE with one end of the bridge supported on one or more legs running on fixed rails or other runway and the other end of the bridge supported by a truck running on an elevated rail or runway.

shaft, (1) A vertical space within a building, generally extending through more than one story, usually enclosed, created to house vertical functions. (2) A passageway through the ground, generally constructed to provide for movement of men or material from one level to another.

shaker conveyor, See oscillating conveyor.

shelf, A horizontal supporting surface above floor level within a rack or storage section.

shelf connector, A device used at each end of a beam to connect the beam to an upright frame.

shipping, The function of recording, reporting, and sending merchandise to the consumer.

shipping area, The space reserved for the shipping function.

ship's conveyor elevator, A deck-mounted pocket or suspended tray conveyor arranged to be lowered into a ship's hold to varying depths for the purpose of loading or unloading packages and objects. Not to be confused with a MARINE LEG.

shop, The space and resources assigned to a functional group, usually artisans engaged in making or repairing a variety of articles.

shuttle conveyor, Any conveyor such as a belt, chain, apron, screw, etc., in a self-contained structure movable in a defined path parallel to the flow of the material.

side loader, A self-loading truck, generally high lift, having load engaging means mounted in such a manner that it can be extended laterally under control to permit a load to be picked up and deposited in the extended position and transported in the retracted position on the truck bed.

side loading attachment, An attachment used to pick up loads and deposit them at right angles to the longitudinal axis of a lift truck. (Primarily intended for narrow aisle operations.)

side-mounted operator compartment, An operator compartment a portion of which is located outside of a longitudinal line connecting between outermost front and rear tires.

side-pull en masse conveyor, An arrangement of horizontal CLOSED CIRCUIT CONVEYOR in which the tension element is at one side and above the moving stream of material.

side pusher conveyor, A TROLLEY CONVEYOR with arms cantilevered to the side to push free trolleys on parallel tracks or with pushers to engage the cantilevered arms of free trolleys.

side shifter, A device to permit lateral movement of the load-engaging means on a lift truck.

signature service, A service designed to provide continuous responsibility for the custody of shipments in transit. A signature is required from each person handling the shipment at each stage of its transit from origin to destination and a record is kept to minimize the possibility of loss or damage.

silo, (1) A tall cylindrical building, usually used for storing bulk commodities. (2) A tall cylindrical building in the ground such as that used for storing missiles.

single arm, Horizontal load member affixed to a column for single side loading.

single rack, A pallet storage rack with only two upright frames and one or more shelves.

SITC, Standard International Trade Classification. (See commodity code. See also standard international trade classification.)

skate-wheel conveyor, A wheel conveyor making use of series of skate wheels mounted on common shafts or axles, or mounted on parallel spaced bars on individual axles. (See wheel conveyor.)

skew table, A live roller conveyor having its rollers skewed for the purpose of moving objects laterally against a guide member. It may have a fixed or an adjustable guide member. (See roller conveyor, skewed.)

skip hoist, A bucket or car operating up and down a defined path receiving, elevating, and discharging bulk materials.

skid support, A pair of upturned channels placed front to back across a pair of shelf beams.

slat conveyor, A conveyor employing one or more endless chains to which nonoverlapping, noninterlocking, spaced slats are attached.

slider bed, A stationary surface on which the carrying portion of a conveyor belt slides.

sliding chain conveyor, One or more endless chains sliding on tracks on which packages or objects are carried.

sliding type switch, A switch with a movable inner frame containing straight and/or curved sections of track. The inner frame can be slid to align these sections of track with other tracks for routing carriers from one track to another.

sling, A flexible means of connecting the load and the hoisting device.

slope conveyor, Usually a troughed belt conveyor used for transporting coal or ore through an inclined passage to the surface from an underground mine. (See apron conveyor, belt conveyor, flight conveyor.)

solid tire, A tire, such as "Cushion Tire" or "Low-Solid Tire," made of resilient solid material.

sorting conveyor, A conveyor which receives mixed unit loads and discharges them to segregated spaces or conveyors in response to an automatic dispatch control. Operator attention is usually required to introduce the dispatch signal into a memory system.

space, One or more areas.

span, Center to center distance between runways.

spark enclosing equipment, Auxiliary units which completely enclose the electrical equipment on the vehicle to prevent the emission of electric sparks. It includes the enclosure of the generator, electric motors, contactors, and protection of certain other electrical equipment.

spindle conveyor, A chain-on-end conveyor in which the chain pins extend in a vertical plane and are usually of enlarged diameter in that portion above the chain on which special revolvable fixtures can be rotated for the purpose of spraying or drying. Outboard rollers or sliding shoes support the chain and products.

spur track, A fixed track arranged to interlock with an adjacent crane girder to permit passage of carriers from the spur track to the crane and vice versa.

squaring shaft, A driven shaft which transmits torque to drive wheels operating on two or more tracks.

stability, A truck's resistance to overturning. (The recognized method of measurement is the "Tilting Platform Tests" covered in Appendix A of ASA Safety Code B56.1-1959 plus additions shown in Section 4 of ITA Recommended Practices Manual.)

stacker conveyor, (1) A CONVEYOR adapted to piling or stacking bulk materials, packages or objects. (See apron conveyor; belt conveyor; flight conveyor; portable conveyor; slat conveyor. (2) A fixed or pivotally mounted BOOM CONVEYOR. (3) With a blending system the stacker operates over the stocking conveyor in a manner similar to a wing tripper to build layered piles or beds of material parallel to the stocking conveyor. (See boom conveyor; portable conveyor.)

stair, A series of steps allowing walking access from one level to another.

stairway, One or more flights of stairs and the necessary landings and PLATFORMS connected therewith to form a continuous and uninterrupted passage from one floor to another or to the roof or to the ground.

standard international trade classification (SITC), A numerical code developed by the United Nations and adopted by the U.S. scheduled airlines and airlines of most other countries as a basis for numerical identification of commodities moving in air freight. (See commodity code. See also SITC.)

station, An established location in which something or someone stands or is assigned to stand or remain. This space includes that area necessary for the operation, material being processed or worked on, and the auxiliary equipment necessary to perform the assigned function.

steel belt conveyor, A BELT CONVEYOR using a steel band belt as the conveying medium.

storehouse, A place or space for storing goods.

story, The space between the surface of two successive floors in a building, or between the top floor and the ceiling or underside of the roof frame.

straddle truck, A self-loading outrigger type industrial truck, generally high-lift, for picking up and handling loads between its outrigger arms.

straight bill of lading, A nonnegotiable document by which a carrier acknowledges receipt of freight and contracts for its movement. The surrender of the original straight bill of lading, is not required by the transportation line upon delivery of the freight except when necessary for the purpose of identifying the consignee.

straight chute, A sloped chute designed to transfer bulk materials, packages, or objects in a straight line from point of entry to point of discharge.

strap slot, A small opening in the STRINGER or on the underside of top DECKMEMBER to facilitate use and prevent movement of unit load fastening material.

stringer, A continuous longitudinal DECK SPACER.

stringermember, A subsurface element on full four-way entry wood pallets beneath and at right angles to the DECK-MEMBER but above the block, leg or post spacers.

supports, conductor, Mechanical supports for CONDUCTOR bars.

suspended tray conveyor, A vertical conveyor having one or more endless chains with suitable pendant trays, cars or carriers which receive objects at one elevation(s) and deliver them to another elevation(s).

suspended tray elevator, See suspended tray conveyor.

suspended tray lift, See suspended tray conveyor.

switch, monorail, A device with a moving section or sections of track which can be moved to permit passage of a carrier from an incoming track to one of various outgoing tracks. (See rotary, sliding type, tongue switch.)

tag line, An electrical conductor system employing flexible cable or cables as the conductors.

tariff (air transport), A document filed with the Civil Aeronautics Board setting forth applicable rules, rates and charges for the movement of goods.

tariff (motor carrier), Documents published and filed with I.C.C. containing rates, fares, rules and regulations of common carriers.

tariff (railroad), A schedule of rates charged with governing rules and regulations. (Part I of I.C.C. Act.)

telescoping conveyor, A conveyor, the length of which may be varied by telescoping frame members. (See extendable conveyor.)

telescoping mast, A multiple mast wherein one member is stationary and the other(s) movable vertically with respect to the stationary member and supporting the FORK CARRIAGE in its vertical movement.

terminal (term.), The terminals of a battery are the parts to which the external electric circuit is connected.

terminal connector (term. conn.), A terminal connector is an electric conductor for carrying current between a cell and a battery terminal.

tethered-electric truck, An electric truck in which the power source is remote from the vehicle and connected by a flexible cable.

thermal overload, automatic, A device which shuts off power when permissible temperatures are exceeded and which reestablishes power when the temperature is reduced.

thermal overload, manual reset type, A device which shuts off power when permissible temperatures are exceeded and which requires manual actuation to reestablish the power.

tierer, pallet load, See pallet load tierer.

tiering, The process of placing one load above another.

tiering guide (portable metal stacking rack), A formed metal piece to serve a dual purpose (1) as a lifting device for crane handling, (2) as a tiering guide.

tilt, Longitudinal angular displacement of mast structure, forks, or carriage.

toe (cantilever rack), The front edge of an extended base on single or double sided upright.

toe board, A vertical projection above a platform or walkway to contain objects and to define the walkway for personnel.

tongue switch, A switch that contains one straight section of track or tongue, pivoted at one end. The tongue can be swung to various positions to connect this section of track to other tracks for transfer of carriers from one track to another.

top cross braces, Diagonal members in the form of a rigid cross connecting the uppermost portion of the rack assembly.

top diagonal brace, A diagonal member connecting uppermost portions of rack assemblies.

top horizontal brace, A member tying two posts together parallel to the depth of the bay at the top of the rack.

top tie, A member connecting rack posts.

torque converter transmission, A hydrodynamic drive which transmits power with ability to change torque. (Torque ratio is a function of speed ratio.)

torque, full load, The torque necessary for a motor to produce its rated horsepower at full-load speed.

tow conveyor, An endless chain supported by trolleys from an overhead track or running in a track at the floor with means for towing floor supported trucks, dollies, or carts.

tractive effort, The motive force, exerted at the circumferences of the driving wheels at the point of contact with the ground.

tractive effort, required, Tractive effort needed to propel a vehicle and any trailers being towed by it, while overcoming (1) ROLLING RESISTANCE of all wheels, including all wheel bearing friction, (2) GRADE RESISTANCE, and (3) inertia of vehicle and any trailers being towed by it.

tractor, motor driven, A motor driven unit, supported from track wheels and propelled by a friction driving wheel or wheels. It is normally attached to, and drives, a carrier or crane end truck.

traffic control, A mechanical or electrical mechanism to prevent collision of objects as they merge from two conveyor lines into a single line.

tramway, A system in which carriers are supported by a track cable and in which the movement of carriers is continuous over one or more spans.

transfer lift section, A section of track which can be moved vertically out of alignment with one stationary track into alignment with another stationary track at a different elevation in order to transfer carriers from one track to the other.

travel speed, The velocity of a vehicle in a horizontal plane.

traveling cranes, Cranes that follow a fixed path. (See overhead, underhung, gantry, semi-gantry, wall cranes.)

tray (battery), A tray of a storage battery is a support or container for one or more storage cells.

trolley conveyor, A series of trolleys supported from or within an overhead track and connected by endless propelling means such as chain, cable, or other linkage with loads usually suspended from the trolleys.

trolley (top running), The unit including the hoisting mechanism which travels on the BRIDGE RAILS.

trolley (underhung), See carrier.

truck, See powered industrial truck.

truck/air/truck service, The surface movement of air freight to and from airports and origin and destination points beyond the usual 25-mile radius of pick-up and delivery service.

truck crane, A PORTABLE CRANE mounted on an automotive vehicle equipped with a power plant for travel.

truck leveler, An hydraulically operated ground level platform over which the rear wheels of a truck or trailer may be driven for the express purpose of raising or lowering the truck body in order to compensate for varying truck bed and delivery platform heights.

tunnel, A below grade passageway.

turning center, The center of the circle described by the truck moving in a turn.

turning radius (inside), Under conditions described for turning radius (OUTSIDE), the distance from center of the smallest circle to the nearest projection of the truck. Primarily used to establish operational clearance.

turning radius (outside), Half the diameter of the largest circle described by the outermost projection of the truck when driving slowly empty or loaded under its own power with the steering mechanism at the optimum steering angle, both forward and reverse, and to the left and the right.

turntable, A device with a movable inner frame containing a straight section of track. The inner frame can be rotated, with a loaded carrier on it, to align the section of track with other tracks for the transfer of carriers from one track to another.

turntable, A horizontal, rotatable conveyor mechanism used for transferring objects between conveyors which are in angular relation to one another.

twist conveyor, An EL CONVEYOR in which the carrying surface and guard gradually exchange their functional duties.

two-way container, A container whose configuration permits retrieving or discharging from opposite directions along the same horizontal axis.

underclearance (portable metal stacking rack), The dimension from the floor to the underside of the base frame.

underdrive, A crane or carrier arrangement which propels the equipment by means of a driving wheel pressing on the underside of the track.

underhung crane, A TRAVELING CRANE with a movable bridge running on the lower flanges of an overhead fixed runway structure and carrying a movable or fixed hoisting mechanism.

undervoltage protection, A device operative on the reduction or failure of voltage which interrupts the main circuit and maintains this condition.

underwriters' laboratories listing (UL), A list of truck models approved as complying with the requirements of and used by the Stock group of insurance companies to determine the premium rate for insurance coverage for various areas of operation.

upright (cantilever rack), A column with the supports necessary to keep it in vertical alignment. The six basic types are as follows:

(Type I) SINGLE SIDES WITH EXTENDED BASE — column with the arms and the base extended in one direction only. (Type II) DOUBLE SIDED WITH EXTENDED BASE — column with arms and base extended from two sides of the column in the same vertical plane. (Type III) IMBEDDED — column with the bottom of the column imbedded into the floor. (Type IV) PINNED — column affixed by pins to the ceiling and floor. (Type V) BOTTOM PINNED — column with the column affixed to the floor and the top of the column is affixed to another column in another row by an aisle tie. (Type VI) FREE STANDING — column with 3 or more extended base legs to make column rigid without being permanently affixed to the floor.

upright frame, An assembly comprised of posts, horizontal braces, and/or cross braces and diagonal braces in one plane.

unsafe act, A departure from an accepted, normal, or correct procedure or practice, an unnecessary exposure to a hazard, or conduct minimizing the degree of safety normally present.

unsafe condition, Any condition that if left uncorrected may lead to an accident.

valuation charges, Transportation charges assessed shippers whose shipments have declared values higher than the historically assumed value to 50 cents a pound or 50 dollars, whichever is greater.

variable reach lift truck, A high lift truck with the additional capability of extending and retracting the forks (and load) in a longitudinal direction.

vault, A room designed to provide security of contents.

vertical belt conveyor, A vertical conveyor in which sections of articulated slat conveyor apron form rigid platforms for vertical movement in continuous flow. The platforms are flexible in but one direction and they assume a vertical position on the noncarrying run to minimize space requirements.

vertical chain conveyor, opposed shelf type, Two or more vertical elevating-conveying units opposed to each other. Each unit consists of one or more endless chains whose adjacent facing runs operate in parallel paths. Thus each pair of opposing shelves or brackets receive objects (usually dish trays) and deliver them to any number of stations.

vertical clearance, The dimension from the floor to the underside of the rail assembly; or from the topside of the rail assembly to the underside of the next rail assembly.

vertical reciprocating conveyor, A reciprocating power or gravity actuated unit which receives only inanimate objects on a track, roller, conveyor, or power conveyor forming the bed of the carrier and transmits these inanimate objects vertically from one elevation to another. This type of conveyor is never designed to carry an operator or passengers and no person may ever be permitted to ride upon it. The carrier must never be furnished with flooring of any type which might provide safe footing for any person. The conveyor shaft must be enclosed and where manual loading and unloading is employed, doors at each operating level must be installed and so interlocked with the carrier that doors at any level can be opened only when the carrier has stopped at that level and carrier cannot be moved until such open doors are closed. No controls for movement of carrier may be located on carrier, or within the enclosure, or within reach of any person standing on the carrier. For further reference see the current B 20 Safety Code for "Conveyors & Related Equipment."

vibrating conveyor, A trough or tube flexibly supported and vibrated at a relatively high frequency and small amplitude to convey bulk material or objects. (See oscillating conveyor.)

vibrating conveyor, balanced, A vibrating conveyor having a secondary mass operating in a direction opposite to the movement of the troughed pan for the purpose of minimizing the dynamic force normally directed into the supports.

vibrating conveyor, natural frequency, A vibrating conveyor in which the rate of free vibration of the trough on its resilient support is approximately the same as the rate of vibration induced by the driving mechanism.

walking beam conveyor, A conveyor employing a multiple arrangement of walking beams with associated rails and/or rollers.

wall tie, A member connecting the rack to a wall.

wall crane, A TRAVELING CRANE having a jib with a movable or fixed hoisting mechanism and operating on a runway attached to the side walls or columns of a building.

warehouse, A building where wares, and/or goods, are stored.

water muffler, A special muffler utilizing a water trap to prevent the emission of sparks or flames from the exhaust system.

waybill, A document prepared by the transportation line at the point of origin giving a description of the shipment and shipping instructions and sent with the shipment or mailed to the agent at the waybill destination or transfer point.

wheel base, Distance from center to center of outermost wheels.

wheel conveyor, A series of wheels supported on axles in a frame over which objects are moved manually or by gravity. (See skate wheel conveyor.)

wheel, driving, Wheel to which tractive effort is applied to cause equipment to move.

wheel load, The load on any wheel with the trolley and lifted load (rated capacity) positioned on the bridge to give maximum loading.

wheel spiral, Helically wound, curved sections of wheel conveyor that lower objects by gravity.

wicket conveyor, A conveyor comprised of two or more endless chains connected by cross bars to which vertical rods are attached at spaced intervals. The cross bars are also provided with spaced projections at level of same to form, in effect, a continuous carrying surface through which the loads cannot fall. Used for handling such products as painted sheets, wall, or composition board, etc., on edge through dryers or bake ovens.

Materials Processing 7

Z94.7

The definitions developed for this section represent a revised version of the original 1972 edition, Z94.7: MATERIALS PROCESSING. Like the original, these definitions cover the processing of metals, plastics, ceramics, and rubber-like materials.

Where applicable, the MATERIALS PROCESSING SUBCOMMITTEE used as a basis the already accepted definitions such as those published by THE SOCIETY OF THE PLASTICS INDUSTRY (e.g., from the PLASTICS ENGINEERING HANDBOOK, Van Nostrand-Reinhold), THE AMERICAN WELDING SOCIETY (especially its STANDARD A 3.0), THE AMERICAN SOCIETY FOR METALS (METALS HANDBOOK), FORGING INDUSTRY ASSOCIATION (FORGING INDUSTRY HANDBOOK), AMERICAN FOUNDRYMEN'S SOCIETY (AFS GLOSSARY OF FOUNDRYMEN'S TERMS), the Industrial Division, Minneapolis-Honeywell Regulator Company (AUTOMATION DICTIONARY OF, and the ANATOMY OF AUTOMATION by Amber and Amber (Prentice-Hall).

The definitions in this section appear in the following subsections: AUTOMATION, FORGING, FOUNDRY, METAL FORMING, METAL MACHINING, PLASTICS, and WELDING.

Although the work of the earlier subcommittee must be acknowledged, this current section was the result of the time, effort, expertise, and devotion of:

CHAIRMAN:

Prof. Benjamin W. Niebel
  Emeritus Professor and Head, Department of Industrial and Management Systems Engineering, The Pennsylvania State University

MEMBERS:

Prof. A. B. Draper
Prof. D. G. Freark
Prof. G. L. Goodrich
Prof. R. E. Keith
  All of Department of Industrial and Management Systems Engineering, The Pennsylvania State University
Prof. G. Kane
  Department of Industrial Engineering, Lehigh University

AUTOMATION

accuracy, The difference between the target and the average from the target plus or minus three standard deviations.

adaptive control, The ability of a control system to change its own parameters in response to a measured change in operating conditions.

address, A character or group of characters that identifies a register, a particular part of storage, or some other data source or destination.

algorithm, A prescribed set of well-defined rules for the solution of a problem in a finite number of steps.

amplifier, Device for controlling power from a source so that more is delivered at the output than is supplied at the input. Source of power may be mechanical, hydraulic, pneumatic, electric, etc.

analog computer, A computer which operates on the principle of creating a physical (usually electrical) analogy of mathematical problem to be solved. Variables such as temperature or flow are represented by the magnitude of a physical phenomenon such as voltage or current. These variables are manipulated by the computer in accordance with mathematical formulas "analoged" on it.

analog-to-digital-converter (ADC), Refers to a device which produces a digital output from an input in the analog form of physical motion or electrical voltages.

ANSI, American National Standards Institute, formerly known as ASA and USASI, its purposes include:

(1) serving as the national institution for voluntary standardization and certification
(2) furthering voluntary standards
(3) assuring the interests of the public, including consumers, labor, industry, and government
(4) providing the need for standards and certification programs
(5) establishing, promulgating, and administering procedures for recognition and approval of standards as American National Standards
(6) encouraging existing organizations to prepare and submit certification programs for accreditation
(7) cooperating with government agencies
(8) promoting knowledge and use of American National Standards
(9) representing USA interests in international nontreaty standardization and certification organizations
(10) serve as a clearinghouse for information on standards, standardization, and certification in the USA and abroad.

APT (Automatically Programmed Tools), A language for programming numerically controlled machine tools. There are many other similar languages with different names.

artificial intelligence, Research and study in methods for the development of a machine that can improve its own operations. The development or capability of a machine that can proceed or perform functions that are normally concerned with human intelligence as learning, adapting, reasoning, self-correction, automatic improvement. In a more restricted sense, the study of techniques for more effective use of digital computers by improved programming techniques.

ASCII (American Standard Code for Information Interchange), A standard data-transmission code that was introduced to achieve compatibility between data devices. It consists of 7 information bits and one parity bit for error-checking purposes, thus allowing 128 code combinations.

assembler language, A source language that includes symbolic machine language statements in which there is a one-to-one correspondence with the instruction formats and data formats of the computer.

attentiation, Refers to the decrease in amplitude of a signal, usually expressed as a ratio or in decibels.

automata theory, Relates to the development of theory which relates the study of principles of operations and applications of automatic devices to various behaviorist concepts and theories.

automation, (1) The implementation of processes by automatic means. (2) The theory, art, or technique of making a process more automatic. (3) Investigation, design, development, and application of methods of rendering processes automatic, self-moving, or self-controlling. (4) The conversion of a procedure, a process, or equipment to automatic operation.

axis, A direction of relative motion between tool and workpiece.

baud, (1) A unit of signalling speed equal to the number of discrete conditions or signal events per second. For example, one baud equals one-half dot cycle per second in Morse Code, one bit per second in a train or binary signals, and one 3-bit value per second in a train or signals each of which can assume one of eight different states. (2) In asynchronous transmission, the unit of modulation rate corresponding to one unit interval per second, i.e., if the duration of the units interval is 20 milliseconds, the modulation rate is 50 baud.

BCD (Binary Coded Decimal), A system of number representation; that is, an information code, in which each decimal digit is represented by 4 binary digits.

binary, A characteristic, property, or condition in which there are but two possible alternatives, e.g., the binary number system using 2 as its base and using only the digits zero (0) and one (1).

binary digit (bit), (1) A numeral in the binary scale of notation. This digit may be zero (0), or one (1). It may be equivalent to an on or off condition, a yes, or a no. Often abbreviated to (bit). (2) The kind of number that computers are internally. There are only two binary digits, 1 and 0, otherwise known as "on" and "off."

bit, (1) Bit is an abbreviation for binary digit. Most commonly a unit of information equalling one binary decision, or the designation of one of two possible and equally likely values or states, usually conveyed as 1 or 0 of anything used to store or convey information (such as 1 or 0, which may also mean "yes" or "no"). (2) A single character in a binary number. (3) A single pulse in a group of pulses. (4) A unit of information, capacity of a storage device. The capacity in bits is the logarithm to the base two of the number of possible states of the device.

bode diagram, A unique plot of log amplitude ratios and phase angle values on a log frequency base for a transfer function. This may be an element, output or loop transfer function.

Boolean algebra, A mathematical system of logic named for George Boole (1815-1864), the famous English mathematician and logician. It deals with classes, propositions, on-off circuit elements, etc. Associated by operators as AND, OR, NOT, EXCEPT, IF . . . THEN . . . which permits computations and demonstrations as in any other mathematical system.

bootstrap, (1) A technique or device designed to bring itself into a desired state by means of its own action, e.g., a machine routine whose first few instructions are sufficient to bring the rest of itself into the computer from an input device. (2) To use a bootstrap. (3) That part of a computer program used to establish another version of the computer program.

breadboard, Usually refers to an experimental or rough construction model of a process, device, or construction.

bridge, In a system of measurement, the instrument in which part or all of a bridge circuit is used in measuring one or more electrical quantities. In relation to a fully electronic stringed instrument, the bridge converts the mechanical vibrations produced by the strings into electrical signals.

bubble memory, Such memories are actually tiny cylinders of magnetization whose axes lie perpendicular to the plane of the single-crystal sheet that contains them. Magnetic bubbles arise when two magnetic fields are applied perpendicular to the sheet. A constant field strengthens and fattens the regions of the sheet whose magnetization lies along it. A pulsed field then breaks the strengthened regions into isolated bubbles, which are free to move within the plane of the sheet. Because the presence or absence of bubbles can represent digital information, and because other external fields can manipulate this information, magnetic-bubble devices could find uses in future data-storage systems.

buffer, A "machine" designed to be inserted between other "machines" or program elements to match impedances, peripheral equipment speeds, to prevent mixed interactions, to supply additional drive or relay capability or simply to delay the rate of information flow. Buffer types are classified as inverting or non-inverting.

buffer storage, (1) A synchronizing element between forms of storage, usually between internal and external. (2) An input device in which information is assembled from external or secondary storage and stored ready for transfer to internal storage. (3) An output device into which information is copied from internal storage and held for transfer to secondary or external storage. Computation continues while transfers between buffer storage and secondary or internal storage or vice versa takes place. (4) Any device which stores information temporarily during data transfers.

bug, (1) A program defect or error. Also refers to any circuit fault due to improper design or construction. (2) A mistake or malfunction. (3) An integrated circuit.

bus, (1) As applied to computer technology, one or more conductors used as a path over which information is transmitted. (2) A circuit over which data or power is transmitted. Often one which acts as a common connection among a number of locations. *Syn:* trunk. (3) A path over which information is transferred, from any of several sources to any of several destinations.

byte, An IBM developed term used to indicate a specific number of consecutive bits treated as a single entity. A byte is most often considered to consist of eight bits which as a unit can represent one character or two numerals.

carrier, (1) A particular wave which has constant amplitude and frequency, and a phase which can be modulated by changing amplitude, frequency, or phase. Also, an entity which has the ability to carry an electric charge through a solid. For example, holes and conduction electrons in semiconductors. (2) A continuous frequency capable of being modulated or impressed with a second (informative carrying) signal.

cascade control, An automatic control system in which the control units, linked in sequence, feed into one another in succession, each regulating the operation of the next in line.

Central Processing Unit (CPU), (1) A unit of a computer that includes the circuits controlling the interpretation and execution of instructions. *Syn:* mainframe. (2) The central processor of a computer system contains main storage, arithmetic unit, control registers, and scratchpad memory.

chad, The tiny piece of paper removed when a hole is punched into a card or paper tape.

channel, (1) That portion of a computer's storage medium which is also accessible to a given reading station. (2) That part of a communication system that connects the message source with the message sink. (3) A path along which signals can be sent, e.g., a data channel, output channel. (4) The portion of a storage medium that is accessible to a given reading or writing station e.g., track, band. (5) In communication, a means of transmission. Several channels may share common equipment. For example, in frequency multiplexing carrier systems, each channel uses a particular frequency band that is reserved for it.

character, One of a set of elements which may be arranged in ordered groups to express information. Each character has two forms: (1) a man-intelligible form, the graphic, including the decimal digits 0-9, and letters A-Z, punctuation marks, and other formatting and control symbols; and (2) its computer intelligible form, the code, consisting of a group of binary bits.

circuit breaker, Refers to various devices for opening electric circuits under abnormal operating conditions; e.g., excessive current, heat, high ambient radiation level, etc. Also called constant breaker.

circuit, integrated (IC), Refers to one of several logic circuits, gates, flip-flops which are etched on single crystals, ceramics or other semiconductor materials and designed to use geometric etching and conductive ink or chemical deposition techniques all within a hermetically sealed chip. Some chips with many resistors and transistors are extremely tiny, others are in effect "sandwiches" of individual chips.

circuit, printed, Refer to resistors, capacitors, diodes, transistors and other circuit elements which are mounted on cards and interconnected by conductor deposits. These special cards are treated with light-sensitive emulsion and exposed. The light thus fixes the areas to be retained and an acid bath eats away those portions which are designed to be destroyed. The base is usually a copper clad card.

circular interpolation, A machine control unit with a small computer module that is able to direct a tool in a circle or arc when given basic information such as the coordinates of the center point, direction of travel and coordinates of the end point.

clock, (1) The most basic source of synchronizing signals in most electronic equipment, especially computers. (2) That specific device or unit designed to time events. (3) A data communications clock which controls the timing of bits sent in a data stream, and controls the timing of the sampling of bits received in a data stream.

closed loop, (1) A circuit in which the output is continuously fed back to its source for constant comparison. Also a group of indefinitely repeated computer instructions. (2) The complete signal path in a control system represented as a group of units connected in such a manner that a signal started at any point follows a closed path and can be traced back to that point.

closed-loop system, Refers to a feedback control system involving one or more feedback control loops, which combine functions of controlled signals and of commands, in order to keep relationships between the two stable.

CMOS (Complementary Metal-Oxide-Semiconductor), Technology uses both p- and n-channel devices on the same silicon substrate. Basic CMOS construction uses n- and p-channel devices connected in series. Only one device is turned on at one time, keeping power dissipation low. Switching of devices through the active region and charging and discharging of capacitance are main causes of dissipation. Major advantages are (1) low power dissipation, (2) good noise immunity (45% of drain voltage), (3) high fanout to other CMOS, (4) allowance for very wide power supply variations, (5) shorter propagation delay than with p-MOS (approximately 60 ns), and (6) full temperature range capabilities. Chip size, however, is larger than p-MOS. CMOS is used in battery-operated systems, aerospace logic systems, and portable digital communications equipment. It is also used for components in digital instruments which operate in noisy environments.

CMOS: Universal Asynchronous Receiver — Transmitter (UART), These UARTs are CMOS/LSI subsystems for interfacing computers or microprocessors to an asynchronous serial data channel. The receiver converts serial start, data, parity and stop bits to parallel data verifying proper code transmission, parity, and stop bits. The data word length can be 5, 6, 7 or 8 bits. Parity may be odd or even. Parity checking and generation can be inhibited. The CMOS UARTs can be used in a wide range of applications including modems, printers, peripherals and remote data acquisition systems. CMOS/LSI technology permits operating clock frequencies up to 3.2 MHz (200K Baud) an improvement of 10 to 1 over previous PMOS UART designs. Power requirements, by comparison, are reduced from 300 mw to 10mw. Status logic increases flexibility and simplifies the user interface.

code, (1) A system of characters and rules for representing information. (2) Relates to transformations or representations of information in different forms according to a preassigned convention. (3) Digital codes may represent numbers, letters of the alphabet, control signals, and the like, as a group of discrete bits rather than as a continuous signal.

command, (1) A pulse, signal or set of signals that initiate a performance. (2) A signal from a machine control unit that initiates one step in a complete program.

communication interface circuit, A communications interface circuit, (USART), is a peripheral device programmed by the CPU to operate using virtually any serial data transmission technique currently in use. It will typically have a speed of 4 megabits per second for synchronous operation and 250 kilobits per second for asynchronous operation. (A USART refers to Universal Synchronous/Asynchronous Receiver/Transmitter.)

Computer Aided Design (CAD), Refers to the capability of a computer to be used for automated industrial, statistical, biological, etc., design through visual devices.

control, (1) Those parts of a computer which carry out instructions in proper sequence, interpret instructions, and apply proper signals. (2) Frequently, it is one or more of the components in any mechnism responsible for interpreting and carrying out manually initiated directions. Sometimes it is called manual control. (3) In some business applications, a mathematical check. (4) In programming, instructions which determine conditional jumps are often referred to as control instructions, and the time sequence of execution of instructions is called the flow of control.

controller, (1) An element or group of elements that take data proportional to the difference between input and output of a device or system and convert this data into power used to restore agreement between input and output. (2) A module or specific device which operates automatically to regulate a controlled variable or system.

core, Refers to tiny "doughnuts" of magnetizable metal that can be in either an on or off state and can represent either a binary 1 (on) or binary 0 (off). Commonly called magnetic core and used as the basic type of main memory for many computers.

criterion, Refers to a value used for testing, comparing, or judging; e.g., in determining whether a condition is plus or minus, true or false; also, a rule or test for making a decision in a computer or by humans.

cross assembler, (1) Refers to a program run on one computer for the purpose of translating instructions for a different computer. (2) Programs are usually assembled by the same assembler or assembly program contained within or used by the processor on which they will be run. Many microprocessor programs, however, are assembled by other computer processors whether they be standard, time-shared, mini or other microcomputers. This process is referred to as cross-assembly, and the programs are not designed for specific microprocessors but are to be used on other computers. They are known as cross-assemblers.

cross talk, Generally cross talk occurs when signals on one circuit emerge on another circuit as interference. The circuit which is the source of the signals is known as the disturbing circuit, and that on which the signals are heard is the disturbed circuit.

cutter compensation, A method of adjusting for the difference between the actual and programmed size of a tool.

cybernetics, (1) The science of systems of control and communication. (2) The diverse field encompasses a) integration of communication, control and systems theories; b) development of systems engineering technology; and c) practical applications at both the hardware and software levels. The term was originated by Norbert Wrener, American mathematician (1894-1964) from the Greek word for "steersman." Recent and projected developments in cybernetics are taking place in at least five important areas: technological forecasting and assessment, complex systems modeling, policy analysis, pattern recognition, and artificial intelligence.

cycle, (1) The functioning of an alternating wave from zero to negative and positive peaks, and back to zero. The number of cycles per second (CPS) is called frequency or alternation. (2) A complete sequence of a wave pattern that recurs at regular intervals. The number of cycles occurring in 1 second is the frequency of the wave, now expressed in hertz. (3) An interval of space or time in which one set of events or phenomena is completed. In alternating current, the time for a change of state from a value through a positive and negative maximum, back to the same value.

damping, Refers to a characteristic built into electrical circuits and mechanical systems to prevent rapid or excessive corrections which may lead to instability or oscillatory conditions, i.e. connecting a register on the terminals of a pulse transformer to remove natural oscillations or placing a moving element in oil or sluggish grease to prevent mechanical overshoot of the moving parts.

data, (1) A general term that is used to denote any or all facts, numbers, letters, symbols, etc., which can be processed or produced by a computer. (2) In a more restricted sense, "data" refers to numerical information as contrasted with non-numerical information. (3) Implying "source data" or "raw data" as contrasted with information which is then defined to mean the knowledge obtained by the processing of data.

dead band, A specific range of values in which the incoming signal can vary without changing the output.

deadtime, A definite delay between two related actions.

debugging, The process of finding and eliminating errors.

decibel, A standard unit expressing a loss or gain in transmission power levels. Abbreviated db. The term "dbm" is also used, when a power of one milliwatt is the reference level. Db indicates the ratio of power output to power input:

$$db = 10 \log_{10}\frac{P_1}{P_2}$$

decoder, (1) Refers to a specific device which determines the meaning of a set of signals and initiates a computer operation based thereon. (2) A matrix of switching elements which selects one or more output channels according to the combination of input signals present.

dedicated computer, A computer devoted exclusively to a singular machine or application.

diagnostic routine, A special routine for a computer or machine control unit that is designed to locate and identify errors and malfunctions.

differential transducer, A type of device which can simultaneously measure two separate stimuli and provide an output proportionate to the difference between them.

digital, Refers to the use of discrete integral numbers in a given base to represent all the quantities that occur in a problem or calculation. It is possible to express in digital form all information stored, transferred, processed, or transmitted by a dual-state condition (i.e., on-off, true-false, and open-closed).

Digital-to-Analog-Converter (DAC), Converts digital signals into a continuous electrical signal suitable for input to an analog computer.

digital computer, A computer which processes information represented by combinations of discrete or discontinuous data as compared with an analog computer for continuous data. More specifically, it is a device for performing sequences of arithmetic and logical operations, not only on data but its own program. Still more specifically it is a stored program digital computer capable of performing sequences of internally stored instructions, as opposed to calculators, such as card programmed calculators, on which the sequence is impressed manually. Related to machine, data processing.

digitizer, Refers to a specific device which converts an analog measurement into digital form. *Syn:* quantizer.

DIP (Dual In-line Package), (1) The most popular IC packaging in the mid-1970's is the plastic, dual-in-line case, using plastic for economic reasons and the dual-in-line package (DIP) configuration for manufacturing efficiently. (2) Chips are enclosed in dual in-line packages which take their names from the double parallel rows of leads which connect them to the circuit board. DIPs are sometimes called "bugs."

direct memory access (DMA), (1) Direct Memory Access, sometimes called data break, is the preferred form of data transfer for use with high-speed storage devices such as magnetic disk or tape units. The DMA mechanism transfers data directly between memory and peripheral devices. The CPU is involved only in setting up the transfer; the transfers take place with no processor intervention on a "cycle stealing" basis. The DMA transfer rate is limited only by the bandwidth of the memory and the data transfer characteristics of the device. The device generates a DMA Request when it is ready to transfer data. (2) High-speed data transfer operation in which an I/O channel transfers information directly to or from the memory. Also called, "data break" or "cycle stealing."

disk drives, A typical unit consists of a floppy disk drive, power supply (100-125 v AC, 60 Hz), cooling fan, disk buffer and address select electronics. It is capable of storing up to 300,000 words on a flexible disk. Up to 16 disk drives can be controlled by one disk controller.

disk storage, A computer memory device capable of storing information magnetically on a disc similar in appearance to a phonograph record.

dynamic RAM, Data is stored capacitively, and must be recharged (refreshed) periodically (every 2 ms. or so) or it will be lost. (See random access memory.)

dynamic response, The specific behavior of the output of a device as a function of the input, both with respect to time.

edit, To modify the form, format, or content of data.

EIA (Electronic Industries Association), A trade association of the electronics industry which formulates technical standards, disseminates marketing data, and maintains contact with government agencies in matters relating to the electronics industry.

Electrically Programmable Read Only Memory (EPROM), Most of these ROMs are 256 words by 8 bit electrically programmable and are ideally suited for uses where fast turnaround and pattern experimentation are important. They generally undergo complete programming and functional testing on each bit position prior to shipment insuring 100% programmability. Some types are packaged in a 24 pin dual in-line package with a transparent quartz lid. The transparent quartz lid allows the user to expose the chip to ultraviolet light to erase the bit pattern. A new pattern can then be written into the device. This procedure can be repeated as many times as required. The circuitry of some types is entirely static; no clocks are required. A pin-for-pin metal mask programmed ROM is ideal for large volume production runs of systems; most are fabricated with silicon gate technology. This low threshold technology allows the design and production of higher performance MOS circuits and provides a higher functioning density on a monolithic chip than many conventional MOS technologies.

end-of-block signal, A code that indicates the end of a block of data.

energy, The capacity for doing work.

equation solver, Usually an analog device used to solve linear simultaneous nondifferential equations, etc.

error, (1) Refers to any incorrect step, process, or result in a computer or data-processing system. The term also refers to machine malfunctions or "machine errors," and to human mistakes or "human errors." (2) Any discrepancy between a computed, observed, recorded, or measured quantity and the true, specified, or theoretically correct value or condition. Contrast with mistake.

feedback, When applied to a transmission, feedback is the return of a fraction of the output to the input. In a closed-loop system, it is the part of the system which brings back information about the condition under control.

feedback control, A type of system control obtained when a portion of the output signal is operated upon and fed back to the input in order to obtain a desired effect.

feedback control loop, A closed transmission path which includes an active transducer and consists of a forward path, a feedback path, and one or more mixing points arranged to maintain a prescribed relationship between the loop input and output signals.

feeler, A device used to sense a "two-state condition," such as on-off, go-no-go, open-closed.

field, A set of one or more characters which is treated as a unit of information.

firmware, (1) A term usually related to microprogramming and those specific software instructions that have been more or less permanently burned into a ROM control block. (2) An extension to a computer's basic command (instruction) repertoire to create microprograms for a user-oriented instruction set. This extension to the basic instruction set is done in read-only memory and not in software. The read-only memory converts the extended instructions to the basic instructions of the computer.

fixed sequential format, A numerical control format where each word in the format is identified by its position.

flip-flop, (1) A type of circuit having two stable states and usually two input terminals (or signals) corresponding to each of the two states. The circuit remains in either state until the corresponding signal is applied. Also, a similar bistable device with an input which allows it to act as a single-stage binary counter. (2) A bistable device; a device capable of assuming two stable states; a bistable device which may assume a given stable depending upon the pulse history of one or more input points and having one or more output points. The device is capable of storing a bit of information; controlling gates, etc.; a toggle.

floppy disk systems, A typical floppy disk provides random access program/data storage. Hard-sector formatted, each disk holds over 300,000 data bytes. Because many floppy controllers have all of their intelligence in microcode, some microcontrollers offer features not practical in designs implemented with hard-wired logic. The host-computer driver need only issue a small sequence of commands to write or read data from the disk.

Fourier analysis, The determination of the harmonic components of a complex waveform either mathematically or by a waveanalyzer device.

gain, Refers to the ratio between the output signal and the input signal of a device.

G code, A command in manufacturing process control changing the mode of operation of the control such as from positioning to contouring or calling for a fixed cycle of the machine.

graphic panel, Master control panel in automation and remote control systems, which shows the relation and functioning of the different parts of the control equipment by means of colored block diagrams.

gray code, (1) A positional binary number notation in which any two numbers whose difference is one are represented by expressions that are the same except in one place or column and differ by only one unit in that column or place. (2) A binary code in which sequential numbers are represented by expressions which are the same except in one place and in that place differ by one unit.

handshaking, (1) A descriptive term often used interchangeably with "buffering" or "interfacing," implying a direct connection or matching of specific units or programs. Some computer terminal programs are called "handshaking" if they greet and assist the new terminal operator to interface with or use the procedures or programs of the system. Other handshaking relates to direct package-to-package connections as regards circuits, programs, or procedures. (2) Exchange of predetermined signals when a connection is established between two data set devices.

hardware, Refers to the metallic or "hard" components of a computer system in contrast to the "soft" or programming components. The components of circuits may be active, passive, or both.

hard-wire logic, Refers to logic designs for control or problem solutions that require interconnection of numerous integrated circuits formed or wired for specific purposes and relatively unalterable. A hard-wired diode matrix is hard-wired logic whereas a RAM, ROM, or CPU can be reprogrammed with little difficulty to change the purpose of operation. Hard-wired interconnections are usually completed by soldering or by printed circuits and are thus hard-wired in contrast to software solutions achieved by programmed microcomputer components.

hexadecimal, Refers to whole numbers in positional notation with 16 as the base. Hexadecimal uses 0 through 15, with the first ten represented by 0 through 9 and the last six digits represented by A, B, C, D, E, and F.

host computer, (1) The primary or controlling computer in a multiple operation. (2) A computer used to prepare programs for use on another computer or on another data processing system; for example, a computer used to compile link edit, or test programs to be used on another system.

hunting, Refers to a continuous attempt on the part of an automatically controlled system to seek a desired equilibrium condition, the system usually contains a standard, a method of determining deviation from this standard and a method of influencing the system such as the difference between the standard and the state.

hybrid computer, Various specially-designed computers with both digital and analog characteristics, combining the advantages of analog and digital computer when working as a system. Hybrid computers are being used extensively in simulation process control systems where it is necessary to have a close representation with the physical world. The hybrid system provides good precision that can be attained with analog computers and greater control than is possible with digital computers, plus the ability to accept input data in either form.

hysteresis, The difference between the response of a unit or system to an increasing signal and the response to a decreasing signal.

impulse, (1) A pulse that begins and ends within so short a time that it may be regarded mathematically as infinitesimal. However, the change in the medium, is usually of a finite amount. (2) A change in the intensity or level of some medium, usually over a relatively short period of time; e.g., a shift in electrical potential of a point for a short period of time compared to the time period, i.e., if the voltage level of a point shifts from −10 to 20 volts with respect to ground for a period of two microseconds, one says that the point received a 30-volt, 2-microsecond pulse. (3) A change in the intensity or level of some medium over a relatively short period of time.

incremental coordinates, Coordinates measured from an origin defined by the preceding value in a sequence of values.

index, A point on a continuous function at which a specific action is desired; or the pulse count of a digital function at which a specific action is desired.

information, The knowledge of facts, measurements and requirements necessary for accomplishing useful work.

information theory, Mathematical analysis of efficiency with which communication channels are employed — the aim being to find the most efficient system of coding for any channel.

input, (1) An adjective referring to a device or collective set of devices used for bringing data into another device. (2) A channel for impressing a state on a device or logic element. (3) Pertaining to a device, process, or channel involved in an input process or to the data or states involved in an input process. In the English language, the adjective "input" may be used in place of "input data," "input signal," "input terminal," etc., when such usage is clear in a given context. (4) Pertaining to a device, process, or channel involved in the insertion of data or states, or to the data or states involved. (5) One or a sequence of, input states.

input equipment, (1) The equipment used for transferring data and instructions into an automatic data processing system. (2) The equipment by which an operator transcribes original data and instructions to a medium that may be used in an automatic data processing system.

inspection, The operation of comparing actual conditions of a workpiece to standards, manually or automatically.

instrument, A device that is capable of converting usable intelligence into electrical signals or a device that is capable of recording, measuring or controlling.

instrumentation, The application of devices for the measuring, recording and/or controlling of physical properties and movements.

integrator, (1) Any device which integrates a signal over a period of time. (2) Unit in a computer which performs the mathematical operation of integration, usually with reference to time. (3) A resistor-condensor circuit at the input to the vertical oscillator. (4) A device whose output is proportional to the integral of the input variable with respect to time.

interface, (1) Refers to instruments, devices or a concept of a common boundary or matching of adjacent components, circuits, equipment, or system elements. An interface enables devices to yield and/or acquire information from one device or program to another. Although the terms adapter, handshake, and buffer have similar meaning, interface is more distinctly a connection to complete an operation. (2) A common boundary — e.g., physical connection between two systems or two devices. (3) Specifications of the interconnection between two systems or units.

interpolation, (1) The process of finding a value between two known values on a chart or graph. (2) General procedure for determining values of a function between known or observed values, e.g., entries in a table of logarithms. There are various procedures, depending on the assumption of a curve (line or parabola) which fit localized values, imitated electrically in some controls for machine tools.

interrupt, Various interrupts relate to the suspension of normal operations or programming routines of microprocessors and are most often designed to handle sudden requests for service or change. As peripheral devices interface with CPUs, various interrupts occur on frequent bases. Multiple interrupt requests require the processor to delay or prevent further interrupts; to break into a procedure; to modify operations, etc., and after completion of the interrupt task, to resume the operation from the point of interrupt.

inverter, (1) A circuit which takes in a positive signal and puts out a negative one, or vice versa. (2) A device that changes AC to DC or vice versa. It frequently is used to change 6 volt or 12 volt direct current to 110 volt alternating current. (3) Arrangement of modulators and filters for inverting speech or music for privacy.

lag, A delay in the response of a system or device.

limit, An index point corresponding to the extreme condition allowable for the system or process.

linear displacement transducer, Commonly referred to as LVDT (Linear Variable Displacement Transformer). An electromechanical device which produces an electrical output proportional to the displacement of a separate movable core.

linear interpolation, Refers to a mode of machine tool contouring control which uses the information contained in a block to produce constant velocities proportional to the distance moved in two or more axes simultaneously.

load, (1) The power consumed by a machine or circuit in performing its function. (2) A resistor or other impedance which can replace some circuit element. (3) To fill the inner storage of a computer with information obtained from auxiliary or external storage. (4) The process of reading the beginning of a program into virtual storage and making necessary adjustments and/or modifications to the program so that it may have control transferred to it for the purpose of execution. (5) To take information from auxiliary or external storage and place it into core storage.

logger, (1) Colloquialism for recorder or printout device in control system. (2) A type of instrument which automatically scans conditions such as pressure, temperature, humidity and records (logs) the findings on a chart. (3) A device which automatically records physical processes and events, usually with respect to time.

logic, (1) As regards microprocessors, logic is a mathematical treatment of formal logic using a set of symbols to represent quantities and relationships that can be translated into switching circuits or gates. Such gates are logical functions such as, AND, OR, NOT, and others. Each such gate is a switching circuit that has two states, on or off (open or closed). They make possible the application of binary numbers for solving problems. The basic logic functions electronically performed from gate circuits is the foundation of the often complex computing capability. (2) The science dealing with the criteria of formal principles of reasoning and thought. (3) The systematic scheme which defines the interactions of signals in the design of an automatic data processing system. (4) The basic principles and application of truth tables and interconnection between logical elements required for arithmetic computation in an automatic data processing system. Related to symbolic logic.

lowpass, Concerns a type of pass which relates to the operation of a circuit or a device, as a filter which permits the passage of low frequency signals and attenuates high frequency signals.

LSI (large-scale integration), Refers to a component density of more than 100 per chip.

machine language, The final language all computers must use is binary. All other programming languages must be compiled or translated ultimately into binary code before entering the processor. Binary language is machine language.

mainframe, Refers to the basic or main part of the computer, i.e., the arithmetic or logic unit. The central processing unit.

manipulated variable, That quantity or condition which is altered by the controller and applied to the controlled system.

manual control, Machine or process control performed by a person.

manuscript, A form used by a part programmer to organize his workpiece machining instructions.

measurement, The process of determing the numerical value of dimensions or factors, determining the quality, extent or degree of the dimensions or factors of a process of workpiece.

memory, (1) One of the three basic components of a CPU, memory stores information for future use. Storage and memory are interchangeable expressions. Memories accept and hold binary numbers or images. (2) In order to be effective, a computer must be capable of storing the data it is to operate on as well as the program that dictates which operations are to be performed. Not only must the memory unit of a computer store large amounts of information, the memory must be designed to allow rapid access to any particular portion of that information. Speed, size, and cost are the critical criteria in any storage unit. Various types are: disc, drum, semi-conductor, magnetic core, charge-coupled devices, bubble domain, etc.

memory cycle, (1) A computer operation consisting of reading from and writing into memory. (2) The time required to complete this process.

microcomputer, A general term referring to a complete tiny computing system, consisting of hardware and software, that usually sells for less than $300 and whose main processing blocks are made of semiconductor integrated circuits. In function and structure it is somewhat similar to a minicomputer, with the main difference being price, size, speed of execution, and computing power. The hardware of a microcomputer consists of the microprocessing unit (MPU) which is usually assembled on a PC board with memory and auxiliary circuits. Power supplies, control console, and cabinet and separate.

microcontroller, Another all-purpose word, this can mean a microprogrammed machine, a microprocessor or a microcomputer used in a control operation — that is, to direct or make changes in a process of operation. But there's a narrower definition, more in keeping with the prefix, in which microcontroller refers to any device or instrument that controls a process with high resolution, usually over a narrow region.

microprocessor, The semiconductor central processing unit (CPU) and one of the principal components of the microcomputer. The elements of the microcomputer. The elements of the microprocessor are frequently contained on a single chip or within the same package but sometimes distributed over several separate chips. In a microcomputer with a fixed instruction set, it consists of the arithmetic logic unit and control logic unit. In a microcomputer with a microprogrammed instruction set, it contains an additional control memory unit.

minicomputer, (1) Generally a minicomputer is a mainframe that sells for less than $25,000. Usually it is a parallel binary system with 8, 12, 16, 18, 24 or 36-bit word length incorporating semiconductor or magnetic core memory offering from 4K words to 64K words of storage and a cycle time of 0.2 to 8 microseconds or less. A bare minicomputer (one without cabinet, console and power supplies) consisting of a single PC card can sell for less than $1,000 in OEM quantities. (2) These units are characterized by higher performance than microcomputers or programmable calculators, richer instruction sets, higher price and a proliferation of high level languages, operating systems, and networking methodologies. Examples HP 21MX Series, DEC PDP 11 Series, Data General Nova Series, General Automation Inc., Computer Automation Inc., products and those from scores of other innovative companies.

mirror image programming, A feature of some machine control units that provides for the reversing of all instructions programmed for a specific axis by means of a switch. This means that a single program can produce two parts that are mirror images.

modem, Refers to a MODulation/DEModulation chip or device that enables computers and terminals to communicate over telephone circuits.

nanosecond, A measurement equivalent to one billionth of a second (10^{-9} s). (Often abbreviated as "n sec" or "ns" in the computer field.)

natural frequency, The frequency of free oscillation of a system.

neutral zone, A range of values for which no control action occurs.

noise, (1) Any unwanted disturbance within a dynamic electrical or mechanical system. (2) Any unwanted electrical disturbance or spurious signal which modifies the transmitting, indicating, or recording or desired data. (3) In a computer, extra bits or words which have no meaning and must be ignored or removed from the data at the time it is used. (4) Random variations of one or more characteristics of any entity such as voltage, current, or data. (5) A random signal of known statistical properties of amplitude, distribution, and spectral density.

normally closed, A designation applied to the contacts of a switch or relay when they are connected so that the circuit will be completed when the switch is not activated or the delay coil is not energized. Symbolized by NC.

normally open, (1) A designation applied to the contacts of a switch or relay when they are connected so that the circuit will be broken when the switch is not activated or the relay coil is not energized. Symbolized by NO. (2) Specific pairs of contacts on relays which are closed only when the relay coil is energized.

numerical control (NC), (1) A manufacturing technique controlled automatically by orders (called commands). These are introduced in the form of numbers — which may be entered by a decade switch, or by a dial switch like the one on a telephone. (2) Refers to the control of machine tools, drafting machines, and the like, by punched paper or magnetic tapes suitably encoded with directive information. As most numerically controlled devices have very limited logical or arithmetic capability (to keep costs low), they rely on their input tapes for detailed and explicit guidance. This may mean 8 bits for every 0.001 of motion, or a great amount of data on the tape. It is common for a computer to prepare the control tapes, using information presented in a more manageable and concise form.

numerical control, direct (DNC), A system connecting a set of numerically controlled machines to a common memory for part program or machine program storage, with provision for on-demand distribution of data for the machines. Direct numerical control systems typically have additional provisions for collection, display or editing of part programs, operator instructions, or data related to the numerical control process.

octal, A numbering system basic to computer operation. A positional notation system using 8 as a base, instead of 2, as in binary, 10, as in decimal, etc.

offset, The difference between the value or condition desired and that actually attained.

open loop, (1) Refers to a control system in which there is no self-correcting action for misses of the desired operational condition, as there is in a closed loop system. (2) A family of automatic control units, one of which may be a computer, linked together manually by operator action.

operating system, (1) A basic group of programs with operation under control of a data processing monitor program. (2) An integrated collection of service routines for supervising the sequencing and processing of programs by a computer. Operating systems may perform debugging, input-output, machine accounting, compilation and storage assignment tasks.

overshoot, (1) Extent to which a servo system carries the controlled variables past its final equilibrium position. (2) For a step change in signal amplitude, undershoot and overshoot are the maximum transient signal excursions outside range from initial to final mean amplitude levels.

parabolic interpolation, Refers to a procedure, usually connected with numerical control of a machine tool, to control the center line of a cutter path; also smooth blended surfaces can be constructed from parabolic segments (spans), each defined by three programmed points (also of which each span shares with its neighbor). Spans may vary in length and skill blend.

parity check, A computer checking method in which the total number of binary 1's (or 0's) is always even or always odd. Either an even-parity or odd numbers of bit patterns to signify a character; thus, all characters would be represented by an even word, are added, modulo 2, and the sum checked against a single, previously computed parity digit, i.e., a check which tests whether the number of ones in a word is odd or even. Synonymous with odd-even check and related to check, redundant and to check, forbidden combination.

phase shift, (1) A time difference between the input and output signals of a system. (2) A change in the phase of a periodic quantity. (3) A change in time relationship of one part of a signal waveform with another, with no change in the basic form of the signal. The degree of change varies with frequency as a signal passes through a channel.

polling, (1) Refers to an important multiprocessing method used to identify the source of interrupt requests. When several interrupts occur simultaneously, the control program makes the decision as to the one that will be serviced first. (2) Refers to a technique by which each of the terminals sharing a communications line is periodically interrogated to determine whether it requires servicing. The multiplexor or control station sends a poll which, in effect asks the terminal selected (do you have anything to transmit). (3) A flexible, systematic, centrally-controlled method of permitting terminals on a multi-terminal line to transmit without contending for the line. The computer contacts terminals according to the order specified by the user, and each terminal contacted is invited to send messages. (4) Refers to centrally controlled method of calling a number of points to permit them to transmit information.

postprocessor, Refers to a set of computer instructions which transform tool centerline data into machine motion commands using the proper tape code and format required by a specific machine control system. Instructions such as feed rate calculations, spindle speed calculations, and auxiliary function commands may be included.

printed circuit, (1) A circuit in which interconnecting wires have been replaced by conductive strips printed, etched, etc., onto an insulating board. It may also include similarly formed components on the baseboard. (2) Refers to resistors, capacitors, diodes, transistors and other circuit elements which are mounted on cards and interconnected by conductor deposits. These special cards are treated with light-sensitive emulsion and exposed. The light thus fixes the areas to be retained and an acid bath eats away those portions which are designed to be destroyed. The base is usually a copper clad card.

process control, (1) Automatic control of industrial processes in which continuous material or energy is produced. (2) Pertaining to systems whose purpose is to provide automation of continuous operations. This is contrasted with numerical control, which provides automation of discrete operations.

program, (1) The plan or procedure for the operation of a machine or process especially when the machine is a computer. (2) A set of instructions arranged in proper sequence to cause the desired operations.

Programmable Controller (PC), Quite often now PCs replace solid-state logic modules. They are used for industrial controls as direct replacements for electromechanical control relays. PCs are designed to provide the user with the benefit of solid-state reliability while avoiding its pitfalls. PCs are provided with shielding or noise-immune logic — i.e., designed-in isolation from high voltage inputs and outputs. In addition, the control logic — rather than being point-to-point wired as with solid-state modules — is programmable. PCs can be programmed in a familiar manner — often directly from a ladder diagram. Therefore, minimal training is required to implement a PC. More effort can be spent on designing the control logic rather than designing a control panel.

PROM, (1) Programmable Read-Only Memory is generally any type which is not recorded during its fabrication but which requires a physical operation to program it. Some PROMs can be erased and reprogrammed through special physical processes. (2) A semiconductor diode array which is programmed by fusing or burning out diode junctions.

proportional control action, Refers to designed control action in which there is a continuous linear relation between the output and the input. Such a condition applies when both the output and the input are within their normal operating ranges and when an operation is at a frequency below a limiting value.

RAM (Random Access Memory), This type memory is random because it provides access to any storage location point in the memory immediately by means of vertical and horizontal co-ordinates. Information may be "written" in or "read" out in the same very fast procedure.

random access, (1) Access to a computer storage under conditions whereby there is no rule for predetermining the position from where the next item of information is to be obtained. (2) Describes the process of obtaining information from or placing information into a storage system where the time required for such access is independent of the location of the information most recently obtained or placed in storage. (3) Describing a device in which random access can be achieved without effective penalty in time.

Read Only Memory (ROM), The ROM is used to store the microprogram or a fixed program depending upon the microprogrammability of the CPU. The microprogram provides the translation from the higher-level user commands, such as ADD, SUBTR, etc., down to a series of detailed control codes recognizable by the microprocessor for execution. The size of the ROM varies according to user requirements within the maximum allowed capacity dictated by the addressing capability of the microprocessor.

real time, (1) In solving a problem, a speed sufficient to give an answer within the actual time the problem must be solved. (2) Pertaining to the actual time during which a physical process transpires. (3) Pertaining to the performance of a computation during the actual time that the related physical process transpires in order that results of the computation can be used in guiding the physical process.

reed switch, Refers to a special switching device which consists of magnetic contactors which are sealed into a glass tube. The contactors are actuated by the magnetic field of an external solenoid, electromagnet or a permanent magnet.

relay, An electromagnetic switching device in which contacts are opened and/or closed by variations in the conditions of one electric circuit and thereby affect the operation of these devices in the same or other electric circuits.

remote control, Any system of control performed from a distance. The control signal may be conveyed by intervening wires, sound (ultrasonics), light, or radio.

repeatability, The closeness of agreement among a number of consecutive measurements of a constant signal, approached from the same direction. Repeatability is expressed as maximum non-repeatability in per cent of span or counts of error.

resolution, (1) The smallest incremental step in separating a measurement into its constituent parts. In a digital system, resolution is one count in its least significant digit. (2) The ratio of maximum to minimum readings of a measuring system. (3) The process of separating the parts which compose a mixed body.

resolver, (1) A means for resolving a vector into two mutually perpendicular components. (2) A transformer, the coupling between primary and secondary of which can be varied. (3) A small section with a faster access than the remainder of the magnetic-drum memory in a computer. (4) A device which separates or breaks up a quantity, particularly a vector, into constituent parts or elements, e.g., the mutually perpendicular components of a plane vector. (5) A small section of storage, particularly in drum, tape or disk storage units, that has much faster access than the remainder of the storage.

response time, (1) The time (usually expressed in cycles of the power frequency) required for the output voltage of a magnetic amplifier to reach 63% of its final average value in response to a step-function change of signal voltage. (2) The time required for the pointer of an instrument to come to apparent rest in its new position after the measured quantity abruptly changes to a new, constant value. (3) The amount of time elapsed between generation of an inquiry at a data communications terminal and receipt of a response at that same terminal. Response time, thus defined includes: transmission time to the computer, processing time at the computer, including access time to obtain any file records needed to answer the inquiry, and transmission time back to the terminal.

robot, Refers to a specific device equipped with sensing instruments for detecting input signals or environmental conditions but with a reacting or guidance mechanism, which can perform sensing, calculations, etc., and with stored programs for resultant actions, i.e., a machine running itself.

RS-232 interface, Refers to the interface between a modem and the associated data terminal equipment, it is standardized by ETA Standard RS232. For voice-band modems the interface leads are single leads with a common ground return. Polartype signals are specified with a minimum amplitude of +3 V at the terminating end. The maximum allowable voltage is +25V. A ground potential difference between equipment of up to 2V is allowed for by specifying that the driver source must provide a +5V signal. The terminating impedence is required to be in the 3000-7000 Ω range. The drivers typically provide voltages in the range +6V to +10V with a source impedance of a few hundred ohms. A negative polarity indicates the binary state "1," marking, or an OFF control state. The positive polarity indicates the binary state "0," spacing, or an ON control state.

sampled data, Data in which the information content is ascertained as discrete points in time.

scanner, Refers to an instrument which automatically samples or interrogates the state of various processes, files, conditions, or physical states and initiates action in accordance with the information obtained.

sense, (1) Refers to the activity to examine, particularly relative to a criterion. (2) To determine the present arrangement of some element of hardware, especially a manually-set switch.

sensitivity, (1) Ratio of the response of a measuring device, to the magnitude of the measured quantity. It may be expressed directly in divisions per volt, milliradians per microampere, etc., or indirectly by stating a property from which sensitivity can be computed (e.g., ohms per volt for a stated deflection). (2) The degree of response of an instrument of control unit to a change in the incoming.

sensor, Refers to a transducer or other device whose input is a quantitative measure of some external physical phenomenon and whose output can be read by a computer.

sequential logic, Refers to a circuit arrangement in which the output state is determined by the previous states of the input.

servomechanism, (1) Closed-cycle system in which a small input power controls a much larger output power in a strictly proportionate manner, e.g., movement of a gun turret may be accurately controlled by movement of a small knob or wheel. (2) Any closed-loop, feedback type of control system.

set point, In a feedback control loop, the point which determines the desired value of the quantity being controlled.

signal, (1) A visible, audible, or other conveyor of information. (2) The intelligence, message, or effect to be conveyed over a communication system. (3) A signal wave. (4) The physical embodiment of a message.

signal-to-noise ratio, Also called signal-noise ratio. (1) Ratio of the magnitude of the signal to that of the noise (often expressed in decibels). (2) The ratio of the amount of signals conveying information to the amount of signals not conveying information.

single-ended, Unit or system designed for use with unbalanced signal, having one input and one output terminal permanently grounded.

software, All of the program manuscripts, tapes, decks of cards, methods sheets, flow charts and other programming documentation associated with computers and numerical control systems. Its counterpart is the physical hardware comprising the computer or NC system.

static behavior, The behavior of a control system or an individual unit under fixed conditions (as contrasted to dynamic behavior, under changing conditions).

stepping motor, Refers to one in which rotation occurs in a series of discrete steps controlled electromagnetically by individual (digital) input signals.

storage, Pertaining to a device into which data can be entered, in which they can be held, and from which they can be retrieved at a later time.

straight cut control system, A system in which the controlled cutting action occurs only along a path parallel to linear, circular, or other machine ways.

switch, (1) A mechanical or electrical device that completes or breaks the path of the current or sends it over a different path.

system, (1) Refers to a collection of parts united by some form of regulated interaction, an organized whole.

time-sharing, (1) A computing technique by which more than one terminal device can use the input, processing and output facilities of a central computer simultaneously. (2) A specific method of operation in which a computer facility is shared by several users for different purposes at (apparently) the same time. Although the computer actually services each user in sequence, the high speed of the computer makes it appear that the users are all handled simultaneously.

transient, A phenomenon caused in a system by a sudden change in conditions, and which persists for a relatively short time after the change.

triac, Relates to two SCRs in parallel and oppositely orientated to provide bidirectional control of A.C. power.

FORGING

air-lift hammer, A type of gravity drop hammer where the ram is raised for each stroke by an air cylinder. Since length of strike may be controlled, ram velocity and thus energy delivered to the workpiece may be varied.

anvil (base), Extremely large, heavy block of metal which supports entire structure of conventional gravity or steam driven forging hammers. Also, the block of metal on which hand (or smith) forgings are made.

batch-type furnace, A furnace for heating materials in which the loading and unloading is done through a single door or slot.

bender, Term denoting a die impression, tool or mechanical device designed to bend forging stock to conform to the general configuration of die impressions subsequently to be used.

billet, A semi-finished, cogged, hot-rolled or continuous-cast metal product of uniform section, usually rectangular with radiused corners. Billets are relatively larger than bars.

blank, A piece of stock (also called a "slug" or "multiple") from which a forging is to be made.

blocking, A forging operation often used to impart an intermediate shape to a forging, preparatory to forging of the final shape in the finishing impression of the dies. Blocking can assure proper "working" of the material and contribute to greater die life.

blocker (blocking impression), The impression in the dies (often one of a series of impressions in a single die set) which imparts to the forging an intermediate shape, preparatory to forging of the final shape.

blocker-type forging, A forging which approximates the general shape of the final part with relatively generous finish allowance and radii. Such forgings are sometimes specified to reduce die casts where only a small number of forgings are desired and the cost of machining each part to its final shape is not excessive.

blow, The impact or force delivered by one workstroke of the forging equipment.

board hammer, A type of gravity drop hammer where wood boards attached to the ram are raised vertically by action of contra-rotating rolls, then released. Energy for forging is obtained by the mass and velocity of the freely falling ram and the attached upper die.

boss, A relatively short protrusion or projection on the surface of a forging often cylindrical in shape.

buster (pre-blocking impression), A type of die impression sometimes used to combine preliminary forging operations such as edging and fullering with the blocking operation to eliminate blows.

check, Crack in a die impression, generally due to forging pressure and/or excessive die temperature. Die blocks too hard for the depth of the die impression have a tendency to check or develop cracks in impression corners.

coining, The process of applying necessary pressure to all or some portion of a forging's surface in order to obtain closer tolerances, smoother surfaces or to eliminate draft. Coining may be done while forgings are hot or cold and is usually performed on surfaces parallel to the parting line of the forging.

cold heading, Working metal at room temperature in such a manner that the cross-sectional area of a portion or all of the stock is increased.

cold inspection, A visual (usually final) inspection of the forgings for visible defects, dimensions, weight, and surface condition at room temperature. The term may also be used to describe certain non-destructive tests such as magnetic particle, dye penetrant and sonic inspection.

cold shut, A defect characterized by a fissure or lap on a forging's surface which has been closed without fusion during the forging operation.

cold trimming, Removing flash or excess metal from the forging in a trimming press when the forging is at room temperature.

cold working, Permanent plastic deformation of a metal at a temperature below its recrystallization point — low enough to produce strain hardening.

counterblow forging equipment, A category of forging equipment wherein two opposed rams are activated simultaneously, striking repeated blows on the workpiece at a midway point. Action may be vertical, as in the case of counterblow forging hammers, or horizontal as with the "Impacter."

die lubricant, A compound sprayed, swabbed or otherwise applied on die surfaces or forgings during forging to reduce friction between the forging and the dies. Lubricants may also ease release of forgings from the dies and provide thermal insulation.

die match, The condition where dies, after having been set up in the forging equipment, are in proper alignment relative to each other.

die set, A guided die holder to assure alignment of the dies during processing.

die shift, A condition requiring correction where, after dies have been set up in the forging equipment, displacement of a point in one die from the corresponding point in the opposite die occurs in a direction parallel to the fundamental parting line of the dies.

dies (die blocks), The metal blocks into which forging impressions are machined and from which forgings are produced.

dowel, A metal insert placed between mating surfaces of the die shank and die holder in the forging equipment to assure lengthwise die match.

draft, The amount of taper on the sides of the forging necessary for removal of the workpiece from the dies. Also, the corresponding taper on the side walls of the die impressions.

draft angle, The angle of taper, expressed in degrees, given to the sides of the forging and the side walls of the die impression.

drawing, A forging operation in which the cross section of forging stock is reduced and the stock lengthened between flat or simple contour dies. (See fuller.)

drop forging, A forging produced by hammering metal in a drop hammer between dies containing impressions designed to produce the desired shape. (See impression die forging.)

drop hammer, A term generally applied to forging hammers wherein energy for forging is provided by gravity, steam or compressed air. (See air-lift hammer, board hammer, steam hammer.)

edger (edging impression), The portion of the die impression which distributes metal, during forging, into areas where it is most needed to facilitate filling the cavities of subsequent impressions to be used in the forging sequence. (See fuller.)

end loss (crop end), Bar end left over after cutting bar lengths of stock into forging multiples. (See multiple.)

extrusion, The process of forcing metal to flow through a die orifice in the same direction in which energy is being applied (forward extrusion); or in the reverse direction (backward extrusion) in which case the metal usually follows the contour of the punch or moving forming tool. The extrusion principle is used in many impression die forging applications.

fillet, The concave intersection of two surfaces. In forging, the desired radius at the concave intersection of two surfaces is usually specified.

flash, Necessary metal in excess of that required to completely fill the finishing impression of the dies. Flash extends out from the body of the forging as a thin plate at the line where the dies meet and is subsequently removed by trimming. Cooling faster than the body of the component during forging, flash can serve to restrict metal flow at the line where dies meet, thus assuring complete filling of the finishing impression.

flash extension, Portion of flash remaining after trimming. Flash extension is measured from the intersection of the draft and flash at the body of the forging to the trimmed edge of the stock.

flash land, Configuration in the finishing impression of the dies designed either to restrict or encourage growth of flash at the parting line, whichever may be required in a particular instance in order to assure complete filling of the finishing impression.

flat die forging (open die forging), Forging worked between flat or simple contour dies by repeated strokes and manipulation of the workpiece. Also known as "hand" or "smith" forging.

flow lines, Patterns in a forging resulting from the elongation of nonhomogeneous constituents and the grain structure of the material in the direction of working during forging; usually revealed by macroetching. (See grain flow.)

forgeability, Term used to describe the relative ability of material to deform without rupture.

forging, The product of work on metal formed to a desired shape by impact or pressure in hammers, forging machines (upsetters), presses, rolls and related forming equipment. Forging hammers, counterblow equipment and high-energy-rate forging machines impart impact to the workpiece, while most other types of forging equipment impart squeeze temperature, but the majority of metals are made more plastic for forging by heating.

forging machine (upsetter or header), A type of forging equipment, related to the mechanical press, in which the main forming energy is applied horizontally to the workpiece which is gripped and held by prior action of the dies.

forging plane, The plane which includes the principal die face and which is perpendicular to the direction of the ram travel. When parting surfaces of the dies are flat, the forging plane coincides with the parting line. (See parting plane.)

forging quality, Term describing stock of sufficiently superior quality to make it suitable for commercially satisfactory forgings.

fuller (fullering impression), Portion of the die which is used in hammer forging primarily to reduce the cross section and lengthen a portion of the forging stock. The fullering impression is often used in conjunction with an edger (or edging impression).

gate (sprue), A portion of the die which has been removed by machining to permit a connection between multiple impressions or between an impression and the bar of stock.

gathering stock, Any operation whereby the cross section of a portion of the forging stock is increased above its original size.

grain flow, Fiber-like lines appearing on polished and etched sections of forgings which are caused by orientation of the constituents of the metal in the direction of working during forging. Grain flow produced by proper die design can improve required mechanical properties of forgings.

grain size, The average size of the crystals or grains in a metal as measured against an accepted standard.

gravity hammer, A class of forging hammer wherein energy for forging is obtained by the mass and velocity of a freely falling ram and the attached upper die. Examples: board hammers and air-lift hammers.

gutter, A slight depression machined around the periphery of an impression in the die which allows space for the excess metal (flash during forging.)

hammer forging, Shaping of metal by impact between dies in one of several types of equipment known as forging hammers. (See air-lift hammer, board hammer, counterblow forging equipment, steam hammers.)

handling holes, Holes drilled in the ends of a forging die block to provide for handling by a crane or bar.

heat (forging), Amount of forging stock placed in a batch-type furnace at one time.

heat of metal, The quantity of material manufactured from one melt at the metal producer's facility. Metal from a single heat is extremely uniform in chemical analysis.

high-energy-rate forging (high velocity or high speed forging), The process of producing forgings on equipment capable of extremely high ram velocities resulting from the sudden release of a compressed gas against a free piston.

hot stamp, Impressing markings in a forging while the forging is in the heated, plastic condition.

hot trim, Removing flash or excess metal from the forging in a trimming press while the forging is in the heated state.

hot working, The mechanical working of metal at a temperature above its recrystallization point — a temperature high enough to prevent strain hardening.

impression, A cavity machined into a forging die to produce a desired configuration in the workpiece during forging.

impression die forging, A forging that is formed to the required shape and size by machined impressions in specially prepared dies which exert 3-dimensional control on the workpiece.

ironing, (1) A press operation used to obtain a more exact alignment of the various parts of a forging, or to obtain a better surface condition. (2) An operation to increase the length of a tube by reduction of wall thickness and outside diameter. (See coining, swaging.)

lap, A surface irregularity appearing as a seam, caused by the folding over of hot metal, fins or sharp corners and by subsequent rolling or forging (but not welding) of these into the surface.

locks, Changes in the plane of the mating faces of the dies. Locks aid in holding die alignment during forging by counteracting lateral thrust which is present to an extent dependent on the shape of the workpiece.

match, A condition in which a point in one die-half is aligned properly with the corresponding point in the opposite die-half within specified tolerance.

matched edges (match lines), Two edges of the die face which are machined exactly at 90 degrees to each other, and from which all dimensions are taken in laying out the die impress and aligning the dies in the forging equipment.

mechanical working, Subjecting metal to pressure, exerted by rolls, hammers or presses, in order to change the metal's shape or physical properties.

no-draft forging, A forging with extremely close tolerances and little or no draft, requiring a minimum of machining to produce the final part. Mechanical properties can be enhanced by closer control of grain flow and retention of surface material in the final component.

parting line, The line along the surface of a forging where the dies meet, or the line along the corresponding edge of the die impression.

parting plane, The plane which includes the fundamental parting line of the dies; the dividing plane between dies. (See forging plane.)

platter, The entire workpiece upon which the forging equipment performs work, including the flash, sprue, tonghold and as many forgings as are made at one time.

press forging, The shaping of metal between dies by mechanical or hydraulic pressure. Usually this is accomplished with a single workstroke of the press for each die station.

proof, Any reproduction of a die impression in any material, frequently a lead or plaster cast. (See die proof.)

punch, (1) The movable die in a trimming press or forging machine. (2) A tool used in punching holes in metal.

pusher furnace, A continuous type furnace where stock to be heated is charged at one end, carried through one or more heating zones and discharged at the opposite end.

quantity tolerance, Allowable variation of quantity to be shipped on a purchase order. This tolerance is properly agreed to by forging producer and purchaser when order is placed. (A schedule of suggested standard quantity tolerances for various order quantities is available through members of FIA.)

restriking, Striking a trimmed forging an additional blow in the dies in order to align or size its several components or sections. The operation can be performed hot or cold.

ring rolling, The process of shaping weldless rings from pierced discs or thick-walled, ring-shaped blanks between rolls which control wall thickness, ring diamter, height, and contour.

roller (rolling impression), The portion of a forging die where cross sections are altered by hammering or pressing while the workpiece is being rotated.

roll forging, The process of shaping stock between power driven rolls bearing contoured dies. The workpiece is introduced from the delivery side of the rolls, and is reinserted for each succeeding pass. Usually used for pre-forming, roll forging is often employed to reduce thickness and increase length of stock.

rotary furnace, A circular furnace constructed so that the hearth and workpieces rotate around the furnace's axis during heating.

saddling (mandrel forging), The process of rolling and forging a pierced disc of stock over a mandrel in order to produce a weldless ring.

scale, The oxide film that is formed on forgings, or other heated metal, by chemical action of the surface metal with the oxygen in the air.

scale pit, A surface depression formed on the forging due to scale remaining in the dies during the forging operation.

seam, A crack or inclusion on the surface of forging stock which may carry through forging and appear on the finished product.

semi-finisher (semi-finishing impression), An impression in the forging die which only approximates the finish dimensions of the forging. Semi-finishers are often used to extend die life of the finishing impression; to assure proper control of grain flow during forging; and assist in obtaining desired tolerances.

shank, The portion of the die or tool by which it is held in position in the forging unit.

shoe, A holder used as a support for the stationary portions of trimming and forming dies; sometimes termed sow block.

shrinkage, The contraction of metal during cooling after forging. Die impressions are made oversize according to precise shrinkage scales to allow forgings to shrink to design dimensions and tolerances.

shrink scale, A measuring scale or rule, used in die layout, on which graduations are expanded to compensate for thermal contraction (shrinkage) of the forging during cooling.

sizing, A process employed to control precisely a diameter of rings or tubular components.

slug, (1) Metal removed when punching a hole in a forging. Also termed "punch out". (2) Forging stock for one workpiece cut to length. (See blank.)

snag grinding (snagging), The process of removing portions of forgings not desired in the finished product, by grinding.

sow block, Metal die holder employed in a forging hammer to protect the hammer anvil from shock and wear. Also called anvil cap or shoe.

steam hammer, A type of drop hammer where the ram is raised for each stroke by a double-action steam cylinder and the energy delivered to the workpiece is supplied by the velocity and weight of the ram and attached upper die driven downward by steam pressure. Energy delivered during each stroke may be varied.

swage (swedge), Operation of reducing or changing the cross sectional area by revolving the stock under rapid impact blows.

tonghold, The portion of the stock by which the operator grips the stock with tongs during forging.

trimming, The process of removing flash or excess metal from a forging.

FOUNDRY

abrasion resistance, Degree of resistance of a material to abrasion or wear. White iron is extremely abrasion resistant because of its large amount of combined carbon in the form of iron carbide, Fe_3C.

abrasives, Hard granular materials for grinding, polishing, blasting either bonded together in the form of wheels or bricks or bonded to paper or cloth belts or discs by glue or resins; or used loose and propelled by certrifugal force from a wheel or an entrained in an air stream. Metallic shot, sand or grit are used in the latter application. Natural abrasives include emery, garnet and silica; synthetic abrasives are corundum and silicon carbide.

accelerator, A material which increases the reaction rate of organic binders. It usually acts as a catalyst in the reaction.

acid, A term applied to refractories and minerals or slags which have a high silica content. Most frequently used with reference to the characteristic reactions between a slag and the refractory lining of a cupola or furnace used for ferrous alloys.

addition agent, Any material added to a charge of molten metal in the bath or ladle to bring the alloy to the required specification.

additive, Any material added to a molding sand to promote casting "peel", to improve flowability or, to reduce expansion, e.g., sea coal, pitch, cereal graphite, plumbage.

aeration, The fluffing of a molding sand just before dropping it into the flask.

A.F.S., American Foundrymen's Society. This abbreviation will be found in connection with various tests, numbers, etc., among the foundry definitions.

air belt, Chamber, surrounding the cupola at the tuyeres, to equalize volume and pressure of blast and deliver it to the tuyeres. Commonly called a windbox.

air dried, Surface drying of cores in open air before baking in an oven; also applied to molds which air-dry when left open, which may cause crumbling or crushing when the mold is closed; a core or mold dried in air, without application of heat.

air-dried strength, Tenacity (compressive, shear, tensile or transverse) of a sand mixture after being air dried at room temperature.

air furnace, A reverbatory type of furnace which is fired by fuel burning at one end of the hearth; the metal is melted by the hot gases which pass over the charge and move toward the stack at the other end. Heat is reflected from the parabolic roof and side walls.

air quenching (normalizing), Cooling of alloys in air from above the transformation range.

alloy, A metallic mixture composed of two or more chemical elements of which at least one must be a metal. Usually an alloy has properties which are different from its components.

annealing, A heat treating process in which castings are heated in a furnace above the transformation temperature and then are allowed to cool to room temperature before removing from the furnace. A fully annealed casting has few residual stresses and is in its softest condition.

apron feeder, A continuous pan conveyor, carried by two strands of chain, used for carrying castings or feeding sand at a controlled rate from a storage hopper.

arc cutting, Cutting a ferrous metal with an electric arc.

as cast, Referring to a casting which has received no heat treatment other than normal gate and riser removal and sand blasting.

back draft, A reverse taper which prevents removal of a pattern from the mold.

backing sand, Reconditioned sand used for ramming main part of mold after pattern has been covered with facing sand.

baghouse, Large chamber for holding bags used in filtration of gases from a furnace to recover metal oxides and other solids suspended in the gases; one form of dust collector.

baked core, A core which has been heated through sufficient time and temperature to produce the desired physical properties attainable from its oxidizing or thermal-setting binders.

baked permeability, The property of a molded mass of sand which when baked at a temperature above 230°F (110° C) and then cooled to room temperature permits passage through it of the gases resulting when molten metal is poured in a mold.

baked strength, Compressive, shear, tensile or transverse strength of a molded sand mixture when baked at a temperature above 230°F (110° C) and then cooled to room temperature.

balanced blast, Arrangement of tuyeres in a cupola which provides for distributing or balancing the blast as required between upper and lower levels of the melting zone.

balanced core, One with the core-seat portion so shaped and dimensioned that it will overbalance that part of the core extending into the metal cavity.

bank sand, Sedimentary deposits, usually containing less than 5 per cent clay, occurring in banks or pits, used in core making and in synthetic molding sands.

bar, rapping, A bar for rapping a pattern to assist withdrawal of pattern from mold.

bars (cleats), Ribs of metal or wood placed across the cope portion of a flask to help support the sand in the cope flask.

base permeability, That physical property which permits gases to pass through packed dry sand grains containing no clay or other bonding substance.

basin, The enlarged mouth of the sprue into which the molten metal is first poured. (See pouring basin.)

bed or bed charge, Initial charge of fuel in cupola upon which the melting is started.

bed coke, First layer of coke placed in the cupola. Also the coke used as the foundation in constructing a large mold in a flask or pit.

bed height, The measured height of the cupola bed above the tuyeres before the first metal charge is added.

bench molder, A craftsman who makes molds for smaller type castings, working at the molder's bench only.

bentonite, A colloidal clay derived from volcanic ash and employed as a binder in connection with synthetic sands, or added to ordinary natural (clay-bonded) sands where strength is required; western or sodium bentonite is found in Wyoming and South Dakota; also in the south central states which have quantities of southern or calcium.

binder, Artifically added bond (usually used to indicate other than clay) to foundry sand, such as cereal, pitch, resin, oil, sulphite by-product, etc.

black heart malleable, American type malleable iron, normal fracture showing a velvety black appearance with a gray rim. This type of malleable does not depend on decarburization for its malleable properties.

blacking, Carbonaceous materials such as plumbago, graphite or powdered carbon, usually mixed with a binder and frequently carried in suspension in water or other liquid; used as a thin facing applied to surfaces of molds or cores to improve casting finish.

blast furnace, In ferrous metallurgy, a shaft furnace supplied with an air blast (usually hot) and used for producing pig iron by smelting iron ore in a continuous operation. The raw materials (iron ore, coke and limestone) are charged at the top, and the molten pig iron and slag which collect at the bottom, are tapped out at intervals. In nonferrous metallurgy, a shaft type of vertical furnace is used for smelting coarse copper, lead and tin ores; similar to the type used for smelting iron, but smaller.

blast gate, Sliding plate in the cupola blast pipe to regulate flow of air.

blast pressure, Pressure of air in blast pipe or wind belt of cupola, depending on location of indicating instrument; usually given in ounces per sq. in. or inches of water.

blast volume, The volume of air introduced governs the melting rate of the cupola. Approximately 30,000 cu. ft. of air is required per ton of metal melted.

bleeder, A defect wherein a casting lacks completeness due to molten metal draining or leaking out of some part of the mold cavity after pouring has stopped.

blow, Forcing of air or gas evolved from either the mold or core material, through the molten metal due to insufficient venting; also commonly used to indicate starting up of a cupola.

blowhole, Irregular shaped cavities with smooth walls produced in a casting when gas, entrapped while the mold is being filled, or evolved during solidification of the metal, fails to escape and is held in pockets.

bob, A riser or feeder, usually blind, to provide molten metal to the casting during solidification, thereby preventing shrinkage cavities.

body core, The main core.

bond, bonding substance or bonding agent, Any material other than water, which, when added to foundry sands, imparts bond strength.

bonding clay, Any clay suitable for use as a bonding material.

bond strength, Property of a foundry sand by virtue of which it offers resistance to deformation.

boss, A projection of circular cross-section on a casting. Usually intended for thickening a section so that it can be drilled and tapped.

bot, A mass of clay used to stop the flow of metal from the taphole of the cupola.

break-off core, A thin core connecting the riser and casting which, while not impeding the flow of metal, serves as a notch to assist in riser removal.

breast, Part of furnace lining connecting spout with bottom and made up with taphole for every heat.

BHN, Brinell hardness number, The number results from measurement of the diameter of the impression left in the surface of a casting by a hardened steel ball of a given diameter pressed against the surface with a known load.

buckle, An indentation in a casting, resulting from expansion of the sand, may be termed the start of an expansion defect.

bumper, Machine for ramming sand in a flask by repeated jarring or jolting.

burned sand, Sand from which the binder or bond has been wholly or partially lost by contact with molten metal.

burned in sand, A layer of adherent sand on the surface of a casting which is attached by glassy iron silicate. In this case the adherent sand layer is difficult to remove. Several trips through a sand blast unit are required to remove the sand layer.

burned on sand, A thin layer of adherent sand on the surface of a casting. The sand grains are attached to the surface of the casting by fayalite, which is crystalline and brittle. The sand grains are easily removed by sand blasting.

burnt-on-sand, A misnomer usually indicating metal penetration into sand resulting in a mixture of sand and metal adhering to the surface of a casting.

butt-off, Operation performed at times to supplement ramming by jolting, either hand or air rammer.

butt rammer, The flat end of the molder's rammer.

"C" or croning process, See shell molding.

captive foundry, A foundry that is an element of a manufacturing establishment where foundry operations are a part of the regular operation of the plant.

carbide, A compound of carbon with one or more metallic elements.

carbon dioxide process (CO_2 process), A process for hardening cores or molds in which carbon dioxide gas is blown for a few seconds through a sand bonded by sodium silicate. The sand mass instantly becomes rigid because of the formation of a silica gel.

carbon equivalent, A relationship of the total carbon, silicon and phosphorus content in a gray iron expressed by the formula C.E. = % T.C. + 0.3 (% Si + % P); i.e., the carbon content having the same effect as the combination of C + Si + P.

carbon steel, Steel that owes its properties chiefly to the presence of carbon, without substantial amounts of alloying elements; also termed "ordinary steel," "straight carbon steel," "plain carbon steel," etc.

casting (noun), Metal object cast to the required shape by pouring or otherwise injecting liquid metal into a mold, as distinct from one shaped by a mechanical process.

casting (verb), Act of pouring metal into a mold.

casting, centrifugal, A process of filling molds by (1) pouring the metal into a sand or permanent mold that is revolving about either its horizontal or its vertical axis; or (2) pouring the metal into a mold that is subsequently revolved before solidification of the metal is complete.

casting, continuous, A process in which the ingot is continuously solidified while it is being poured, and the length is not determined by mold dimensions.

casting, die (verb), The injection of molten (primarily nonferrous) metal into a metal mold under pressure.

casting, machine (verb), Process of casting by machine.

casting, open sand (noun), Casting poured into an uncovered mold.

casting, permanent mold, A casting produced in a re-usable metal or refractory mold.

casting, plaster, A casting made in a plaster mold.

casting, precision, A casting of high dimensional accuracy produced by such processes as investment casting, plaster mold.

casting, sand, A casting produced in a mold made of green sand, dried sand, or a core sand.

cast iron, Essentially an alloy of iron, carbon, and silicon in which the carbon is present in excess of the amount which can be retained in solid solution is austenite at the eutectic temperature. When cast iron contains a specially added element or elements in amounts sufficient to produce a measurable modification of the physical properties of the section under consideration, it is called alloy cast iron. Silicon, manganese, sulphur, and phosphorus, as normally obtained from raw materials, are not considered as alloy addition.

cast iron, white, Cast iron in which substantially all the carbon is present in the form of iron carbide. Such a material has a white fracture.

cast-weld assembly, Welding one casting to another to form a complete unit.

cavity, (1) The impression in a mold produced by withdrawal of the pattern and to be filled by the casting metal. (2) A hollow or sunken space, or a void in the interior of a casting.

cement sand, A synthetic sand that is bonded with Portland cement.

cereal binder, A binder used in core mixtures and molding sands, derived principally from corn flour.

chamotte, A highly calcined flint clay which, when mixed either with raw clay or with sand and raw clay, is used as a molding material for steel castings.

chaplets, Metal supports or spacers used in molds to maintain cores, or parts of the mold which are not self-supporting in their proper positions during the casting process.

charge, A given weight of metal or fuel introduced into the cupola or furnace.

charging deck (or floor), The floor from which furnace charging is performed, located at or just below the charging doors.

cheek, Intermediate section of a flask that is inserted between cope and drag to decrease the difficulty of molding unusual shapes or to fill a need for more than one parting line.

chill, (1) Addition of solid metal to molten metal in ladle to reduce temperature before pouring. (2) Depth to which chilled structure penetrates a casting.

chill block, A cast iron test block in which the depth of chill, as determined by fracture, is used as an estimate of the cast iron's quality.

chills, Metal inserts in molds or cores at the surface of a casting or within the mold to hasten solidification of heavy sections and cause the casting to cool at a uniform rate.

choke, A restriction in the gating system for the purpose of keeping dirt, dross or slag from entering the casting proper.

clamp-off, An indentation in the casting surface due to displacement of sand in the mold.

clay, An earthy or stony mineral aggregate consisting essentially of hydrous silicates of alumina, plastic when sufficiently pulverized and wetted, rigid when dry, and vitreous when fired at a sufficiently high temperature. Clay minerals most commonly used in the foundry are montmorillonites and kaolinites.

clay substance (A.F.S. clay), That portion of foundry sand which, when suspended in water, fails to settle 1 in. per minute and which consists of particles less than 20 microns (0.02 mm or 0.0008 in.) in diameter.

clay-wash, Thin emulsion of clay and water for coating gaggers and the inside of flasks. Also used as grout.

coke, A porous, gray infusible product resulting from the dry distillation of bituminous coal, petroleum or coal tar pitch, which drives off most of the volatile matter. Used as a fuel in cupola melting.

cold box process, A two part organic resin binder system, mixed in conventional mixers and blown into shell or solid core shapes at room temperature. A vapor mixed with air is blown through the sand, resulting in instant curing, stripping, setting and immediate pouring of metal around it.

cold chamber machine, A die casting machine which does not have the metal chamber and plunger immersed in hot metal.

cold setting binders, Term used to describe any binder that will harden the core sufficiently at room temperature so that the core can be removed from its box without distortion. Commonly used in reference to oil-oxygen types of binders.

cold setting process, Any of several systems for bonding mold or core aggregates by means of organic binders, relying on the use of catalysts rather than heat for polymerization (curing).

cold shot, Small globule of metal embedded in but not entirely fused with the casting.

cold shut, A casting defect caused by imperfect fusing of molten metal coming together from opposite directions in a mold, or due to folding of the surface.

collapsible sprue, A sprue pattern of flexible material or of spring tube design for use in the squeeze molding of match plate patterns. Frequently a pouring cup is incorporated in the design.

colloidal clay, Finely divided clay of the montmorillonite (bentonite), kaolinite (fire clay) or illite class; prepared for use as sand binders.

colloidal material, Finely divided materials which are less than 0.5 micron (0.000015 in.) such as albumin, glue, starch, gelatin and bentonite.

columnar structure, A coarse structure of parallel columns of elongated grains which is caused by a highly directional solidification pattern which results from steep thermal gradients across the mold metal interface.

combined carbon, Carbon in iron and steel which is combined chemically with other elements; not in the free state as graphitic or temper carbon.

combined water, That water in mineral matter which is chemically combined and drives off only at temperatures above 110° C (230° F).

compressive strength, sand, Maximum stress in compression which an AFS Standard compacted sand specimen is able to withstand without significant dimensional deformation.

contraction cracks, Cracks formed by restriction of the metal while it is contracting in the mold; may occur just after solidification (called a hot tear) or a short time after the casting has been removed from the mold.

conveyor, belt, A continuously moving belt used in an automated or semi-automated foundry to move material from one station to another.

conveyor, pallet, A material handling conveyor that holds one or more molds per section and transports them from the molding station, through pouring to shake out.

conveyor, pneumatic tube, A means of moving small specimens or orders from one place to another by special carriers which are propelled through metal tubes by pneumatic suction. Also can apply to finely divided materials such as sand binders.

cope, Upper or topmost section of a flask, mold or pattern.

cope, false, Temporary cope used only in forming the parting and therefore not a part of the finished mold.

coping out, The extension of sand of the cope downward into the drag, where it takes an impression of a pattern.

core, A preformed sand aggregate inserted in a mold to shape the interior or that part of a casting which cannot be shaped by the pattern.

core assembly, Putting together a complex core made of a number of sections.

core binder, Any material used to hold the grains of core sand together.

core blow, A gas pocket in a casting adjacent to a cored cavity and caused by entrapped gases from the core.

core box, Wood, metal or plastic structure, the cavity of which has the shape of the desired core which is to be made therein.

core break-off, A core designed to produce a sharp break line in the gate for case of removal of the feeder.

core cavity, The interior form of a core box that gives shape to the core. Also, the cavity produced in a casting by use of a core.

core collapsibility, The rate of disintegration of a core at elevated temperature.

core, cover, A core set in the place during the ramming of a mold to cover and complete a cavity partly formed by the withdrawal of a loose part of the pattern. Also used to form part or all of the cope surface of the mold cavity. A core placed over another core to created a flat parting line.

core driers, Supports used to hold cores in shape while being baked; constructed from metal or sand for conventional baking, or from plastic material for use with dielectric core baking equipment.

core, drop, A type of core used in forming comparatively small openings occurring above or below the parting; the seat portion is so shaped that the core is easily dropped into place.

core extruder, A special shell core making machine that produces a continuous length of cores, usually of cylindrical cross section.

core filler, Material used to replace sand in the interior of large cores — coke, cinder, saw dust, etc. Usually added to aid collapsibility to reduce weight and to save core binders.

core, green sand, (1) A core formed from the molding sand and generally an integral part of the pattern and mold. (2) A core made of unbaked molding sand. (See core sand.)

coremaker, A craftsman skilled in the production of cores for foundry use.

core oil, Linseed-base or other oil used as a binder for baked cores.

core ovens, Low-temperature ovens used for baking cores. Maximum operation temperature is 600° F.

core paste, A prepared adhesive for joining sections of baked or cured cores.

core plates, Heat resistant plates used to support cores while being baked; may be metal or non-metallic, the latter being a requisite with dielectric core baking.

core print, Projections attached to a pattern in order to form recesses in the mold at points where cores are to be supported.

core, ram-up, A core inserted into the pattern and rammed up in the mold.

coreroom, Department of the foundry in which cores are made.

core, sag, A decrease in the height of a core, usually accompanied with an increase in width, as a result of insufficient green strength of the sand to support its own weight.

core sand, Sand for making cores to which a binding material has been added to obtain good cohesion and porosity after curing. (See core, green sand.)

core, sectional, A core made in two or more parts and pasted or wired together.

core setting, Operation of placing the cores in a jig or mold.

core shift, A variation from specified dimensions of a cored section because of a change in position of the core or misalignment of cores in assembling.

core shooter, A device using low air pressure to fluidize a core sand mix. The air is released quickly so that it forces the mix into a core box.

core sprayer, A device for spraying a viscous coating on cores.

core, stop-off, A core used to simplify the parting line of a pattern; that is, to make it unnecessary to carry the parting line above or below its normal position in order to provide for lugs or cored holes.

core, strainer, A baked sand or a refractory disc with holes of a uniform size through its thickness. Used to control the discharge of metal from pouring basins or to regulate the flow of metal in gating systems of molds; also to prevent entrance of dross or slag into the mold cavity.

core, suspended, A core having the core seat so formed that it may be suspended from above the mold.

core vents, (1) A wax product, round or oval in cross section, used to form the vent passage in a core. Also refers to a metal screen of slotted piece used to form the vent passage in the core box employed in a core blowing machine. (2) Holes made in a core to facilitate the escape of gases.

core wash, A suspension of fine clay or graphite applied to cores by brushing, dipping or spraying to improve the cast surface of the cored portion of the casting.

coring, Variable composition in solid-solution dendrites; the center of the dendrite is richer in one element, as shown by the pertinent solidus-liquidus lines in a phase diagram.

cover half, In die casting, the stationary half of the die.

crack, hot tear, A rupture occurring in a casting at or just below the solidifying temperature by a pulling apart of the soft metal; caused by thermal contraction stresses.

cracking strip, A fin of metal molded on the surface of a casting to prevent cracking.

Croning process (c process), Now called shell molding or shell core practice. The process was named after its German inventor Johannes Croning. (See shell molding.)

crucible, A ceramic pot or receptacle made of materials such as graphite or silicon carbide, with relatively high thermal conductivity, bonded with clay or carbon, and used in melting metals; sometimes applied to pots made of cast iron, steel, or wrought steel. The zone in the cupola between the bottom and the tuyere is also known as the cruicible zone. The name is derived from the cross (crux) with which ancient alchemists adorned it.

crucible furnace, A furnace fired with coke, oil, gas or electricity in which metals are melted in a refractory crucible.

crush, Buckling or breaking of a section of mold due to incorrect register when closing. Also, an indentation in the casting surface due to displacement of sand in the mold when the mold was closed.

cupola, A cylindrical shaft furnace lined with refractories for melting metal in direct contact with the fuel by forcing air under pressure through openings near its base.

cupola blower, A machine which compresses a large volume of air at low pressure for operation of the cupola.

cupola drop, The sand bottom, bed, and unmelted charges dropped from the cupola at the end of a heat.

cupola tender, Person responsible for operating a cupola.

curing time (no bake), That period of time needed before a sand mass reaches maximum hardness and rigidity.

cutoff machine, abrasive, A device which uses a thin, bonded abrasive wheel rotating at high speed to cut off gates and risers from castings.

cuts, Defects in a casting resulting from erosion of the sand by metal flowing over the mold or cored surface.

cutter, gate, A scoop or other form for cutting gates in the mold.

cyclone (centrifugal collector), In air pollution control, a controlled descending vortex created to spiral objectionable gases and dust to the bottom of a collector cone.

De Lavaud process, A centrifugal process employed chiefly for making cast iron pipe.

decarburization, Loss of carbon from the surface of a ferrous alloy as a result of heating in a medium, usually including oxygen, which reacts with carbon.

decibel (db), A unit for measuring sound level or amount of acoustical power at a given location within a plant; one-tenth of a bel.

deformation test, An AFS sand property test which is determined on an instrument such as the Dietest Universal Sand Strength Testing Machine. The test determines the amount of deformation in thousandths of an inch that occurs before a sand specimen ruptures.

degasser, A material for removing gases from molten metals and alloys (usually nonferrous metals).

degassing, The operation of removing all or most of the gas from a crucible or ladle of molten metal prior to pouring.

degree of ramming, A measure of the amount of compaction which has occurred in a molding operation. (See ramming.)

deoxidation, Removal of dissolved oxygen from molten ferrous alloys, usually accomplished by adding materials with a high affinity for oxygen, the oxides of which are either gaseous or which readily form slags.

dewaring, The process of melting out the expendable wax pattern from an investment mold by heating, usually at temperatures less than 121° C (250° F).

diatomaceous earth, A hydrous form of silica which is soft, light in weight and consists mainly of microscopic shells of diatoms or other marine organisms. Widely used for furnace insulation.

die, A metal block used in forming materials by casting, molding, stamping, threading or extruding.

die casting, A high production casting process in which the molten metal (usually a nonferrous alloy) is forced under greater than atmospheric pressure into a water cooled metal mold cavity.

die coating, See release agent.

dielectric oven, A rapid curing high frequency electric oven used to cure cores which are bonded with non conducting materials.

dip coat, In solid and shell mold investment casting, a fine ceramic coating applied as a slurry to the pattern to produce maximum surface smoothness; followed by a cheaper investment material.

direct-arc furance, An electric arc furnace in which the metal being melted is one of the poles.

directional solidification, The solidification of molten metal in a casting in such a manner that feed metal is always available for that portion that is just solidifying.

dispersed shrinkage, Small shrinkage cavities dispersed through a casting, which are not necessarily a cause for rejection.

distribution, sand grain, Variation or uniformity in the particle size which comprise sand aggregate when properly screened by U.S. Standard Sieves.

downcomer, In air pollution control, a pipe for conducting gases down into a conditioner for subsequent cleaning.

downtime, Time lost from normal casting activities, caused by unscheduled interruptions or breakdowns. (As used in foundry operations.)

draft, pattern, See pattern draft.

drag, Lower or bottom section of a mold or pattern. Originally called a nowel.

draw, (noun), A term sometimes used to denote a shrink appearing on the surface of a casting or a riser.

draw, (verb), To remove the cope or drag from a pattern plate or the pattern from the cope or drag.

draw bar, A bar used for lifting the pattern from the sand mold.

draw plate, A plate attached to a pattern to facilitate drawing of the pattern from the mold.

dried sand, Sand which has been dried by a mechanical drier prior to its use in core making.

drop, A casting defect caused by the dropping of sand from the cope or other overhanging section, which results in a casting defect known as a drop after the mold has been poured.

drop off or drop out, Sand falling from the cope of a mold. (See drop.)

dropping the bottom, See cupola drop.

dross, Metal oxides in or on the surface of molten metal.

dry permeability, The property of a molded mass of bonded or unbonded sand, dried at (105 to 110° C) (200 to 230° F) and cooled to room temperature, that allows passage of gases resulting during pouring of molten metal into a mold.

dry sand mold, A mold from which the moisture content has been removed by suitably heating the mold prior to pouring molten metal therein.

duplexing, A method of producing molten metal of desired analysis, the metal being melted in one furnace and refined in a second.

durville process, A casting process that involves a rigid attachment of the mold in an inverted position above the crucible. The melt is poured by tilting the entire assembly, causing the metal to flow along a connecting launder and down the side of the mold.

dust, Small solid particles created by breaking up larger particles by any process.

electric furnace, A furnace for industrial purposes, either melting or heat-treating, in which the heat source is an electric current.

erosion scab, A casting defect which occurs when the molten metal has been agitated, boiled or partially eroded the sand in the mold, leaving a solid mass of sand and metal at that particular spot.

exothermic riser sleeve, A riser sleeve which is made of a material which liberates heat upon contact with a molten alloy. Riser toppings are frequently composed of similar materials.

expansion scab, Rough thin layer or metal partially separated from the body of the casting by a thin layer of sand, and held in place by a thin vein of metal, usually resulting in an indentation in the casting, caused by spalling of the mold face.

facing, facing material, Coating material applied to the surface of a mold to protect the sand from the heat of the molten metal; also to impart smooth surface to casting.

facing sand, Specially prepared molding sand mixture used in the mold adjacent to the pattern to produce a smooth casting surface.

feeder, feeder head, A reservoir of molten metal to compensate for the contraction of metal as it solidifies. Molten metal flowing from the feed head, also known as a riser, prevents voids in the casting.

feeder, sand, A device for discharging a uniform thickness or volume of sand onto a belt or other conveying equipment to maintain uniform delivery.

feeding, Pouring additional molten metal into a freshly poured mold to compensate for volume shrinkage while the casting is solidifying. Also the continuous supply of molten metal, as from a riser, to the solidifying metal in the casting. Also refers to keeping risers open by manipulation of feeding rods.

ferritic steels, Steels in which ferrite is the predominant phase. These steels are magnetic.

ferritic matrix, In iron castings, the term refers to cast irons which have a large percentage of ferrite in the microstructures.

fettle, A British term which refers to the process of removing all runners and risers and cleaning off adhering sand from the casting surface.

fillet, Concave corner piece usually used at the intersection of right-angled surfaces (that would otherwise meet at an angle) on patterns and core boxes. A struck fillet is one that is dressed to shape in place, usually of wax. A planted fillet is one made separately and affixed in place. Fillets used at recentrant angles in cast shapes lessen the danger of cracks and avoid "fillet shrinkages."

fin, A thin projection of metal from the casting, formed as a result of imperfect mold or core joints.

fineness, sand, The extent of subdivision of a foundry sand, as determined by the A.F.S. fineness test.

fines, A term the meaning of which varies with the type of foundry or the type of work. It refers to those sand grain sizes substantially smaller than the predominating grain size.

finish allowance, Amount of stock left on the surface of a casting for machine finish.

fireclay, A clay with a high fusion temperature having a P.C.E. of not less than cone 19 (1520° C, 2768° F). Clays with a P.C.E. from 19 to 26, inclusive, are called "low heat-duty fireclays"; those with a P.C.E. of cone 27 or higher are designated as "refractory"; and P.C.E. of 31032 denotes "high heat-duty" fireclays.

fire sand, A refractory sand which resists high temperatures.

flask, Metal or wood frame without top and without fixed bottom used to hold the sand of which a mold is formed; usually consists of two parts, cope and drag.

flask clamp, A device for binding together the parts of a flask.

flask pins, Guides to assure proper alignment of the cope and drag of the mold after the pattern is withdrawn.

flask, slip, A tapered flask which depends on a movable strip of metal to hold the sand in position. After closing the mold the strip is withdrawn and the flask can be removed.

flask, snap, A hinged flask. Can be removed after completion of the mold.

flask, tight, A type of flask which remains on mold during pouring. Lugs are normally provided for clamping cope and drag together for pouring.

flaskless molding, Sand is blown into a slip flask and then it is squeezed so firmly around the pattern that it is difficult to strip the mold without a special stripping station. Once made the mold is so strong that no flasks are needed. The making time is about 5 seconds per cycle. With a shuttle, twice that speed can be achieved.

flat back, A pattern with a flat surface at the drag face. The pattern lies wholly in the drag and the cope has a plane surface with only risers and the down sprue.

floor molder, Craftsman employed on the main foundry floor in making molds for large castings generally from loose patterns.

flow-off (pop-off), A large vent, usually located at the high point in a cope cavity, the flow-off fills with molten metal and helps to relieve the surge of pressure which accompanies the completion of filling.

flowability, The property of a foundry sand mixture which enables it to fill pattern recesses and move in any direction against pattern surfaces under pressure.

fluidity, The ability of molten metal to flow readily as measured by the length of a standard spiral casting.

follow-board, A board which conforms to the form of the pattern and defines the parting surface of the drag.

founding, The science of melting and casting of metals into useful objects to serve the needs of man and industry.

foundry, A building, establishment, or works where metal castings are produced.

foundry facing, Material, usually carbonaceous, applied to the surface of a sand mold to prevent the molten metal from penetrating and reacting with the molding sand.

foundryman, Craftsman employed in the production of metal castings.

foundry returns, Metal in the form of gates, sprues, runners, risers, and scrapped castings returned to the furnace for remelting.

foundry scrap, Consists of gates, sprues, risers, defective castings, and borings, all of known composition, which are charged back into the melting cycle.

front slagging, A process wherein both slag and molten metal flow out through the taphole. The slag is then skimmed off the surface of the molten metal.

gaggers, Metal pieces of irregular shape used to reinforce and support sand in deep pockets of molds.

gas holes, Rounded cavities, either spherical, flattened, or elongated, in a casting, caused by the generation and/or accumulation of gas or entrapped air during solidification of the casting.

gate, End of the runner in a mold where molten metal enters the casting or mold cavity; sometimes applied to entire assembly of connected channels, to the pattern parts which form them or to the metal which fills them, and sometimes is restricted to mean the first or main channel.

gated patterns, One or more patterns with gates or channels attached.

gate, finger, Gate used on thin castings to allow rapid filling of mold. If is wedge-shaped with thin edge divided vertically to produce several members or fingers.

gate, horn, A semicircular gate to convey a molten metal over or under certain parts of a casting so that it will enter the mold at or near the center; also used as a skim gate.

gate, pencil, A series of small round gates entering the mold cavity from above and coming from a common pouring basin.

gate, ring, A gate so formed that a number of small gates conduct the metal from a circular runner to a mold in the center.

gate, runner, A horizontal channel for running metal into the mold cavity.

gate, slot, A gate used on vertical cylindrical castings in which the down sprue and casting are connected over a large part or all of the height of the casting.

gate, step, A gate which allows entrance of the metal into the mold in steps.

gate strainer, A gate designed to prevent slag and dirt from entering the mold and also to control the rate at which metal enters the mold cavity.

gate, swirl, A gate used with a feeder and runner designed to swirl the metal in the feeder to remove impurities.

gate wedge, A gate the shape of a wedge feeding directly into the mold cavity.

gating system, The complete assembly of sprues, runners, gates and individual casting cavities in the mold. Term also applies to similar portions of master patterns, pattern die, patterns, investment mold and the finished casting.

grain fineness number, A.F.S., Approximately the number of mesh per inch of that sieve which would just pass the sample if its grains were of a uniform size, that is, the average of the sizes of grains in the sample. It is approximately proportional to the surface area per unit weight of sand, exclusive of clay.

grain, sand, The granular material of molding or core sand, left after removing the clay substance in accordance with the A.F.S. fineness test.

gray cast iron, Cast iron which contains a relatively large percentage of its carbon in the form of graphite, and substantially all of the remainder of the carbon in the form of eutectoid carbide. Such material has a gray fracture.

green casting, A casting in the as-cast unheat-treated, or unaged condition.

green permeability, The ability of a molded body of sand in a tempered condition to permit passage of gases through its mass.

green sand, A naturally-bonded sand or a compounded molding sand mixture which has been tempered with water for use while still in the damp or wet condition.

halloysite, A massive, clay-like or earthy material used as a bond in certain types of green sand molding. It is isotropic, has a conchoidal fracture and often resembles unglazed porcelain.

hardness, A measure of the ability to resist surface penetration. There are several methods for measuring or evaluating this. The Brinell hardness number measures the resistance of a metal to permanent indentation by another body. The Brinell machine is a device whereby a hardened steel ball is forced into the flat surface of the specimen under definite pressure for a definite length of time: the number is found by dividing the applied load (kg) by spherical area of indentation (mm^2). Rockwell hardness test for soft metals is similar, but the number is read from an arbitrary B scale on the Rockwell machine. In Rockwell C test, used for alloy and other steels, the penetrating point is a diamond cone, with numbers read from a C scale. The Rockwell superficial hardness tester is similar to the standard instrument but employs a smaller load making a shallower indentation: it is used for thin sheets and for parts having only surface hardness (e.g., casehardened steel). The Shore Scelroscope determines hardness by measuring the height of rebound of a diamond-pointed weight from a flat surface of the material in question. The hardness of the rebound is proportional to the hardness, with the number being read from an arbitrary scale on the instrument.

hardness, Brinell, See hardness.

hardness, Rockwell, See hardness.

hand shank, A pouring ladle carried and used by one man.

head, The pressure exerted by a fluid as a head of steam or of molten metal.

heap sand, Also referrred to as system sand or unit sand; usually regarded as sand "heaped" on the foundry floor after it has been reclaimed and placed in "heaps" for re-use; chiefly used as "backing sand."

heat, A stated tonuage of metal obtained from a period of continuous melting in a cupola or furnace; or the melting period required to obtain that tonuage.

heel, Metal left in ladle after pouring has been completed; metal kept in channel type induction furnaces during stand-by periods.

high pressure molding, A mold which has been compacted by hydraulic squeezing at a pressure greater than 100 pounds per square inch on the pattern plate. The resulting mold must have a hardness greater than 85 on this AFS B scale meter.

hot blast, Blast which has been heated prior to entering into the combustion reaction.

hot box process, A furan resin based core-making process similar to shell core-making. Cores are solid throughout unless a mandrel is used.

hot spots, Localized areas of a mold or casting where high temperatures are reached and maintained for a period of time.

hot strength, Strength of a sand mix as determined at any temperature above room temperature.

hot tear, Surface discontinuity of fracture caused by either external loads or internal stresses or a combination of both acting on a casting during solidification, and subsequent contraction at temperatures near the solidus.

illite, A mineral, typically $K\ A1_3\ Si_3\ O_{10}\ (OH)_2$ found in many clays; large clay deposits of illite are found in Illinois and Michigan.

impregnation, The treatment of castings with a sealing medium to stop pressure leaks, such as soaking under pressure with or without prior evacuation but usually in heated baths. Media include sodium silicate solutions, drying oils or styrenes, plastics or proprietary compounds.

inclusions, Particles of impurities (usually oxides, sulphides, silicates and such) that are held mechanically, or are formed during solidification or by subsequent reaction within the solid metal.

inconel, Any of a series of oxidation resistant alloys usually containing up to 80% or more nickel, 8 to 14% chromium and 5% or so of iron.

indirect-arc furnace, An electric-arc furnace in which the metal bath is not one of the poles of the arc.

induction furnace, A melting furnace which uses the heat developed by an electrical induction coil.

ingot, A mass of metal cast in a heavy cast iron permanent mold. The ingot is cast in a convenient size and shape for remelting or hot working operations.

inoculant, Materials which when added to molten metal modify the structure, and thereby change the physical and mechanical properties to a degree not explained on the basis of the change in composition resulting from their use.

inoculation, Addition to molten metal of substances designed to form nuclei for crystallization.

insulating pads and sleeves, Insulating material such as gypsum, diatomaceous earth, etc. used to lower the rate of solidification. As riser sleeves and topping they keep the riser liquid thus increasing feeding efficiency.

internal shrinkage, A void or network of voids within a casting caused by improper feeding of that section during solidification.

inverse chill, The condition in a casting section where the interior is mottled or white while the outer sections are gray iron.

investment, A flowable mixture of a graded refractory filler, a binder and a liquid vehicle which when poured around the patterns conforms to their shape and subsequently sets hard to form the investment mold.

investment casting, The process of casting metal into an investment mold.

iron, ferritic, A cast iron (gray ductile or malleable) having a ferritic matrix.

iron, hard or white, Irons possessing white fractures because substantially all of the carbon is in the combined form. Irons to be malleabilized are cast white, as are many abrasion-resistant irons.

iron, malleable, A mixture of iron and carbon, including smaller amounts of silicon, manganese, phosphorus, and sulphur, which after being cast (white iron, carbon in combined form as carbides) is converted structurally by heat-treatment into a maxtrix of ferrite containing nodules of temper carbon.

iron, pearlitic, A cast iron (gray ductile or malleable) having a more or less pearlitic matrix.

iron oxide, A core sand additive to increase the high temperature and penetration resistance of core sand mixes. It contains about 85% of pulverized ore.

iserine, A black sand which consists mainly of magnetic iron ore, but which also contains a considerable amount of titanium.

isocure, A proprietary name for a binder system which was developed for use in the proprietary Ashland cold box process.

isocyanate acid, Isomeric cyanic acid HNCO.

izod test, A pendulum type impact test, in which the specimen is supported at one end as a cantilever beam so the energy required to break the specimen is a measure of impact strength.

jacket, mold, Wooden or metal form which is slipped over a mold made in a snap or slip flask, to support the four sides of the mold during pouring; maintains alignment of cope and drag halves of mold. After pouring jackets and mold weights are shifted to another row of molds.

jet scrubber, In air pollution control, a high velocity water jet directed into the throat of a venture section of a cupola to separate out particulates.

jobbing foundry, A foundry engaged in the manufacture of numerous types of castings not intended for use in its own product. Usually refers to a foundry making castings for many other companies.

jolt squeeze molding, Compaction of a sand mold which weighs 75 pounds or less by first using jolt action to compact the sand around the pattern and then using a squeezing action to complete the molding operation.

kaolin, Fire clay which is the purest form of china clay which consists of a silicate of aluminum. The name is derived from a hill in China.

kip, A term sometimes used to represent a unit load of 1000 pounds or 1000 psi.

kish, Free graphite which has separated from molten hypereutectic iron.

kiss gating, Gating with a minimum of metal left at the casting breakoff point.

knockout pins (ejector pins), Pins of a small diameter affixed to die casting or permanent mold dies to eject the casting upon opening the die. Also used with shell pattern plates to strip the shell after curing.

ladle, Metal receptacle frequently lined with refractories used for transporting and pouring molten metal. Types include hand, bull, crane, bottom-pour, holding, teapot, trolley, shank, lip-pour, buggy, truck, mixing reservoir.

ladle, bottom pour, Ladle from which metal, usually steel, flows through a nozzle in the bottom.

ladle, bull, A large ladle for carrying molten metal. Frequently used to designate a transfer ladle.

ladle liner, A preformed shape made of the same materials as graphite crucibles and widely used in steel ladles in non-ferrous foundries for carrying metal from the melting unit to the flasks.

ladle, teapot, A ladle in which, by means of an external spout, iron is removed from the bottom rather than the top of the ladle.

leakers, Foundry term for castings which fail to meet liquid or gas pressure tests.

lifters, Metal tool for removing loose sand from cope or drag before closing mold. Also, a device which is attached to the cope, to hold the sand together when the cope is lifted.

light-up, The act of igniting the fuel in a melting furnace.

lining, Inside refractory layer of firebrick, clay, sand or other material in a furnace or ladle.

lining, monolithic, A lining made without the customary layers and joints of a brick wall. Usually made by tamping or casting refractory material into place, drying and then burning in places on the job.

linseed oil, The most common of all drying oils and used in the foundry principally as a base for core oil in the foundry. Produced by applying pressure to the seeds of flax, the liquid constituent being linseed oil.

loam, a mixture of sand, silt and clay particles in about equal proportions; i.e., roughly 50% sand and 50% silt and sand.

loose piece, (1) Core box; part of a core box which remains embedded in the core, and is removed after lifting off the core box.(2) Pattern; laterally-projecting part of a pattern so attached that it remains in the mold until the body of the pattern is drawn. Back-draft is avoided by this means. (3) Part of a permanent mold which remains on the casting, and is removed after casting is ejected from the mold.

lost wax process, A pattern casting process in which a wax or thermoplastic pattern is used. The pattern is invested in a refractory slurry; after the mold is dry, the pattern is melted or burned out of the mold cavity.

malleable iron, A mixture of iron and carbon, including smaller amounts of silicon, manganese, sulphur and phosphorus which, after being cast as white iron, is converted structurally by heat-treatment into a matrix of ferrite containing nodules of temper carbon, and substantially free of all combined carbon.

master pattern, A pattern embodying a contraction allowance in its construction, used for making castings to be employed as patterns in production work.

matched parting, Forming of a projection on the parting surface of a cope half of a pattern, and a corresponding depression in the surface of the drag.

matchplate, A plate of metal or other materials on which patterns and gating systems split along the parting line, are mounted back to back to form an integral piece.

metalloid, An element intermediate between metals and nonmetals possessing both metallic and non-metallic properties, as arsenic.

metallostatic pressure, The pressure developed within a molten metal while it is still liquid. The metal head can exert considerable lifting force on a cope particularly if the cope is deep and the metal is poured rapidly.

melting loss, Loss of metal in the charge during the operation of melting, usually due to oxidation or volatilization.

melting point, The temperature at which a metal begins to liquify. Pure metals, eutectics and some intermediate phases, melt as a constant temperature. Alloys generally melt over a range of temperature.

melting rate, The tonnage of metal melted per hour.

melting ratio, The proportion of the weight of metal to the weight of fuel used in melting.

metal penetration, A casting surface defect which appears as if the metal had filled the voids between the sand grains without displacing them.

microhardness, The hardness of microconstituents of a material as determined by using a diamond indentor and a special machine equipped with a microscope.

microshrinkage, Tiny cavities in a cast structure. They are a fraction of a millimeter in diameter, with irregular outlines. Microscopic analysis shows that they occur at intersections of convergent dendritic arms.

microstructure, The structure of polished and etched specimens of casting sections as revealed by a microscope at more than 90 diameters.

mishmetal, An alloy of rare earth metals containing about 50% cerium with the balance lanthanum, neodymium and similar elements.

misrun, Casting not fully formed resulting from metal poured so cold that is solidified before filling the mold completely.

modification, A process in which the eutectic temperature, structure and composition of aluminum-silicon alloys are apparently altered by the addition of small amounts of a third element, such as sodium. A similar effect can be achieved by chill casting.

moisture content, Amount of water contained in a substance that can be driven off by heating at 104-110° C (220-230° F).

moisture teller, A patented apparatus for the rapid determination of the moisture content of molding sands using the reaction between calcium carbide and water to produce a gas pressure which is then measured.

mold, The form, made of sand, metal or any other investment material, which contains the cavity into which molten metal is poured to produce a casting of definite shape and outline.

mold board, The board upon which the pattern is placed to make the mold.

mold blower, A device for filling a flask with molding sand by means of compressed air.

mold coating, Coating on sand molds to prevent metal penetration and to improve casting finish.

mold cavity, In a mold, the hole which, when filled with metal becomes the casting. Gates and risers are not considered part of the mold cavity.

mold jacket, A wooden or metal form slipped over a mold made in a snap or slip flask to support the sides and maintain alignment of the cope and drag during pouring.

mold shift, A casting defect which results when the mold halves did not match at the parting line.

mold wash, Usually an aqueous emulsion containing various compounds, such as graphite, silica flour and the like, used to coat the face of the cavity in the casting mold.

mold weights, Weights placed on top of molds to offset internal or ferrostatic pressure during pouring.

moldability, Ability of sand to flow into a flask and around a pattern, measured by the amount of sand falling through an inclined screen or slot.

moldability controller, A patented device for controlling moisture additions to a sand mix to maintain a consistent moldability index.

molding, floor, Making sand molds from loose or production patterns of such size that they cannot be satisfactorily handled on a bench or molding machine, the equipment being located on the floor during the entire operation of making the mold.

molding, machine, May refer to squeezer or jolt-squeezer machines on which one operator makes the entire mold, or to similar or larger machines including jolt-squeeze-strippers, and jolt and jolt-rollover pattern draw machines on which the cope and drag halves of molds are made.

molding material, A material that is suitable for making molds into which molten metal can be cast.

molding, pit, Molding method in which the drag is made in a pit or hole in the floor.

molding sands, Sands containing over 5 per cent natural clay, usually between 8 and 20 per cent.

montmorillonite, A very plastic clay called bentonite. It is more siliceous than kaolinite. (See bentonite.)

mottled cast iron, Cast iron which consists of a mixture of variable proportions of gray cast iron and white cast iron; such a material has a mottled fracture.

muller, A type of foundry sand mixing machine.

mulling, Process of mixing sand and clay particles either by compressing with a heavy roller in preparation for molding or by forcing the sand mixture to flow between wheels mounted on a vertical axis and the side walls of the container. Also, the process of mixing sands with a rubbing action as well as stirring.

multiple mold, A composite mold made up of stacked sections, each of which produces a complete gate of castings and poured from a central down-gate.

natural sand, One derived from a rock, in which the grains separate along their natural boundaries. This includes unconsolidated sand, or a soft sandstone where little pressure is required to separate the individual grains.

naturally-bonded molding sand, A sand containing sufficient bonding material as mined to be suitable for molding purposes.

neck down, (knock-off, washburn, or cameron core), A thin core or tile used to restrict the riser neck, making it easier to break or cut off the riser from the casting.

NI-hard, Hard white cast iron containing 4% Ni and 2% Cr.

no bake binder, A synthetic liquid resin sand binder that hardens completely at room temperatures without baking. Used in the cold setting process.

nodular iron, Iron of a normally gray cast iron type that has been suitably treated with a nodularizing agent so that all or the major portion of its graphitic carbon has a nodular or spherulitic form as cast. It is malleable as cast.

noise radiator, In OSHA terminology a device which creates noise.

noise spectrum, The various frequencies which occur in a noise.

nondestructive testing (nondestructive inspection), Testing or inspection that does not destroy the object being tested. These include radiography, magnetic particle testing, dye penetrant testing and ultrasonic testing.

oil core or mold, A core or mold in which the sand is held together by an oil binder.

oil-oxygen binder (cold-setting, air-setting binders), A synthetic auto-oxidizing liquid, oil-based binder that partially hardens at room temperature, using an oxygen releasing agent. Baking is needed to complete the curing cycle.

olivine sand, $Mg_2Fe_2SiO_4$, a naturally occuring mineral composed of fosterite and fayalite, crushed and used as a molding sand.

one-piece pattern, Solid pattern, not necessarily made from one piece of material. May have one or more loose pieces.

open hearth furnace, A furnace for melting metal, in which the bath is heated by the combustion of hot gases over the surface of the metal and by radiation from the roof.

open sand casting, A casting which is produced in a mold which is all poured in the drag, and which has no cope or other top covering.

optical pyrometer, A temperature measuring device through which the operator sights the heated object and compares its incandescence with that of an electrically heated filament whose brightness can be regulated to match the intensity of the source. Proper calibration and practice are needed to read consistent temperatures to ± 30° C.

orange peel, A surface defect often found in low carbon steel castings poured in shell molds, sometimes referred to as fish eyes.

orifice plate, In a cupola, a device used to measure the volume of air delivered by the blower.

Ottowa sand, A largely round grained silica sand of high purity originating near Ottawa, IL. Also known as St. Peter sand stone.

overflows (overflow wells), Small cavities cut into the face of a die casting die adjacent to the main cavity and connected to it by a shallow output. Overflow wells ensure cavity fill by providing for gas ejection and the removal of the first cold metal flow.

pH, A symbol denoting the negative logarithm of the concentration of the hydrogen ion in gram atoms per liter. Used to express both acidity and alkalinity. pH = log 1/H per liter. An important factor in foundry sand control where a pH = 7 is neutral, values less than 7 are acidic and greater than 7 are basic. (The symbol is derived from the French "pouvoir hydrogene" or "hydrogen power".)

padding, The addition of a wedge of excess metal to the cross section of a casting wall, usually extending downward from the riser to ensure adequate feed metal to a localized area where a shrink would occur if the added metal were eliminated.

panoramic analyzer, An instrument for detecting the level of sounds in a foundry to assure compliance with OSHA standards. The output is either on an oscilloscope or chart.

particulate matter, In air pollution control, solid or liquid particles, except water, visible with or without a microscope, that make up the obvious portion of smoke.

parting compound, A material dusted or sprayed on patterns or mold halves to prevent adherence of sand and to promote easy separation of cope and drag parting surfaces when cope is lifted from drag.

parting line, A line on a pattern or casting corresponding to the separation between the cope and drag portions of a sand mold.

parting sand, Fine sand for dusting on sand mold surfaces that are to be separated.

patching, Repair of a furnace lining, repair of a mold or core.

pattern, A form of wood metal or other materials around which molding material is placed to make a mold for casting metals.

pattern board, Board having a true surface upon which a pattern is placed preparatory to making a mold of the pattern.

pattern die, A body composed of two or more mating parts, containing one or more negative impressions of the master pattern and used for production of expendable patterns.

pattern draft, Taper allowed on vertical faces of a pattern to permit easy withdrawal of the pattern from the compacted mold.

pattern, grand master, Pattern constructed with triple shrinkage allowance from which metal master patterns are made.

pattern, investment molding, A reproduction with an expendable material of the object to be cast and usually formed in a pattern die.

pattern layout, Full-sized drawing of a pattern showing its arrangement and structural features.

patternmaker's shrinkage, Shrinkage allowance made on all patterns to compensate for the change in dimensions as the solidified casting cools in the mold from freezing temperature to the metal to room temperature. Patterns are made larger by the amount of shrinkage characteristic of the particular metal in the casting and the amount of resulting contraction to be encountered. Rules or scales are available for use.

patternmaking, Skilled craft of modeling in wood, metal, or other materials objects to be cast in metal.

pattern, master, Pattern constructed with double shrinkage allowance from which production patterns are made.

pattern, plaster, A pattern made from plaster of Paris.

pattern, split, Pattern made in two or more parts. Also called "parted pattern."

pearlitic malleable, The product obtained by a heat treatment of white cast iron which converts some of the combined carbon into graphite nodules but which leaves a significant amount of combined carbon in the product. Pearlitic malleable may also be obtained through the use of addition alloys which retard the breakdown of pearlite.

peen, (1) Flat-pointed end of a rammer used in ramming sand into a mold. (2) Process of repairing slight leaks in castings by repeated impacts of a blunt tool or peening hammer. (3) Peening action obtained by impact of metal shot.

pencil core, A core projecting to the center of a blind riser to admit atmospheric pressure to force out feed metal.

pencil gate, Gating directly into the mold cavity through the cope by means of one or more small vertical gates connecting the pouring basin and mold cavity.

penetration, metal, Condition where molten metal has penetrated into the face of a sand mold beyond the midpoint of the first layer of sand grains, resulting in a mixture of metal and sand adhering to the cast surface.

perlite, A highly siliceous volcanic rock which can be expanded into a porous mass of particles by heating. Perlite can be used as an insulation in foundry sand mixes or as a riser topping or sleeves in steel casting.

permanent mold, A metal mold of two or more parts that is used repeatedly for the production of many castings of the same form. Liquid metal is poured in by gravity. Not an ingot mold.

permeability, As applied to sand molds, permeability means the property of the sand which permits passage of gases. The magnetic permeability of a substance is the ratio of the magnetic induction of the substance to the intensity of the magnetizing field to which it is subjected.

phenolic resin, A resin made by the polymerization of a phenol with an aldehyde. Used as a binder for cores and sand molds.

pig, An ingot of virgin or secondary metal to be remelted for use.

pig iron, Cast iron generally produced by the reduction of iron ore in the blast furnace; also the over-iron in the foundry poured into pig molds.

pinhole porosity, Very small holes scattered through a casting, possibly caused by microshrinkage or gas evolution during solidification.

pipe, A cavity formed by shrinkage of the metal during solidification, usually occurring in a riser.

pit, A sharp depression in the surface of metal.

plaster molds, With plaster as the mold material, liquid plaster slurry is poured over the pattern, allowed to harden and the pattern is then removed. The hardened mold is then heated to drive off the moisture. Smooth finishes and close tolerances are the advantages. Aluminum and copper-base alloys are cast by this process.

plaster of Paris, A semi-hydrated form of calcium sulphate made by sintering gypsum at 120-130° C (248-266° F).

plumbago, Graphite in powdered form. Plumbago crucibles are made from this graphite plus clay.

porosity (sand), The ratio of volume of the pores or voids to volume of the entire mass, usually expressed as a percentage. Generally the volume of pores is obtained by determining the amount of water or gas absorbed by the mass, and therefore does not include the volume of the sealed pores. The result so obtained is "apparent" rather than "true" porosity. (Not synonymous with permeability.)

poured short, Casting which lacks completeness due to the mold not being filled.

pouring, Transfer of molten metal from furnace to ladle, ladle to ladle, or ladle to molds.

pouring basin, The enlarged mouth of the sprue into which the molten metal is first poured.

pouring cup, The flared section at the top of the downsprue. It may also be a baked core sand core cup placed on top of the cope to line up with the down sprue.

print back, After drawing the pattern, the surface of the mold is dusted with a carbonaceous material such as graphite, coal, etc., and the pattern replaced, rapped slightly into position, and again removed, leaving a smooth mold surface.

primary choke (choke), That part of the gating system which most restricts or regulates the flow of metal into the mold cavity.

production foundry, A foundry engaged in manufacture of castings in large quantities, usually highly mechanized to minimize manual labor. May be either a jobbing or captive foundry.

quartz, A form of silica which occurs in hexagonal cyrstals. It is the most common of all solid minerals.

radiant heat, Heat which is transmitted by electromagnetic waves.

ram, Process of packing sand in a mold. (See ramming.)

ramaway, A casting defect resulting from a section of the mold being forced away from the pattern by ramming sand after it has conformed to the pattern contour.

ramming, The operation of packing sand around a pattern in a flask to form a mold.

ramming, jolt or jar, Packing sand in the mold by raising and dropping upon a table the sand, pattern and flask, the sand itself being the ramming medium. Jolt squeezers, jarring machines, and jolt rammers are molding machines using the principle of jar ramming.

rapping, Knocking or jarring the pattern to loosen it from the sand in the mold before withdrawing the pattern.

rapping bar, Metal rod used to rap the draw spike to loosen the pattern in the sand for drawing.

rat, A lump on the surface of a casting caused by a portion of the mold face sticking to the pattern.

rat-tail, A minor buckle, occurring as a small irregular line or lines on a cast surface.

receiving ladle (forehearth), A large ladle placed in front of the cupola into which all metal is tapped. It acts as a mixer and reservoir and permits homogenization of the cupola output.

release agent (parting agent), A material such as silicone, stearate, oil or wax for lubricating a die, pattern or core box to facilitate easy removal of a casting, mold or core.

relief sprue, In a mold a vertical channel, the approximate size of the downsprue connected to the runner to relieve pressure surge during pouring. It functions like a standpipe in a plumbing system.

remelt, Sprues, gates, risers and defective castings returned directly to the crucible furnace.

residual stress, See stress, residual.

respirator, A filtering device which covers the nose and mouth and prevents inhalation of dust or fumes; should have the U.S. Bureau of Mines certificate of approval for the specific contaminant being filtered out.

riddle, A screening device, manually or mechanically operated, for removing coarse particles and contamination from molding sand.

riddled sand, Sand that has been passed through a riddle or screen, mounting patterns on boards integrally with skimmers, gates, risers, etc.

Ringelmann's scale, In air pollution control, a black and white mesh scale reading from all clear to solid black. Used to measure the density of smoke. Observer normally uses the chart comparator 50 feet from the point of emission.

riser, A reservoir of molten metal provided to compensate for the internal contraction of the casting as it solidifies.

riser, blind, A riser which does not break through the top of the cope mold, and is entirely surrounded by sand; often combined with spin gates, and together forming an efficient method of gating and feeding a casting.

riser, gating, Practice of running metal for the casting through the riser to help directional solidification.

riser height, The distance from the top of the riser when liquid to the top of the riser neck. Riser height when solid may be several inches less than when liquid because of loss of feed metal to the casting.

riser neck, The connecting passage between the riser and casting. Only the height and width, or diameter, of the riser neck usually are reported, although the shape may be equally important.

riser, open, Conventional form of riser usually located at the heaviest section of the casting and extending through the entire height of the cope.

riser pad (riser contact), An enlargement of the riser neck where it joins the casting. The purpose of the pad is to prevent "breaking in" of the riser when it is struck or cut from the casting.

riser, side, A riser attached to the side of a casting.

riser, top, A riser attached to the top surface of a casting.

rolling over, The operation of reversing the position of a flask in which the drag part of the pattern has been rammed with the parting surface downward.

roll-over machine, A molding machine with which the flask is rolled over before the pattern is drawn from the mold.

runner, A channel through which molten metal or slag is passed from one receptacle to another; in a mold, the portion of the gate assembly that connects the downgate or sprue with the casting ingate or riser. The term also applies to similar portions of master patterns, pattern die, patterns, investment mold and the finished casting.

runner box, Device for distributing molten metal around a mold by dividing it into several streams.

runner extension, That part of a runner which extends beyond the farthest ingate as a blind end. The end should be enlarged to at least twice the runner volume to act as an effective dirt trap and to prevent bounce back of the flowing stream.

runner riser, A conventional runner, usually in the horizontal plane, which permits flow of molten metal to the ingate and is large enough to act as a reservoir to feed the casting.

runout, A casting defect caused by incomplete filling of the mold due to molten metal draining or leaking out of some part of the mold cavity during pouring; escape of molten metal from a furnace, mold or melting crucible.

sag, A decrease in metal section in a casting due to sagging of the cope or core.

sand, A loose, granular material resulting from the disintegration of rock. The name sand refers to the size of grain and not to mineral composition. Diameter of the individual grains can vary from approximately 6 to 270 mesh. Most foundry sands are made up principally of the mineral quartz (silica). Reason for this is that it is plentiful, refractory and cheap.

(1) *bank sand*, Sedimentary deposits, usually containing less than 5% clay.
(2) *dune sand,* Wind-blown deposits of sand found near large bodies of water.
(3) *molding sands*, Sands which contain over 5% natural clay; usually between 10 and 20%.
(4) *silica sand,* Although most foundry sands contain a high percentage of silica, the term silica sand is generally reserved for those that show a minimum of 95% silica content. Many high grade silica sands will analyze better than 99% pure silica.
(5) *miscellaneous sands,* Include zircon, olivine, $CaCO_3$, black sands (lava grains), titanium minerals, etc.

sand, backing, An inexpensive sand used for mold filler.

sand, bank, Sand from a bank or pit which usually contains less than 5% clay-like material.

sand, black, Sand with a carbonaceous coating or a skin deposit which usually occurs after the sand has been used at least once.

sand blast, Sand driven by a blast of compressed air (or steam); used to clean castings; to cut, polish or decorate glass or other hard substances; for cleaning building fronts, etc.

sand, burned, Condition in which the claying portion of the sand has been destroyed by the heat of the metal.

sand castings, Metal castings produced in sand molds.

sand, coated, Any sand whose grains have been coated so as to improve results obtained; non-thermosetting resins are commonly used.

sand control, Procedure whereby various properties of foundry sand, such as fineness, permeability, green strength, moisture content, etc., are adjusted to obtain castings free from blows, scabs, veins and similar defects.

sand cutting, Preparing sand for molding, either by hand or by machine, the operation also aerating the sand.

sand expansion, Maximum dimensional increase which a sand undergoes when heated to a given temperature.

sand inclusions, Sand which has loosened from the mold and become entrapped in the molten metal.

sand mulling, A method of evenly distributing the bond around the sand grain by a rubbing action.

sand, naturally-bonded, A sand containing sufficient bonding material as mined to be suitable for molding purposes.

sand plow, A bladed device used to divert sand from a belt conveyor into a sand hopper.

sand porosity, Volume of the pore spaces or voids in a sand. (Not synonomous with permeability.)

sand, rebonded, Used or reclaimed molding sand restored to usable condition by the addition of new bonding material.

sand reclamation, Processing of used foundry sand by thermal or hydraulic methods, so that it may be used in place of new sand without substantially changing current foundry sand practice.

sand slinger, Mechanical device which throws sand into a flask or core-box, by centrifugal action.

sand spun process, A centrifugal casting technique for coating the mold with a thin layer of unbonded refractory parting material. Particularly useful for short lengths of cylindrical shapes.

sand, tempering, Dampening and cutting over or otherwise mixing sand to produce uniform distribution of moisture.

sand toughness number, The product of deformation, times green compressive strength, times 1000. It is an indication of the workability of the sand mixture, and is usually expressed STN.

scab, A defect on the surface of a casting which appears as a rough, slightly raised surface blemish, crusted over by a thin, porous layer of metal under which is a honeycomb or cavity that usually contains a layer of sand; defect common to thin-walled portions of the casting, or around hot areas of the mold.

scaling (scale), Surface oxidation, partially adherent layers of corrosion products, left on metals by heating or casting in oxidizing atmospheres.

scrap, Metal for remelting; includes scrapped machinery, rail or structural steel and rejected castings.

screen analyses, See sieve analysis.

screen, A sieve or riddle with openings of a definite size. Sieves are used to separate the different fractions of a sand distribution or to remove agglomerated particles or lumps.

seam, A surface defect on a casting related to but of lesser degree than a cold shut; a ridge on the surface of a casting caused by a crack in the mold face.

segregation, (1) A sand problem which occurs during transportation in bulk or in sand silos or hoppers. It is difficult to reblend the various fractions into a homogeneous mix. (2) A casting defect which is evidenced by the concentration of alloying elements in specific regions usually as a result of the primary crystallization of one phase with the subsequent concentration of other elements in the remaining liquid.

self curing binder, A material used in core making that sets up to form cured cores without heat or additional assistance.

separator, A mechanical device which separates or grades materials into constituent parts, used in the foundry to remove fines from system sands and dust from the air.

set gates, Gating systems formed by patterns, in contrast to gates cut in the sand by hand.

shake-out, The operation of removing castings from the mold. A mechanical unit for separating the molding material from the solidified metal casting.

shank, The handle which is attached to a ladle.

sharp sand, A sand that is substantially free of bond. The term has no reference to the grain shape.

shaw process, A precision casting process which uses ceramic molds in five stages and using conventional metal patterns. The patented mold material microcrazes to produce sufficient permeability to pour castings weighing up to 1500 pounds (682 kg).

shell molding, A process for forming a mold from resin-bonded sand mixtures brought in contact with pre-heated (300-500° F; 150-260° C) metal patterns, resulting in a firm shell with a cavity corresponding to the outline of the pattern. Also called Croning process.

shift, A casting defect caused by mismatch of cope and drag or of cores and mold.

shot blasting, A casting cleaning process which uses a metal abrasive (grit or shot) propelled by centrifugal or pneumatic force.

shrinkage, Change in size as the metal passes from the fluid to the solid state in the mold.

shrinkage cavity, A void left in cast metals as a result of solidification shrinkage and the progressive freezing of metal toward the center.

shrinkage cracks, Cracks that form in metal as a result of the pulling apart of grains by thermal contraction before complete solidification.

shrinkage, patternmaker's, A scale divided in excess of standard measurement to allow for the difference in size between the casting and the corresponding mold cavity. Used by patternmakers to avoid calculations for shrinkage.

shrink hole, A hole or cavity in a casting resulting from contraction and insufficient feed metal, and formed during the time the metal changed from the liquid to the solid state.

sieve analyses, Determination of the partial size distribution of a molding sand expressed in terms of the weight retained on each of a series of standard sieves stacked according to the decreasing mesh sizes.

silica, Silicon dioxide, the prime ingredient of sharp sand and acid refractories.

silica brick, Refractory material of Ganister bonded with hydrated lime and fired at high temperature.

silica flour, Material commonly produced by pulverizing pure grains of quartz sand in large ball mills. It is available in several mesh sizes, generally from 80 to 325 mesh. In analysis, it is 99.% pure silica, and is commonly used in foundry practice for improving finish and increasing hot strength. It is also used in production of core and mold washes.

silica gel, A colloidal form of silica used as a drying agent.

silica sand, Sand with a minimum of 95% silica.

silt, Very fine particles that pass a No. 270 mesh sieve, but which are not plastic nor sticky when wet. (Between $-50 + 5$ microns in size.)

sintering, The bonding of adjacent surfaces of particles or a compact by heating to a suitable temperature and cooling.

sintering point, The temperature at which a molding material begins to adhere to a casting or in a test when sand adheres to a platinum ribbon under controlled conditions, or the temperature at which sand grains cohere to one another.

skeleton pattern, A framework or skeleton of ribs representing the interior and exterior form and the metal thickness of the required casting.

skin core, A flat core or tile placed in a mold to skim a flowing stream of metal. Commonly used in pouring basins, it holds back slag and dirt while clean metal flows underneath to the downsprue.

slag, A nonmetallic covering which forms on the molten metal as a result of the flux action in combining impurities contained in the original charge, some ash from the fuel, and any silica and clay eroded from the refractory lining. It is skimmed off prior to tapping the heat.

slag inclusion, Nonmetallic solids entrapped in solid metal.

slag trap, An enlargement, dam, or extrusion in the gating or runner system is a mold for the purpose of preventing molten slag particles from entering the mold cavity.

slick, slicker, smoother, A tool used for mending and smoothing the surfaces of a mold after withdrawal of pattern and before closing mold.

slip casting, In ceramics, the pouring of slip (a water suspension of finely ground clay) into a plaster of paris mold. After hardening, it is dried and fired.

slip jacket, A wooden or metal form which is slipped over a mold after removal of the snap or slip flask, to support the four sides of the mold during pouring. Jackets are shifted, with the mold weights, from one row of molds to another during pouring.

slurry, A term loosely applied to any clay-like dispersion. It may be used to wash ladles or other refractory linings to impart a smoother surface; as a bonding addition to molding sand; as a thin loam over specially made molds; or as a mixture to fill in joints or cracks of a core, etc.

slush casting, A casting made from an alloy that has a low melting point and freezes within a wide range of temperatures. The metal is poured into the mold, and brought into contact with all surfaces so as to form a shell of frozen metal; then the excess metal is poured out. Castings that consist of completely enclosed shells may be made by using a definite quantity of metal and a closed mold.

smoke, A type of emission resulting from incomplete combustion. It consists predominantly of small gas-borne particles of combustible materials which are present in sufficient quantities to be observable independently of the presence of other solids in the gas stream.

snagging, The process of rough cleaning castings by grinding.

soaking, Prolonged heating of a metal furnace or ladle at a selected temperature.

sodium silicate, See water glass.

soldiers, Thin pieces of wood used to strengthen a body of sand or to hold it in place.

solidification, The physical process of change from a liquid to a solid state.

solidification shrinkage, The decrease in size accompanying the freezing of a molten metal.

sonic testing, Using sound waves above audible frequency via a supersonic reflectoscope to measure time sound waves taken as they return from opposite sides of the casting. Defects return the waves in more or less time.

spark test, A method for determining the approximate composition of steel by producing sparks on a grinding wheel.

spiegeleisen, An alloy of iron and manganese used in basic and acid open hearth steel-making practices. Also can be used in cupola charges. A high manganese pig iron containing 15 to 20% Mn and 4.5 to 6.5% carbon.

spiral fluidity test, A quantitative test for determining the fluidity of an alloy by pouring molten metal into a mold with a long narrow spiral channel. The fluidity index is determined by the length of flow which the alloy achieved.

splash core, A core or tile placed in a mold to prevent erosion of the mold at places where metal impinges with more than normal force. Splash cores are commonly used at the bottom of large rammed pouring basins, at the bottom of long downsprues, or at the ingates of large molds.

spout, A refractory lined trough through which metal flows from the furnace to the ladle.

sprue, The vertical channel connecting the pouring basin with the skimming gate, if any, and the runner to the mold cavity, all of which together may be called the gate. In top-poured castings the sprue may also act as riser. Sometimes used as a generic term to cover all gates, risers, etc. returned to the melting unit for re-melting. Also applies to similar portions of master patterns, pattern die, patterns, investment mold and the finished casting.

sprue base, An enlargement or rounded section at the bottom of the downsprue, used to help streamline the flow of metal into the runner and thus lowering its viscosity.

sprue cutter, A metal tool used in cutting the pouring aperture.

sprue hole, The opening through which the metal is poured into the cope to run the casting.

sprue plug, Wood or metal tapered roll or pin used to form a sprue opening, or a metal plug left in the sprue after pouring has begun to allow filling of the pouring basin before the metal is allowed to enter the sprue.

spruing, Removing gates from castings after the metal has solidified.

squeeze board, A board used on the cope half of the mold to permit squeezing it on a jolt squeeze machine.

squeeze head, In certain type of molding machines, a stationary or moveably plate against which a filled mold is compressed in order to complete the ramming of the sand.

squeeze foot, One of a number of 4″ × 4″ (0.101 600 m) feet which squeeze a mold by hydraulic means. The feet are independently loaded so the entire mold can be uniformly compacted regardless of the contour of the pattern. This action is called molded with compensating hydraulic squeeze.

stack molding, A molding method which makes use of both sides of one half mold, one side acting as the drag and the other half as the cope, all the castings being poured through a common down gate.

stack molding, Use of composite molds made up of stacked sections, each of which produces a complete gate of castings which is poured from a central down gate.

standard pattern, A pattern of high-grade material and workmanship in daily use or used at frequent intervals. A pattern used as a master to make or check production patterns.

sticker, A lump on the surface of a casting caused by a portion of the mold face sticking to the pattern. Also, a forming tool used in molding.

stool, Support for a crucible in a lift out furnace.

stopping-off, Filling in a portion of a mold cavity which is not to be cast.

streamline flow, Steady flow of liquid metal with a minimum of changes in direction and subsequent eddying. True streamlined flow cannot be achieved in metal casting but disruptive turbulence can be eliminated.

stress, residual, Those stresses set up in a metal as a result of non-uniform plastics deformation or the unequal cooling of a casting.

strickle or strike, A tool with a straight or curved edge used for removing excess and from a flask or pattern.

strike-off, Operation of removing excess sand from core box or flask.

stripper pins, A series of pins (usually 4) which support a rammed flask at the parting surface so that the mounted pattern may be drawn by lowering the pattern plate.

stripping plate, The plate which holds the sand in place while the pattern is drawn vertically through the plate.

stripping time, In oil-oxygen and no bake mixtures, the moment when the core box may be satisfactorily drawn from the core or the pattern from the sand.

styrofoam pattern, Expendable pattern of foamed plastic, especially polystyrene used in producing castings by the full mold process. Especially useful for large automotive body dies or for art sculpture castings.

superheat, Any increment of temperature above the melting point of a metal or alloy.

swab, A sponge or piece of waste, hemp, or other material used in dampening sand around a pattern before withdrawing it. Sometimes used in blacking molds which might be broken by a brush.

sweep, A template cut to the profile of the desired mold shape, which when revolved around a stake or spindle, produces that mold.

swell, A casting defect consisting of an increase in metal section due to the displacement of sand by metal pressure.

swing frame grinder, A device for grinding large castings which must remain stationary. The grinder is suspended by a hoist because of its weight.

synthetic molding sand, Any sand compounded from selected individual materials which, when mixed together, produce a mixture of the proper physical and mechanical properties from which to make foundry molds.

system sand, Foundry sand which is recirculated back to the mulling station each time it is used. There small percentages of new binder and sand are added, the sand is remulled and sent to the molding station again.

talc, Hydrated magnesium silicate; it is soft, greasy to the touch and very pliable; widely used in parting compounds, core compounds and in core paste; often called red talc or yellow talc.

taphole, Opening in the furnace breast through which the molten metal is tapped into the spout.

temper (noun), The moisture content of a sand at which any certain physical test value is obtained, i.e., temper with respect to green compressive strength, permeability, retained compressive strength, etc.

temper (verb), Mixing sand with sufficient water or other liquid to develop desired molding properties.

temper carbon, The free or graphitic carbon that precipitates from solution, usually in the form of rounded or equiaxed nodules in the structure, during the graphitizing or malleabilizing of white cast iron.

temperature, holding, (1) Temperature above the critical phase transformation range at which castings are held as a part of the heat treatment cycle. (2) The temperature maintained when metal is held in a furnace, usually prior to pouring.

temperature, pouring, The temperature of the metal as it is poured into the mold.

ternary alloy, An alloy that contains three principal elements.

thixotrophy, In sand testing, the proportion of the volumes of liquid and solid matters which is required, one minute after shaking, for the suspension to just flow out of an 8-mm diameter test tube, when the test tube is changed from a vertical to a horizontal position.

tooling points, Reference points on drawings of a pattern or a casting to guide patternmaker or machinist in subsequent finishing operations and in dimension checks.

transfer ladle, A ladle that may be supported on a monorail or carried in a shank and used to transfer metal from the melting furnace to the holding furnace or from furnace to pouring ladles.

tucking, Pressing sand with the fingers under flask bars, around gaggers, and other places where the rammer does not give the desired density.

tumbling, Cleaning castings by rotating them in a container in the presence of cleaning materials.

tuyere, An opening in the cupola shell and refractory lining through which the air blast is forced.

ultrasonic cleaning, Immersion cleaning aided by ultrasonic waves which cause microagitation.

ultrasonic testing, A nondestructive method of testing metal for flaws based on the fact that ultrasonic waves are reflected and refracted at the boundaries of a solid medium.

unkilled steel, A mild steel insufficiently deoxidized so that it evolves gas and blowholes during solidification.

urea formaldehyde resin, A thermosetting product of condensation from urea or thiourea and formaldehyde, soluble in water and used as a sand binder in core and mold compounds.

"V" process, A Japanese patented process to produce molds using only unbonded sand. The vented pattern plate is covered with a heated plastic film and a mild vacuum draws the film close to the pattern, the flask is located on the plate and loosely bonded sand is added to fill the flask. The plate is rapped to achieve compaction and a second sheet of plastic is placed over the top of the cope. Next a vacuum is drawn through a special port in the flask side. Now the vacuum on the pattern plate is released and the mold is stripped and ready to be assembled to a drag. The mold is poured and the vacuum is released. Clean up recycling of the sand are simple routines.

vacuum casting, A casting process in which metal is melted and poured under very low atmospheric pressure; a form of permanent mold casting where the mold is inserted into liquid metal, vacuum is applied and metal drawn up into the cavity.

vacuum degassing, The use of vacuum technique to remove dissolved gases from molten alloys.

vacuum refining, Melting in a vacuum, usually by electrical induction, to remove gaseous contaminants from the metal.

veining, A defect on the surface of a casting appearing as veins or wrinkles and associated with excessive thermal movement of the sand, especially core sands.

vent, A small opening or passage in a mold or core to facilitate escape of gases when the mold is poured.

venting, Perforation with a venting wire of the sand over and around a mold cavity to assist in the escape of the gases.

vent rods, Rods used to make a passageway in the mold for escaping gases.

vent wax, Wire or rod shapes of wax placed during making of the core. During baking the wax is melted, leaving an opening through which gases may escape.

vent wire, A wire used to make vents or small holes in the mold to allow gas to escape.

vibrator, A device, operated by compressed air or electricity, for loosening and withdrawing patterns from a mold, or for vibrating a hopper or chute to promote the flow of material from the hopper or chute.

virgin metal (primary metal), Metal extracted directly from the ore; not previously used.

viscosity, The resistance of a fluid substance to flowing, quantitatively characteristic for an individual substance at a given temperature and under other definite external conditions.

void, A shrinkage cavity produced in castings during solidification.

warpage, Deformation other than contraction that develops in a casting between solidification and room temperature; also, distortion occurring during annealing, stress relieving and high-temperature service.

wash, A casting defect resulting from erosion of sand by metal flowing over the mold or cored surfaces. They appear as rough spots and excess metal on the casting surface. Also called cuts.

washburn core, A thin core which constricts the riser at the point of attachment to the casting. The thin core heats quickly and promotes feeding of the casting. Riser removal cost is minimized.

water test, To subject a casting to water pressure in such a manner that any porous areas will show leakage.

water glass, Sodium silicate, a viscous liquid which, when mixed with powdered fireclay, forms a refractory cement; used in CO_2 molding.

wax pattern, (1) A precise duplicate, allowing for shrinkage, of the casting and required gates, usually formed by pouring or injecting molten wax into a die or mold. (2) Wax molded around the parts to be welded by a thermit welding process.

weak sand, Sand lacking in the proper amount of clay, bond or moisture. May be intentionally prepared to avoid cracking of intricate castings.

weep hole, An opening placed in a casting to allow drainage of moisture.

wet scrubber (gas washer), In air pollution control, a liquid (usually water) spray device for collecting pollutants in escaping foundry gases.

whirl gate, A gate and sprue arrangement which tangentially introduces molten metal into a riser so the centrifugal action forces dirt or slag to the center of the riser and away from the riser connection as the metal enters the casting cavity.

Williams core, A core placed in a blind riser so that the metal within the riser is subject to atmospheric pressure, thereby eliminating establishment of partial vacuum within the riser as the metal begins to solidify, the core acting as a passageway for the ingress of air, thus promoting free flow of metal to feed the casting.

wind box, The chamber surrounding a cupola through which air is conducted under pressure to the tuyeres.

wood flour, A pulverized wood product used in the foundry to furnish a reducing atmosphere in the mold, help overcome sand expansion, increase flowability, improve casting finish and provide easier shakeout.

x-ray, Form of radiant energy with wavelength shorter than that of visible light and with the ability to penetrate materials that absorb or reflect ordinary light. X-rays are usually produced by bombarding a metallic target with electrons in a high vacuum. In nuclear reactions it is customary to refer to photons originating in the nucleus as gamma rays and to those originating in the extranuclear part of the atom as x-rays.

yield, Comparison of casting weight to total weight of metal poured into mold.

yield strength, The stress at which a material exhibits a specified limiting permanent strain.

Young's Modulus (E), The Modulus of Elasticity for pure tension with no other stress activity; it has, in most metals, practically the same compression. Denoted by E it is the tensile (or compressive) stress per unit of linear strain or tensile stress intensity/tensile strain.

zircon, The mineral zircon silicate, $ZrSiO_4$, a very high melting point acid refractory material used as a molding material in steel foundries.

zircon sand, A very refractory sand of unusual fineness, low thermal expansion and high thermal conductivity, exceedingly useful and efficient in foundry practice. It consists principally of zirconium silicate and is widely used for improving finish and preventing "burning-in" of sand on the casting surface.

zirconia, ZrO_2 an acid refractory up to 2500° C (4532° F) having good thermal shock resistance and low electrical resistivity.

METAL FORMING

age hardening, A process of aging that increases hardness and strength and ordinarily decreases ductility. Age hardening usually follows rapid cooling or cold working.

air-hardening steel, An alloy steel that is hardened by cooling in air from a temperature higher than the transformation range. Also called self-hardening steel.

alloy, A substance that has metallic properties and is composed of two or more chemical elements of which at least one is a metal.

alpha iron, The form of iron that is stable below Ae_3 (910° C, 1970° F) and characterized by a body-centered cubic crystal structure.

annealing, A heat treatment designed to effect: (1) softening of a cold worked structure by recrystallization or grain growth or both; (2) softening of an age hardened alloy by causing a nearly complete precipitation of the second phase in relatively coarse form; (3) softening of certain age hardenable alloys by dissolving the second phase and cooling rapidly enough to obtain a supersaturated solution (this usuage is generally applied to nickel-base and copper-base alloys; the treatment should be called more precisely a solution heat treatment); (4) relief of residual stress and change in properties induced during the thermal flattening of magnesium alloy sheet at slightly elevated temperatures.

artificial aging, An aging treatment above room temperature.

austempering, A trade name for a patented heat treating process that consists in quenching a ferrous alloy from a temperature above the transformation range, in a medium having a rate of heat abstraction sufficiently high to prevent the formation of high-temperature transformation products; and in maintaining the alloy, until transformation is complete, at a temperature below that of pearlite formation and above that of martensite formation.

austenite, A solid solution in which gamma iron is the solvent; characterized by a face-centered cubic crystal structure.

bending rolls, Three staggered rolls adjusted to put the desired curvature in plate or used for coiling and uncoiling strip or wire.

bend radius, The radius corresponding to the curvature of a bent specimen or bent area of a formed part, and measured on the inside of a bend.

bend tests, Various tests used to determine the ductility of sheet or plate that is subjected to bending. These tests may include determination of the minimum radius or diameter required to make a satisfactory bend and the number of repeated bends that the material can withstand without failure when it is bent through a given angle and over a definite radius.

blank, A pressed presintered or fully sintered compact, usually in the unfinished condition and requiring cutting, machining or some other operation to produce the final shape.

blank holder, The tool that prevents the rim of a sheet metal blank from wrinkling while it is being deep drawn.

blanking, Shearing out a piece of sheet metal in preparation for deep drawing.

blue brittleness, Reduced ductility occurring as a result of strain aging, when certain ferrous alloys are worked between 300 and 700° F. This phenomenon may be observed at the working temperature or subsequently at lower temperatures.

board hammer, A type of drop hammer in which boards attached to the ram are lifted between rollers.

brake, A piece of equipment used for bending sheet; also called a "bar folder." If operated manually, it is called a "hand brake"; if power driven, it is called a "press brake."

brale, A diamond penetrator, conical in shape, used with a Rockwell hardness test for hard metals.

buckling, The bending of sheets or plates under a compressive load.

buff, A polishing wheel usually consisting of a large number of treated or untreated muslin disks sewed together.

bulging, Local expansion of tubes or shells by internal hydraulic pressure or by compression of a rubber cylinder.

bull block, A power-driven reel for drawing heavy-gage wire through a die.

burnishing, Plastic smearing such as may occur on metallic surfaces during buffing.

burnt, A term applied to a metal permanently damaged by having been heated to a temperature close to the melting point.

burr, A rough or sharp edge left on metal by a cutting tool.

carbide, A coumpound of carbon with one or more metallic elements.

carbonitriding, A process in which a ferrous alloy is case hardened by first being heated in a gaseous atmosphere of such composition that the alloy absorbs carbon and nitrogen simultaneously, and then being cooled at a rate that will produce desired properties.

carbon steel, Steel that owes its properties chiefly to the presence of carbon without substantial amounts of other alloying elements; also termed "ordinary steel," "straight carbon steel," "plain carbon steel."

carburizing, A process that introduces carbon into a solid ferrous alloy by heating the metal in contact with a carbonaceous material — solid, liquid, or gas — to a temperature above the transformation range and holding at that temperature.

case hardening, A process of hardening a ferrous alloy so that the surface layer or case is made substantially harder than the interior or core.

charpy test, A pendulum type of impact test in which a specimen, supported at both ends as a simple beam, is broken by the impact of the falling pendulum.

closed dies, Forging dies in which the compressive force is applied to the whole surface of the forging. In open dies, there is no constraint to lateral flow in some directions.

coining, A process of impressing images or characters of the die and punch onto a plane metal surface.

cold pressing, Forming a compact at room temperature.

cold work, Plastic deformation at such temperatures and rates that substantial increases occur in the strength and hardness of the metal. Visible structural changes include changes in grain shape and, in some instances, mechanical twinning or banding.

cold working, Deforming a metal plastically at such a temperature and rate that strain hardening occurs. The upper limit of temperature for this process is the recrystallization temperature.

cupping, The breaking of wire with a cup fracture accompanied by very little reduction of area; observed during cold drawing. Also the forming of sheet into cuplike objects such as shells, by deep drawing.

curling, Rounding the edge of sheet metal into a closed or partly closed loop.

decarburization, The loss of carbon from the surface of a ferrous alloy as a result of heating in a medium that reacts with the carbon.

deformation bands, Bands within individual cold worked crystals. The bands and matrix differ in orientation.

die radius, The radius on the exposed edge of a deep drawing die, over which the sheet flows in forming drawn shells.

die sinking, Forming or machining a depressed pattern in a die.

dimpling, A process of cupping sheet so as to permit the use of rivets that have countersunk heads. Dimpling differs from counters in that it is a cupping operation and no metal is cut away after the rivet hole have been drilled.

direct extrusion, Extrusion through a die placed at the opposite end of the billet from the ram.

draft, (1) The amount of taper in the side walls of die impressions. (2) The taper given to the sides of a pattern to enable it to be withdrawn easily from the mold.

drawbench, The stand that holds the die and drawhead used in drawing wire, rod and tubing.

drop forging, Forming metal, usually under impact, by compression within dies designed to produce the required shape. The term is ordinarily used synonymously with hot die forging.

drop hammer, A forging machine that employs the impact resulting from the action of gravity, with or without added steam or air pressure, on a falling ram.

ear, A wavy projection formed in the course of deep drawing, as a result of directional properties or anisotropy of the sheet.

electroforming, Production or reproduction of articles by electrolytic deposition.

embossing, Raising a design in relief against a surface.

extrusion, Shaping metal into a chosen continuous form by forcing it through a die of appropriate shape.

ferrite, A solid solution in which alpha iron is the solvent, and which is characterized by a body-centered cubic crystal structure.

finished steel, Steel that is ready for the market without further work or treatment.

finishing temperature, The temperature at which hot mechanical working of metal is completed.

forging rolls, Rolling mills that forge comparatively uniform shapes by using variable radii around the circumference of rolls that rotate in the opposite direction from those ordinarily used for rolling.

forging strains, Differential strains that result from forging or from cooling from the forging temperature, and that are accompanied by residual stresses.

Guerin forming, A method of forming in which the metal is forced to conform to the shape of a male die by the application of a hydrostatic force of confined rubber.

hammer forging, A forging process in which the work is deformed by repeated blows.

hammering, Beating metal sheet into a desired shape either over a form or on a high-speed mechanical hammer, in which the sheet is moved between a small curved hammer and a similar anvil to produce the required dishing or thinning.

hard drawn, A temper produced in wire, rod, or tube by cold drawing.

hardness, Defined in terms of the method of measurement. (1) Usually the resistance to indentation. (2) Stiffness or temper of wrought products. (3) Machinability characteristics.

heat treatable alloy, An alloy hardenable by heat treatment.

heat treatment, A combination of heating and cooling operations, timed and applied to a metal or alloy in the solid state in a way that will produce desired properties.

hold down, The tool that exerts pressure normal to a sheet blank during deep drawing, in order to prevent wrinkling and to cause uniform circumferential compression while the large circumference of the blank is reduced to that of the shell.

Hooker process, A process of cold extrusion in a crank press, employed in making small arms cartridge cases and small thin-walled seamless tubes.

hot forming, Working operations such as bending and drawing sheet and plate, forging, pressing, and heading, performed on metal heated to temperature above room temperature.

hot pressing, The simultaneous forming and heating of a compact.

hot shortness, Brittleness in hot metal.

hot working, Plastic deformation of metal at such a temperature and rate that strain hardening does not occur. The lower limit of temperature for this process is the recrystallization temperature.

indirect extrusion (inverted), An extrusion process in which the metal is forced back inside a hollow ram that pushes the die.

ingot, A casting intended for subsequent rolling or forging.

ironing, Thinning the walls of deep drawn articles by reducing the clearance between punch and die.

killed steel, Steel deoxidized with a strong deoxidizing agent such as silicon or aluminum in order to reduce the oxygen content to a minimum so that no reaction occurs between carbon and oxygen during solidification.

Mannesmann process, A process used for piercing tube billets in making seamless tubing. The billet is rotated between two heavy rolls mounted at an angle, and is forced over a fixed mandrel.

martempering, The process of quenching an austenitized ferrous alloy in a medium at a temperature in the upper portion of the temperature range of martensite formation, or slightly above that range, and holding in the medium until the temperature throughout the alloy is substantially uniform. The alloy is then allowed to cool in air through the temperature range of martensite formation.

mechanical working, Subjecting metal to pressure exerted by rolls, dies, presses, or hammers, to change its form or to affect the structure and consequently the mechanical and physical properties.

natural aging, Spontaneous aging of a supersaturated solid solution at room temperature.

necking down, Reduction in area concentrated at the subsequent fracture when a ductile metal is tested in tension.

normalizing, A process in which a ferrous alloy is heated to a suitable temperature above the transformation range and is subsequently cooled in still air at room temperature.

overaging, Aging under conditions of time and temperature greater than those required to obtain maximum strength.

overbending, Allowance for spring-back when bending metal to a desired angle.

overheated, A term applied when, after exposure to an excessively high temperature, a metal develops an undesirable coarse grain structure but is not permanently damaged.

overstressing, Permanently deforming a metal by subjecting it to stresses that exceed the elastic limit.

pearlite, The lamellar aggregate of ferrite and carbide. *Note:* It is recommended that this word be reserved for the microstructures consisting of thin plates or lamellae — that is, those that may have a pearly luster in white light. The lamellae can be very thin and resolvable only with the best microscopic equipment and technique.

plastic deformation, Permanent distortion of a material under the action of applied stresses.

press forging, The forging process in which metal stock is formed between dies, usually by hydraulic pressure. Press forging is an operation that employs a single, slow stroke.

punch, The moveable part that forces the metal into the die in equipment for sheet drawing, blanking, coining, embossing and the like.

punching, Shearing holes in sheet metal with punch and die.

punch radius, The radius on the bottom end of the punch, over which the sheet is bent in deep drawing. This radius is independent of the shape of the punch.

push bench, Equipment used for drawing heavy-gauge tubes by cupping sheet and forcing it through a die by pressure against the bottom of the cut.

quench hardening, A process of hardening a ferrous alloy of suitable composition by heating within or above the transformation range and cooling at a rate sufficient to increase the hardness substantially.

quenching, A process of rapid cooling from an elevated temperature by contact with liquids, gases or solids.

recrystallization, A process whereby the distorted grain structure of cold worked metals is replaced by a new, strain-free grain structure during annealing above a specific minimum temperature.

red shortness, Brittleness in steel when it is red hot.

restriking, Cold striking of a finished forging to remove distortions after it has been quenched.

roller straightening, A process involving a series of staggered rolls of small diameter, between which rod, tubing and shapes are passed for the purpose of straightening. The process consists of a series of bending operations.

roll forming, (1) An operation used in forming sheet. Strips of sheet are passed between rolls of definite settings that bend the sheet progressively into structural members of various contours, sometimes called "molded sections." (2) A process of coiling sheet into open cylinders.

roll threading, Applying a thread to a bolt or screw by rolling the piece between two grooved die plates, one of which is in motion, or between rotating grooved circular rolls.

rotary shear, A cutting machine with sharpened circular blades or disk-like cutters used for trimming edges and slitting sheet and foil.

rotary swager, A machine used for pointing wire rod and tubing and for initial working of moderately brittle metals and alloys in which a pair of tapered dies is used. The dies are rotated and are pressed intermittently against the work material by an eccentric action.

scaling, Surface oxidation caused on metals by heating in air or in other oxidizing atmospheres.

shaving, Trimming heavy-gauge blanks to remove uneven sheared edges.

shear, (1) A type of cutting operation in which the metal object (sheet, wire, rod, or such) is cut by means of a moving blade and fixed edge or by a pair of moving blades that may be either flat or curved. (2) A type of deformation in which parallel planes in metal crystals slide so as to retain their parallel relation to one another, resulting in block movement.

single-acting hammer, A forging hammer in which the head is raised by a steam cylinder and piston and the blow is delivered by the free fall of the head.

single-action press, A forming press that operates with a single function, such as moving a punch into a die with no simultaneous action for holding down the blank or ejecting the formed work.

sizing, A final pressing of a sintered compact to secure the desired size.

skelp, A plate of steel or wrought iron from which pipe or tubing is made by rolling the skelp into shape longitudinally and welding the edges together.

spinning, The procedure of making sheet metal disks into hollow shapes by pressing a tool against a rotating form (spinning chuck).

spring-back, An indicator of elastic stresses, frequently measured as the increase in diameter of a curved strip after removing it from the mandrel about which it was held. The measurement is employed as an indicator of the extent of recovery or relief of residual stresses that has been achieved by the transformation of elastic strain to plastic strain during heating or stress relieving.

stamping, A process used to cut lines of letters, figures and decorations on smooth metal surfaces. The impact of a punch with comparatively sharp projecting outlines impresses the characters into the surface of the metal.

strain hardening, An increase in hardness and strength caused by plastic deformation at temperatures lower than the recrystallization range.

stress relieving, A process of reducing residual stresses in a metal object by heating to a suitable temperature and holding for a sufficient time. This treatment may be applied to relieve stresses induced by casting, quenching, normalizing, machining, cold working, or welding.

stretcher straightening, A process for straightening rod, tubing and shapes by the application of tension at the ends of the stock. The products are elongated a definite amount to remove warpage.

stretch forming, A process of forming panels and cowls of large curvature by stretching sheet over a form of the desired shape. This method is more rapid than hammering and beating.

stripper punch, A punch that serves as top or bottom of the die cavity and later moves further into the die to eject the compact.

temper, A condition produced in a metal or alloy by mechanical or thermal treatment and having characteristic structure and mechanical properties.

temper brittleness, Brittleness that results when certain steels are held within, or are cooled slowly through a certain range of temperature below the transformation range. The brittleness is revealed by notched-bar impact tests at room temperature of lower temperatures.

tempering, A process of reheating quench-hardened or normalized steel to a temperature below the transformation range, and then cooling at any rate desired.

upsetting, (1) A metal working operation similar to forging. (2) The process of axial flow under axial compression of metal, as in forming heads on rivets by flattening the end of wire.

wire rod, A semifinished product from which wire is made. It is generally of circular cross section approximately ¼ in. in diameter.

work hardness, Hardness developed in metal as a result of cold working.

METAL MACHINING

allowance, The least intentional difference or the maximum interference between two mating parts.

axial rake angle, Applies to the angular (not helical or spiral) cutting faces: (1) The angle of a plane containing the cutting face, or tangent to the cutting face at a given point, and the reamer axis. (2) The angle between the tool face, on angular teeth, perpendicular to a tangent between the cutting edge and the cutter axis, measured in an axial direction.

axial relief angle, The angle between the relieved surface, or flank of the tooth, and a plane perpendicular to a line between the axis of the cutter and the cutting edge, measured in an axial direction.

back rake angle, The angle between the face of a tool and a line parallel to the base of the tool or holder, measured in a plane parallel to the centerline of the point and at right angles to the base. The angle is positive if the tool face slopes toward the tool or holder base, in a direction away from the point, and is negative if the contrary is true.

basic size, The theoretically perfect size from which variation is permitted.

bilateral tolerance, The form in which variation is permitted in both directions from the basic size. (See tolerance.)

boring, The production of internal surfaces.

brazed shank tool, One in which the cutting tool is a different material than the shank to which it has been brazed.

broaches, (1) Back off angle — the angle between the tooth flank and a plane parallel to the tool axis tangent to the cutting edge. (2) Pace angle — the angle between the tool face and a plane perpendicular to the tool axis and tangent to the cutting edge.

built-up edge, The welding of work material to the face of the cutting tool.

centerless internal grinding, The generation of finished internal surfaces by rotating the workpiece between two rollers and the regulating wheel, with a grinding wheel mounted on a shaft operating inside the cylindrical section.

center mounted grinding, The generation of finished external cylindrical surfaces on work-pieces mounted between centers.

chamfer angle, The angle between the reamer axis and the cutting edge of the chamfer measured in an axial plane at the cutting edge.

chamfer relief angle, The axial relief angle at the outer corner of the chamfer, measured by projection into a plane tangent to the periphery at the outer corner of the chamfer.

chipbreaker, A means of providing an assistance to the operation that will result in breaking the chip into small pieces.

chisel edge angle, The angle included between the chisel edge and the cutting edge in a plane parallel to the end of the drill.

chucking internal grinding, The generation of finished internal surfaces on workpieces mounted in chucking devices that provide rotation of the workpiece.

clearance fit, A fit between mating parts having limits of size so prescribed that a clearance always results in assembly.

common broaching operations, See surface broaching, internal broaching.

common drilling operations, See drilling, reaming, counterboring, countersinking, tapping, and fly cutting.

common grinding operations, See tool and cutter grinding, center mounted grinding, through feed centerless grinding, in-fed centerless grinding, end-feed centerless grinding, centerless internal grinding, chucking internal grinding, vertical surface grinding, horizontal surface grinding, crush form grinding, snap grinding.

crush form grinding, The grinding of irregular cross sections by a grinding wheel whose face has been formed by feeding a crusher roll against the grinding wheel while rotating at a considerably slower speed than those employed in grinding.

cutting force, The force acting normal to the tool base during a cutting operation.

cutting ratio, The ratio of the chip thickness prior to entering the shear zone and its thickness after passing through the shear zone.

dimensions and tolerances, See tolerance, allowance, basic size, limits, standard size, bilateral tolerance, unilateral tolerance, standard hole practice, standard shaft practice, selective assembly, fit, clearance fit, interference fit, transition fit.

drilling, (1) The origination or enlargement of a hole using a fluted end-cutting tool. (2) The origination or enlargement of a hole utilizing the tailstock to provide an axial feed to the tool.

drills, See helix angle, point angle, lip relief angle, chisel edge angle.

economic tool life, The period of time, usually given in minutes, during which the tool performs the function required, at cutting conditons that result in least cost per piece.

electric discharge machining, The removal of material by the high frequency discharge of DC current through a controlled gap between a tool and the workpiece submerged in a dielectric material.

electrochemical machining, The removal of material by a high frequency discharge of DC current through a controlled gap submerged in an electrolyte that assists in the removal of material from the workpiece.

electrolytic grinding, The removal of material by utilizing DC current flow from workpiece to wheel in the presence of electrolyte. The deplating operation results in the bulk of the material removal.

end cutting edge angle, The angle between the trailing edge of the tool and the line formed by the surface generated, measured in a place parallel to the tool or holder base.

end-feed centerless grinding, The generation of finished external surfaces by feeding the workpiece, either manually or mechanically, past the regulating and grinding wheels to an end stop. Used primarily on tapered sections.

end milling, The production of a flat or slotted surface when employing a shank-mounted cutter.

end relief angle, The angle between the end flank and a perpendicular to the tool or holder base.

experimental terms, See flank land wear, crater wear, shear plane, shear angle, cutting ratio, tool-ship interface, tool temperature, cutting force, thrust force, radial force, friction force, tool-chip thermocouple.

facing, The production of a flat surface when the cutting tool moves perpendicular to the axis of rotation of the workpiece.

face milling, The production of a flat or slotted surface when employing a shank-mounted cutter.

fit, The relationship existing between two mating parts with respect to the amount of clearance or interference which is present when they are assembled.

fixture, A device for holding the workpiece.

flank land wear, The abrasion on the flank of the cutting tool, usually measured by the height of the relatively uniform area of wear.

flaws, The irregularities of any sort that occur at relatively infrequent intervals.

fly cutting, The enlargement of a hole utilizing a single point cutting tool.

form milling, The production of a nonlinear surface when employing an arbor-mounted cutter.

friction force, The force acting along the tool face, caused by the chip sliding across the tool.

helical rake angle, Applies to helical and spiral cutting faces only, not angular. The angle between a plane tangent to the cutting face at a given point on the cutting edge, and the reamer axis.

helix angle (drills), The angle between a line tangent to the land of the flute and the axis of rotation at their point of intersection.

helix angle (milling cutters), The angle between a line tangent to a point on the helical tooth and the axis of the cutter measured at their point of intersection.

horizontal surface grinding, The generation of finished flat surfaces by mounting workpieces on a traversing work table below a grinding wheel that rotates about a horizontal axis.

in-fed centerless grinding, The generation of finished external cylindrical surfaces accomplished by closing the distance between the grinding wheel and the regulating wheel, with the width of grinding limited to the width of the grinding wheel.

interference fit, A fit between mating parts having limits of size so prescribed that an interference always results in assembly.

internal broaching, The production of internal surfaces other than the origination of holes by employing an elongated tool of increasing size teeth, that is actuated by pulling or pushing.

jig, A device for holding the workpiece and guiding the tool.

knurling, The production of a pattern-roughened surface by the transverse feed of the tool.

lay, The direction of the predominant surface pattern.

limits, The extreme permissible values of a dimension.

lip relief angle, The angle between a tangent on the surface back of the cutting edge at the periphery and a plane at right angles to the axis of the drill.

machinability, That qualitative property of a work material, under conditions or production, that reflects on the ease or difficulty with which the operation takes place.

macro-machining, processes, See common lathe, drilling, milling, broaching and grinding operations.

maximum production tool life, The period of time, usually given in minutes, during which the tool performs the function required, at cutting conditions that result in the least practical time to process a piece.

microinch, This is the unit of measurement used to describe surface irregularity quantitatively. A microinch is one millionth of an inch.

micro-machining operations, See electric discharge machining, electrochemical machining, electrolytic grinding, ultrasonic machining, plasma-jet machining.

milling cutters, See axial relief angle, radial relief angle, axial rake angle, radial rake angle and helix angle.

nose radius, The radius generated at the intersection of the leading and trailing edges of the tool measured in the plane of the tool face.

parting, The severing of a section from a workpiece by moving the cutting tool perpendicular to the axis of rotation of the workpiece.

plasma-jet machining, The removal of material by directing a stream of electrons, of sufficient energy, at a surface that will result in melting and vaporization of the solid material.

point angle, The angle included between the lips projected upon a plane parallel to the drill axis and parallel to the two cutting lips.

production applications, See machinability, tool life, economic tool life, maximum production tool life, chipbreaker, throwaway insert, solid shank tool, brazed shank tool, jig, fixture, build-up edge.

profile, This describes the contour of any specified cross section of the surface. It discloses waviness, roughness, a combination of both, as well as flaws.

radial force, The force acting on the tool to resist the depth of cut.

radial rake angle, (1) The angle between the cutting face and a line that passes through the tool axis and is tangent to the cutting edge. (2) The angle between the tool face and a tangent to the cutting edge originating at the axis of the cutter, measured in a radial direction.

radial relief angle, The angle between the relieved surface or flank of the tooth, and a plane perpendicular to a line between the axis of the cutter and the cutting edge, measured in a radial direction.

reamers, See relief angle, radial rake angle, axial rake angle, helical rake angle, chamfer angle, chamfer relief angle.

reaming, The slight enlargement of a hole using a multiple point cutting tool.

relief angle, The angle measured in a transverse plane, between the relieved surface and a plane tangent to the periphery at the cutting edge.

roughness-width cutoff, The distance over which the surface irregularities are to be averaged.

selective assembly, The segregation of parts into groups in order to effect particular matings of component parts.

shear angle, The angle inscribed between the shear plane and the newly formed surface.

shear plane, The area over which the chip is removed from the parent material.

side cutting edge angle, The angle between the leading edge of the tool and a line perpendicular to the line formed by the surface generated, measured in a plane parallel to the tool or holder base.

side relief angle, The angle between the side flank and a perpendicular to the tool or holder base.

side rake angle, The angle between the face of a tool and a line parallel to the base of the tool or holder measured in a plane at right angles to the base and at right angles to the centerline of the point. The angle is positive if the tool face slopes toward the tool or holder base in the direction away from the cutting edge, and is negative if the contrary is true.

side milling, The production of two or three perpendicular flat surfaces utilizing a single cutter.

single point tools, See back rake angle, side rake angle, end relief angle, side relief angle, side cutting edge angle, nose radius.

slab milling, The production of a flat surface utilizing an arbor-mounted cutter.

slotting, The production of a straight-sided groove in the workpiece.

snag grinding, The nonprecision removal of unwanted material, such as sprues and gates in foundry operations by means of very high-speed grinding operation.

solid shank tool, One in which the shank and the cutting edge of the tool are one and the same material.

standard hole practice, The method of allocating dimensions and tolerances in which the basic hole size is the nominal size and the minimum hole size. Tolerance and allowance is then applied as determined from the appropriate class of fit.

standard shaft practice, The method of allocating dimensions and tolerances in which the basic shaft size is the nominal size and the maximum shaft size. Tolerance and allowance is then applied as determined from the appropriate class of fit.

standard size, The sizes corresponding to the subdivisions of recognized units of length.

straddle milling, The production of more than one flat surface while employing more than one cutter.

surface, The surface of an object is represented by the boundary line which separates that object from its surroundings.

surface broaching, The production of external surfaces of either straight or curvilinear form by employing an elongated tool of increasing size teeth, that is actuated by pulling or pushing.

surface requirements, See surface roughness, surface roughness width, waviness, flaws, lay, roughness-width cutoff, surface, microinch, profile.

surface roughness, The measure of the deviation from the nominal in the form of the finely spaced irregularities.

surface roughness width, The distance between successive peaks of the finely spaced irregularities that form the roughness pattern.

tapping, The production of screw threads on an internal surface.

throw-away insert, A small piece of cutting tool material prepared for mechanical mounting in a tool holder.

through feed centerless grinding, The generation of finished external cylindrical surfaces by passing the workpiece between the grinding wheel and the regulating wheel, with a longitudinal movement imparted by the regulating wheel axis inclined to the axis of the grinding wheel.

thrust force, The force acting on the tool, resisting the advanced or feed of the tool.

tolerance, The unavoidable variation permitted in a dimension for economical production.

tool and cutter grinding, The generation of finished surfaces on cutting tools.

tool-chip interface, The contact surface between the chip and the tool face during a cutting operation.

tool-chip thermocouple, The thermocouple made of the tool and chip material and utilized as a means of measuring average temperature at points of contact between chip and tool.

tool geometry, See single point tools, drills, reamers, milling cutters, broaches.

tool life, The period of time, usually given in minutes, during which the tool performs the function required.

tool temperature, The average tool-chip interface temperature.

transition fit, A fit between mating parts having limits of size so prescribed as to partially or wholly overlap, so that either a clearance or interference may result in assembly.

turning, The production of a curvilinear surface when the cutting tool moves in direction parallel to the axis of rotation of the workpiece.

ultrasonic machining, The removal of material by high frequency vibration of a tool that imparts energy to abrasive grains introduced between the tool and the workpiece by a slurry. The abrasive crystals then impart hammer blows to the workpiece in removing minute particles of unwanted material.

unilateral tolerance, The form in which variation is permitted in only one direction from the basic size. (See tolerance. bilateral tolerance.)

vertical surface grinding, The generation of finished flat surfaces by mounting workpieces on a slowly rotating table beneath a grinding wheel that rotates about a vertical axis.

waviness, This consists of recurrent widespread irregularities in the form of waves on which is superimposed the roughness.

PLASTICS

A-stage, An early stage in the reaction of certain thermosetting resins, in which the material is still soluble in certain liquids and fusible. Sometimes referred to as resol. (See B-stage, C-stage.)

absorption, (1) The penetration into the mass of one substance by another. (2) The process whereby energy is dissipated within a specimen placed in a field of radiation energy. Since processes other than absorption occur, e.g., scattering, in which only a fraction of the energy removed from a beam is retained in the specimen, the total amount of energy removed from a well-collimated beam will be greater than the amount actually dissipated within the sample.

acid-acceptor, A compound which acts as a stabilizer by chemically combining with acid which may be initially present in minute quantities in a plastic, or which may be formed by the decomposition of the resin.

acrylic plastics, Plastics based on resins made by the polymerization of acrylic monomers, such as ethyl acrylate and methyl methacrylate.

activation, The process of inducing radioactivity in a specimen by bombardment with neutrons or other types of radiation.

adherend, A body which is held to another body by an adhesive.

adhesion, The state in which two surfaces are held together by interfacial forces which may consist of valence forces or interlocking action or both. (See adhesion, mechanical; adhesion, specific.)

adhesion, mechanical, Adhesion between surfaces in which the adhesive holds the parts together by interlocking action. (See adhesion, specific.)

adhesion, specific, Adhesion between surfaces which are held together by valence forces of the same type as those which give rise to cohesion. (See adhesion, mechanical.)

adhesive, A substance capable of holding materials together by surface attachment. Adhesive is the general term and includes among others cement, glue, mucilage, and paste. All of these terms are loosely used interchangeably. Various descriptive adjectives are applied to the term adhesive to indicate certain characteristics, as follows:

physical form, e.g., liquid adhesive, tape adhesive
chemical type, e.g., silicate adhesive, resin adhesive
materials bonded, e.g., paper adhesive, metal-plastic adhesive, can-label adhesive
conditions for use, e.g., hot-setting adhesive.

adhesive, assembly, An adhesive which can be used by bonding parts together, such as in the manufacture of a boat, airplane, furniture, and the like. The term assembly adhesive is commonly used in the wood industry to distinguish such adhesives (formerly called joint glues) from those used in making plywood (sometimes called veneer glues). It is applied to adhesives used in fabricating finished structures of goods, or subassemblies thereof as differentiated from adhesives used in the production of sheet materials for sale as such, for example, plywood or laminates.

adsorption, A concentration of a substance at a surface or interface of another substance.

aging, (1) The effect of exposure of plastics to an environment for an interval of time. (2) The process of exposing plastics to an environment for an interval of time. Aging, according to the first definition, may be very great or not, depending on the environmental conditions, such as radiation, moisture, temperature, surrounding gases, etc., and the period of exposure.

air vent, Small outlet, usually a groove, to prevent entrapment of gases.

alkyd plastics, Plastics based on resins composed principally of saturated polymeric esters, in which the recurring ester groups are an integral part of the main polymer chain, and in which ester groups occur in any cross-links that may be present between chains. (See polyester plastics.)

all-veneer construction, Plywood without lumber cores, more frequently multiple for strength requirements, often 7-ply or 9-ply, to equal the thickness of conventional member-core plywood. The maximum thickness of any single sheet of veneer seldom exceeds ¼ in.

allyl plastics, Plastics based on resins made by additional polymerization of monomers containing allyl groups, such as diallyl phthalate.

ambient temperature, Temperature of the medium surrounding an object.

amino plastics, Plastics based on resins made by the condensation of amines such as urea and melamine, with aldehydes.

anchorage, Part of the insert that is molded inside of the plastic and held fast by the shrinkage of the plastic.

anneal, (1) To heat a molded plastic article to a predetermined temperature and slowly cooled to relieve stresses. (2) To heat steel to a predetermined temperature above the critical range and slowly cool it, to relieve stresses and reduce hardness.

assembly, The collection of and placing together in proper order of the layers of veneer, lumber and/or other materials, with the adhesive, ready to be pressed and bonded into a single unit.

assembly time, In adhesives, refers to the elapsed time after the adhesive is spread and until the pressure becomes effective.

atactic, A random arrangement of the unsymmetrical group with respect to the carbon-carbon backbone plane.

autoclave, A closed vessel for conducting a chemical reaction or other operation-made pressure.

B-stage, An intermediate stage in the reaction of certain thermosetting resins, in which the material swells when in contact with certain liquids and softens when heated, but may not entirely dissolve or fuse. The resin in an uncured thermosetting molding compound is usually in this stage. Sometimes referred to as resitol. (See A-stage, C-stage.)

back-pressure-relief port, An opening from an extrusion die for escape of excess material.

back taper, Reverse draft used in mold to prevent molded article from drawing freely. (See undercut.)

backing plate, A plate which backs up the cavity blocks, guide pins, bushings, etc. (Sometimes called support plate.)

balanced construction, Plywood which has an odd number of plies, and is symmetrical on both sides of its center line.

bale; bundle, A bundle or package of freshly glued (cold-pressed) plywood, held together by clamp irons (q.v.) after removal from the cold press. The bale is kept under pressure until initial adhesion is accomplished.

bastard-sawn, Lumber or veneer, usually hardwood, in which the annual growth rings make angles of 30 to 60 degrees with the surface of the piece.

binder, A component of an adhesive composition which is primarily responsible for the adhesive forces which hold two bodies together. (See extender, filler.)

blanket, Veneers which have all been laid up on a flat table. The complete assembly is placed on or in the mold all at one time; useful only on simple curved surfaces to be molded by the flexible-bag process. Also used to denote a form of bag made of rubber in which the edges are sealed against the mold by clamps.

bleeding, (1) The flow of color from one region into an adjoining region. (2) In the manufacture of plywood, the escape of a portion of the steam-air mixture during cooking, to permit mixing of the steam and air and to maintain uniform temperature at all levels in the autoclave, when molding with flexible pressure.

blind hole, Hole that is not drilled entirely through.

blind-hole partial thread, Thread counterbored from the front for terminal or other assembly fit.

blister, (1) Undesirable rounded elevation of the surface of a plastic, whose boundaries may be indefinitely outlined, somewhat resembling in shape a blister on the human skin. A blister may burst and become flattened. (2) An elevation of the surface of an adherend, somewhat resembling in shape a blister on the human skin; its boundaries may be indefinitely outlined and it may have burst and become flattened. A blister consists of trapped air, water or solvent, and can be caused by insufficient adhesive, inadequate curing time, temperature, or pressure.

blocking, An adhesion between touching layers of plastics, such as that which may develop under pressure during storage or use.

bloom, A visible exudation or efflorescence on the surface of a plastic. Bloom can be caused by lubricant, plasticizer, etc.

bolster, Spacer or filler in a mold.

bond, The attachment at an interface between an adhesive and an adherend. (See joint.)

bond, To attach materials together by adhesives.

bubble, Internal void or a trapped globule of air or other gas.

bulk density, The density of a molding material in loose form (granular, nodular, etc.) expressed as a ratio of weight to volume (e.g., g/cm^3 or lb/ft^3).

bulk factor, The ratio of the volume of the loose molding compound to the volume of the same quantity in a molded solid piece. Also, the ratio of the density of the solid plastic object to the apparent density of the loose molding powder.

burned, Showing evidence of thermal decomposition through some discoloration, distortion, or destruction of the surface of the plastic.

C-stage, The final stage in the reaction of certain thermosetting resins, in which the material is relatively insoluable and infusible. The resin in a fully cured thermosetting molding is in this stage. Sometimes referred to as resite. (See A-stage, B-stage.)

carburize, To increase carbon content of surface of any steel.

case harden, To harden surface of a piece of steel to a relatively shallow depth.

cast film, A film made by depositing a layer of liquid plastic onto a surface and stabilizing this form by evaporation of solvent, by fusing after deposition, or by allowing a melt to cool. Cast films are usually made from solutions or dispersions.

catalyst, A substance which when added in minor proportion accelerates a chemical reaction.

caul, In plywood manufacture, a sheet the size of the platens used in hot pressing. Aluminum sheet approximately 1/16 in. thick is generally used. Plywood assemblies are inserted between pairs of cauls, to facilitate loading the press, and to protect plywood faces from contact with the steel plates of the hot press. At one time plywood cauls. 1/16 to 1/4 in. thick were used; they were replaced by aluminum for quicker heating and better durability.

caul, canvas-covered plywood, A special type of plywood caul used in progressive gluing, to control heat and moisture. Also used for two-plying fragile veneer faces in hot presses.

caul, plywood, Used in cold-pressing with conventional adhesives, to assure undamaged faces and to prevent transmission of defects to adjacent assemblies. Usually ¼ to ⅜ in. thick, with waxed surfaces to prevent adhesion.

cavity, That portion of the mold which forms the outer surface of the molded article.

cavity-retainer plate, See chase.

cell, A single cavity formed by gaseous displacement in a plastic material. (See cellular plastic.)

cellular plastic, A plastic whose density is decreased substantially by the presence of numerous cells disposed throughout its mass. (See cell, foamed plastic.)

cellular striation, A layer of cells within a cellular plastic which differs greatly from the characteristic cell structure of the material.

cellulosic plastics, Plastics based on cellulose compounds, such as esters (cellulose acetate) and ethers (ethyl cellulose).

cement, See adhesive, bond.

centipoise, A unit of viscosity, conveniently and approximately defined as the viscosity of water at room temperature. The following table of approximate viscosities at room temperature may be useful for rough comparisons:

liquid	viscosity in centipoises
water	1
kerosene	10
motor oil SAE 10	100
castor oil; glycerine	1,000

corn syrup	10,000
molasses	100,000

Note: Under SI, centipoise has been replaced by pascal-second (Pa-s) where 1 centipoise = 0.001 Pa-s).

chalking-dry, Chalk-like appearance or deposit on the surface of a plastic. (See haze, bloom.)

channel, See port.

chase, The main body of the mold which contains the molding cavity or cavities, or cores, the mold pins, the guide pins (or the bushings), etc.

chase ring, A ring used in hobbing to restrain the blank against spreading during the sinking of the hob.

chemically foamed plastic, A cellular plastic whose structure is produced by gases generated from the chemical interaction of its constituents.

clamp irons, In plywood manufacture, the pressure-maintenance equipment, which includes the "I" beams or double channel irons, together with clamp screws or turnbuckle rods, to hold bales under pressure after cold gluing.

clamping plate, A mold plate fitted to the mold and used to fasten the mold to the machine.

closed-cell foam, A cellular plastic in which there is a predominance of non-interconnecting cells.

cohesion, The forces holding a single substance together.

cold flow, See creep.

cold molding, The shaping of an unheated compound in a mold under pressure, followed by heating the article to cure it.

cold pressing, A bonding operation in which an assembly is subjected to pressure without the application of heat.

cold slug, The first material to enter an injection mold; so called because in passing through the sprue orifice it is cooled below effective molding temperature.

cold-slug well, Space provided directly opposite the sprue opening of the injection mold to trap the cold slug.

compregnated wood, A consolidation of the term compressed-impregnated wood, referring usually to an assembly of layers of veneer impregnated with a liquid resin and bonded under very high pressures. More commonly, but not always, the veneer layers have parallel grain, i.e., laminated wood construction.

condensation, A chemical reaction in which two or more molecules combine, with the separation of water or some other simple substance. If a polymer is formed, the process is called polycondensation. (See polymerization.)

consistency, The resistance of a material to flow or permanent deformation when shearing stresses are applied to it. The term is generally used with materials whose deformations are not proportional to applied stresses. Viscosity is generally considered to be a similar internal friction that results in flow in proportion to the stress applied. (See viscosity, viscosity coefficient.)

cooling fixture, Equipment used to hold molded article while cooling after removal from the mold; it may be channeled to be cooled with water.

copolymer, See polymer.

copolymerization, See polymerization.

core, That portion of the mold that forms the required inner surfaces of the molded article.

cores, centers, A term usually applied to the central layer of plywood, which in lumber-core construction is the principal strength factor. It is also applied when the central layer is a veneer. The terms are sometimes used in the Pacific Northwest to designate the layer that is spread with adhesive, which agrees with the above in 3-ply construction but is inconsistent when applied to 5-ply. Under this latter usage, center is used to indicate the middle ply. Core may refer also to the remaining part of the log that is too small to be cut into rotary veneer on a lathe.

core and separator, The center section of an extrusion die.

core pin, Pin used to mold a hole.

core-pin plate, Plate holding core pins.

core-retainer plate, See chase.

crazing, Development of fine cracks on the surface of a plastic, sometimes extending into the body of the material.

creep, The dimensional change with time of a plastic under load, following the instantaneous elastic or rapid deformation. It is the permanent deformation resulting from prolonged application of a stress below the elastic limit. Creep at room temperature is sometimes referred to as cold flow.

crossbanding, The transverse veneer layers that distinguish plywood from laminated wood. Their presence counteracts the tendency of wood to split, as well as to shrink and swell. In standard 5-ply construction, it is the layer between the face and the core and between the back and the core, sometimes called face crossing and back crossing, respectively.

cross-linking, The development of chemical linkages between long-chain molesters.

cull, Material remaining in the transfer pot after the mold has been filled.

cure, To change the properties of a plastic by chemical reaction, which may be condensation, polymerization, or addition; usually accomplished by the action of heat or catalyst or both, with or without pressure.

dead fold, A fold which does not spontaneously unfold.

degradation, A deleterious change in the chemical structure of a plastic. (See deterioration.)

delamination, The separation of the layers in a laminate.

deterioration, A permanent change in the physical properties of a plastic evidenced by impairment of these properties.

diaphragm gate, Gate used in molding annular or tubular articles.

die-adaptor, That part of an extrusion die which holds the die block.

die block, That part of an extrusion die which holds the forming bushing and core.

die body, That part of an extrusion die used to separate and form material.

dielectric heating, See high-frequency heating.

diffusion, The movement of a material in the body of a plastic.

dilatant, A dilatant fluid, or inverted pseudoplastic, is one whose apparent viscosity increases instantaneously with increasing rate of shear; i.e., the act of stirring creates instantly an increase in resistance to stirring.

diluent, In an organosol, a liquid component which has little or no solvating action on the resin, its purpose being to modify the action of the dispersant.

disc gate, See diaphragm gate.

dished, Showing a symmetrical distortion of a flat or curved section of a plastic object, so that, as normally viewed, it appears concave, or more concave than intended. (See warp.)

dispersant, In an organosol, a liquid component which has a solvating or peptizing action on the resins, so as to aid in dispersing and suspending it.

dispersion, A heterogeneous system in which a finely divided material is distributed in a matrix of another material. In plastics technology, a dispersion is usually the distribution of a finely divided solid in a liquid or a solid. For example, (1) pigments or fillers in molded plastics, (2) plastisols, or (3) organosols. (See plastisol, organosol.)

doctor roll; doctor bar, A device for regulating the amount of liquid material on the rollers of a spreader.

domed, Showing a symmetrical distortion of a flat or curved section of a plastic object, so that, as normally viewed, it appears more convex than intended. (See warp.)

doping, Coating the mold or mandrel with a substance which will prevent the molded plywood part from sticking to it, and will facilitate easy removal.

dowel, Pin used to maintain alignment between two or more parts of a mold.

draft, The degree of the taper of a side wall or the angle of clearance, designed to facilitate removal of molded article from the mold.

dry spot, Area of incomplete surface film on laminated plastics; in laminated glass, an area over which the interlayer and the glass are not bonded.

duplicate cavity-plate, Removable plate that retains cavities, used where two-plate operation is necessary for loading inserts, etc.

durometer hardness, The hardness of a material as measured by the Shore Durometer.

ejector pin, Pins used to eject the finished article from the mold.

ejector-pin-retainer plate, Retainer into which ejector pins are assembled.

ejector plate, A plate which backs up the ejector pins and holds the ejector assembly together.

ejector rod, A rod which actuates the ejector assembly when the mold is opened.

elasticity, That property of a material by virtue of which it tends to recover its original size and shape after deformation. If the strain is proportional to the applied stress, the material is said to exhibit Hookean or ideal elasticity.

elastomer, A material which at room temperature can be stretched repeatedly to at least twice its original length and, upon immediate release of the stress, will return with force to its approximate original length.

envenomation, The process by which the surface of a plastic close to or in contact with another surface is deteriorated. Softening, discoloration, mottling, crazing, or other effects may occur.

epoxy plastics, Plastics based on resins made by the reaction of epoxides or oxiranes with other materials such as amines, alcohols, phenols, carboxylic acids, acid anhydrides, and unsaturated compounds.

ethylene plastics, Plastics based on resins made by the polymerization of ethylene or copolymerization of ethylene with other unsaturated compounds.

exotherm; exothermic heat, Heat generated by a chemical reaction; in polymers a particular problem since the dissipated heat must be dissipated.

expandable plastic, A plastic which can be made cellular by thermal, chemical, or mechanical means.

expanded plastic, See cellular plastic.

extender, A substance, generally having some adhesive action, added to an adhesive to reduce the amount of the primary binder required per unit area. (See binder, filler.)

extraction, The transfer of a material from a plastic to a liquid in contact with it.

extrusion, A method whereby heated or unheated plastic forced through a shaping orifice becomes one continuously formed piece.

extrusion, autothermal, A method of extrusion in which the conversion of the drive energy, through friction, is the sole source of heat.

fabricating; fabrication, The manufacture of plastic products form molded parts, rods, tubes, sheeting, extrusions, or other forms by appropriate operations such as punching, cutting, drilling, and taping. Fabrication includes fastening plastic parts togedther or to other parts by mechanical devices, adhesives, heat-sealing, or other means.

face, In plywood manufacture, the veneer on the exposed surface of the plywood.

false body, The deceptively high apparent viscosity of a pseudoplastic fluid at a low rate of shear.

family mold, A multicavity mold containing individual cavities of different sizes or designs.

filler, A relatively inert material added to a plastic or an adhesive to modify its strength, permanence, working properties or other qualities, or to lower its cost. (See reinforced plastics binder, extender.)

filler-specks, Visible specks of a filler, such as woodflour or asbestos, which stand out in color contrast against the background.

film, (1) Sheeting having a nominal thickness not greater than 0.010 in. (2) In adhesives, a thin, dry sheet of paper impregnated with a phenol-formaldehyde resin adhesive. Film sometimes is used to refer to a liquid coating of adhesive.

fin, That part of the flash which remains attached to the molded article.

finishing, The removal of defects or the development of desired surface characteristics on plastic products.

fisheye, Small globular mass which has not blended completely into the surrounding material; particularly evident in a transparent or translucent material.

flash, That portion of the charge which flows from or is extruded from the mold cavity during the molding.

flash groove, A groove ground in a form to allow the escape of excess material.

flash ridge, That part of a flash mold along which the excess material escapes until the mold is closed.

flat grain, Grain produced in approximately a tangential direction, or plain-cs veneers.

floating chase, Mold member, free to move vertically, which fits over a lower plug or cavity, and into which an upper plug telescopes.

flow mark, A visible mark resulting from the solidification of a pattern of flow.

fluxing temperature, See fusion temperature.

foam, See cellular plastic.

foamed plastic, See cellular plastic.

force; force plug, The force is the member which transmits the pressure from the press unit to the top of the molding charge. In positive molds, there is an extension to or protuberance beneath the force which telescopes within the molding cavity (or loading chamber) and which is termed a force plug (compression molding). (When used in reference to injection molding, see core.)

force plate, Plate for holding force plug or plugs in compression molding.

forming, A process in which the shape of plastic pieces such as sheets, rods, or tubes, is changed to a desired configuration. The use of the term forming in plastics technology does not include such operations as molding, casting, or extrusion, in which shapes or articles are made from molding materials or liquids.

furane plastics, Plastics based on resins in which the furane ring is an integral part of the polymer chain, made by the polymerization or polycondensation of furfual, furfuryl alcohol, or other compounds containing a furane ring, or by the reaction of these furane compounds with not more than an equal weight of other compounds. Furane is also spelled furan.

fusion, In vinyl dispersions, the heating of a dispersion to produce a homogeneous mixture. There is an apparent mutual solvation of the resin and plasticizer.

fusion temperature, In vinyl dispersions the temperature at which fusion occurs; also called fluxing temperature.

gamma transition; glassy transition, The change in an amorphous polymer or in amorphous regions of a partially crystalline polymer from (or to) a viscous or rubbery condition to (or from) a hard and relatively brittle one. This transition generally occurs over a relatively narrow temperature region and is similar to the solidification of a liquid to a glassy state; it is not like a first order phase transformation where discontinuities occur. Not only do hardness and brittleness undergo rapid changes in this temperature region, but other properties such as thermal expansibility and specific heat also change rapidly. This phenomenon has been called second-order transition, glass transition, rubber transition, and rubbery transition. The word transformation has also been used instead of transition.

gamma-transition temperature; glassy-transition temperature, The temperature region in which the gamma or glassy transition occurs. The measured value of gamma- or glassy-transition temperature depends to some extent on the details of the method of test. (See gamma or glassy transition.)

gate, The short, usually restricted section of the runner at the entrance to the cavity of an injection or transfer mold.

gel, (1) A semisolid system consisting of a network of solid aggregates, in which liquid is held. (2) The initial jelly-like solid phase which develops during the formation of a resin from a liquid. Both types of gel have very low strengths and do not flow like a liquid. They are soft and flexible, and will rupture under their own weight unless supported externally.

gel point, The stage at which a liquid begins to exhibit pseudo-elastic properties. This stage may be conveniently observed from the inflection point on a viscosity-time plot. (See gel.)

gelation, (1) Formation of a gel. (2) In vinyl dispersions, formation of gel in the early stages of fusion.

glass, An inorganic product of fusion which has cooled to a rigid conditon without crystallizing. Glass is typically hard and relatively brittle, and has a conchoidal fracture.

glass finish, A material applied to the surface of a glass reinforcement to improve its effect upon the physical properties of the reinforced plastic.

glassy transition, See gamma transition; glassy transition.

glue, Originally, a hard gelatin obtained from hides, tendons, cartilage, bones, etc., of animals. Also an adhesive prepared from this substance by heating with water. Through general use the term is now synonymous with the term adhesive. (See adhesive, sizing.)

glue, See bond.

glue, assembly, See adhesive, assembly.

glue joint, That part of an aggregated product which comprises the adhesive (or glue) and the parts in contact therewith.

granular structure, Apparent incomplete fusion of and at least partial retention of their original form by the particles from which a plastic is formed.

grinding-type resin, A vinyl resin which requires grinding to effect dispersion in plastisols or organosols.

guide pin, Pin which assures alignment of the mold halves.

guide-pin bushing, A guiding bushing through which the leader pin moves.

grid, Channel shaped mold-supporting members.

gum, Any of a class of colloidal substances, exuded by or prepared from plants sticky when moist, composed of complex carbohydrates and organic acids, which are soluble or swell in water. The term gum is sometimes used loosely to denote various materials that exhibit gummy characteristics under certain conditions, for example, gum balata, gum benzion, and gum asphaltum. Gums are included by some in the category of natural resins. (See adhesive, glue, resin.)

halocarbon plastics, Plastics based on resins made by the polymerization of monomers composed only of carbon and a halogen or halogens.

hardener, A substance or mixture of substances added to an adhesive to promote or control the curing reaction by taking part in it. The term is also used to designate a substance added to control the degree of hardness of the cured film. (See catalyst.)

haze, Indefinite cloudy appearance within or on the surface of a plastic, not describable by the terms chalking, bloom.

head block; retainer board, A thick (3 to 5 in.) large piece of laminated lumber, usually with veneer crossings, used for bottom and top of a bale of plywood, during pressing and clamping.

heat forming, See thermoforming.

heat-treat, Term used to cover annealing, hardening, tempering, etc., of metals.

heater-adapter, That part of an extrusion die around which heating medium is held.

high-density plywood, Plywood of special construction, made at high specific pressure, usually 500 psi and up. With the increase in pressure comes a corresponding increase in density, or specific gravity.

high-frequency heating, The heating of materials by dielectric loss in a high-frequency electrostatic field. The material is exposed between electrodes, and by absorption of energy from the electrical field is heated quickly and uniformly throughout.

hob, A master model used to sink the shape of a mold into a soft steel block.

hobbing, A process of forming a mold by forcing a hob of the shape desired into a soft steel blank.

hold-down groove, A small groove cut into the side wall of the molding surface to assist in holding the molded article in that member while the mold opens.

holder block, See chase.

hot-short, Inelastic, nonstretchable and easily broken by tension when hot.

hub, See hob.

hubbing, See hobbing.

hydrocarbon plastics, Plastics based on resins composed of carbon and hydrogen alone.

inhibitor, A substance which prevents or retards a chemical reaction.

insert, An object molded or cast into a plastic article for a definite purpose.

insert, eyelet-type, Insert having a section which protrudes from the material and is used for spinning over in assembly.

insert, floating of, When pressure is applied on the mold and the material softens, it flows upwards. If the insert is loose on the retaining pin, some of the material flows under the insert or into the anchorage points and carries or floats the insert off the retaining pin.

insert, flow of material into, If a threaded open-hole insert is being molded, the retaining pin does not prevent the material from flowing into the insert unless the retaining pin is threaded, which would increase the cost of molding.

insert, open-hole, One having a hold drilled completely through it.

insert, protruding, One having a part which protrudes from the molded material.

insert, rivet, One having a protruding part which is riveted in assembly.

insert, through-type, One which is exposed on both sides of the molded article.

inventory, In injection molding or extrusion, the amount of plastic contained in the heating cylinder or barrel.

irradiation, The subjection of a material to radiant energy for the purpose of producing a desired effect or of determining the effect of the radiant energy on the material

isocyanate plastics, Plastics based on resins made by the condensation of organic isocyanates with other compounds. (See urethane plastics.)

isotactic, Pertaining to a type of polymeric molecular structure containing sequences of regularly spaced asymmetric atoms arranged in like configuration in the main polymer chain. Materials containing isotactic molecules may exist in highly crystalline form, because of the high degree of order that may be imparted to such structure.

joint, The location at which two adherends are held together with a layer of adhesive. (See bond.)

joint, butt, A type of edge joint in which the edge faces of the two adherends are at right angles to the other faces of the adherends.

joint, edge, A joint made by bonding the edge faces of two adherends with adhesives.

joint, lap, A joint made by placing one adherend partly over another and bonding together the overlapped portions.

joint, scarf, A joint made by cutting away similar angular segments of two adherends and bonding the adherends with the cut areas fitted together.

joint, starved, A joint which has an insufficient amount of adhesive to produce a satisfactory bond. This condition may result from too thin a spread to fill the gap between the adherends, excessive penetration of the adhesive into the adherend, too short an assembly time, or the use of excessive pressure.

knit line, See weld line.

knockout, Any part or mechanism of a mold used to eject the molded article.

knockout bar, A bar or plate in a knockout frame used to back up a row or rows of knockout pins.

knockout pin, See ejector pin.

knockout-pin plate, See ejector plate.

laminate, A product made by bonding together two or more layers of material or materials.

laminated, cross, Pertaining to a laminate in which some of the layers of materials are oriented at right angles to the remaining layers with respect to the grain or strongest direction in tension. (See laminated, parallel.) Balanced construction of the laminations about the center line of the thickness of the laminate is normally assumed.

laminated, parallel, Pertaining to a laminate in which all the layers of material are oriented approximately parallel with respect to the grain or strongest direction in tension.

laminated wood, An assembly of wood layers in which the wood grain or the fillets of the adjacent layers are parallel, contrasted with plywood, which is characterized by cross layers or crossings, usually alternated with the parallel face, core and back layers.

land, (1) The portion of a mold which provides the separation or cutoff of the flash from the molded article. (2) In the screw of an extruder, the bearing surface along the top of the flights. (3) In an extrusion die, the surface parallel to the flow of material.

landed force, Force with shoulder which seats on land in landed positive mold.

landed plunger, See landed force.

latch, Device used to hold two members of a mold together.

leader pin, See guide pin.

leader-pin bushing, See guide-pin bushing.

lignin plastics, Plastics based on resins made by the treatment of lignin with heat or by reaction with chemicals or with not more than an equal weight of synthetic resins.

loading board, A device for holding preforms spaced to correspond with the positions of multiple cavities of a compression mold and for dropping the preforms simultaneously into the cavities.

loading space, Space provided in a compression mold or in the pot used with a transfer mold to accommodate the molding material before it is compressed.

locating ring, A ring which serves to align the nozzle of an injection cylinder with the entrance of the sprue bushing and the mold to the machine platen.

lumber-core construction, A type of plywood in which the center layer or core is lumber, usually edge-glued together from narrow strips and seldom less than ⅜ in. thick.

lyophilic, In vinyl dispersions, having affinity for the dispersing medium.

lyophobic, In vinyl dispersions, not having affinity for the dispersing medium.

mechanically foamed plastic, A cellular plastic whose structure is produced by physically incorporated gases.

melamine plastics, Plastics based on resins made by the condensation of melamine and aldehydes.

metastable, Describing an unstable condition of a plastic evidenced by changes of physical properties not caused by changes in composition or in environment. Metastable refers, for example, to the temporarily more flexible condition of some plastics after molding. No physical tests should be made while the plastic is in a metastable condition, unless data regarding this condition are desired.

migration, The transfer of a material from a plastic to another solid in contact with it.

mold base, The assembly of all parts making up an injection mold, other than the cavity, cores, and pins.

mold insert (removable), Part of a mold cavity of force which forms undercut or raised portions of a molded article.

molding, bag, A method of molding or laminating which involves the application of fluid pressure, usually by means of air, steam, water, or vacuum, to a flexible material which transmits the pressure to the material being molded or bonded.

molding, blow, A method of forming objects from plastic masses by inflating with compressed gas.

molding, compression, A method of forming objects from plastics by placing the material in a confining mold cavity and by applying pressure, and usually heat.

molding, contact-pressure, A method of molding or laminating, in which the pressure used is only slightly more than that necessary to hold the materials together during the molding operation. This pressure is usually less than 10 psi (68.948 kPa).

molding, high-pressure, Molding or laminating in which the pressure used is greater than 200 psi (1378.96 kPa).

molding, injection, A method of forming plastic objects from granular or powdered plastics by fusing the plastic in a chamber with heat and pressure and then forcing part of the mass into a cooler chamber where it solidifies. This method is commonly used to form objects from thermoplastics.

molding, low-pressure, Molding or laminating in which the pressure used is 200 psi (1378.96 kPa).

molding, transfer, A method of forming molded objects from granular, powdered or preformed plastics by fusing the plastic in a chamber by heat, and then forcing essentially the whole mass into a hot chamber, where it solidifies. This method is commonly used to form objects from thermosetting plastics.

monomer, A relatively simple chemical which can react to form a polymer.

mounting plate, See clamping plate.

Newtonian liquid, A liquid in which the rate of flow is directly proportional to the force applied. The viscosity is independent of the rate of shear, and there is no yield value in Newtonian flow.

nonrigid plastic, A plastic which has a stiffness or apparent modulus of elasticity of not over 10,000 psi (68.948 MPa) at 23°C, when determined in accordance with the Standard Method of Test for Stiffness in Flexure of Plastics (ASTM Designation: D 747).

normalize, To heat ferrous alloys above the critical temperature range and then air-cool.

novolak, A phenolic-aldehydic resin which, unless a source of methylene groups is added, remains permanently thermoplastic. (See resinoid, thermoplastic.)

nozzle, Restricted orifice at the end of the heating cylinder of an injection or transfer machine.

nylon plastics, Plastics based on a resin composed principally of a long-chain synthetic polymeric amide which has recurring amide groups as an integral part of the main polymer chain.

open-cell foam, A cellular plastic in which there is a predominance of interconnected cells.

orange peel, Uneven surface somewhat resembling an orange peel.

organosol, A suspension of a finely-divided resin in a volatile organic liquid. The resin does not dissolve appreciably in the organic liquid at room temperature, but does at elevated temperatures. The liquid evaporates at the elevated temperature and the residue on cooling is a homogenous plastic mass. Plasticizers may be dissolved in the volatile liquid.

overflow groove, Small groove used in molds to allow material to flow freely to prevent weldlines and low density, and to dispose of excess material.

pack hardening, Heating of steel for hardening in a closed container packed with either an inert or an active material.

parallels, Spaces or supports used under the top and/or bottom halves of the mold to prevent deflection, or to provide space for an ejector mechanism.

part, In its proper literal meaning, a component of an assembly. However, the word is widely misused to designate any individual manufactured article, even when (like a cup, a comb, a doll) it is complete in itself, not part of anything.

permanence, Resistance of a plastic to appreciable changes in characteristics with time and environment.

phenolic plastics, Plastics based on resins made by the condensation of phenols such as phenol and cresol, with aldehydes.

pickup groove, See hold-down groove.

plastic, The adjective plastic indicates that the noun modified is made of, consists of, or pertains to plastic. The above definition may be used as a separate meaning to the definitions contained in the dictionary for the adjective plastic.

plastic, A material that contains as an essential ingredient an organic substance of large molecular weight, is solid in its finished state, and at some stage in its manufacture or its processing into finished articles, can be shaped by flow.

plastic, rigid, A plastic which has a stiffness or apparent static modulus of elasticity greater than 100,000 psi (689.48 MPa) at 23°C, when determined in accordance with the Standard Method of Test for Stiffness in Flexure of Plastics (ASTM Designation: D 747).

plastic, semirigid, A plastic which has a stiffness or apparent modulus of elasticity of between 10,000 and 100,000 psi (68.948 and 689.48 MPa) at 23°C, when determined in accordance with the Standard Method of Test for Stiffness in Flexure of Plastics (ASTM Designation: D 747).

plastic welding, The joining of two or more pieces of plastic by fusion of the material in the pieces at adjoining or nearby areas, either with or without the addition of plastic from another source.

plasticate, To soften by heating or kneading.

plasticity, A property of plastics or adhesives which allows the material to be deformed continuously and permanently without rupture upon the application of a force that exceeds the yield value of the material.

plasticize, To soften by addition of plasticizer.

plasticizer, A material incorporated in a plastic or an adhesive to increase its workability and its flexibility or distensibility. The addition of the plasticizer may lower the melt viscosity, the temperature of the glassy transition, or the elastic modulus of the plastic.

plastify, See plasticate.

plastigel, A plastisol exhibiting gel-like flow properties; one having an effective yield value.

plastisol, A suspension of finely-divided resin in a plasticizer. The resin does not dissolve appreciably in the plasticizer at room temperature, but does at elevated temperature. On cooling, a homogeneous plastic mass (plasticized resin) results.

plunger, See force and pot plunger.

plywood, A cross-bonded assembly made of layers of veneer or veneer in combination with a lumber core or plies joined with an adhesive. Two types of plywood are recognized, namely (1) veneer plywood and (2) lumber-core plywood.

poise, A unit of viscosity in which the shearing stress is expressed in dynes per square centimeter to produce a velocity gradient of one centimeter per second per centimeter. Under SI, viscosity is measured in pascal-seconds, where 1 poise = 0.100 Pa.s. (See viscosity, centipoise.)

polyamide plastics, See nylon plastics.

polycondensation, See condensation.

polyester plastics, Plastics based on resins composed principally of polymeric esters in which the recurring ester groups are an integral part of the main polymer chain and which are cross-linked by carbon-to-carbon bonds.

polyethylene, A plastic or resin prepared by the polymerization of ethylene as essentially the sole monomer.

polymer, A compound formed by the reaction of simple molecules having functional groups that permit their combination to proceed to high molecular weights under suitable conditions. Polymers may be formed by polymerization (addition polymer) or polycondensation (condensation polymer). When two or more monomers are involved, the product is called a copolymer.

polymerization, A chemical reaction in which the molecules of a monomer are linked together to form large molecules. When two or more monomers are involved, the process is called copolymerization or heteropolymerization. (See condensation.)

polypropylene, A plastic or resin prepared by the polymerization of propylene as essentially the sole monomer.

polystyrene, A plastic based on a resin made by polymerization of styrene as the sole monomer. Polystyrene may contain minor proportions of lubricants, stabilizers, fillers, pigments, and dyes. (See styrene plastics.)

polyvinyl acetate, A resin prepared by the polymerization of vinyl acetate alone.

polyvinyl alcohol, A polymer prepared by the hydrolysis of polyvinyl esters.

polyvinyl chloride, A resin prepared by the polymerization of vinyl chloride alone.

polyvinyl chloride-acetate, Copolymer of vinyl chloride and vinyl acetate.

port, Inlet or outlet of oil, water or steam channel.

postforming, A method of shaping substantially cured thermoset plastic sheets or other forms by heating and stressing into other configurations.

pot, Chamber to hold and heat molding material for a transfer mold.

pot life, See working life.

pot plunger, A plunger used to force softened molding material into the closed cavity of a transfer mold.

pot-retainer, Plate channeled for heat and used to hold pot of transfer mold.

preform, A coherent block of granular or fibrous plastic molding compound, or of fibrous material with or without resin. A preform is made by compressing the material sufficiently to produce a coherent block for convenience in handling.

prepolymer, A chemical structure intermediate between that of the monomer or monomers and the final polymer or resin.

pressure pad, Hardened steel reinforcements distributed around a mold to help the land to absorb the final closing pressure.

primer, A coating applied to a surface, prior to the application of an adhesive or lacquer, enamel or the like, to improve the performance of the bond.

primary gluing, See secondary gluing.

progressive gluing, A method of curing a resin adhesive in successive steps or stages by application of heat and pressure, between the platens of a hot press. Used only for a wood of larger area than the press platens.

pseudoplastic, A pseudoplastic fluid is one whose apparent viscosity or consistency decreases instantaneously with increase in rate of shear; i.e., an initial relatively high resistance to stirring decrease abruptly as the rate of stirring is increased.

pushback pins, See return pins.

reinforced plastics, Plastics with some strength properties greatly superior to those of the base resin, resulting from the presence of high-strength fillers embedded in the composition. The reinforcing fillers are usually fibers, fabrics, or mats made of fibers. The plastic laminates are the most common and strongest type of reinforced plastics. (See filler.)

resin, A solid, semisolid, or pseudosolid organic material which has an indefinite and often high molecular weight, exhibits a tendency to flow when subjected to stress, usually has a softening point or melting range, and usually fractures conchoidally.

resin, liquid, An organic polymeric liquid which, when converted to its final state for use, becomes a solid.

resinoid, See novolak.

retaining pin, Pin on which insert is placed and located prior to molding.

retarder, See inhibitor.

return pins, Pins which return the ejector mechanism to molding position.

ring gate, Annular opening for entrance of material into cavity of injection or transfer mold.

runner, (1) The secondary feed channel that runs from the inner end of the sprue in an injection or transfer mold to the cavity gate. (2) The piece formed in a secondary feed channel or runner.

saran plastics, Plastics based on resins made by the polymerization of vinylidene chloride or copolymerization of vinylidene chloride with not more than an equal weight of other unsaturated compounds.

sealing diameter, Portion of the insert which is free of knurl and is allowed to enter into the mold to prevent the flow of material.

secondary gluing, The process of gluing together wood and plywood parts in assembling wood products, such as aircraft. Contrasted with primary gluing, when veneers are glued into plywood. (See assembly in plastics.)

set, To convert into a fixed or hardened state by chemical or physical action, such as condensation, polymerization, oxidation, vulcanization, gelation, hydration, or evaporation of volatile constitutents. (See cure.)

sheet, A piece of plastic sheeting produced as an individual piece rather than in a continuous length, or cut as individual pieces from a continuous length. (See sheeting, film.)

sheeting, A form of plastic in which the thickness is very small in proportion to length and width, and in which the plastic is present as a continuous phase throughout, with or without filler. (See film.)

shelf life, The length of time over which a product will remain fit for use during storage under specific conditions (especially temperature).

shim, In the manufacture of plywood, a long narrow patch glued into the panel, or into the lumber core.

shoe, See chase.

silicone plastics, Plastics based on resins in which the main polymer chain consists of alternating silicon and oxygen atoms, with carbon-containing side groups.

sizing, The process of applying a material on a surface in order to fill pores and thus reduce the absorption of the subsequently applied adhesive or coating, or to otherwise modify the surface properties of the substrate to improve adhesion. Also, the material used for this purpose. The latter is sometimes called size.

sleeve ejector, Bushing-type knockout.

sliding plate, See duplicate cavity plate.

slip joint, The method of laying up veneers in flexible-bag molding, wherein the edges are beveled and allowed to overlap over part or all of the scarfed area.

softening range, The range of temperatures in which a plastic changes from a rigid to a soft state. Actual values will depend on the method of test. Sometimes erroneously referred to as softening point.

solid-piled, Sometimes called dead-piled or bulked-down. Plywood fresh from clamps or hot press is piled on a solid, flat base, without stickers, and weighted down while reaching normal temperature and moisture content.

solcation, The process of swelling, gelling, or solution of a resin by a solvent or plasticizer as a result of mutual attraction.

solvency, Solvent action, or strength of solvent action.

spacers, See parallels.

split cavity, Cavity made in sections.

split-cavity blocks, Blocks which when assembled contain a cavity for molding articles having undercuts.

spread, The quantity of adhesive per unit joint area applied to an adherend. It is preferably expressed in pounds of liquid or solid adhesive per thousand square feet of joint area. Single spread refers to application of adhesive to only one adherend of a joint. Double spread refers to application of adhesive to both adherends of a joint.

sprue, (1) The primary feed channel that runs from the outer face of an injection or transfer mold to the mold gate in a single-cavity mold or to the runners in a multiple-cavity mold. (2) The piece formed in a primary feed channel or sprue.

sprue bushing, In an injection mold, a hardened steel insert which contains the tapered sprue hole and has a suitable seat for making close contact with the nozzle of the injection cylinder.

sprue-ejector pin, See sprue-puller.

sprue-puller, A pin having a Z-shaped slot undercut in its end, by means of which it serves to pull the sprue out of the sprue bushing.

stir-in resin, A vinyl resin which does not require grinding to effect dispersion in a plastisol or an organosol.

stop buttons, Multiple islands limiting travel of ejector mechanism when returned to molding position.

stress-crack, External or internal crack in a plastic caused by tensile stresses less than that of its short-time mechanical strength. The development of such cracks is frequently accelerated by the environment to which the plastic is exposed. The stresses which cause cracking may be present internally or externally or may be combinations of these stresses. The appearance of a network of fine cracks is called crazing.

stripper plate, A plate which strips the molded article from mold pin, force or cores.

stripping fork, Tool, usually of brass or laminated sheet, used to remove articles from the mold (also called comb).

styrene plastics, Plastics based on resins made by the polymerization of styrene or copolymerization of styrene with other unsaturated compounds.

styrene rubber plastics, Plastics consisting of at least 50 percent of a styrene plastic combined with rubbers and other compounding ingredients.

support plate, See backing plate.

support post or pillar, Post used to resist deflection under pressure.

syndiotactic, Alternating placement of the group on either side of the chain with respect to the carbon-carbon backbone plane.

syneresis, The contraction of a gel. This is usually evidenced by the separation of a liquid. See first definition of gel.

tack, Stickiness of an adhesive. This property is measured as the pull resistance to effect division without failure or deformation occurring in the adherend surroundings or at the interface while the adhesive still exhibits viscous or plastic flow. The measured value may vary with time, temperature, film thickness, etc.

temper, To reheat after hardening to some temperature below the critical temperature, followed by air cooling to obtain desired mechanical properties and to relieve hardening strains.

thermally foamed plastics, Cellular plastics produced by applying heat to effect gaseous decomposition or volatilization of a constituent.

thermoelasticity, Rubber-like elasticity exhibited by a rigid plastic and resulting from an increase of temperature.

thermoforming, Forming with aid of heat. Retention of the desired shape may be achieved by cooling in place after forming. In the case of thermosetting materials, prolonged heating may be necessary to effect cure in place.

thermoplastic, Capable of being repeatedly softened by increase of temperature and hardened by decrease of temperature. Thermoplastic applies to those materials whose change upon heating is substantially physical rather than chemical. Thermoplastic polymers do not contain cross-links.

thermoset, Pertaining to the state of a resin in which it is relatively infusible.

thermosetting, Capable of being changed into a substantially infusible or insoluble product when cured by application of heat or chemical means. Thermosetting polymers form bridges between molecule chains referred to as cross-links.

thixotropic, A thixotropic fluid is one whose apparent viscosity decreases with time, to some constant value at any constant rate of shear; i.e., its apparent viscosity can be gradually decreased, to that limit, by stirring; when stirring is discontinued, the apparent viscosity increases gradually back to the original value.

undercut, Any depression on the mold which prevents or hinders withdrawal of the molded article.

urea plastics, Plastics based on resins made by the condensation of urea and aldehydes.

urethane plastics, Plastics based on resins made by the condensation of organic isocyanates with compounds or resins that contain hydroxyl groups. Urethane plastics are a type of isocyanate plastics. (See isocyanate plastics.)

vacuum forming, A forming process in which a plastic sheet is changed to a desired shape by causing it to flow by reducing the air pressure on one side of the sheet.

veneer, A thin sheet or layer of wod, sliced, rotary-cut, half-round or sawn from a log, block or flitch. There is no sawkerf waste in a knife-cut veneer. Veneer is the raw material from which plywood and laminated wood are assembled. Thicknesses may range from 1/100 to ¼ in. and are seldom greater.

vent, See air vent.

vinyl acetate plastics, Plastics based on resins made by the polymerization of vinyl acetate or copolymerization of vinyl acetate with no more than an equal weight of other saturated compounds.

vinyl alcohol plastics, Plastics based on resins made by the hydrolysis of polyvinyl esters or copolymers of vinyl esters.

vinyl chloride plastics, Plastics based on resins made by the polymerization of vinyl chloride with not more than equal weights of other unsaturated compounds.

vinyl plastics, Plastics based on resins made from vinyl monomers except those specifically covered by other classification, such as acrylic and styrene plastics. Typical vinyl plastics are polyvinyl chloride, polyvinyl acetate, polyvinyl alcohol and polyvinyl butyral, and copolymers of vinyl monomers with unsaturated compounds.

vinylidene plastics, See saran plastics.

viscosity, The property of resistance to flow exhibited within the body of a material. This property can be expressed in terms of relationship between applied shearing stress and resulting rate of strain in shear. Viscosity is usually taken to mean Newtonian viscosity, in which case the ratio of shearing stress to the rate of shearing strain is constant. In non-Newtonian behavior, which is the usual case with plastic materials, the ratio varies with the shearing stress. Such ratios are often called the apparent viscosities at the corresponding shearing stress. (See viscosity coefficient).

viscosity coefficient, The shearing stress necessary to induce a unit velocity flow gradient in a material. In actual measurement, the viscosity coefficient of a material is obtained from the ratio of shearing stress to shearing rate. This assumes the rates to be constant and independent of the shearing stress, a condition which is satisfied only by Newtonian fluids. Consequently, in all other cases, values obtained are apparent and represent one point on the flow curve. In the metric system, the viscosity coefficient is expressed in poises. Under SI, the unit is pascal-seconds.

warp, warpage, Dimensional distortion in a plastic object after molding or other fabrications. (See dished, domed.)

weathering, The exposure of plastics outdoors.

weathering, artificial, The exposure of plastics to cyclic laboratory conditions comprising high and low temperatures, high and low relative humidities, and ultraviolet radiant energy, with or without direct water spray, in an attempt to produce changes in their properties similar to those observed on long-time continuous exposure outdoors. The laboratory exposure conditions are usually intensified beyond those encountered in actual outdoor exposure in an attempt to achieve an accelerated effect.

web, A textile fabric, paper, or a thin metal sheet of continuous length handled in roll form as contrasted with the same material cut into sheets.

weld line, In a molded article, a zone of defective appearance and/or strength resulting from defective union of two fronts of plastic flowing together within the mold. Sometimes called weld mark.

working life, The period during which a compound, after mixing with catalyst, solvent, or other compounding ingredients, remains suitable for its intended use.

yield value, The force which must be applied to a plastic to initiate flow also called yield stress.

WELDING

air carbon-cutting, An arc-cutting process wherein the severing of metals is effected by melting with the heat of an arc between an electrode and the base metal and an air stream is used to facilitate cutting.

arc blow, The deflection of an electric arc from its normal path because of magnetic forces.

arc-time, The length of time the arc is maintained in making an arc weld.

arc voltage, The voltage across the welding arc.

arc welding, A group of welding processes wherein coalescence is produced by heating with an arc or arcs, with or without the application of pressure and with or without the use of filler metal.

automatic welding, Welding with equipment which performs the entire welding operation without constant observation and adjustment of the controls by an operator. The equipment may or may not perform the loading and unloading of the work.

backhand welding, A welding technique wherein the welding torch or gun is directed opposite to the progress of welding.

backing, Material (metal, weld metal, asbestos, carbon, granular flux, gas, etc.) backing up the joint during welding.

bare electrode, A filler-metal electrode, used in arc welding, consisting of a metal wire with no coating other than that incidental to its manufacture or preservation.

base metal, The metal to be welded, soldered or cut.

bevel angle, The angle formed between the prepared edge of a member and a plane perpendicular to the surface of the member.

braze welding, A method of welding whereby a weld is made using a filler metal, having a liquidus above 800°F (427°C) and below the solidus of the base metals. The filler metal is not distributed in the joint by capillary attraction.

buildup sequence, The order in which the weld beads of a multiple-pass weld are deposited with respect to the cross-section of the joint.

butt joint, A joint between two members lying approximately in the same plane.

carbon arc-cutting, An arc-cutting process wherein the severing of metals is effected by melting with the heat of an arc between a carbon electrode and the base metal.

carbon arc-welding, An arc-welding process wherein coalescence is produced by heating with an arc between a carbon electrode and the work and no shielding is used. Pressure may or may not be used and filler metal may or may not be used.

cold welding, A solid state welding process wherein coalescence is produced by the external application of mechanical force alone.

cool time, The time interval between successive heat times in multiple-impulse welding or in the making of seam welds by resistance welding.

corner joint, A joint between two members located approximately at right angles to each other in the form of an L.

covered electrode, A filler-metal electrode, used in arc welding, consisting of a metal core wire with a relatively thick covering which provides protection for the molten metal from the atmosphere, improves the properties of the weld metal and stabilizes the arc.

cover glass, A clear transparent material used in goggles, hand shields and helmets to protect the filter lens from spattering material.

crater, In arc welding, a depression at the termination of a weld bead or in the weld pool beneath the electrode.

crater crack, A crack in the crater of a weld bead.

cutting torch, A device used in oxygen cutting for controlling and directing the gases used for preheating and the oxygen used for cutting the metal.

cylinder, A portable cylindrical container used for transportation and storage of a compressed gas.

deposited metal, Filler metal that has been added during a welding operation.

deposition efficiency, The ratio of the weight of deposited metal to the net weight of electrodes consumed, exclusive of stubs.

deposition rate, The weight of metal deposited in a unit of time.

depth of fusion, The distance that fusion extends into the base metal or previous pass from the surface melted during welding.

dip brazing, A brazing process in which the heat required is furnished by a molten chemical or metal bath. When a molten chemical bath is used, the bath may act as a flux. When a molten metal bath is used, the bath provides the filler metal.

edge joint, A joint between the edges of two or more parallel or nearly parallel members.

edge preparation, The contour prepared on the edge of a member for welding.

electrode holder, A device used for mechanically holding the electrode and conducting current to it.

electron beam welding, A welding process wherein coalescence is produced by the heat obtained from a concentrated beam composed primarily of high velocity electrons impinging upon the surfaces to be joined.

electronic heat control, A device for adjusting the heating value (rms value) of the current in making a resistance weld by controlling the ignition or firing of the tubes in an electronic contractor. The current is initiated each half-cycle at an adjustable time with respect to the zero point on the voltage wave.

electroslag welding, A welding process wherein coalescence is produced by molten slag which melts the filler metal and the surfaces of the work to be welded. The weld pool is shielded by this slag which moves along the full cross section of the joint as welding progresses. The conductive slag is maintained molten by its resistance to electric current passing between the electrode and the work.

face of weld, The exposed surface of a weld on the side from which welding was done.

faying surface, That surface of a member which is in contact or in close proximity with another member to which it is to be joined.

filler metal, The metal to be added in making a welded, brazed or soldered joint.

fillet weld, A weld of approximately triangular cross section joining two surfaces approximately at right angles to each other in a lap joint.

fillet weld size, For equal leg fillet welds, the leg length of the largest isosceles right-triangle which can be inscribed within the fillet-weld cross section. For unequal leg fillet welds, the leg length of the largest right-triangle which can be inscribed within the fillet-weld cross section.

filter lens, A filter, usually colored glass, used in goggles, helmets and handshields to exclude harmful light rays.

flash, The molten metal which is expelled, or which is squeezed out by the application of pressure, and solidifies around the weld.

flashback, A recession of the flame into or back of the mixing chamber of the torch.

flashing time, The time during which the flashing action is taking place in flash welding.

flash welding, A resistance-welding process wherein coalescence is produced, simultaneously over the entire area of abutting surfaces, by the heat obtained from resistance to electric current between the two surfaces, and by the application of pressure after heating is substantially completed. Flashing and upsetting are accompanied by expulsion of metal from the joint.

flat position, The position of welding wherein welding is performed from the upper side of the joint and the face of the weld is approximately horizontal.

flux, Material used to prevent, dissolve or facilitate removal of oxides and other undesirable substances.

flux cored arc-welding, An arc-welding process wherein coalescence is produced by heating with an arc, between a continuous filler metal (consumable) electrode and the work. Shielding is obtained from a flux contained within the electrode. Additional shielding may or may not be obtained from an externally supplied gas or gas mixture.

forehand welding, A welding technique wherein the welding torch or gun is directed toward the progress of welding.

forge welding, A solid state welding process wherein coalescence is produced by heating and by applying pressure or blows sufficient to cause permanent deformation at the interface.

friction welding, A solid state welding process wherein coalescence is produced by the heat obtained from mechanically induced sliding motion between rubbing surfaces. The work parts are held together under pressure.

furnace brazing, A brazing process in which the heat required is obtained from a furnace.

fusion, The melting together of filler metal and base metal, or of base metal only, which results in coalescence.

fusion zone, The area of base metal as determined on the cross section of a weld.

gas metal arc-cutting, Method of metal arc-cutting wherein the severing of metals is effected by melting with an arc between a continuous metal (consumable) electrode and the work. Shielding is obtained entirely from an externally supplied gas or gas mixture.

gas metal arc-welding, An arc-welding process wherein coalescence is produced by heating with an arc between a continuous filler metal (consumable) electrode and the work. Shielding is obtained entirely from an externally supplied gas, or gas mixture. Some methods of this process are called MIG or CO_2 welding.

gas-shielded stud welding, A stud welding method wherein coalescence is produced by heating with an arc drawn between a metal stud or similar part, and the other work part, until the surfaces to be joined are properly heated, when they are brought together under pressure. Shielding is obtained from an inert gas such as helium or argon.

gas tungsten arc-cutting, An arc-cutting process wherein the severing of metals is effected by melting with an arc between a single tungsten (nonconsumable) electrode and the work. Shielding is obtained from a gas or gas mixture.

gas tungsten arc-welding, An arc-welding process wherein the coalescence is produced by heating with an arc between a single tungsten (nonconsumable) electrode and the work. Shielding is obtained from a gas or gas mixture. Pressure may or may not be used and filler metal may or may not be used. (This process has sometimes been called TIG welding.)

gas welding, A group of welding processes wherein coalescence is produced by heating with a gas flame or flames, with or without the application of pressure, and with or without the use of filler metal.

groove angle, The total included angle of the groove between parts to be joined by a groove weld.

groove radius, The radius of a J- or U-groove.

groove weld, A weld made in the groove between two members to be joined.

hand shield, A protective device, used in arc welding, for shielding the face and neck. A hand shield is equipped with a suitable filter lens and is designed to be held by hand.

heat-affected zone, That portion of the base metal which has not been melted, but whose mechanical properties or microstructure have been altered by the heat of welding, brazing, soldering or cutting.

heat time, The time that current occurs during any one impulse in multiple-impulse welding or when making welds by resistance welding.

helmet, A protective device, used in arc welding, for shielding the face and neck. A helmet is equipped with a suitable filter lens and is designed to be worn on the head.

hold time, The time during which force is applied at the point of welding after the last impulse of current ceases in resistance welding.

induction brazing, A brazing process in which the heat required is obtained from the resistance of the work to induced electric current.

induction welding, A welding process wherein coalescence is produced by the heat obtained from resistance of the work to induced electric current, with or without the application of pressure.

intermittent weld, A weld wherein the continuity of the weld is broken by recurring unwelded spaces.

interpass temperature, In a multiple-pass weld, the temperature (minimum or maximum as specified) of the deposited weld metal before the next pass is started.

joint design, The joint geometry together with the required dimensions of the welded joint.

joint geometry, The shape and dimensions of a joint in cross section prior to welding.

joint penetration, The minimum depth a groove of flange weld extends from its face into a joint, exclusive of reinforcement.

kerf, The space from which metal has been removed by a cutting process.

lap joint, A joint between two overlapping members.

laser beam cutting, A cutting process wherein the severing of metals is effected by the heat obtained from the impingement upon the workpiece of a concentrated coherent light beam emitted from a laser, with or without an externally supplied gas.

laser beam welding, A welding process wherein coalescence is produced by the heat obtained from the application of a concentrated coherent light beam impinging upon the surfaces to be joined.

layer, A stratum of weld metal, consisting of one or more weld beads.

lead angle, The angle that the electrode makes in advance of a line perpendicular to the weld axis at the point of welding, taken in a longitudinal plane.

leg of a fillet weld, The distance from the root of the joint to the toe of the fillet weld.

local preheating, Preheating a specific portion of a structure.

local stress-relief heat treatment, Stress-relief heat treatment of a specific portion of a structure.

longitudinal resistance-seam welding, The making of a resistance-seam weld in a direction essentially parallel to the throat depth of a resistance-welding machine.

machine oxygen cutting, Oxygen cutting with equipment which performs the cutting operations under the constant observation and control of an operator. The equipment may or may not perform the loading and unloading of the work.

machine welding, Welding with equipment which performs the welding operation under the constant observation and control of an operator. The equipment may or may not perform the loading and unloading of the work.

manifold, A multiple header for interconnection of gas of fluid sources with distribution points.

manual welding, Welding wherein the entire welding operation is performed and controlled by hand.

melting rate, The weight or length of electrode melted in a unit of time.

metal arc-cutting, An arc-cutting process wherein the severing of metals is effected by melting with the heat of an arc between a metal electrode and the base metal.

metal powder cutting, An oxygen-cutting process wherein the severing of metals is effected by the use of powder, such as iron, to facilitate cutting.

neutral flame, A gas flame wherein the portion used is neither oxidizing nor reducing.

nugget, The weld metal joining the parts in spot, seam or projection welds.

nugget size, The diameter or width of the nugget measured in the plane of the interface between the pieces joined.

open-circuit voltage, The voltage between the output terminals of the welding machine when no current is in the welding circuit.

overhead position, The position of welding wherein welding is performed from the underside of the joint.

overlap, Protrusion of weld metal beyond the toe or root of the weld.

oxidizing flame, A flame having an oxidizing effect (excess oxygen).

oxyacetylene welding, A gas-welding process wherein coalescence is produced by heating with a gas flame or flames obtained from the combustion of acetylene with oxygen, with or without the application of pressure and with or without the use of filler metal.

oxy-fuel gas cutting, An oxygen cutting process wherein the severing of metals is effected by means of the chemical reaction of oxygen with the base metal at elevated temperatures, the necessary temperature being maintained by means of gas flames obtained from the combustion of a specified fuel gas and oxygen.

oxygen cutting, A group of cutting processes wherein the severing or removing of metals is effected by means of the chemical reaction of oxygen with the base metal at elevated temperatures. In the case of oxidation-resistant metals the reaction is facilitated by the use of a chemical flux or metal powder.

oxygen gouging, An application of oxygen cutting wherein a bevel or groove is formed.

oxygen lance, A length of pipe used to convey oxygen to the point of cutting in oxygen-lance cutting.

pass, A single longitudinal progression of a welding operation along a joint or weld deposit. The result of a pass is a weld bead.

peening, The mechanical working of metals by means of impact blows.

percussion welding, A resistance-welding process wherein coalescence is produced simultaneously over the entire abutting surfaces by the heat obtained from an arc produced by a rapid discharge of electrical energy with pressure percussively applied during or immediately following the electrical discharge.

plasma arc-cutting, An arc-cutting process wherein severing of the metal is obtained by melting a localized area with a constricted arc and removing the molten material with a high velocity jet of hot, ionized gas issuing from the orifice.

plasma arc-welding, An arc-welding process wherein coalescence is produced by heating with a constricted arc between an electrode and the work piece (transferred arc) or the electrode and the constricting nozzle (non-transferred arc). Shielding is obtained from the hot, ionized gas issuing from the orifice which may be supplemented by an auxiliary source of shielding gas. Shielding gas may be an inert gas or a mixture of gases. Pressure may or may not be used, and filler metal may or may not be supplied.

platen, A member with a substantially flat surface to which dies, fixtures, backups or electrode holders are attached, and which transmits the electrode force or upsetting force in a resistance-welding machine.

platen force, The force available at the movable platen to cause upsetting in flash or upset welding. This force may be dynamic, theoretical or static.

plug weld, A circular weld made through a hole in one member of a lap or tee joint joining that member to the other. The walls of the hole may or may not be parallel and the hole may be partially or completely filled with weld metal. (A fillet-welded hole or a spot weld should not be construed as conforming to this definition.)

porosity, Gas pockets or voids in metal.

postheat current, The current through the welding circuit during postheat time in resistance welding.

postheating, The application of heat to an assembly after a welding, brazing, soldering or cutting operation.

preheating, The application of heat to the base metal immediately before welding, brazing, soldering or cutting.

preheat temperature, The temperature specified that the base metal must attain in the welding, brazing, soldering or cutting area immediately before these operations are performed.

pressure gas welding, Gas-welding process wherein coalescence is produced, simultaneously, over the entire area of abutting surfaces, by heating with gas flames obtained from the combustion of a fuel gas with oxygen and by the application of pressure, without the use of filler metal.

procedure qualification, The demonstration that welds made by a specific procedure can meet prescribed standards.

projection welding, A resistance-welding process wherein coalescence is produced by the heat obtained from resistance to electric current through the work parts held together under pressure by electrodes. The resulting welds are localized at predetermined points by projections, embossments or intersections.

quench time, The time from the end of weld time to the beginning of temper time in resistance welding.

reducing flame, A gas flame having a reducing effect. (Excess fuel gas.)

regulator, A device for controlling the delivery of gas at some substantially constant pressure regardless of variation in the higher pressure at the source.

residual stress, Stress remaining in a structure or member as a result of thermal or mechanical treatment or both.

resistance brazing, A brazing process in which the heat required is obtained from the resistance to electric current in a circuit of which the work is a part.

resistance seam-welding, A resistance-welding process wherein coalescence at the faying surfaces is produced by the heat obtained from resistance to electric current through the work parts held together under pressure by electrodes. The resulting weld is a series of overlapping resistance-spot welds made progressively along a joint by rotating the electrodes.

resistance spot-welding, A resistance-welding process wherein coalescence at the faying surfaces is produced in one spot by the heat obtained from the resistance to electric current through the work parts held together under pressure by electrodes. The size and shape of the individually-formed welds are limited primarily by the size and contour of the electrodes.

reverse polarity, The arrangement of direct current arc-welding leads wherein the work is the negative pole and the electrode is the positive pole of the welding arc.

root crack, A crack in the weld or heat-affected zone occurring at the root of a weld.

root face, That portion of the groove face adjacent to the root of the joint.

root opening, The separation between the members to be joined, at the root of the joint.

semiautomatic arc welding, Arc welding with equipment which controls only the filler metal feed. The advance of the welding is manually controlled.

shielded metal arc-cutting, A method of metal-arc cutting wherein the severing of metals is effected by melting with the heat of an arc between a covered metal electrode and the base metal.

shielded metal arc-welding, An arc-welding process wherein coalescence is produced by heating with an arc between a covered metal electrode and the work. Shielding is obtained from decomposition of the electrode covering. Pressure is not used and filler metal is obtained from the electrode.

slag inclusion, Nonmetallic solid material entrapped in weld metal or between weld metal and base metal.

soldering, A group of joining processes wherein coalescence is produced by heating to a suitable temperature and by using a filler metal having a liquidus not exceeding 800° F (427° C) and below the solidus of the base metals.

solid state welding, A group of welding processes wherein coalescence is produced essentially at temperatures below the melting point of the base metals being joined, without the addition of a brazing filler metal. Pressure may or may not be used.

spatter, In arc and gas welding, the metal particles expelled during welding and which do not form a part of the weld.

spot weld, A weld made between or upon overlapping members wherein coalescence may start and occur on the faying surfaces or may have proceeded from the surface of one member. The weld cross section (plan view) is approximately circular.

squeeze time, The time interval between the initial application of the electrode force on the work and the first application of current in making spot and seam welds by resistance welding and in projection or upset welding.

stack cutting, Oxygen cutting of stacked metal plates arranged so that all the plates are severed by a single cut.

straight polarity, The arrangement of direct current arc-welding leads wherein the work is the positive pole and the electrode is the negative pole of the welding arc.

stress-relief heat treatment, Uniform heating of a structure or portion thereof to a sufficient temperature, below the critical range, to relieve the major portion of the residual stresses, followed by uniform cooling. (Note: terms normalizing, annealing, etc., are misnomers for this application.)

stud welding, An arc-welding process wherein coalescence is produced by heating with an arc drawn between a metal stud, or similar part, and the other work part until the surfaces to be joined are properly heated, when they are brought together under pressure. Partial shielding may be obtained by the use of a ceramic ferrule surrounding the stud. Shielding gas or flux may or may not be used.

submerged arc welding, An arc-welding process wherein coalescence is produced by heating with an arc or arcs between a bare metal electrode or electrodes and the work. The arc is shielded by a blanket of granular, fusible material on the work. Pressure is not used and filler metal is obtained from the electrode and sometimes from a supplementary welding rod.

surfacing, The deposition of filler metal on a metal surface to obtain desired properties or dimensions.

tack weld, A weld made to hold parts of a weldment in proper alignment until the final welds are made.

tee joint, A joint between two members located approximately at right angles to each other in the form of a T.

temper time, That part of the postweld interval following quench time to the beginning of hold time in resistance welding.

thermit welding, A group of welding processes wherein coalescence is produced by heating with super heated liquid metal and slag resulting from a chemical reaction between a metal oxide and aluminum, with or without the application of pressure. Filler metal, when used, is obtained from the liquid metal.

throat of a fillet weld, Theoretical — the distance from the beginning of the root of the joint perpendicular to the hypotenuse of the largest right-triangle that can be inscribed within the fillet-weld cross section. Acutal — the shortest distance from the root of a fillet weld to its face.

toe of weld, The junction between the face of a weld and the base metal.

torch brazing, A brazing process in which the heat required is furnished by a gas flame.

tungsten electrode, A nonfiller-metal electrode, used in arc welding, made principally of tungsten.

undercut, A groove melted into the base metal adjacent to the toe or root of a weld and left unfilled by weld metal.

upset, The localized increase in volume in the region of a weld, resulting from the application of pressure.

upset welding, A resistance-welding process wherein coalescence is produced, simultaneously over the entire area of abutting surfaces or progressively along a joint, by the heat obtained from resistance to electric current through the area of contact of those surfaces. Pressure is applied before heating is started and is maintained throughout the heating period.

vertical position, The position of welding wherein the axis of the weld is approximately vertical.

weave bead, A type of weld bead made with transverse oscillation.

weld, A localized coalescence of metal wherein coalescence is produced either by heating to suitable temperatures, with or without the application of pressure, or by the application of pressure alone, and with or without the use of filler metal. The filler metal either has a melting point approximately the same as the base metals or has a melting point below that of the base metals but above 800° F (427° C).

weldability, The capacity of a metal to be welded under the fabrication conditions imposed into a specific, suitably designed structure and to perform satisfactorily in the intended service.

weld bead, A weld deposit resulting from a pass.

weld gauge, A device designed for checking the shape and size of welds.

weld metal, That portion of a weld which has been melted during welding.

welder, One who is capable of performing a manual or semiautomatic welding operation. (Sometimes erroneously used to denote a welding machine.)

welder certification, Certification in writing that a welder has produced welds meeting prescribed standards.

welder qualification, The demonstration of a welder's ability to produce welds meeting prescribed standards.

welding current, The current in the welding circuit during the making of a weld. In resistance welding, the current used during a preweld or postweld interval is excluded.

welding goggles, Goggles with tinted lenses, used during welding, brazing or oxygen cutting, which protect the eyes from harmful radiation and flying particles.

welding machine, Equipment used to perform the welding operation. For example, spot-welding machine, arc-welding machine, seam-welding machine, etc.

welding operator, One who operates machine or automatic welding equipment.

welding procedure, The detailed methods and practices including all joint welding, procedures involved in the production of a weldment.

welding rod, A form of filler metal used for welding or brazing wherein the filler metal does not conduct the electrical current.

welding transformer, A transformer used for supplying current for welding.

weldment, An assembly whose component parts are joined by welding.

Applied Mathematics

Z94.8

8

The APPLIED MATHEMATICS TERMINOLOGY provides definitions for about 400 terms frequently encountered by industrial engineers in the course of their work. This TERMINOLOGY was compiled by the Applied Mathematics Subcommittee and is a revision of the originally issued Z94.8 APPLIED MATHEMATICS-1972.

Since glossaries covering quality control and statistics are addressed by the Z1 Committee on Quality Assurance, most of these terms have been deleted from the 1972 Z94 TERMINOLOGY — leaving only basic terms for continuity. (The Z1 Committee glossaries can be obtained separately.)

The Subcommittee acknowledges with grateful appreciation the permission granted by the many authors and publishers to quote from their published works, and trusts that the direct reference in each case to the Source Code will constitute deserved recognition. If the definition is largely a direct quote, reference is made to that source by the Source Code Number, which immediately follows the definition. In some cases, page numbers follow the source code numbers separated by a colon. Where more than one code and source numbers are listed for a term, the first is the source of the direct quotation and the others are references of additional information concerning the term defined. Of course, there are no source code numbers when the term is not a direct quote.

The following lists those who have so generously, faithfully, and devotedly donated their time, effort, and expertise:

SUBCOMMITTEE ON APPLIED MATHEMATICS:
PAST and PRESENT

THOMAS W. CALVIN, Chairman, IBM Corporation
GEORGE O. HAWLEY, Past Chairman, Sandia Laboratories (Retired)
DR. RICHARD L. CARTER (deceased), Rensselaer Polytechnic Institute
GEORGE CHANDLER, General Atomic Company
JULIAN EDELMAN, Sylvania Electric Products, Inc.
DR. SAUL I. GASS, Mathematica, Inc. & University of Maryland
DR. ISAAC N. GIBRA, Stevens Institute of Technology
DR. SEBASTIAN B. LITTAUER (deceased), Columbia University
DR. RICHARD R. PRAIRIE, Sandia Laboratories
R. J. STOKELY, General Atomic Company
DR. ROBERT L. WILLIAMS, Ohio University

Sources or Cross References for Glossary of Terms

Code **Source or Cross Reference**

1 American Society for Quality Control, Standard A1-1971, ANSI Z1.5-1971, *Definitions, Symbols, Formulas and Tables for Control Charts,* American Society for Quality Control, 161 W. Wisconsin Avenue, Milwaukee, Wisconsin 53203.

2 American Society for Quality Control, Standard A2-1978, ANSI Z1.6-1978, *Terms, Symbols, and Definitions for Acceptance Sampling,* 161 W. Wisconsin Avenue, Milwaukee, Wisconsin 53203.

3 American Society for Quality Control, Standard A3-1979, ANSI Z1.7-1979, *Quality Systems Terminology,* American Society for Quality Control, 161 W. Wisconsin Avenue, Milwaukee, Wisconsin 53203.

4 Arley, N. and Buch, K.R., *Introduction to Theory of Probability and Statistics,* Wiley, 1950.

5 Bailey, N. T. J., *The Elements of Stochastic Processes,* Wiley, 1964.

6 Bizley, M. T. L., *Probability,* Cambridge, 1957.

7 Brownlee, K. A., *Statistical Theory and Methodology in Science and Engineering,* Second Edition, Wiley, 1965.

8 Burington, R. S. and May, D.C., *Handbook of Probability and Statistics,* Handbook Publishers, 1953.

9 Charnes, A. and Cooper, W. W., *Management Models and Industrial Applications of Linear Programming,* Volumes I & II, John Wiley and Sons, New York, 1961.

10 Cramer, H., *The Elements of Probability Theory,* Wiley, 1955.

11 Dantzig, G. B., *Linear Programming and Extensions,* Princeton University Press, Princeton, New Jersey, 1963.

12 *Decomposition Principles for Solving Large Structured Linear Programs,* Mathematica, Princeton, New Jersey, 1963.

13 Feller, W., *An Introduction to Probability Theory and Its Applications,* Volume I, Second Edition, Wiley, 1957.

14 Ford, L. R. and Fulkerson, D. R., *Flows in Networks,* Princeton University Press, Princeton, New Jersey, 1962.

15 Gass, S. I., *Linear Programming,* McGraw-Hill, New York, New York, Second Edition, 1964.

16 *Glossary and Abbreviations for Quality Assurance Program,* Quality Assurance Department, Sandia Corporation, Albuquerque, New Mexico, 1962.

17 Hadley, G., *Linear Programming,* Addison-Wesley, Reading, Massachusetts, 1962.

18 Hadley, G., *Nonlinear and Dynamic Programming,* Addison-Wesley, Reading, Massachusetts, 1964.

19 IBM, *An Introduction to Linear Programming,* IBM Corporation, White Plains, New York.

20 International Electrotechnical Commission (EC), Technical Committee 56 (Reliability), Draft Copy, *Definitions of Reliability Terms.*

21 James, Glenn and James, Robert C., *Mathematics Dictionary, Multi-lingual Edition,* D. Van Nostrand Company, Inc., 1959 (1966 Reprint).

22 Kendall, Maurice G. and Buckland, William R., *Dictionary of Statistical Terms,* Second Edition, Hafner Publishing Company, New York, 1960.

23 McElrath, Gayle W.; Astrachan, Max; Leone, Fred C.; and Bocke, Robert W., *Significance Tests,* four papers presented at the Twelfth Annual Convention of the American Society for Quality Control, Boston, Massachusetts, May 26-28, 1958.

24 Nemhauser, G. L., *Introduction to Dynamic Programming,* John Wiley and Sons, New York, 1966.

25 Parzen, E., *Modern Probability Theory and Its Application,* Wiley, 1960.

26 Stuart, Alan, *Basic Ideas of Scientific Sampling,* Number Four of Griffin's Statistical Monographs and Courses, Hafner Publishing Company, New York, 1962.

27 United States Department of Defense, Mil.-Std.-109A, 30 October 1961, *Quality Assurance Terms and Definitions,* United States Government Printing Office, Washington, D.C., 1961.

28 United States Department of Defense, Mil.-Std.-721B, 25 August 1966, *Definitions of Effectiveness Terms for Reliability, Maintainability, Human Factors, and Safety,* United States Government Printing Office, Washington, D.C., 1966.

29 Uspensky, J. V., *Introduction to Mathematical Probability,* McGraw-Hill, 1937.

30 Wilks, S. S., *Elementary Statistical Analysis,* Princeton, 1951.

31 Wilks, S. S., *Annuals of Mathematical Statistics,* Vol. 12 (1941), p. 91 and Vol. 13 (1942), p. 400.

32 Morse, Phillip M., *Queues, Inventories and Maintenance,* John Wiley and Sons, Inc., New York, New York.

33 Saaty, Thomas L., *Elements of Queueing Theory,* McGraw-Hill, New York, New York, 1959.

34 Bellman, Richard E., *Dynamic Programming,* Princeton University Press, Princeton, New Jersey, 1957.

35 Eisenhart, Churchill, *Realistic Evaluation of the Precision and Accuracy of Instrument Calibration Systems,* Journal of Research of the National Bureau of Standards 67C, No. 2, 161-187, 1963.

36 Murphy, R. B., *On the Meaning of Precision and Accuracy,* Material Research and Standard, Vol. 1, No. 4, 264-267, 1961.

absolute maximum, The function f(X) defined over a set D is said to take on its absolute maximum over D at a point X* in D if $f(X) \leqslant f(X^*)$ for every point X in D.

accelerated test, A test in which the applied stress level is chosen to exceed that stated in the reference conditions in order to shorten the time required to observe the stress response of the item, or magnify the response in a given time. To be valid, an accelerated test must not alter the basic modes and/or mechanisms of failure. (20)

acceleration factor, The ratio between the times necessary to obtain a stated proportion of failures for two different sets of stress conditions involving the same failure modes and/or mechanisms. (20)

acceptance sampling, Sampling inspection in which decisions are made to accept or reject product; also, the science that deals with the procedures by which decisions to accept or reject are based on the results of the inspection of samples. **Note 1:** The alternative to acceptance is termed "rejection" for the purpose of the definition, although, in practice, the alternative may take some form other than outright rejection. **Note 2:** In lot-by-lot sampling, acceptance and rejection relate to individual lots. In continuous sampling, acceptance and rejection relate to individual units, or to blocks of consecutive units, depending on the stated procedure. (2)

access, random, When units in the waiting line are chosen for service by some rule for random choice among them, the procedure is one of random access to service — or service by random access.

accounting prices, Internal prices assigned to company-owned resources as an aid of decision-making and control. In certain linear programming applications, the dual variables at optimality may be regarded as including a set of accounting prices for the various scarce resources. Accounting prices in the linear-programming format are called shadow prices (q.v.). (12)

accuracy, Degree of agreement of individual measurements with an accepted reference value. Normally, the degree of correspondence is obtained by comparing the discrepancy between the average of a number of observed values (q.v.) of a measurable magnitude and its standard value. (3)

active redundancy, See redundancy, active. (20)

activity, A variable whose level is to be computed in a programming problem. (19)

activity analysis problem, A linear programming problem whose constraints represent available resources and variables represent commodities to be manufactured. The i-th coefficient of an activity vector "j" represents the number of units of resources "i" required to produce a unit of commodity "j." The right-hand-side coefficients indicate the amount of each resource available, while the objective function coefficients represent the profit per unit of activity. (15)

activity level, The value taken by a variable in an intermediate or final solution to a programming problem. (19)

activity vector, A column of the constraint matrix (q.v.) associated with a variable in a programming problem.

adjacent extreme point methods, Two extreme points of a convex set (q.v.) lying on the same boundary edge are called adjacent. All methods, such as the simplex method (q.v.) which move from one extreme point (q.v.) to an adjacent one are called extreme point methods. (9)

admissible basis, A set of m linearly independent activity vectors associated with a basic feasible solution (q.v.). (See basis, basic solution.) (15)

algorithm, A mathematical procedure or set of rules whose routine application yields the solution to a particular class of problems.

alternate optima, Distinct solutions to the same optimization problem. (19)

analysis of variance, This technique partitions the total variability of a set of observations into parts, each part associated with defined sources of variation used as criteria of classification for the observations.

angular matrix, A block-triangular matrix in which all blocks are zero except those down the diagonal and along the bottom row. That is, it is a matrix of the form

$$\begin{bmatrix} A_{11} & & & \\ & A_{22} & & \\ & & \ddots & \\ A_{m1} & A_{m2} & \cdots & A_{mm} \end{bmatrix}$$

where the Aij are submatrices (or "blocks") and all omitted entries are matrices consisting entirely of zeroes. Linear-programming problems with angular matrices lend themselves to special types of computational treatment, and in particular to that of decomposition. (12)

arc, In a connected network with more than two nodes, an arc is an ordered pair (n_i, n_j), where n_i and n_j are nodes of the network.

arithmetic mean, The sum of a set of values divided by the number of values in the set.

arrival rate, mean, In a waiting line, is the expected number of arrivals occurring in a specified time unit. (12)

arrival rate distribution, constant, The term constant applied to rates and times implies that a given rate or given time varies throughout its history under study in accordance with one probability density function. (12)

arrival time, constant, See arrival rate distribution, constant. (12)

artificial basis, A set of m artificial vectors which may be used to initiate the simplex method (q.v.). (15)

artificial variables, Auxiliary variables introduced into the equality constraints, at the start of the simplex method of linear programming, in order to provide an identity basis required to initiate the procedure. (12)

artificial vector, The column vector associated with an artificial variable. It is a unit vector, with +1 in the row where the artificial variable occurs and zero elsewhere. (12)

assembly line balancing problem, An assembly line consists of a number of work stations. To assemble the product under consideration, a number of jobs must be performed subject to certain sequencing requirements concerning the order in which they are performed. Given the desired production rate of the product, management wishes to accomplish two objectives: assign jobs to balance the work among stations and minimize the total number of these stations. The times required to do each job are specified. The problem can be formulated as integer linear programming (q.v.). (18)

assignable cause, A factor contributing to the variation in quality that it is possible to detect and identify. (1)

assignment problem, The problem of placing n elements into n cells, one element to a cell, at minimum cost, where the individual cost of putting element i in cell j is a given constant c_{ij}. The problem can be stated as a linear programming problem, or a transportation problem. The problem is usually interpreted as one of assigning n persons to n jobs, such that the value of person i in job j is c_{ij}. The problem is to determine the assignment with maximum total value. (19)

attributes inspection, A term used to designate a method of inspection whereby units of product are examined to determine for each unit whether it does or does not conform to a requirement. Example: Go and Not-Go gaging of a dimension. **Note:** Inspection by attributes may be either one of two kinds — inspection whereby either the unit of product is classified simply as defective or nondefective, or the number of defects in the unit of product is counted, with respect to a given requirement or set of requirements.

augmented matrix, The coefficient matrix augmented by the column of right-hand-side constants; the same as the constraint matrix (q.v.). (19)

availability, A measure of the degree to which an item is in the operable and commitable state at the start of the mission, when the mission is called for at an unknown (random) point in time. (28)

availability (achieved), The probability that an item when used and maintained under stated conditions in an ideal support environment will be in a satisfactory state at any given time. (20)

It may be expressed as:

$$A_a\,(t) = \frac{MTBM}{MTBM + \overline{M}}$$

where

MTBM = mean time between maintenance — is a function of both preventive and corrective maintenance requirements

$\overline{M}$ = mean active maintenance downtime — a value which is a function of the repair times associated with corrective and preventive maintenance.

availability (intrinsic), The probability that an item is in satisfactory state at a stated instant of time (t) when it is used and maintained under stated conditions. It may be expressed as:

$$A_i\,(t) = \frac{MTBF}{MTBF + MTTR}$$

MTBF = mean time between failures

MTTR = mean time to repair (excludes preventive maintenance downtime, supply downtime, and administrative downtime). (20)

availability (operation), The probability that an item is in a satisfactory state at a stated instant of time (t) when it is used and maintained under stated conditions in actual support environment. It may be expressed as:

$$A_o\,(t) = \frac{MTBM}{MTBM + MDT}$$

where

MTBM = mean time between maintenance

MDT = mean downtime (includes preventive maintenance, supply and administrative downtime). (2)

average, A measure of central tendency (mean, mode, median, etc.).

basic feasible solution, A basic solution to the m equations of the linear programming problem which also satisfies the non-negativity constraints. (19)

basic solution, Given a system of m simultaneous linear equations in "n" unknowns, $Ax = b$ $(m < n)$ and with rank m. If an m x m nonsingular matrix is chosen from A, and if all the n-m variables associated with the remaining columns of this matrix are set equal to zero, the solution to the resulting system of equations is called a basic solution. (17)

basic variables, The m variables not initially set equal to zero in a basic solution (q.v.). (19)

basis, A basis in n-dimensional Euclidean space is a set of n linearly independent vectors. In linear programming, a basis is an (m x m) nonsingular matrix composed of activity vectors, i.e., a basis matrix. (15)

basis inverse, The inverse of the basis matrix. (19)

basis matrix, See basis. (19)

basis vector, A column of a basis matrix. (19)

bidding, competitive, Each firm seeking a particular contract must submit a sealed price and the firm which submits the lowest price is awarded the contract.

birth-and-death process, A stochastic process which attempts to describe the growth and decay of a population the members of which may die or generate new individuals. The types mainly studied have relatively simple laws of reproduction and mortality. (22:29 5:91 13:402 25:265)

birth process, A stochastic process describing the population of a system in which individual members may generate new members. The expression is often confined to the case where the variate (population) increases only by jumps of amount + 1, the probability of a jump from n to n + 1 in time dt being asymptotically $\lambda_n dt$. Here λ_n may also depend on t. (See also Poisson process, branching process.) (22:29 5:84 13:402 13:267)

black box, Describes a system (organism or mechanism) whose structure is unknown either because it cannot be observed or it is too complex to be understood.

blemish, An imperfection that occurs with a severity sufficient to cause awareness but that should not cause any real impairment with respect to intended normal, or reasonably foreseeable, usage requirements.

block pivot, By introducing several columns at once into the basis, the notion of pivot element may be generalized to that of block pivot. (11)

block-triangular matrix, A matrix that can be written in the form

$$\begin{bmatrix} A_{11} & & & \\ A_{21} & A_{22} & & \\ & & \cdots & \\ _{m1} & A_{m2} & \cdots & A_{mm} \end{bmatrix}$$

where the Aij are submatrices or blocks and all entries not shown are submatrices consisting entirely of zeros. The Aij need not be square, but the vertical dimensions of all blocks in one row must coincide, as well as the horizontal dimensions of all blocks in one column. (12)

bounded variable, A variable xj in a linear programming problem whose values are limited from above or below or both.

bounded variable problem, A linear programming problem in which all or most of the variables have bounds of the type $0 \leq xj \leq bj$. (15)

branching process, A stochastic process describing the growth of a population in which the individual members may have offspring, the lines of descent "branching out" as new members are born. (22:32, 5:58, 13:272)

burn-in, The operation of items prior to their ultimate application intended to stabilize their characteristics and to identify early failures. (20)

capability, A measure of the ability of an item to achieve mission objectives given the conditions during the mission. (28)

capacitated transportation problem, A bounded variable transportation problem (q.v.). (15)

catastrophic failure, See failure, catastrophic. (20)

caterer problem, A caterer requires for the next number of days r_j fresh napkins on the jth day. j = 1, 2, . . . n. Two types of laundry service are available. A slow service which requires p days and costs b cents per napkin; a fast service requiring q days $q < p$ but costing c cents per napkin $c > b$. Beginning with no usable napkins on hand or in the laundry, the caterer meets the demands by purchasing napkins at a cents per napkin. How does the caterer purchase and launder napkins to meet requirements so as to minimize the total cost of n days? (19)

chain (in a graph), A sequence of arcs $(i,i_1), (i_1,i_2), (i_2,i_3) \ldots (i_k,j)$ connecting nodes i and j is called a chain. (11)

chance causes, Factors, generally many in number but each of relatively small importance, contributing to variation, which have not been identified.

chance constrained programming, A method of treating the stochastic constraints in stochastic linear programming. A vector d_S with all components greater than 0 and less than or equal to 1 for all s in a given set S is defined. Then the stochastic linear program may be reformulated as

$$\text{MIN } cx$$

$$\text{Subject to Probability} \left\{ \begin{matrix} A^{(s)} x \geq b_S \\ x \geq 0 \end{matrix} \right\} \geq d_S \quad \text{for all s in S.}$$

d_S is a vector chosen prior to solution of the chance constrained problem. (See stochastic programming.) (9)

channels, multiple, A waiting line is said to have multiple channels when there is more than one station at which service is provided.

characteristic, A property that can serve to differentiate between items. The differentiation may be either quantitative (by variables) or qualitative (by attributes). (3)

characteristic root, The characteristic root of a square matrix A is a value λ such that $|A - \lambda I| = 0$, where I is the identity matrix. For a p x p matrix there are, in general, p such roots. They are also known as Latent Roots and Eigenvalues.

The corresponding row-vectors u or column-vectors v for which

$$uA = \lambda u \text{ or } Av = \lambda v$$

are called characteristic vectors. (22)

checkout, Tests or observations of an item to determine its condition or status. (28)

cluster sampling, Cluster sampling is simple random sampling applied to groups of population members, each group being considered as a single unit in the selection process. (26)

coefficient matrix, The matrix of left-hand-side coefficients in a system of linear equations. It is to be distinguished from the matrix obtained by appending the right-hand side, which is called the "augmented matrix" of the system. (19)

coefficient of loss of service, The ratio of the average number of idle service stations (servers) to the total number of service stations (servers).

column vector, One column of a matrix, or a matrix consisting of a single column. The elements of the column are interpreted as the components of the vector. (19)

complementary slackness theorem, For symmetric primal and dual linear programs (q.v.) the following theorem holds:

Whenever inequality occurs in the ith relation of the primal (or dual) constraints for an optimizing solution, then the ith dual (or primal) variable of an optimizing solution vanishes. Conversely, if the ith variable of the dual system is positive, then the ith relation of the primal system is an equality. (15)

composite algorithms, For linear program, if neither the basic solution nor the dual solution generated by its simplex multipliers, remain feasible, the corresponding algorithm is called composite. (19)

concave function, Geometrically, a function whose graph is never below the chords joining any two points on the graph. A function f such that

$$\alpha_1 f(x_1) + \alpha_2 f(x_2) \leq f(\alpha_1 x_1 + \alpha_2 x_2) \text{ for all}$$

$$\alpha_1 \geq 0, \alpha_2 \geq 0, \alpha_1 + \alpha_2 = 1 \text{ and all } x_1 \text{ and } x_2$$

in the convex domain of definition of f.

A function for which the inequality always holds when $\alpha_1 > 0$ and $\alpha_2 > 0$ is called strictly concave. (19)

cone, A set of vectors s is called a cone if for every vector X in S, λ X is in S for all $\lambda \geq 0$. (15)

confidence interval, An interval, estimated from sample values, expressed in terms of its probability (confidence) of including the parameter under estimate. The assertion that the parameter lies in the interval will be true in a proportion (confidence) of the cases when the assertion is made. (15)

constant arrival rate distribution, See arrival rate distribution, constant.

constant failure period, That period during which failure of units of a particular item occurs at an approximately uniform rate. (20)

constant failure rate period, That possible period during which the failures occur at an approximately uniform rate. (20)

constant vector, The right-hand side of a set of linear inequalities (equalities). By convention the linear relations of a linear programming problem are so arranged that all variables and their coefficients appear on the left, and the column of constants appears on the right. (19)

constraint, An equation or inequality relating the variables in an optimization problem. (19)

constraint matrix, In linear programming, the augmented matrix of the constraint equations. It is the matrix formed by the coefficient columns, or left-hand sides, and the column of constants. (19)

constraint qualification, This qualification places restrictions on the nature of the vicinity of a point satisfying certain conditions for optimality. These restricitons insure that the conditions are indeed, properly related to optimality.

control chart, A graphical representation with control limits and plotted values of some statistical measure for a series of samples or subgroups. A central line is commonly shown. (3)

convex combination, For the points $X_1, X_2, \ldots, X_n$ a convex combination is a point

$$X = \alpha_1 X_1 + \alpha_2 X_2 + \ldots + \alpha_n X_n$$

where

$$\alpha_i \geqslant 0 \text{ and } \Sigma_i \alpha_i = 1.$$

convex cone, A cone is convex if the positive sum of any two vectors in the cone is also in the cone.

convex function, Geometrically, a function whose graph is never above the chords joining two points on the graph. A function f such that

$$\alpha_1 f(x_1) + \alpha_2 f(x_2) \geqslant f(\alpha_1 x_1 + \alpha_2 x_2) \text{ for all}$$

$$\alpha_1 \geqslant 0, \alpha_2 \geqslant 0, \alpha_1 + \alpha_2 = 1 \text{ and all } x_1 \text{ and } x_2$$

in the convex domain of definitions of f. (19)

convex hull, The convex hull of a set of points S is the set of all convex combinations of sets of points from S. (15)

convex polyhedral cone, A cone generated by a convex polyhedron. (9)

convex polyhedron, The convex hull of a finite number of points. (15)

convex programming, Optimization of a convex function over a convex region. Linear programming and quadratic programming are special cases of convex programming. (19)

convex set, Geometrically, a set of points that contains all the points on the line segment joining any two points of the set; i.e., a set is convex if and only if each convex combination of any two points in the set is also in the set. (15)

convolution, Let $F_1(x), F_2(x), \ldots F_n(x)$ be a sequence of distribution functions. The distribution

$$f(x)=\int_{-\infty}^{\infty} dF_1(x_1) \ldots \int_{-\infty}^{\infty} dFn(x - x_1 \ldots - x_{n-1}(x_{n-1})$$

is called the convolution of the distributions. The relationship is sometimes written

$$F(x) = F_1(x)^* F_2(x)^* \ldots F_n(x).$$

If the associated variates are independent, F(x) is the distribution function of their sum. (22:65, 5:8, 13:250, 25:404)

correlation, A measure of the interdependence between two variates. It is usually a pure number which varies between − 1 and 1 with the intermediate value of zero indicating the absence of correlation, but not necessarily the independence of the variates. The limiting values indicate perfect negative or positive correlation, respectively. (22)

cost coefficient, The coefficient of a variable in the cost function of a linear program. The elements of any objective function are sometimes referred to as cost coefficients. (19)

cost function, The objective function of a cost minimization program. (19)

cost range, The range of values of a cost coefficient of a basic solution variable for which the basis stays optimum. (19)

cost row, The row of objective function coefficients of a linear program matrix. (19)

coupling equations, When some of the constraints involve only the variables X and some others involve only the variables Y, the remaining equations involving both X and Y are called coupling equations. (19)

cumulants, These are values that are given by the coefficients in the expansion of a power series formed from the logarithms of the characteristic function of a variable.

cumulative distribution function, F(x) of a variate x is the total frequency of members with variate values less than or equal to x. As a general rule the total frequency is taken to be unity, in which case the distribution function is the proportion of members bearing values $\leqslant$ x. (22:87, 9:38, 10:58, 13:168)

cut (in a network), A collection of oriented arcs such that every oriented path from source to sink contains at least one arc in the collection. Also a partition of the nodes into two sets such that the source node belongs to one set and the sink node to the other.

cutting plane, The hyperplane boundary of an added linear inequality constraint in a programming problem. It may be thought of as "cutting off" part of the original convex region of feasible solutions in order to form a new convex region. (11)

cycling (linear programming), The repeating of a basic feasible solution during application of the Simplex Algorithm (q.v.). (15)

debugging, The check out of equipment, complex items, or computer programs prior to use in order to detect design and/or operational difficulties.

decision function, A decision function is a rule of conduct which, at any stage of a sampling investigation, tells the statistician whether to take further observations or whether enough information has been collected, and in the latter case, what decision to make upon it. At each stage beyond the first the decision function is a function of the preceding observations. (22)

decision under certainty, If a choice must be made between two or more actions, decision under certainty involves that realm where each action is known to lead invariably to a specific outcome.

decision under risk, Decision under risk is concerned with that realm of choice where each action leads to one of a set of possible specific outcomes, each outcome occurring with a known probability.

decision under uncertainty, Decision under uncertainty involves that realm of choice where each action or combination has as its consequence a set of possible specific outcomes, but where the probabilities of these outcomes are completely unknown or are not even meaningful.

decomposition principle, A method for subdividing a large linear programming problem into smaller, more manageable linear programs; solving the smaller problems in the standard way and then combining these solutions to solve the original problem. In particular linear programs which qualify for decomposition are those for which the variables can be separated into classes X, Y, Z, . . . and the constraints into classes C_0, C_1, C_2, . . . such that: The constraints in class C_0 can involve all the variables; constraints in class C_1 involve only the variables in X; the constraints in class C_2 involve only the variables Y; the constraints in class C_3 involve only the variables Z; and so on. The subsets of constraints each involving only a part of the variables are called subprograms; while the subset of constraints in C_0 is called the "master program." (12)

defect, An imperfection that occurs with a severity sufficient to cause a product or service not to satisfy intended normal, or reasonably foreseeable, usage requirements.

degenerate solution, For linear programs, a basic solution in which at least one basic variable is zero. (9)

degradation failure, See failure, degradation. (20)

delay, See waiting time. (33)

dependability, A measure of the item operating condition at one or more points during the mission, including the effects of reliability, maintainability, and survivability, given the item condition(s) at the start of the mission. It may be stated as the probability that an item will (a) enter or occupy any one of its required operational modes during a specified mission, and (b) perform the functions associated with those operational modes. (28)

dependent variable, See independent variable.

derating, (1) Using an item in such a way that applied stresses are below rated values, or (2) The lowering of the rating of an item in one stress field to allow an increase in rating in another stress field. (28)

designated imperfection, A category of imperfections which, because of their type and/or severity, are to be treated as an event for control purposes.

deterministic model, As opposed to a stochastic model, a model which contains no random elements and for which, therefore, the future course of the system is determined by its state at present (and/or in the past).

diet problem, A linear programming problem in which the constraints represent minimum daily requirements for specified nutrients and variables represent foods. The ith coefficient of the jth activity vector represents the amount of the ith nutrient in one unit of the jth food. The right-hand side coefficients indicate the daily requirement of each nutrient, while the objective function coefficients represent the cost per unit of each food. (15)

diophantine programming, See integer linear programming. (9)

discrete variable problem, See integer linear programming. (19)

distribution-free method, A method, e.g., of testing a hypothesis or of setting up a confidence interval which does not depend on the form of the underlying distribution.

downtime, See time, down. See also Work Measurement and Methods. (28)

dual linear programming problems, A pair of linear programs, called "primal" and "dual" of the following form:

Unsymmetric Case:

Primal	Min cx $AX = b$ $x \geq 0$	subject to
Dual:	Max Wb $WA \leq c$	subject to

Symmetric Case:

Primal	Min cx $AX \geq b$ $x \geq 0$	subject to
Dual:	Max Wb $WA \leq c$ $W \geq 0$	subject to

(15)

dual problem, See dual linear programming problems. (19)

dual simplex algorithm, Starting with a dual-feasible solution, the algorithm selects the most infeasible vector to leave the primal basis and then computes which vector must enter to maintain dual feasibility. As a consequence of this change of basis, other variables may become primal infeasible, but since the objective value changes monotonically toward optimum, the algorithm is finitely convergent. It is exactly the primal algorithm applied to the dual problem. (19)

duality theorems for linear programming, (1) Main theorem: If both the primal and dual problems have a finite optimum, then the optimum values of the objective functions are equal. (2) Corollary: If either problem has a feasible finite optimum, then so does the other, and the optimum values are equal. (3) Corollary: A feasible but unbounded solution to one problem implies no feasible solution for the other.

(4) Corollary: No feasible solution to one problem implies that the other is either unbounded or infeasible. (5) Weak theorems of the alternative: A variable and its complementary slack are not both nonzero. (6) Strong theorem of the alternative: Among all alternate optima, at least one solution exists in which a variable and its complementary slack are not both zero, and the one of the pair that is zero in this solution is zero in all alternate solutions. (19)

dynamic programming, A method for optimizing a set of decisions which may be made sequentially. Characteristically, each decision may be made in the light of the information embodied in a small number of observables called state variables. The incurred cost for each period is a mathematical function of the current state and decision variables, while future states are functions of these variables. The aim of the decision policy is to minimize the total incurred cost, or equivalently the average cost per period. (19)

early failure period, That possible early period, beginning at some stated time and during which the failure rate is decreasing rapidly in comparison with the subsequent period.

elementary matrix, A unit or identity matrix except for one unique column or row. The inverse of a nonsingular elementary matrix is also an elementary matrix. (19)

elimination method, An alogrithm for solving a set of m x m independent linear equations. (See Gaussian elimination.) (19)

equation, A proposition (or relation) that a given mathematical function shall be equal to another function (often a constant). (19)

error mean-square, The residual or error sum-of-squares (q.v.) divided by the number of degrees of freedom on which the sum is based. It provides an estimator of the residual or error variance. (22)

error sum of squares, In the analysis of variance and covariance, and regression analysis, it is the sum of squares that remains after subtracting from the total sum of squares all sums of squares associated with the factors of the experiment. It also is referred to as the residual sum of squares.

estimation, Inference about the numerical value of unknown population values from incomplete data such as a sample. If a single figure is calculated for each unknown parameter the process is called point estimation. If an interval is calculated within which the parameter is likely, in some sense, to lie, the process is called interval estimation. (22)

event, An occurrence of some attribute.

exponential service time, When the probability density of the service time has an exponential distribution, this is referred to as exponential service time.

extremal problem, See master program.

extreme point, A point of a convex set which does not lie on a line segment joining any other two points of the set; also called a "vertex." (19)

factor analysis, A branch of multivariate analysis in which the observed variates x_i ($i = 1, 2, \ldots, p$) are supposed to be expressible in terms of a number $m < p$ factors f_j together with residual elements.

failure, The termination of the ability of an item to perform its required function. (3)

failure analysis, The logical, systematic examination of an item or its diagram(s) to identify and analyze the probability, causes, and consequences of potential and real failures. (28)

failure, catastrophic, Failures which are both sudden and complete. (20)

failure, complete, Failure resulting from deviations in characteristic(s) beyond specified limits such as to cause complete lack of the required function. Note: The limits referred to in this category are special limits specified for this purpose. (20)

failure criteria, Rules for failure relevancy such as specified limits for the acceptability of an item. (20)

failure, degradation, Failures which are both gradual and partial. (20)

failure, dependent, One which is caused by the failure of an associated item(s). Not independent. (28)

failure, gradual, Failures that could be anticipated by prior examination. (20)

failure, independent, One which occurs without being related to the failure of associated items. Not dependent. (28)

failure, inherent weakness, Failures attributable to weakness inherent in the item itself when subjected to stresses within the stated capabilities of that item. (20)

failure mechanism, The physical, chemical or other process which results in a failure. (20)

failure, misuse, Failures attributable to the application of stresses beyond the stated capabilities of the item. (20)

failure mode, The effect by which a failure is observed; for example, an open or short circuit condition, or a gain change. (20)

failure, partial, Failures resulting from deviations in characteristic(s) beyond specified limits but not such as to cause complete lack of the required function. (20)

failure, random, Any failure whose occurrence is unpredictable in an absolute sense but which is predictable only in a probabilistic or statistical sense. (28)

failure rate, The number of failures of an item per unit measure of life (cycles, time, miles, events, etc., as applicable for the item). (28)

failure rate acceleration factor, The ratio of the accelerated testing failure rate to the failure rate under stated reference test conditions and time period. (20)

failure rate, assessed, The failure rate of an item determined as a limiting value or values of the confidence interval with a stated confidence level, based on the same data as the observed failure rate of nominally identical items.

failure rate, extrapolated, Extension by a defined extrapolation or interpolation of the observed or assessed failure rate for durations and/or conditions different from those applying to the conditions of that observed or assessed failure rate.

failure rate, observed, The ratio of the total number of failures in a sample to the total cumulation observed time on that sample. The observed failure rate is to be associated with particular, and stated time intervals (or summation of intervals) in the life of the items, and with stated conditions.

failure rate, predicted, For the stated conditions of use and the design considerations of an item, the failure rate computed from the observed, assessed or extrapolated failure rates of its parts. (20)

failure, secondary, Failure of an item caused either directly or indirectly by the failure of another item. (20)

failure, sudden, Failures that could not be anticipated by prior examination. (20)

failure, wear-out, A failure which occurs as a result of deterioration processes or mechanical wear and whose probability of occurrence increases with time. (20)

fair game, In probability theory, a game consisting of a sequence of trials is deemed to be a "fair" game if the cost of each trial is equal to the expected value of the gain from each trial. In game theory, a game which with proper play neither adversary has an advantage.

Farkas' lemma, If for every solution of $WA \leq 0$ it is also true that $Wb \leq 0$, then there exists a vector $X \geq 0$ such that $Ax = b$; and thus $(WA)\,X = Wb$. (If a linear homogeneous inequality $Wb \leq 0$ holds for all W satisfying a system of homogeneous inequalities $WA \leq 0$, then the inequality can be expressed as a non-negative combination of the inequalities of the system $WA \leq 0$.) (11)

fat solution, See permanently feasible set. (19)

feasible basis, A basis yielding a basic feasible solution (q.v.). (19)

feasible solution, A nonnegative solution satisfying the constraints of a mathematical programming problem. (19)

finite and infinite games, A game is finite if each player has only a finite number of possible pure strategies; it is infinite if at least one player has an infinite number of possible pure strategies (e.g., a pure strategy might ideally consist of choosing an instant from a given interval of time at which to fire a gun). (21)

fixed charge problem, Nonlinear programming problem of the form.

$$\begin{aligned} &\text{Minimize} && \Sigma f_j(x_j), \\ &\text{Subject to} && AX \geq 0, \\ & && X \geq 0, \end{aligned}$$

$$\text{Where } f_j(x_j) = c_j x_j + d_j \delta_j,\ d_j > 0,$$

$$\text{And } \delta_j = \begin{cases} 0 \text{ if } x_j = 0, \\ 1 \text{ if } x_j > 0. \end{cases}$$

The d_j are called fixed charges since d_j is incurred only if $x_j > 0$. (18)

fractional programming, A class of mathematical programming problems in which the objective function is the quotient of linear functions. (19)

frame, The list of units, or items, accessible for test. Each unit has a serial number associated with it, actually or conceptually. If there are units in the population that are not covered by the frame, statistical inferences (estimates, confidence limits, etc.) refer to the frame, not the population. Generalizations from the frame to the population must be based on judgment. (3)

gambler's ruin, The name given to one of the classical topics in probability theory. A game of chance can be related to a series of Bernoulli trials at which a gambler wins a certain predetermined sum of money for every success and loses a second sum of money for every failure. The play may proceed until his initial capital is exhausted and he is ruined. The statistical problems involved are concerned with the probability of the ruin of a player, given the stakes, initial capital and chances of success, and with such matters as the distribution of the length of play. There are many variations to this classical problem, which is closely associated with problems of the random walk (q.v.) in particular, of sequential sampling. (22:117, 5:25, 13:334, 25:144)

game theory, The study of the following problem: If n players P_1, P_2, . . . P_n play a given game G, how must the ith player P_i play to achieve the most favorable result? Special two-person games can be solved by linear programming methods. More generally, the mathematical study of cooperative/competitive situations. (15)

Gaussian elimination, A reduction method for systems of simultaneous linear equations, in which one of the equations is solved for one of the variables in terms of the remaining variables. When these expressions for the solved variable are substituted into the remaining equations, the result is an equivalent system with one less equation and one less variable. A Gaussian elimination step is exactly equivalent to a pivot step. It is a single change of basis and can be expressed functionally as premultiplying by the inverse of an appropriate elementary column matrix. Sufficient repetition of this procedure can yield the numerical solution in case of a nonsingular square system, and a solution of parametric form (a linear function of a subset of the variables) if the number of variables exceeds the number of equations. (19)

geometric solution, A graphic method of solving a linear programming problem, by plotting the half-planes determined by the constraints and the lines of constant value for the objective function. (19)

global optimum, A feasible solution (q.v.) which gives a value to the objective function at least as great (small) as any other in the feasible region. It is contrasted with a local optimum, which yields the best objective function value of all points in some subset of the feasible region. In convex programming problems a local optimum is a global optimum. (See absolute maximum.) (19)

gradient of a function, A vector at a point, whose direction is the direction of most rapid change of some function f, and whose magnitude is the rate of change of f in that direction. (19)

gradient methods, A term applied to algorithms which at each iteration in their use seek to improve the current function value by moving in the direction of the gradient of the function f (x) to be optimized. The application of this method in the presence of constraints gives rise to numerous forms depending upon the type of problem and the manner in which the method is modified to handle constraints. (18)

graph (linear), Compare network. A linear graph consists of a number of nodes or junction points, each joined to some or all of the others by arcs or lines. (11)

graph, signal flow, See signal flow graph.

half-space, The set of points whose coordinates satisfy a linear inequality. (11)

holding time, Time that an item is not operational, so that it may be serviced.

Hungarian method, A combinatorial procedure designed for solving an assignment problem. It is based on a proof by the Hungarian mathematician Egervary. (11)

hyperplane, The set of all points X in n-space whose coordinates $(x_1, \ldots, x_n)$ satisfy a linear equation of the form.

$$a_1 x_1 + \ldots + a_n x_n = b. \qquad (19)$$

identity matrix, A square matrix with diagonal elements all unity and all other elements zero. (19)

implicit prices, Same as marginal values, shadow prices, dual variable levels, i.e., numbers giving the incremental worth of a relaxation of one unit in the right-hand side of a constraint. (19)

independent equations, A set of equations none of which can be expressed as a linear combination of the others. With linear equations, the condition for independence is that the matrix (coefficient columns) shall be nonsingular or, equivalently, have rank equal to the number of equations. (19)

independent trials, The successive trials of an event are said to be independent if the probability of outcome of any trial is independent of the outcome of the others. The expression is usually confined to cases where the probability is the same for all trials. In the sampling of attributes, such a series of trials is often referred to as "Bernoullian Trials." It includes all the classical cases of drawing colored balls from urns with replacement after each draw, coin tossing, dice rolling, and the events associated with other games of chance. (22:136, 13:118, 25:95, 29:44)

independent variable, This term is regularly used in contradistinction to "dependent variable" in regression analysis. When a variate y is expressed as a function of variables x_1, x_2, etc., plus a stochastic term the x's are known as "independent variables." The terminology is rather unfortunate since the concept has no connection with either mathematical or statistical dependence. The usage is so convenient, however, and so common that strongly coordinated action would be necessary to change it. (22)

inequality, A proposition which relates the magnitudes of two mathematical expressions or functions A and B. Inequalities are of four types: A is greater than B $(A > B)$; A is less than B $(A < B)$; A is greater than or equal to B $(A \geq B)$; A is less than or equal to B $(A \leq B)$. The first two types are called "strict" and the last two "relaxed" or "weak." (19)

inequality relation, A constraint which is an inequality. (19)

infeasible basis, A basis corresponding to an infeasible solution. Postmultiplying the inverse of such a basis by the constant vector, one obtains a solution in which at least one variable value is negative. (19)

initial basis, The set of column vectors associated with the basic variables for which a starting solution will be obtained in linear programming. It is often an identity matrix. (19)

initial feasible basis, The set of column vectors associated with the basic variables of the first feasible solution to a linear programming problem. (19)

initial solution, The solution used at the beginning of an iterative solution to a mathematical problem.

in-process inspection, Inspection which is performed during the manufacturing or repair cycle in an effort to prevent defectives from occurring and to inspect the characteristics and attributes as early as expedient.

input-out coefficient, A coefficient of an activity vector of the activity analysis problem. The (i, j) coefficient represents the amount of resource i required to produce one unit of activity j. (19)

inspection, The process of measuring, examining, testing, gauging or otherwise comparing the unit with the applicable requirements.

instantaneous failure rate (hazard), At a particular time, the rate of change of the number of items which have failed divided by the number of items surviving. (20)

integer form, A cutting plane used to solve integer linear programs. (11)

integer linear programming, A linear programming problem with the added restriction that some or all of the variables are constrained to be integers. (19)

interarrival time, The time between two successive arrivals. (Mean interarrival time: the arithmetic average of the actual interarrival times or the expected value of the interarrival time distribution; reciprocal of interarrival rate.)

interindustry (input-output) analysis, An interpretation of the economy in terms of the interactions of industrial organizations. Also termed "Leontief Model." (19)

inventory, active, That group of items in storage assigned an operational status. (28)

inventory, inactive, That group of items in storage being held in reserve for possible future commitment to the operational inventory. (28)

inverse of A matrix, An inverse of the square matrix A is another matrix B of the same dimension such that $AB = BA = I$, where I is the identity matrix. (19)

inversion, An operation on a matrix yielding the matrix's inverse (19)

item, interchangeable, One which (1) possesses such functional and physical characteristics as to be equivalent in performance, reliability, and maintainability, to another item of similar or identical purpose; and (2) is capable of being exchanged for the other item (a) without selection for fit or performance, and (b) without alteration of the times themselves or of adjoining items, except for adjustment. (28)

item, replaceable, One which is interchangeable with another item, but which differs physically from the original item in that the installation of the replaceable item requires operations such as drilling, reaming, cutting, filing, shimming, etc., in addition to the normal application and methods of attachment. (28)

item, substitute, One which possesses such functional and physical characteristics as to be capable of being exchanged for another only under specified conditions or for particular applications and without alteration of the items themselves or of adjoining items. (28)

item, unit, An object on which a measurement or observation may be made. (3)

iteration, A single cycle of operations in a solution algorithm made up of a number of such cycles. (19)

Kuhn-Tucker conditions, A set of conditions which are necessary and sometimes sufficient for a given feasible solution of a mathematical programming problem to yield optimum. (19)

knapsack problem, If a_i is the weight of the ith object and b_i is its relative value, then the problem of outfitting a knapsack in order to maximize the total relative value and not exceed a weight limitation c is given by:

$$\text{Max} \quad \sum_i b_i x_i \qquad \text{subject to}$$

$$\sum_i a_i x_i \leqslant c$$

where $x_j = 0$ or 1. It is an integer linear program. (11)

Lagrange multipliers, Auxiliary variables $\lambda_1, \lambda_2, \ldots, \lambda_m$ used in optimizing a function $f(x_1, x_2, \ldots, x_n)$ subject to the m constraints, $\phi = 0, \ldots\ldots, \phi_m = 0$ on the x_j. The Lagrange function $F(\chi, \lambda) = F(\chi) + \sum_i^m \lambda_i \phi_i$ is formed and one set of (m + n) partial derivatives (for each χ_j and λ_i) are set equal to zero and solved for an optimizing set of χ_j and corresponding set of λ. (19)

Laplace transform, If a function g(t) is related to a second function f(x) by the equation

$$g(t) = \int_0^\infty e^{-tx} f(x) dx$$

then g (t) is the Laplace transform of f (x).

In statistical theory it is more customary to use the Fourier transform, which has certain advantages over the Laplace form; e.g., if f (x) is a frequency function the Fourier transform always exists whereas the Laplace transform may not do so for real t. (22)

left-hand side, The mathematical expression to the left of the equality or inequality sign in an equation or inequality. In linear programming, by convention, the left-hand side of each constraint is the complete linear function, while the right-hand side is the constant term. (19)

Leontief system, See interindustry analysis.

lexicographic ordering, Dictionary ordering. (19)

linear combination, An expression of the form

$$a_1 x_1 + a_2 x_2 + \ldots + a_n x_n$$

where the a_i are coefficients and the x_i are variables or vectors. (19)

linear constraint, A constraint which contains variables which are related in terms of linear combinations. (19)

linearly dependent, A set of vectors $\{X_i\}$ is linearly dependent if a set of numbers a_j, not all equal to zero, can be found such that

$$a_1 X_1 + a_2 X_2 + \ldots + a_k X_k = 0.$$

(19)

linear equation, An equation whose left-hand side and right-hand side are both linear functions of the variables. Such an equation can always be put in the form

$$f(x, y, z, \ldots) = c.$$

where f is a linear function and c is a constant. (19)

linear estimator, An estimator which is a linear function of the observations. (22)

linear function, A function of the form

$$a_0 + a_1 x_1 + a_2 x_2 + \ldots + a_n x_n,$$

where the a_i are coefficients, not all 0, and the x_i are variables. The geometrical representation of a linear function is a straight line, plane, or hyperplane.

linear independence, A set of vectors (X_i) is linearly independent if the only set of numbers ai for which

$$a_1 X_1 + a_2 X_2 + \ldots + a_k X_k = 0$$

is $\quad a_1 = a_2 = \ldots = a_k = 0.$

(19)

linear inequality, An inequality whose left-hand side and right-hand side are both linear functions of the variables. (19)

linear model, A model where each dependent variable is a linear function of independent variables.

linear programming, The concept of expressing the inter-relationship of activities of a system in terms of a set of linear constraints in nonnegative variables. A program, i.e., values of the variables, is selected which satisfies the constraints and optimizes a linear objective function in these variables. (17)

linear programming problem, The problem of minimizing or maximizing a linear function in n variables subject to m linear constraints, with the variables restricted to be non-negative. Mathematically, we have

$$\text{Min (max)} \quad cX \qquad \text{subject to}$$

$$AX = b$$

$$X \geqslant 0$$

with A an (mxn) matrix.

The constraints AX = b can also be given in terms of inequalities, i.e., $AX \geqslant b$, $AX \leqslant b$ or a combination of such constraints. (19)

local optimum, See global optimum. (19)

loop (in a graph), A chain of arcs connecting node i to itself in a graph is called a loop (if the arcs are distinct, the loop is a simple loop). (11)

lot tolerance percent defective (LTPD), Expressed in percent defective, the poorest quality in an individual lot that should be accepted. NOTE: The LTPD is used as a basis for some inspection systems, and is commonly associated with a small consumer's risk. (See risk.) (2)

machine-assignment problem, Given m machines or machine types and n products or jobs; let d_{ij} be the time required to process one unit of product j on machine i, x_{ij} the number of units of j produced on machine i in the coming time period, a_i the time available on machine, b_j the number of units of j which must be completed, and c_{ij} the cost of processing one unit of product j on machine i. The machine-assignment problem thus reduces to finding nonnegative x_{ij} which will satisfy the constraints

$$\sum_{j=1}^{n} d_{ij}x_{ij} = a_i \qquad i = 1, \ldots, m$$

$$\sum_{i=1}^{m} x_{ij} = b_j \qquad j = 1, \ldots, n$$

and minimize the cost $Z = \sum_i \sum_j c_{ij}x_{ij}$. (17)

maintainability, A characteristic of design and installation which is expressed as the probability that an item will be retained in or restored to a specified condition within a given period of time, when the maintenance is performed in accordance with prescribed procedures and resources. (28)

maintenance, All actions necessary for retaining an item in or restoring it to a specified condition. (28)

maintenance, corrective, The actions performed, as a result of failure, to restore an item to a specified condition. (28)

maintenance, preventive, The actions performed in an attempt to retain an item in a specified condition by providing systematic inspection, detection and prevention of incipient failure. (29)

man-function, The function allocated to the human component of a system. (28)

master program, That part of a decomposed problem which expresses the extreme point solution of the subproblems as convex combinations and which also satisfy the constraints in the class C_0 (the class of constraints which can include all the variables). (See decomposition principle.) (19)

mathematical programming, The general problem of optimizing a function of several variables subject to a number of constraints. If the function and constraints are linear in the variables and a subset of the constraints restrict the variables to be nonnegative, we have a linear programming problem. (19)

matrix, A rectangular array of mn numbers arranged in m rows and n columns as follows:

$$\begin{vmatrix} a_{11} & \cdots & a_{1n} \\ \cdot & & \\ \cdot & & \\ \cdot & & \\ a_{m1} & \cdots & a_{mn} \end{vmatrix}$$

(19)

matrix element, One of the mn numbers of a matrix. (19)

matrix game, A zero-sum two-person game. The payoff (positive, negative or zero) of player one to player two when player one plays his i^{th} strategy and player two his j^{th} strategy is denoted by a_{ij}. The set of payoffs a_{ij} can be arranged into an m x n matrix. (19)

max-flow min-cut theorem, A theorem which applies to a maximal flow network problem which states that the maximal flow from the source to the sink is equal to the minimal cut capacity of all cuts separating the source and sink nodes. (14)

maximal network flow problem, Consider a connected linear network in which goods must flow from an origin (source) to a final destination (sink) over the arcs of the network in such a fashion as to maximize the total flow that the network can support. The arcs are connected at intermediate nodes and each branch has a given flow capacity that cannot be exceeded. The flow into an intermediate node must equal the flow out of that node. (15)

mean life, The arithmetic mean (q.v.) of the times-to-failure of the units of a given item. (20)

mean-maintenance-time, The total preventive and corrective maintenance time divided by the total number of preventive and corrective maintenance actions during a specified period of time. (28)

mean service rate, The conditional expectation of the number of services completed in one time unit, given that service is going on throughout the entire time unit.

mean-time-between-failures (MTBF), For a particular interval, the total functioning life of a population of an item divided by the total number of failures within the population during the measurement interval. (28)

mean-time-to-failure, This expression applies to components of a system and not to systems. The expected length of time that a component is in *use* in a system in operation from the moment the component is put into the system to the moment it fails. It is the expected length of life of the components. The mean-time-to-failure is sometimes confused with the mean-time-between-failures (MTBF) which applies to systems which experience subsystem or component failures which can be replaced or repaired, putting the system into operation again. The confusion occurs in part because for the Poisson failure distributions the two quantities have the same distribution.

mean-time-to-repair (MTTR), The total corrective maintenance time divided by the total number of corrective maintenance actions during a given period of time. (28)

median, An average with the number of data points above and below it equal; the 50th percentile value. (28)

minimal cost flow problem, A network flow problem in which costs c_{ij} are the cost of shipping one unit of a homogeneous commodity from node i to node j. A given quantity F of the commodity must be shipped from the origin node to the destination at minimum cost. (19)

minimax principle, A principle introduced into decision-function theory by Wald (1939). It supposes that decisions are taken subject to the condition that the maximum risk in taking a wrong decision is minimized. The principle has been criticized on the grounds that decisions in real life are scarcely ever made by such a rule, which enjoins "that one should never walk under a tree for fear of being killed by its falling." In the theory of games it is not open to the same objection, a prudent player being entitled to assume that his adversary will do his worst. (22)

minimax theorem, Let $\bar{X}$ and $\bar{Y}$ represent the mixed strategy sets available to player 1 and player 2, respectively, in a matrix game with payoff matrix A. By definition

$$\max_{X \text{ in } \bar{X}} \min_{Y \text{ in } \bar{Y}} E(X, Y) = \max_{X \text{ in } \bar{X}} \min_{Y \text{ in } \bar{Y}} XAY$$

and

$$\min_{Y \text{ in } \bar{Y}} \max_{X \text{ in } \bar{X}} E(X, Y) = \min_{Y \text{ in } \bar{Y}} \max_{X \text{ in } \bar{X}} XAY$$

where $X \geqslant 0$, $Y \geqslant 0$ and the components of each X and Y sum to unity.

Theorem 1. For any (mxn) matrix A the maximin and minimax values, as defined above, exist and are equal.

Theorem 2. Every matrix game has a saddle point which defines an optimal strategy pair X^* and Y^* and a value of the game $V^* = E(X^*, Y^*)$. Analogous theorems hold, as well, for certain more complicated classes of games.

mission, The objective or task, together with the purpose, which clearly indicates the action to be taken. (28)

mixed strategy, A mixed strategy for a player P is a vector $X = (x_1\ x_2, \ldots, x_m)$ of nonnegative numbers x_i such that

$$\sum_{i=1}^{m} x_i = 1.$$

The elements x_i represent the frequencies with which the player selects his i^{th} strategy.

mode, The mode was originally conceived of as that value of the variate which is possessed by the greatest number of members of the population. Although the idea of the most frequently encountered (or fashionable) value of the variate is probably very old, it was not generally used in statistics until popularized by K. Pearson (1894). The concept is essentially of use only for continuous distributions, although it can be extended to the discontinuous case. More formally, if f (x) is a frequency function, a mode is a value of x for which

$$\frac{df(x)}{dx} = 0, \frac{d^2 f(x)}{dx^2} < 0.$$

There may thus be more than one mode of a distribution, though the practical occurrence of multimodality is comparatively rare. (22)

model, A mathematical or physical representation of a system (q.v.) often used to explore the many variables (q.v.) influencing the system.

monotone, A monotonic (or monotone) increasing quantity is a quantity which never decreases (the quantity may be a function, sequence, etc., which either increases or remains the same, but never decreases). A sequence of sets $E_1, E_2, \ldots$ is monotonic increasing if E_n is contained in E_{n+1} for each n. A monotonic (or monotone) decreasing quantity is a quantity which never increases (the quantity may be a function, sequence, etc., which either decreases or remains the same, but never increases).

A sequence of sets, $E_1, E_2, \ldots$ is monotonic decreasing if E_n contains E_{n+1} for each n. A monotonic (or monotone) system of sets is a system of sets such that, for any two sets of the system one of the sets is contained in the other. A mapping of a topological space A onto a topological space B is said to be monotone if the inverse image of each point of B is a continuum. A mapping of an ordered set A onto an ordered set B is monotone provided x^* precedes (or equals) y^* whenever x^* and y^* are the images in B of points x and y of A for which x precedes y. (21)

monotonic increasing function, A function f is monotonic increasing on an interval (a, b) if $f(y) \geqslant f(x)$ for any two numbers x and y (of this interval) for which $x < y$. (21)

Monte Carlo method, A simulation technique by which approximate evaluations are obtained in the solution of mathematical expressions so as to determine the range or optimum value. The technique consists of simulating an experiment to determine some probabilistic property of a system or population of objects or events by the use of random sampling applied to the components of the system, objects, or events.

moving average, The moving average is the series of arithmetic averages obtained by averaging the last k successive terms in a time series, these terms being spaced at equal intervals. Thus, each average is usually obtained by dropping from the computation the oldest term used in the preceding group of k terms and adding in its place the newest term to be included in the new k group. Weights may be assigned to each term of the group of k successive terms.

mixed-integer programming, Integer programs in which some, but not all, of the variables are restricted to integral values. (19)

MTBF, See mean-time-between-failures. (20)

multicommodity network problem, A network problem in which more than one commodity can flow through the network at the same time. The capacity constraints on the arcs apply to the sum of the flows of the commodities. (19)

multiperson game, A game where a number of players are each competitively striving towards a specific objective which any one player only partially controls.

multiple regression, A statistical procedure to determine a relationship between the average of a dependent variable and the given values of one or more independent variables.

n-person game, A multiperson game in which the success of (pay off to) each player depends not only on his actions but also on those of the other players.

near-optimum solution, A feasible solution to a mathematical programming problem whose value of the objective function can be shown to be near the optimum value. (19)

negative definite or semidefinite quadratic form, X^TAX is negative definite (semi-definite) if $-X^TAX$ is positive definite (semidefinite). (18)

network flow problem, See maximal network flow problem.

node, One of the defining points of a network; a junction point joined to some or all of the others by arcs. (9)

nonbasic variable, A variable in a solution to a linear programming problem obtained by the simplex method and whose value has been arbitrarily set equal to zero. (19)

non-conformity, An imperfection that occurs with a severity sufficient to cause a product or service to fail to meet a specification requirement.

nondegeneracy assumption, The assumption that all basic feasible solutions to a linear programming problem are nondegenerate. This assumption is convenient in some proofs which demonstrate the finiteness and convergence of the simplex algorithm. (15)

nondegenerate feasible solution, A basic feasible solution in which all of the basic variables are positive. (19)

nonlinear constraint, A constraint which contains variables whose relationship cannot be expressed in terms of linear combinations. (19)

nonlinear equation, An equation at least one of whose terms is a nonlinear function of the variables. (19)

nonlinear function, A function defined as something other than a sum of terms consisting of a constant times a single variable plus a final constant; i.e., not a linear function. (19)

nonlinear programming, An inclusive term covering all types of constrained optimization problems except those where the objective function and the constraints are all linear. Special types of nonlinear programming for which some theory has been developed are convex programming, concave programming, and quadratic programming. (19)

nonnegativity constraint, A restriction that a variable can take on only positive or zero values, e.g., $x \geq 0$. (19)

nonrepaired item, See item, nonrepaired. (20)

nonsingular matrix, A square matrix whose determinant is not zero, and thus whose rank (number of linearly independent columns) is equal to its dimension (number of rows). Any nonsingular matrix has an inverse. (19)

northwest corner rule, A procedure for determining a first basic feasible solution to a transportation problem. (15)

objective function, The combination of the variables of a mathematical programming problem whose value is to be maximized or minimized subject to the constraints of the problem. For a linear programming problem, the objective function is a linear combination of variables. (19)

objective function element, One of the coefficients in the objective function. (19)

objective value, The value of the objective function at a particular solution. (19)

observed value, The particular value of a characteristic determined as a result of an observation or test. (3)

operable, The state of being able to perform the intended function. (28)

operational, Of, or pertaining to, the state of actual usage. (28)

operations research, An organized and systematic analysis of complex situations, such as arise in the activities of risk-taking organizations of people and resources. The analysis makes use of certain specific disciplinary methods, such as, probability statistics, mathematical programming, and queueing theory. The purpose of operations research is to provide a more complete and explicit understanding of complex situations, thus leading to an optimum performance of individuals, utilizing the resources available.

opportunity cost, The money or other value sacrificed by choosing one course of action thus ruling out others.

optimize, To maximize or minimize an objective function. (19)

optimum solution, A set of values of the variables which optimizes the objective function, subject to the constraints of the problem. (19)

order-statistics, When a sample of variate values are arrayed in ascending order of magnitude these ordered values are known as order-statistics. Examples are the smallest value of a sample and the median. More generally, any statistic based on order statistics in this narrower sense is called an order-statistic, e.g., the range and the interquartile distance. (22)

parameter, This word occurs in its customary mathematical meaning of an unknown quantity which may vary over a certain set of values. In statistics it most usually occurs in expressions defining frequency distributions (population parameters) or in models describing a stochastic situation (e.g., regression parameters). The domain of permissible variation of the parameter defines the class of population or model under consideration. (22)

parametric programming, A method for investigating the effect on an optimal linear programming solution of a sequence of proportionate changes in the elements of a single row or column of the matrix. Most commonly the method is applied to either the objective function row or the right-hand-side column. (19)

Pareto curve, An empirical relationship describing the number of persons y whose income is x, first advanced by Pareto (1897) in the form

$$y = Ax^{-(1+\alpha)} \quad 0 \leqslant x \leqslant \infty.$$

The expression is now used to denote any frequency distribution of this form whether related to incomes or not. The variable x may be measured from some arbitrary value, not necessarily zero.

The coefficient A in the expression for the Pareto curve is generally referred to as the "Pareto Index." It affords evidence of the concentration of incomes, or, more generally, of the concentration of variate values in distributions of the Pareto type. (22)

partition, To separate a linear program into related subprograms. More generally to divide a set into nonintersecting subsets. (19)

payoff, The amount received by one of the players in a play of a game. (21)

payoff function, For a two-person zero-sum game, the payoff function M is the function for which M (x, y) (positive or negative) is the amount paid by the minimizing player to the maximizing player in case the maximizing player uses pure strategy x and minimizing player uses pure strategy y. (21)

payoff matrix, A two dimensional array $[P_{ij}]$ in which the columns (or rows) represent the available strategies, the row (or columns) represent the states of nature and the entries P_{ij} represent the outcomes (on some scale of measure) for the ith column and the jth row.

permanently feasible set (stochastic programming), Consider a linear program with random parameters. Here one considers only the convex set of those vectors X which will be feasible regardless of the subsequently observed parameters of the linear program. For each (A,b) the X's satisfying $AX = b$, $X \geq 0$ form a convex polyhedron. The set of permanently feasible X's are those that are elements of the intersection of these polyhedra where the intersection is over all A,b. (19)

personnel-assignment problem, See assignment problem. (15)

personnel subsystem, A management concept which considers that functional part of a system which provides, through effective development and implementation of its various elements the specified human performance necessary in the operation, maintenance, support, and control of the system in a specified environment. (28)

perturbation techniques, The process of modifying slightly the right-hand-side coefficients of a linear programming problem to insure against the possibility of cycling. (15)

phase I, The mathematical procedure used by the simplex algorithm, two-phase method, to find a first feasible solution to a linear programming problem. (19)

phase II, The mathematical procedure used by the simplex algorithm, two-phase method, to find an optimal basic feasible solution, given a first basic feasible solution. (19)

phase in queuing theory, There are some service activities which are quite long and whose times do not follow a Poisson distribution, yet which can be subdivided into successive, distinctly separable stages each of which possesses a Poisson distributed service time. Furthermore, by the nature of the service, the first member of the queue cannot enter the first stage until the unit being served leaves the last stage. Each of these stages is referred to in the literature as a phase. The division of long service operations into phases possessing each its own Poisson distribution can contribute to making the mathematical treatment of the queuing process more tractable.

piecewise linear approximation, The division of the domain of definition of a function into subregions, and the replacement of the function by some closefitting linear function in each subregion. (19)

pivot column, The column of the matrix containing the pivot element. In a linear programming iteration, it is the column associated with the entering variable (nonbasic variable picked to become basic). (19)

pivot element, The pivot-row element of the pivot column. (19)

pivot row, The row of the matrix containing the pivot element. In a linear programming iteration, it is the row associated with the departing variable (basic variable picked to become nonbasic). (19)

pivot step, A step consisting of a single transformation of the matrix in a pivotal method of reduction of a set of linear equations. (19)

pivotal method, One of the many methods used in the solution of sets of linear equations, in which some particular coefficient plays a dominant role at each stage of the elimination process. One such method, which forms the base of the ordinary simplex method of linear programming, is called the method of multiplication and subtraction. A step in this method involves transforming the equations so that a designated variable x_s shall appear only in the rth equation and no others. To accomplish this, row r of the matrix is first divided by a_{rs} and suitable multiples of this row are subtracted from all other rows so that the coefficient of x_s in the other equations is zero. The coefficient ars appears in the computation of each new coefficient, it is called the pivot element. (12)

positive definite (or semidefinite) quadratic form, A quadratic form is called positive definite if

$$X^TAX > 0 \text{ for every } X \text{ except } X = 0.$$

It is called positive semidefinite if

$X^TAX \geq 0$ for every X and there exist $X \neq 0$ for which $X^TAX = 0$. (17)

precision, Degree of mutual agreement among individual measurements. Relative to a method of test, precision is the degree of mutual agreement among individual measurements made under prescribed like conditions. (See source code 35.) (3)

predicted, Expected at some future date, on the basis of analysis of past experience. (28)

preemptive priority (preemptive service), A queue discipline in which the arrival of a higher priority while a lower priority is in service requires the return to the queue of the lower priority. There are two cases:

a. Repeat rule: The ejected item returns to service, having lost all service before ejection;
b. Preemptive resume rule: The ejected item resumes at the point of service where it was interrupted. (33)

pricing, In linear programming, the selecting of a nonbasic variable to enter the basis using an appropriate pricing criterion. (19)

pricing criterion, The criterion for choosing an entering variable in the simplex method. (19)

pricing vector, (1) In the modified simplex method this refers to the updated cost row. (2) In the revised simplex method this refers to the "simplex multipliers." (19)

primal dual algorithm, In the ordinary two-phase method of linear programming, the first phase is concerned with finding a basic feasible starting solution, and does this by introducing artificial variables and driving them all to zero. The basic feasible solution so obtained, however, may be far from optimal. In the primal-dual algorithm, the sequence of computations is such that when the artificial variables are finally driven to zero, the resulting basic solution is not only feasible but optimal. The process starts with a solution to the dual problem. Corresponding to this solution, a restricted primal problem is written down and solved in which some but not all of the primal variables are allowed to enter the basic set. Then a new solution to the dual problem is found, with a larger value of the objective function; a new restricted primal problem is written down and solved; and so on. Eventually a solution to the dual is obtained which maximizes its objective function. The corresponding primal solution will be free of artificial variables, and will constitute an optimal solution to the original linear programming problem. (12)

primal problem, See dual linear programming problems.

primal simplex algorithm, Refers to a type of linear programming algorithm in which a basic feasible solution to the primal problem is maintained at each step of the iterative process, with a feasible solution to the dual problem appearing only at optimality. (19)

primal solution, The basic feasible solution of the primal variables at optimum. (19)

principle of optimality, In dynamic programming the principle of optimality is: "An optimal policy has the property that whatever the initial state and decision are, the remaining decisions must constitute an optimal policy with regard to the state resulting from the first decision." (34)

priority, preemptive, See preemptive priority.

prior probability, See posterior probability, Bayes' theorem.

probability, A basic concept which may be taken either as expressing in some way a "degree of belief," or as the limiting frequency in an infinite random series. Both approaches lead to much the same calculus of probabilities. (22:226, 6:220, 25:18, 31:59)

process, Any set of conditions or causes which work together to produce a given result. (32)

process quality, The value of percent defective or of defects per hundred units in product from a given process. **Note:** The symbols p′ and c′ are commonly used to represent the true process average in fraction defective or defects per unit; and 100p′ and 100c′, the true process average in percent defective or in defects per hundred units. (2)

product form of inverse (computing form), A computationally efficient (for sparse matrices) way of updating only the required portion of the inverse; often used in revised simplex (q.v.) codes due to its calculation and input-output characteristics. (19)

profit range, The interval $[c - \Delta_1, c + \Delta_2]$ where c is a profit coefficient in the objective function, and the two endpoints are points where a change of basis must first occur to maintain optimality. (19)

programming, The investigation of the structure and state of a system and the objective to be fulfilled in order to construct a statement of the actions to be performed, their timing, and their quantity (called a program or schedule) which will permit the system to move from a given status toward the defined objective. (11)

programming problems, Programming problems are concerned with the efficient use or allocation of limited resources to meet desired objectives. (15)

pure strategy, A mixed strategy (q.v.) which has all frequencies equal to zero except one which equals unity. (11)

quadratic form, A quadratic form is a numerical function of n variables which can be written

$$Z = \sum_{i=1}^{n} \sum_{j=1}^{n} a_{ij}x_ix_j.$$

If $A = \left[a_{ij}\right], X = (x_1, x_2 \ldots x_n)$

then in matrix notation

$$Z = X^T AX.$$

(19)

quadratic function, A function which contains second-degree terms in the variables, e.g., $a + bx + cx^2$. (19)

quadratic programming, Maximization (minimization), subject to linear constraints, of an objective function which is a concave (convex) quadratic function. (19)

quality, The totality of features and characteristics of a product or service that bear on its ability to satisfy a given need. (3)

quality assurance, All those planned or systematic actions necessary to provide adequate confidence that a product or service will satisfy given needs.

quality audit, A systematic examination of the acts and/or decisions with respect to quality in order to independently verify or evaluate compliance to the operational requirements of the quality program and/or the specification or contract requirements of the product or service.

quality control, The operational techniques and the activities which sustain a quality of product or service that meets the needs; also the use of such techniques and activities.

quality system, The collective activities and events that provide an organized plan to ensure that the quality of a product, process, or service will satisfy the given needs.

quantal response, The response of a subject to a stimulus is said to be quantal when the only observable — or the only recorded — consequence of applying the stimulus is the presence or absence of a certain reaction, e.g., death. A quantal response may be expressed as a two-valued variate taking values 0 and 1. (22)

quantitative data, A term usually used to describe data in which the variables concerned are quantities, as distinct from data derived from qualitative attributes.

queue discipline, The rule for selecting a unit in the waiting line for service.

queuing, multistage, Involves two or more steps in a process, each of which involves waiting in a line. In this case, processing is done in sequence rather than in parallel.

queuing theory, The theory involving the use of mathematical models, theorems and algorithms in the analysis of systems in which some service is to be performed under conditions of randomly varying demand, and where waiting lines or queues may form due to lack of control over either the demand for service or the amount of service required, or both. Utilization of the theory extends to process, operation and work studies. (See source code 33.)

random, This word may be taken as representing an undefined idea, or, if defined, must be expressed in terms of the concept of probability. A process of selection applied to a set of objects is said to be random if it gives to each one an equal chance of being chosen. Generally, the use of the word "random" implies that the process under consideration is in some sense probabilistic. (22:237, 13:29)

random failure, See failure, random. (20)

random walk, The path traversed by a particle which moves in steps, each step being determined by chance either in regard to direction or in regard to magnitude or both. Cases most frequently considered are those in which the particle moves on a lattice of points in one or more dimensions, and at each step is equally likely to move to any of the nearest neighboring points. The theory of random walks has many applications, e.g., to the migration of insects, sequential sampling and, in the limit, to diffusion processes. (22:239, 13:311, 25:393)

range, The difference between the largest observed value and the smallest observed value. (2)

rank of a matrix, The maximum number of linearly independent rows of the matrix. Equivalently, the maximum number of linearly independent columns, or the size of the largest square submatrix with an inverse. (19)

rating, The value of an item parameter which can be attained under specified conditions. (28)

ready rate, operational (combat), Percent of assigned items capable of performing the mission of function for which they were designed, at a random point in time. (28)

redundancy, In a system, the existence of more than one means for performing a given function.

redundancy, active, That redundancy wherein all redundant items are operating simultaneously rather than being switched on when needed. (20)

redundancy, standby, That redundancy wherein the alternative means of performing the function is inoperative until needed, and is switched on upon failure of the primary means of performing the function. (20)

redundant equations, A set of equations, one of which may be expressed as a linear combination of the others. Such an equation can be omitted from the system without affecting the solution. (19)

reliability (of an item, expressed numerically), The probability that an item will perform a required function under stated conditions for a stated period of time. **Note:** This definition is used when defining the characteristic intended by such modified terms as "assessed reliability" and "predicted reliability." (3)

reliability (general definition), Ability of an item to perform a required function under stated conditions for a stated period of time. (3)

reliability, human performance, The probability that man will accomplish all required human functions under specified conditions. (28)

reliability, inherent, The potential reliability of an item present in its design. (20)

repair, See maintenance, corrective. (28)

repairable unit, See unit, repairable. (20)

repeatability (measurement), See precision.

response, The reaction of an individual unit to some form of stimulus. It may be reaction to a drug, as in bio-assay, or the reaction to a request for information, as in sample surveys of human beings. (22)

restriction, An equation or inequality limiting the feasible range of variation of one or more variables. Also used for constraint. (19)

revised simplex method (computing form), A variant of the simplex method, especially suitable for problems in which the number of variables is much larger than the number of equations. The method employs (1) an implicit inverse, "product form of the inverse," or (2) an explicit inverse of the current basis to compute an updated row or column of the tableau as required. The simplex multipliers comprise the profit row of the inverse; they are used to test for optimality and to select a new entering variable. (19)

right-hand side, The mathematical expression on the right of the equality or inequality sign in an equation or inequality. In linear programming, by convention, the right-hand side of each constraint is merely the constant term, with the complete linear function as the left-hand side. (19)

right-hand-side element, An element of the column vector comprising the right-hand sides of the constraints. (19)

right-hand-side range, The interval $[b - \Delta_1, b + \Delta_2]$ where b is the original value of the right-hand-side element and the two endpoints are the first points where a change of basis must occur to maintain feasibility. Within this interval, the optimal objective value is a linear function of the right-hand-side element. (19)

saddle point, A function $F(X, \lambda)$ is said to have a saddle point at the point (X_0, λ_0), if there exists an $\epsilon > 0$ such that for all X, $|X - X_0| < \epsilon$ and all λ $|\lambda - \lambda_0| < \epsilon$ we have

$$F(X, \lambda_0) \leq F(X_0, \lambda_0) \leq F(X_0, \lambda)$$

(18)

saddle point of a game, For any finite two-person zero-sum game, the elements aij of the payoff matrix satisfy the relation

$$\max_i (\min_j a_{ij}) \leq \min_j (\max_i a_{ij}).$$

If the sign of equality holds, then $\max_i (\min_j a_{ij}) = \min_j (\max_i a_{ij}) = v$ and there exist pure strategies i_0 and j_0 for the maximizing and minimizing players, respectively, such that if the maximizing player chooses i_0 then the payoff will be at least v no matter what strategy the minimizing player chooses, and if the minimizing player chooses j_0 then the payoff will be at most v no matter what strategy the maximizing player chooses. Thus

$$v = a_{i_0 j_0} = \max_i a_{ij_0} = \min_j a_{i_0 j}.$$

In this case, the game is said to have a saddle point at (i_0, j_0). There might be more than one saddle point at which the value v is taken on. Similar statements hold for an infinite two-person zero-sum game, for which there might or might not be a saddle point. (21)

sample, A part of a population, or a subset from a set of units, which is provided by some process or other, usually by deliberate selection with the object of investigating the properties of the parent population or set. (22)

screening test, A test or combination of tests, intended to remove unsatisfactory items or those likely to exhibit early failures. (20)

secondary failure, See failure, secondary. (20)

sensitivity, The responsiveness of a solution to changes in the numerical values of the problem parameters. (19)

sensitivity analysis, An analysis of the effect on the solution of a mathematical problem, for example, as the parameters of the problem are varied. (11)

sensitivity testing, A general technique which attempts to determine a relation between quantal response and some stimulus by using a destructive test.

separable constraint, Let

$$g(x_1, x_2, \ldots, x_n) \leqslant 0$$

be the ith constraint in a mathematical programming formulation. Then, if

$$g(x_1, x_2, \ldots, x_n) = g_1(x_1) + g_2(x_2) + \ldots + g_n(x_n),$$

the constraint is said to be separable. (18)

separable objective function, When the objective function of a mathematical programming problem in n variables is capable of being written as a sum of n functions each of which is a function of only a single variable, the objective function is said to be separable. More precisely

$$Z = f(x_1, x_2, \ldots, x_n) = f_1(x_1) + f_2(x_2) + \ldots + f_n(x_n)$$

is separable. (18)

separable programming, A class of nonlinear programming problems in which each function appearing may be expressed as the sum of separate functions of single variables. When all the functions have been separated, it is possible to make a piecewise linear approximation to each one. For convex constraints, this is all that is required, since the LP algorithm will always choose the two nearest points surrounding the desired value, giving accuracy dependent only on the fineness of the approximation. For concave constraints, however, the ordinary LP algorithms will choose from the two most widely separated points, ignoring those in between and giving generally a most unsatisfactory accuracy. However, the separable programming algorithm assures that the points chosen are contiguous, and thus accurate. Of course, only a local optimum solution is assured.

service rate distribution, constant, See arrival rate distribution, constant.

service time, constant, See arrival rate distribution, constant.

serviceability, A measure of the degree to which servicing of an item will be accomplished within a given time under specified conditions. (28)

servicing, The replenishment of consumables needed to keep an item in operating condition, but not including any other preventive maintenance or any corrective maintenance. (28)

service rate, mean, See mean service rate.

set, Any well-defined collection of objects, things or symbols. To be well-defined, it must be possible to tell beyond doubt whether or not a given object, thing or symbol belongs to the collection being considered.

shadow prices, The values of the non-slack dual variables at optimality. This usage is appropriate in linear programming problems where the primary variables $x_1, \ldots x_n$ are quantities and where the right-hand side b_i of the typical constraint

$$\sum_{ij} a_{ij} x_j = b_i$$

is in the same unit and represents the quantity available system-wide of some particular scarce resource. Since, at optimality, the objective function z may be expressed as

$$z = b_1 w_1 + b_2 w_2 + \ldots + b_m w_m$$

where the w_i are the dual variables, we see that the profit accruing from the optimal usage of the resources $b_1, \ldots, b_m$ may be broken up into m imputed portions $b_i w_i$; and, in turn, w_i may be regarded as an imputed unit value (shadow price) for b_i. The shadow price may be interpreted as a fictitious price that society gives to a commodity when the economy is operating in a state of perfect competition. (12)

shortest route problem, Given N nodes numbered $i = 1, 2, \ldots$, N and a set of numbers (d_{ij}) where d_{ij} = the distance required to travel from the ith node to the jth node. Find the minimum distance route from node 1 to N. (19)

signal flow graph, A graphical model of a system in which nodes represent the values of system variables and directed branches between the nodes represent functional relationships between the system variables.

simplex, A simplex is an n-dimensional convex polyhedron having exactly n + 1 vertices. (15)

simplex algorithm, See simplex method.

simplex method, A computational procedure for solving a linear programming problem. The method embodies an algebraic algorithm, termed the simplex algorithm, which transforms by iterative steps a starting solution into an optimum solution (or determines that such a solution does not exist). (19)

simplex multipliers, The components of a vector $(\pi_1, \pi_2, \ldots, \pi_m)$ associated with the current solution. If the objective function is

$$\sum_i c_i x_i,$$

and the current basis is B, and the subscripts of the basic variables are $i_1, i_2, \ldots, i_m$, the simplex multipliers are defined by the equation

$$(\pi_1, \pi_2, \ldots, \pi_m) = \left(c_{i_1}, c_{i_2}, \ldots, c_{i_m}\right) B^{-1}$$

(12)

simulation, The design and operation of a model of a system. Commonly implies the use of a computer programmed and/or designed to accept selected inputs; to treat those inputs in a manner analogous to the way the real system operates, and to read out measurements of the status and change results of the programmed operations affected by those inputs.

singular matrix, A square matrix whose rank is less than its dimension and whose determinant is zero. A singular matrix has no inverse. (19)

sink, That single node in a capacitated network at which all flow is assumed to terminate.

slack, The difference between the larger and smaller members of an inequality. If $u \leq v$, the slack is defined as $s = v - u$. (12)

slack variable, An auxiliary variable introduced to convert an inequality constraint to an equation. (19)

slack vector, The column vector associated with a slack variable. It is a unit vector with + 1 or − 1 in the row in which the slack appears, and zeros elsewhere. (19)

solution, In linear programming, a set of values of the variables which satisfy all the given constraints. If all of these values are nonnegative, the solution is called feasible; if one or more are negative, it is called nonfeasible or infeasible. Solutions are otherwise classified as basic (number of variables with nonzero values less than or equal to the number of constraints) and nonbasic (number of nonzero variables greater than the number of constraints). (12)

solution level, The set of values taken by the variables in a solution.

source, That single node in a capacitated network from which all flow is assumed to originate.

sparse vector or matrix, A vector or matrix whose elements are mostly zeros. (19)

standard deviation, The most widely used measure of dispersion of a frequency distribution. It is equal to the positive square roots of the variance (q.v.). (22)

standby redundancy, See redundancy, standby. (20)

starting basis, The set of column vectors associated with the basic variables of a starting solution in linear programming. (19)

starting solution, The first solution used at the beginning of an iterative-type solution to an optimization problem.

state probability, The state of a system is specified by the number of units in the queue, waiting for service, the number of units in service, etc. The state probability is then the probability that the system is in a given state.

stationary process, A stochastic process $\{x_t\}$ is said to be strictly stationary if the multivariate distribution of

$$x_{t_1} + h, x_{t_2} + h, \ldots x_{t_n} + h$$

is independent of h for any finite set of parameter values $x_{t_1} + h, \ldots x_{t_n} + h, t_1, t_2, \ldots t_n$.

The process is said to be stationary in the wide sense if the mean and variance exist and are independent of t. (22:279, 5:4)

statistic, A summary value calculated from a sample of observations, usually but not necessarily as an estimator of some population parameter; a function of sample values. (22)

statistical quality control, Quality control in which statistical techniques are used. These techniques include the use of frequency distributions, measures of central tendency and dispersion, control charts, acceptance sampling, regression analysis, test of significance, etc. (3)

steady state, A physical condition in which the variables (q.v.) affecting a system (q.v.) are either invariant or periodic functions of time.

steppingstone method, A name for a special form of the simplex method used for solving the transportation problem. Given a feasible solution, the algorithm tells how to improve this solution by introducing a new source-destination allocation and reallocating among existing allocations so that the constraints remain satisfied. Those existing allocations whose values are modified in this way are called steppingstones. If no such reallocation can improve the solution, the current solution is optimal. (19)

step stress test, A test consisting of several stress levels applied sequentially for periods of equal duration to a sample. During each period a state stress level is applied and the stress level is increased from one step to the next. (20)

stochastic, The adjective "stochastic" implies the presence of a random variable; e.g., stochastic variation is variation in which at least one of the elements is a variate and a stochastic process is one wherein the system incorporates an element of randomness as opposed to a deterministic system. (22:279, 4:6, 29:161)

stochastic programming, A generalization of linear programming in which any of the unit costs, the coefficients in the constraint equations, and the right-hand sides may be random variables. The aim of such programming is to choose levels for the variables which will minimize some function of the cost. See permanently feasible set. (19)

storage life (shelf life), The length of time an item can be stored under specified conditions and still meet specified requirements. (28)

strategy, The decision rule used for making the choice from available courses of action.

strategy (mixed), See minimax theorem.

stratified random sampling, The process of selecting a simple random sample from each of the population strata. (26)

stratum, A group of units from a population, a subpopulation, usually defined by relevant population characteristics. (26)

strictly concave function, The negative of a strictly convex function. (19)

strictly convex function, See convex function.

strictly increasing function, A function f is strictly increasing on an interval (a, b) if $f(y) > f(x)$ for any two numbers x and y (of this interval), for which $x < y$. (21)

suboptimal, (1) Not yet optimal. (2) Optimal over a subregion of the feasible region. (19)

supporting hyperplane, Let W be a boundary point of a convex set D. Then $C^T X = Z$ is called a supporting hyperplane at W if $C^T W = Z$ and if all of D lies in one closed half-space produced by the hyperplane. For every boundary point W of D, there exists at least one supporting hyperplane at W. (18)

surplus variable, A slack variable when an inequality is of the "greater than or equal to" form.

surveillance, (1) The continuing analysis and evaluation of records, methods, and procedures including the act of verification to assure conformance with technical requirements. (2) A system whereby supplies and equipment in storage are subjected to, but not limited to, cyclic, scheduled, and special inspection and continuous action to assure that material is maintained in ready for issue condition. (27)

survivability, The measure of the degree to which an item will withstand hostile man-made environment and not suffer abortive impairment of its ability to accomplish its designated mission. (28)

symmetric parametric programming, The simultaneous parameterization of the right-hand side and the objective function. This is useful in economic studies when both costs and requirements change a linear function of a single parameter — for example, time. Another use is as a sort of primal-dual algorithm to go from a pseudo right-hand side and objective function (for which the starting basis is optimal and feasible) to true values of the right-hand side and objective function. In some problems, this approach requires far fewer iterations than a more conventional algorithm. (19)

system, A set of objects with relations between the objects and their attributes.

system effectiveness, A measure of the degree to which an item can be expected to achieve a set of specific mission requirements, and which may be expressed as a function of availability, dependability and capability. (28)

system, linear, A system is said to be linear if an input $f_1(t)$ produces an output $g_1(t)$ and an input $f_2(t)$ produces an output $g_2(t)$ and if then an input $af_1(t) + bf_2(t)$ produces an output $ag_1(t) + bg_2(t)$ for all $f_1(t)$ and $f_2(t)$ and all constants a and b.

tableau, The current matrix, with auxiliary rows and/or columns, as it appears at an iterative stage the standard simplex method.

technology matrix, The coefficient matrix, or (a_{ij}), as distinguished from the costs coefficients (c_j) and the right-hand sides (b_j).

testing, A means of determining the capability of an item to meet specified requirements by subjecting the item to a set of physical, chemical, environmental, or operating actions and conditions. (3:17)

time, active, That time during which an item is in the operational inventory. (28)

time, adjustment or calibration, That element of maintenance time during which the needed adjustments or calibrations are made. (28)

time, administrative, Those elements of delay time that are not included in supply delay time. (28)

time, alert, That element of uptime during which an item is thought to be in specified operating condition and is awaiting a command to perform its intended mission. (28)

time (as used in reliability definitions), Refers to any duration of observations of the considered items — either in actual operation or in storage, readiness, etc., but excludes downtime due to a failure. Note: In definitons where "time" is used, this parameter may be replaced by distance, cycles, or other measures of life as may be appropriate. This refers to terms such as acceleration factor, wear-out failure, failure rate, mean life, mean-time-between-failures, mean-time-to-failure, reliability, and useful life. (20)

time, checkout, That element of maintenance time during which performance of an item is verified to be in specified condition. (28)

time, cleanup, That element of maintenance time during which the item is enclosed and extraneous material not required for operation is removed. (28)

time delay, That element of downtime during which no maintenance is being accomplished on the item because of either supply delay or administrative reasons. (28)

time, down (downtime), That element of time during which the item is not in condition to perform its intended function. (28)

time, fault correction, That element of maintenance time during which a failure is corrected by (a) repairing in place; (b) removing, repairing, and replacing; or (c) removing and replacing with a like serviceable item. (28)

time, fault location, That element of maintenance time during which testing and analysis are performed on an item to isolate a failure. (28)

time, inactive, That time during which an item is in reserve (in the inactive inventory). (28)

time, item obtainment, That element of maintenance time during which the needed item or items are being obtained from designated organizational stockrooms. (28)

time, mission, That element of uptime during which the item is performing its designated mission. (28)

time, modification, The time necessary to introduce any specific change(s) to an item to improve its characteristics or to add new ones. (28)

time, preparation, That element of maintenance time needed to obtain the necessary test equipment and maintenance manuals, and set up the necessary equipment to initiate fault location. (28)

time-series, A time-series is a set of ordered observations on a quantitative characteristic of an individual or collective phenomenon taken at different points of time.

tolerance(s): (industrial production), The amount by which a given characteristic of a manufactured item may differ from its specified-nominal-value and still be acceptable. A tolerance may be one-sided, upper or lower; two-sided tolerances, both upper and lower, are most common. Two-sided tolerances are usually, although not necessarily, symmetrical.

transformation matrix, An elementary matrix representing a single change of basis. It is the inverse of a matrix formed from an identity matrix (which is the present basis in terms of itself) by replacing one unit vector (representing the departing variable) with another vector (representing the entering vector). The entering variable is expressed in terms of the present (old) basis. (19)

transformation of a matrix, A change in the appearance of a matrix which leaves the solutions to the corresponding set of linear equations unchanged; a change of basis. (19)

transient state, In contrast with steady state, a system before some time t such that there is a probability approaching zero that the system can return to that (transient) state.

transportation problem, A homogeneous product is to be shipped in the amounts $a_1, a_2, \ldots, a_m$ respectively from each of m shipping origins and received in amounts $b_1, b_2, \ldots, b_n$ respectively by each of n shipping destinations. The cost of shipping a unit amount from the ith origin to the jth destination is c_{ij} and is known for all combinations (i, j). The problem is to determine the amounts x_{ij} to be shipped over all routes (i, j) so as to minimize the total cost of transportation. (15)

transshipment problem, A generalized transportation problem in which items may pass through intermediate nodes between source and destination. (11)

traveling-salesman problem, A salesman is required to visit each of n cities, indexed by 1, . . . n. He leaves from a home base city indexed by 0, visits each of the n cities exactly once, and returns to city 0. It is required to find an itinerary which minimizes the total distance traveled by the salesman. (15)

tree, A network having n nodes is a tree if it has (n − 1) arcs and no loops. (11)

trim problem, The problem of cutting rolls of paper or other material to meet the orders for a specified list of roll widths. The problem can be formulated as a linear programming problem. (19)

unbounded solution, Solution with some infinite or unbounded values, and therefore with an infinite optimum for the objective function. (19)

unique optimal solution, The optimal solution, in a problem which has one and only one optimal solution. (19)

unit, nonrepaired, An item which is not repaired after a failure.

unit, repairable, A unit which is repaired after a failure.

universe, The totality of the test of items, units, measurements, etc., real or conceptual, that is under consideration. A population. (1)

upper bounded linear programming problem, A linear programming problem in which all the variables x_j are restricted by a corresponding upper bound d_j, i.e., $x_j \leq d_j$. (15)

useful life, The period from a stated time, during which, under stated conditions, an item has an acceptable failure rate, or until an irreparable failure occurs. (20)

utility, A numerical index representing relative preferences assigned among a group of objects, alternatives, prospects, or choices.

utilization parameter, In queuing theory, is the ratio between mean arrival and mean servicing rates.

value of a game, The number v (expected payoff) associated with any two-person zero-sum game for which the minimax theorem (q.v.) holds. (21)

variable, Generally, any quantity which varies. More precisely, a variable in the mathematical sense, i.e., a quantity which may take any one of a specified set of values. It is convenient to apply the same word to denote nonmeasurable characteristics, e.g., "sex" is a variable in this sense since any human individual may take one of two "values," male or female. It is useful, but far from being the general practice, to distinguish between a variable as so defined and a random variable or variate (q.v.). (22:310, 7:422)

variate, In contradistinction to a variable (q.v.) a variate is a quantity which may take any of the values of a specified set with a specified relative frequency or probability. The variate is therefore often known as a random variable. It is to be regarded as defined, not merely by a set of permissible values like an ordinary mathematical variable, but by an associated frequency (probability) function expressing how often those values appear in the situation under discussion. (22:312, 4:26, 13:199)

vector, An ordered set or array of quantities, usually n real numbers (x_1, x_2, . . . x_n). A vector arrayed horizontally is called a row vector; one arrayed vertically is called a column vector. For example, the rows of a matrix are row vectors, the columns of a matrix are column vectors. (19)

Venn diagram, A schematic representation of the universal set and its subsets. A rectangle is usually used to represent the universal set, points to designate elements of the set and circles to depict subsets of the universal set. The diagram is sometimes referred to as a Euler Diagram.

vertex, A point of a convex set which does not lie on a line segment joining any other two points of the set; also called an extreme point. Sometimes used as a synonym for node (q.v.). (19)

waiting line (queue), A line formed by units waiting for service.

waiting time, The time elapsed while a unit is waiting for service.

wear-out failure, See failure, wear-out.

wear-out failure period, That possible period during which the failure rates of units of a particular item are rapidly increasing due to deterioration processes. (20)

zero-sum two-person game (matrix game), A two-person game of strategy in which the amount lost by one player is won by the other. Such a game has an equivalent formulation as a linear programming problem. (15)

Organization Planning and Theory

9

Z94.9

In the Spring of 1977 the Subcommittee for the revision and updating of the 1972 (Z94.9) Standard on Organization Planning and Theory was created. Most of the original members were no longer completely available, so a small new committee was formed. Its small size facilitated the work of updating and revision.

The terms and definitions were selected by the Subcommittee as having value and general usage for those involved in industrial engineering. In preparing this glossary the Subcommittee has avoided, on the one hand, the inclusion of specialized esoteric concepts that have only rare usage and, on the other hand, the inclusion of a great number of ordinary terms for which a dictionary definition will suffice.

As before, reference works, textbooks, and journal material was consulted. The definitions agreed upon represent a consensus derived from the references and the combined judgment and experience of the Subcommittee members.

In 1979, Prof. Dale S. Beach resigned as Co-Chairman, but continued as a regular subcommittee member.

Professor Delmar W. Karger, Co-Chairman
Dean Emeritus and Ford Foundation Professor of Management Emeritus of Rensselaer Polytechnic Institute.

Professor Dale S. Beach, Co-Chairman until 1979
School of Management, Rensselaer Polytechnic Institute.

Dr. Richard T. Barth, Professor
University of British Columbia.

Dr. Joseph Steger, Sr. Vice President and Provost
University of Cincinnati

administration, (1) Usually synonymous with the term *management*. However, it sometimes refers to the portion of management exclusive of establishing goals and policies. (2) That group of people who perform the process of administration.

administrative management theory, A collection of concepts and prescriptions concerning organization and management that emphasizes general principles of administration. These principles cover such issues as the functions of management, specialization, authority, responsibility, span of control, departmentation, special and general staffs, and unity of command. This school of thought was originated by Henri Fayol.

assistant (line assistant), A person or position, directly in the chain of command, who serves in a combined line and staff capacity. He provides advice and service to help his principal carry out the whole range of his responsibilities. He also acts in place of his chief when the latter is absent or when specific delegation is made. The assistant or line assistant is differentiated from the "assistant to" defined below.

assistant to (staff assistant), One who serves as an aide to an executive in the capacity of an investigator, researcher, and analyzer. He provides advice and service to his principal generally in only specified parts of the principal's job. He possesses no authority over other subordinates in the chain of command.

authority, The right to exercise power and to extend jurisdiction over others for the attainment of performance. (1) The "legitimate" right to direct or influence the performance of others under the condition of applying rewards and penalties. Authority lies in the position in an organization, not in the man. This authority derived from institutionalized power should be distinguished from other authority concepts. (2) Authority by subordinate acceptance — authority is a function of the degree to which subordinates accept decisions and direction and is derived from the group of subordinates. (3) Authority of person — authority is derived from superior ability or knowledge or charismatic qualities. (4) Authority by legal decree — authority invested by law to enforce statutes.

autocratic leader, One who tends to control in an absolute manner through the use of personal domination and application of coercive measures.

behavior, Those minimum sequences of actions or movements by an individual to which meaning can be assigned. The mode of conduct that a person exhibits.

behavioral modification, A technique developed from the work of B. F. Skinner that provides for a systematic coupling of desired behavior and desired rewards to achieve desired outcomes through scheduling reinforcements.

board of directors, A group of people that oversees the governance and management of an organization or corporation. Among its functions are those of trusteeship for constituent groups and clients, determination of policies, overall objectives, selection and/or approval of major executives, review of performance, trusteeship of assets, and distribution of earnings. *Syn:* Board of Trustees in non-profit making organizations.

bureaucracy, A large, formal organization characterized by emphasis on form, hierarchy of levels, specialization of labor, established rules and standards of conduct, records and data keeping, and professionalization of administration.

centralization, (1) The process of consolidating authority and decision-making within a single office or person. (2) The act of bringing together, physically or geographically operations or organizational units related by nature or function to form a central grouping.

chain of command, The prescribed line connecting the hierarchy of offices or persons through which authority and responsibility flow.

change agent, A (behavioral science) consultant who has responsibility for initiating, designing, and implementing organizational development (OD) efforts. While much of the OD efforts conducted until a few years ago placed strong reliance on *external* change agents (outside consultants), there has recently been a steady shift toward the use of *internal* change agents. (However, there is yet no hard evidence as to which is better.) In either case, the change agent can be viewed as a manager leveraging the variables important to OD. (See organization development.)

Chief Executive Officer (C.E.O.), The executive actually responsible for all activities of the firm, his title (in addition to being C.E.O.) can be President and/or Chairman.

Chief Operating Officer (C.O.O.), The person in charge of all operations (day-to-day activities) of a firm. The position reports to the C.E.O. and the holder of the position usually has the additional title of President or Executive Vice President or Senior Vice President.

classical organization theory, A collection of theories of organization and management that were formulated during the first third of the twentieth century and that were derived from personal experience, observation, and descriptive analysis of organizations. Included within the broad range of classical theory are scientific management pioneered by Frederick W. Taylor, bureaucratic theory developed by Max Weber, and administrative management theory, developed by Henri Fayol. Central features of classical theory are prescriptions for managerial actions and emphasis upon formal organization structure.

cognitive dissonance, A concept which holds that the individual strives for consistency within himself, but is faced with conflicts before deciding among alternatives. After making a decision, the individual is often faced with uncertainty as to its wisdom. The post-decision and post-action states are termed "cognitive dissonance." Elimination of the tensions induced by cognitive dissonance lead to "cognitive consonance." From Leon Festinger, *A Theory of Cognitive Dissonance*.

collegiality, A system of organization that is based upon the concept of responsible self-governance among organizational members who possess a strong commitment to the organization, its tasks, and goals. Often collegiality is applied in the governance of organizations and organizational subdivisions staffed principally by professionals who possess intellectual expertise, professional commitment, and a desire for autonomy.

committee, A group of people or a form of organization established to achieve one or more of the following: (a) interchange ideas and information, (b) obtain facts and ideas and synthesize into a report or recommendation, (c) obtain meeting of minds or a consensus.

communication, The transfer of information and understanding from one point or person to another person. The basic elements in the process of communication are an information source, encoding, transmission, reception, and decoding.

communication channel, A pathway or route to transmit information and understanding.

communication network, The system of senders, receivers, and channels for the transmission of information and understanding.

conflict, Interpersonal and social forces, ideas, behaviors, and motivations which are acting in opposition to one another, often involving strife or contest.

consensus, A collective opinion, or general agreement, that a decision has been accepted by the group. This does not necessarily imply unanimity.

consultative supervision (consultative management), That form of supervision which utilizes full interchange and sharing of ideas between superiors and subordinates, often taking place in group meetings. In consultative supervision the superior retains the power of final decision. This is distinguished from democratic supervision where decision is by group consensus and from autocratic supervision where decision is by superior domination.

control, Measurement of performance or actions and comparison with established standards in order to maintain performance and actions within permissible limits of variance from the standard. May involve taking corrective action to bring performance into line with the plan or standards.

criterion, The standard or rule by which a judgment of the effectiveness of a course of action, or of performance, can be made. (See control.)

cybernetics, (1) The field of control and communication theory in general, without specific restriction to any area of application or investigation. (2) The behavior and design of mechanisms, organisms, and/or organizations that receive and generate information and respond to it in order to attain a desired result.

decentralization, authority, Placing the authority and decision-making power as close as possible to the organizational level at which the work is done.

decentralization, physical, The geographical dispersing of facilities and activities. The management control can either be centralized or decentralized.

decision-making, The response to a need or stimulus by means of acquiring and organizing information, processing this information to yield alternative courses of action, and selecting one course of action from among the alternatives.

decision rule, A predetermined criterion which may be used by a decision-maker to select from alternative courses of action.

decoding, To operate on a signal in order to extract the message and present it in usable form at its destination.

delegation, To grant or confer responsibility and commensurate authority from one executive or organizational unit to another in order to accomplish a particular assignment; often used in the sense of a superior in the organization delegating to subordinates.

department, An organizational unit established to operate in, and be responsible for, a specified activity or a physical or functional area.

departmentalization (departmentation), The division of an organization into formal groups according to criteria such as function, process, equipment, products, territory, or customer.

division, One of the groupings into which activities are separated; organizational structure which includes more than two departments within its jurisdiction.

division of work, The separation of tasks into less complex subtasks. This may be to use simpler skills or to make use of special skills.

employee, A person who provides services for an organization and receives compensation and benefits for such services in accordance with general rules established for members of the organization, and who is considered a member of the organization for the period of time during which he serves in the sense that he acknowledges the right of the employer to direct and control his services. Excluded is a self-employed individual, i.e., an individual who carries on a business endeavor as a proprietor or partner, or who renders services as an independent contractor.

encoding, To operate on a message from the source of transmission and convert it to signals which the communication channel will accept and which can be decoded.

executive, (1) An employee — a) whose primary duty consists of the management of the enterprise in which he is employed or of a recognized department thereof; b) who customarily directs the work of two or more employees therein; c) who has the authority to hire or fire other employees, or whose suggestions and recommendations as to hiring, firing, advancement promotion, demotion, or other change in status of other employees will be given particular weight; d) who customarily and regularly exercises discretionary power; e) who does not devote more than 20 percent of his time to activities other than those described in a) through d); f) whose salary and other compensation reflects the intellectual and discretionary content of the position. (Adapted from Explanatory Bulletin, Regulations Part 541, 1956, defining the terms of Section 13a of the Fair Labor Standards Act.) (2) A manager, usually of a top or middle management level.

executive committee, A committee consisting of top officers who are on the board of directors and outside board members (usually where the Board only meets quarterly) that assist and act to aid the C.E.O. in making very major policy, planning, and operating decisions between board meetings. In other firms the Executive Committee, if it exists, is composed of a small group of executives reporting to the C.E.O. and assisting the C.E.O. in decision-making.

feedback, Information (data) extracted from a process or situation and used in controlling (directly) or in planning or modifying immediate or future inputs (actions or decisions) into the process or situation. Feedback can be either positive or negative, although the field of cybernetics is based upon negative feedback. Negative feedback indicates that the system is deviating from a prescribed course of action and should readjust to a new steady state.

first-line supervisor, An individual at the lowest rank of the managerial hierarchy. He bears general responsibility for the performance of his organizational unit, guides and directs the work of non-supervisory employees, provides fact-to-fact leadership, and handles such personnel responsibilities as performance appraisal, discipline, instruction, and grievance processing.

forecasting, The process of predicting or projecting the future for the purpose of reducing uncertainties. Specific forecasting techniques include Delphi, trend extrapolation, dynamic modeling, scenario writing, mapping, and the use of relevance or perspective trees, to name some of the more important ones. Among forecasting dimensions often considered are those pertaining to economic, legal, social, political, ecological, and technological conditions.

function, Usually an organizational sub-goal. May be defined in terms of end to be achieved or behavior required to achieve ends. Example: administrative function, protection function, manufacturing function.

functional authority, The right to exercise control as assigned to a particular organizational unit or person. Conventionally used to designate specific staff authority exercised-over line organizational units.

functional organization, A form of organization structure in which activities are grouped according to the purpose, service or utility they possess.

general staff, A form of organizational unit, most common in the military and in government, composed of a staff group usually assigned to a high level office and principally concerned with developing policies and plans.

group cohesion, The tendency of a group of people to maintain themselves as a group; the attractiveness which the group holds for its members; the unity and solidarity of the group. It is the total field of forces acting on members to remain in the group.

group dynamics, The behavior and relationships of people derived from their interactions within a group.

group executive, Term usually used in large companies where the Group Executive heads a group of companies and/or plants or operating locations (organizations). This arrangement is frequently used in large manufacturing and/or service organizations. Each component is headed by an executive who has a title such as President (of a wholly or partially owned subsidiary corporation) or Manager. The Group Executive also usually has an additional title such as Senior Vice President, Corporate Vice President or Division president of the parent company.

Hawthorne effect, The novelty or interest in a changed situation which leads, at least initially, to positive results. In the original Hawthorne Studies, this effect was revealed in terms of increased performance resulting from the workers' knowledge that they were being observed with interest by behavioral science researchers, treated as important, and that their inputs were being taken into consideration.

hierachy, A ranking or ordering of those in authority.

homeostatic system, Any cybernetic system so arranged as to maintain one particular state or to maintain equilbrium among its component parts.

human resource accounting, A relatively recent development in management which refers to expanding the accounting procedures of organizations in order to take into consideration the *human* assets along with capital assets. This represents a shift from thinking of organizational members (employees) as costs to regarding them as assets.

incentive, A stimulus which induces action. Any inducement, material or nonmaterial, which impels or encourages a person to behave in a certain way.

information, (1) The content of any meaningful communication. (2) In information theory, information refers to a quantitative measure of the amount of order and certainty which exists in a dynamic information system, without reference to the semantic content of the symbols involved.

information theory, A study of the quantity of information contained in a message, and also the capacity of a communications channel to transmit information.

input, That which is put in. In the organizational context usually taken to mean any resource allocation to a system.

input-output analysis, A matrix which provides a quantitative framework for the description of an economic unit. Basic input-output analysis is a unique set of input-output ratios for each production and distribution process. If the ratios of inputs per unit of output are known for all production processes, and if the total production of each end product of the economy, or of that section being studied is known, it is possible to compute precisely the production levels required at every intermediate stage to supply the total sum of end products. Further, it is possible to determine the effect at every point in the production process of a specified change in the volume and mix of end products.

interaction, Communication, association, and/or contact among people. The way in which a person affects the behavior of others and is in turn affected by others. Interaction is an element of social behavior.

interdependence, The extent to which the behavior of the elements of a system affect each other individually and collectively.

interface, Those connections of one system whereby one system is matched to another system where the systems are distinctly different because of the basic nature of the systems (man-machine, for example), because of origin of design and construction (designed by different subcontractors, for example), or because of basic objectives of each system alone.

job, (1) The combination of tasks, duties and responsibilities assigned to an individual employee and usually considered his normal or regular assignment. (2) The contents of a work order.

job analysis, Determination of the characteristics of a job through detailed observation and evaluation of the activities, facilities required, conditions of work, and the qualifications needed in a worker.

job classification, The grouping of jobs on the basis of the nature of the functions performed or level of pay, or on the basis of job evaluation, historic groupings, collective bargaining, or arbitrary determination.

job description, A summary of the essential activities involved in the performance of a job that is abstracted from a job analysis and used in the classifying of jobs and the selection of employees to fill them.

job design, The specification of job content, work methods, and relationships with other jobs in order to satisfy technological and organizational requirements plus the human concerns of the job holders. Techniques used in job design include industrial engineering techniques, human factor analysis, rotation of tasks within a work unit, job enlargement, job enrichment, and creation of self-managed work teams.

job enlargement, Extending the parameters of the job being performed by an employee in order to make the work more psychologically rewarding. A job may be extended in two dimensions — horizontal and vertical. *Horizontal* job enlargement can take the form of a greater variety of tasks on which the employee works, an increased number of tasks, and job rotation. *Vertical* job enlargement, or *job enrichment*, usually takes the form of more planning, more controlling, or more team participation on the part of the job incumbent. The additional controlling aspects may include self-pacing as well as inspection. This form of enlargement involves building motivators into the job. (See job design.)

job enrichment, See job enlargement.

job evaluation, The systematic appraisal of the value of an individual job in an organization in relation to other jobs. Jobs are commonly graded according to differences in skill, effort, physical demands, and working conditions. It is an aid in setting the money worth of jobs for pay purposes.

job specification, A form of job description listing the mental and physical qualifications and special skills required in an individual worker for a given job in order to facilitate the selection and placement of employees.

job standardization, The procedure of specifying a standard practice or a standard method for a job.

laboratory training, See sensitivity training.

leadership, Interpersonal influence, exercised in a situation and directed, through the communication process, toward the attainment of a specified goal or goals.

line, (1) In the organizational sense line activities are those that contribute most directly to the accomplishment of the organization's primary objective. (2) An authority and responsibility relationship between a superior and subordinate extending vertically in an uninterrupted train from the top to the bottom position in an organization.

management, (1) The process of utilizing material and human resources to accomplish designated objectives. It involves the activities of planning, organizing, directing, coordinating, and controlling. (2) That group of people who perform the functions described above.

management audit, The process of evaluating how effectively management has operated the organization. Typical of criteria often employed in this audit are production efficiency, earnings, utilization of assets, management and executive talent, and fairness to stockholders.

management by committee, The committee is sometimes used when the subjects or problems are broad or involve many functions. A committee is then formed to manage the responsibility area, usually the members are the managers of the involved functions. Typical examples where this management form is used are patents and new products.

management by exception, A style of management in which the manager establishes a systematic pattern of acceptable operations thus freeing himself from routine occurrences in order to devote his talents to the more difficult problems which are exceptions to the routine.

management, lower, The first level of supervision which is directly in charge of a group of employees. In very large organizations, having many levels of management, lower management could include the first two levels above the nonsupervisory employee level. The foreman, general foreman, and supervisory positions comprise lower management.

management, middle, That broad group of managers and administrators which is located below the top policy-making management level and above the level of supervision. It includes both line and staff personnel.

management by objectives, A program or an approach to managing. In consultation with their organizational superiors, individual managers plan their goals and unit objectives for a forthcoming period, carry on operations guided by these goals, and at the end of the period participate with their superiors in evaluating performance in relation to the objectives.

management information system, An organization's structured system of information inputs covering both the internal and external environments and an associated assimilation, storage and analysis system that provides at the time and place required the information outputs (in an easily understandable manner) necessary to manage the organization efficiently and effectively — and only those outputs unless the system is interrogated and requested to deliver additional data and/or analyses. Usually, in the modern sense, a computer is integral with the system; but this is not a requirement.

management, top, The ultimate level of authority consisting of those directors and principal administrative officers of the company who are responsible for the determination of broad policies, procedures, objectives, and goals. It implements these policies and procedures through its continuous function of defining organization, authority, responsibilities, staffing and by coordinating, integrating, measuring and controlling the organization. (See Chief Executive Officer.)

manager, (1) One who plans work and organizes and directs people toward the accomplishment of organizational objectives. (2) One who is responsible for the planning and employment of resources toward the achievement of organizational objectives even though he may not supervise people.

management (manager) development, The application of planned efforts to assist in supplying, maintaining, and improving managers at, or intended for, the middle and top organizational levels in order that they can more efficiently attain the objectives of the enterprise.

managerial grid, A method of analyzing leadership and managerial styles that was formulated by R. R. Blake and J. S. Mouton. These styles of leadership are classified according to the manager's "concern for people" as one broad dimension and "concern for production" as the other broad dimension. The styles are plotted upon a grid with each dimension scaled from 1 to 9. Blake and Mouton identified five principal styles: 1,1–impoverished; 9,1–task; 1,9–country club; 5,5–middle-of-the-road; 9,9–team management.

managerial style, The way in which a manager relates to others in his work place. This includes such characteristics as consideration for people, concern for production, objectivity, work pace, and degree of involvement of subordinates in decision-making.

matrix management, A management system whereby an employee (usually a professional or semi-professional) has two bosses, but each boss is responsible for a different aspect of the work being performed. For example, project managers can be responsible for outputs and budgets but their manpower resides in functional components whose managers retain some responsibility for the functionally oriented activities performed by their personnel.

means-ends analysis, A method of organization planning and program planning that involves (a) starting with the general goal to be achieved, (b) discovering a set of means for accomplishing this goal, and (c) taking each of these means, in turn, as a new subgoal and discovering a set of more detailed means for achieving it and so on to the point where a particular means can be carried out by existing programs of action.

morale, The total satisfactions derived from the job, the work group, one's supervisor, the organization, and the general environment. It pertains to the general feeling of well-being, satisfaction, and happiness of people.

motive, That which is within the individual, rather than without, that incites him to action.

motivation, A willingness to expend energy to achieve a goal or a reward.

new product venture team, A management system or organizational form used to manage the development and exploitation of a new product. Usually the team consists of a representative from manufacturing, development and product engineering, marketing and finance (sometimes other functions are also represented) who work under a Venture Team Manager to (usually) manage the product from the idea stage through initial marketing. The Venture Team Manager usually reports to the head of a New Product Department or an individual with overall authority for the new product activity.

noise, (Communication Theory), Noise is something which tends to obscure the message and render it unintelligible.

objective, A desired end result, condition, or goal which forms a basis for managerial decision-making. (See means-ends analysis.)

office of president (of chairman), An arrangement in which the office of the chief executive consists of more than one executive. This organizational device is sometimes used in very large corporations and typically consists of three to five senior executives who perform collaboratively to make top level decisions and plans. One of these is the chief executive who is accountable to the board of directors.

organization, (1) The classification or groupings of the activities of an enterprise for the purpose of administering them. Division of work to be done into defined tasks along with the assignment of these tasks to individuals or groups of individuals qualified for their efficient accomplishment. (2) Determining the necessary activities and positions within an enterprise, department, or group, arranging them into the best functional relationships, clearly defining the authority, responsibilities and duties of each, and assigning them to individuals so that the available effort can be effectively and systematically applied and coordinated.

organization, formal, A planned and established structure of relationships among people, designed to achieve specified objectives. Thus one may speak of the formal organization of a company, government agency, school or a hospital.

organization, informal, The type of organization structure which comprises the authority, responsibility, communicative, and associative relationships among functions, physical factors, and personnel that are supplemental to the "formal" organization structure and may be "for," "against" or "neutral" with regard to the achievement of organizational objectives. It develops spontaneously.

organization chart, A diagram or graphic representation of an organization which shows, to varying degrees, functions, responsibilities, people, authority and relationships among these.

organization development (OD), The planned, organization-wide process of change designed to improve organization effectiveness and adaptation (to changing environmental demands). This is accomplished through planned interventions by an internal or external agent (change agent) using theory and techniques of applied behavioral science. Although the planned interventions or strategies depend on the circumstances and diagnosis, the focus has usually been on the attitudes, norms, values, relations, and organizational climate, rather than on the goals, structure, and technology of the organization. Lately, however, more and more attention has been given to the interplay of structure, technology, and people variables. Techniques employed by OD change agents include sensitivity training, managerial grid applications, goal-setting sessions, team building, confrontation meetings, and interpersonal peacemaking, to name some of the more important ones. Note that OD is much more comprehensive than management development, which is focused only on a particular manager or group of managers in order to change individual managerial behavior. (See change agent.)

organization planning, The process of planning and designing the organization structure and providing for appropriate interrelationships among people and constituent units. Often involves analysis of goals, grouping of activities, making audits, planning organizational changes, and preparation of policy manuals. May involve manpower planning, training and development of management personnel.

organization structure, A framework within which the dynamics of human relations forces take form and come into interaction. A plan under which the totality is subdivided, and job duties, personal relations and lines of authority are specified.

organization theory, That branch of study concerned with identification and analysis of the strategic variables and relationships of structure and process involving groups of people engaged in coordinated, purposeful activities over meaningful spans of time.

organizational behavior (OB), A field of knowledge and inquiry concerned with the systematic study of organizations: their origins and growth, and effect upon individual members, constituent groups, and other organizations. Organizational behavior is equally concerned with individual behavior, group processes, attitudes and motivations, communication, and the effect of these upon organizational performance.

organizational climate, The multidimensional set of properties of the work environment, perceived directly or indirectly by employees, and assumed to be a major force in influencing their attitudes, motivation, and behavior. The perceived quality and configuration of organizational climate is usually assessed along dimensions such as structure, responsibility, team spirit, standards, warmth-support, rewards, organizational clarity, etc.

organizing, The process of determining the work to be done, grouping work into appropriate units, and defining the desired relationships among people so that the entire body can pursue identified goals.

participation, The process by which people contribute ideas toward the solution of problems affecting the organization and their jobs. The actions by which managers involve their subordinates in the decision-making process.

participative management, A philosophy and system of management in which employees at one or many hierarchical levels of an organization share in setting goals, making decisions, and solving problems. The degree of employee influence in the decision-making process may range from simply advice-giving to full authority for decisions. Various organizational arrangements may be used such as advisory committees, labor-management committees, workers' councils, self-managed work teams, and employee representatives on governing boards.

personnel management, The function of acquiring, developing, motivating, and maintaining a competent work force so that the objectives of the organization are properly achieved and so that the members of this work force obtain satisfaction from their participation in their organization.

plan, A predetermined course of action over a specified period of time which represents a projected response to an anticipated environment in order to accomplish a specific set of adaptive objectives.

planning, Planning is the process whereby an individual or an organization identifies opportunities, needs, strategies, objectives and policies that are used to guide and manage the organization through future periods. All planning consists of (a) accumulation of information, (b) sorting and relating bits of information and beliefs, (c) establishing premises, (d) forecasting future conditions, (e) establishing needs, (f) identifying opportunities, (g) establishing objectives and policies, (h) structuring alternative courses of action, (i) ranking or selecting total systems of action which will achieve the best balance of ultimate (future) and immediate objectives, (j) establishing criteria and means for measuring adherence to the selected program of action and (k) so managing the organization to achieve the objectives. Some kinds of planning include short and long range planning, product planning, financial planning, etc.

policy, The verbal, written, or implied general plans of action that guide the members of the organization in the conduct of its operation, and incorporating in them broad premise and limitations within which further planning activities take place.

politics, The network of human behaviors and interactions by which social power is acquired, transferred, and exercised.

power, The ability to influence the behavior of other persons by any or all of the following means: (a) coercion — the application or threat of use of physical sanctions, or use of force to control satisfactions of basic human wants such as food, shelter, comfort and the like, (b) control of material rewards such as pay and benefits, (c) use of symbolic rewards and penalties via leadership, communication, allocation of prestige symbols, ritual, persuasion, and suggestion. (Adapted from Amitai Etzioni, *A Comparative Analysis of Complex Organizations,* the Free Press of Glencoe, Inc., 1961, p. 4-5).

productivity, The quantitative and qualitative result of the input of all resources. The most widely used productivity measure is one-dimensional (one measure of input and one measure of output) that defines it as PRODUCTIVITY — is output per labor input (e.g., number of trees planted per employee hour, etc.). A broader and more modern view involves value measures, e.g., LABOR PRODUCTIVITY — equals value added per employee. VALUE ADDED is defined as the net contribution of the business to the value of the output.

profession, A profession is a type of occupation whose work, values, and members ideally conform to the following criteria: (a) it requires advanced, specialized formal education and training as distinguished from general academic education or an apprenticeship, (b) professional work requires the consistent exercise of discretion, judgment, and personal responsibility, (c) it is based upon a deep and organized body of knowledge. Efforts are continually made to expand the knowledge through research, (d) the profession if advanced and regulated by a national level association of its members which helps establish standards for entry into the profession, establishes standards of ethical practice by members of the profession, and applies sanctions where these standards are violated, (e) the members of the profession ideally maintain a social consciousness and sense of trusteeship toward their clients, employers and the general public. They seek to maintain and update their knowledge and skills in keeping with advances in their field of work.

project management, An organizational form that is generally superimposed upon a traditional functional organizational structure. Integrated teams of specialists work under the coordination and direction of a project manager to accomplish a project of limited duration. The project manager coordinates and manages across functional and organizational lines to complete a specific project or program. (See matrix management.)

responsibility, Responsibility originates when one accepts the assignment to perform assigned duties and activities. The acceptance creates a liability for which the assignee is held answerable for and to ther assignor. It constitutes an obligation or accountability for performance.

role, The concept of role can be used in three distinct ways: (a) the demands of a given social position as determined by the pressures of the social group on the individual, (b) an individual's own conception of the part someone in his position is supposed to play in an organization, (c) the actions of individuals in terms of their relevance or implications for a social structure and its norms. It refers to the ways members of a position act with or without conscious design, in accord with or in violation of organizational norms.

role expectation, Role expectations organize expectations and responses to those expectations in the interaction system between the individual and others in the system. Role expectations refer to the expectations of others that the individual will behave in a certain way and also to the individual's expectations of what his behavior should be to meet the approval of others. The formal demands made upon an individual by his employer and the informal demands made by persons and groups associated with the individual in the work situation.

satisficing, Human behavior and decision-making based upon discovery and selection of satisfactory alternatives rather than optimal alternatives.

scalar chain, The chain of direct authority relationships from superior to subordinate and the grading of duties according to degrees of authority and responsibility. Equivalent to chain of command.

semantics, The science dealing with the relationship between signs and symbols (including words), their meaning, and human behavior.

sensitivity training, A method of training that predominately utilizes experiential learning in order to help people gain greater understanding of their own behavior, the behavior of others, the effect of one's behavior on others, and of group processes. Sensitivity training also seeks to develop specific communication and behavioral skills. The heart of sensitivity training occurs in unstructured groups (T-groups) in a face-to-face setting under the general guidance of a trainer who is relatively non-directive in his style. *Syn:* laboratory training.

social responsibility, The obligation an organization has as a member of society, especially in the areas of ecology, consumerism, and equal employment opportunity.

sociometry, The technique of analysis and decision-making for action in small groups whereby the individuals are identified, paired, or clustered according to their expressed preferences for others on such criteria as liking, disliking, leadership, efficiency, and frequency of communication.

span of control, The number of subordinates that a given person supervises.

specialization, Also referred to as division of work. This concerns dividing or grouping all the work to be done into small homogeneous packages which may be performed by individuals, groups, or departments.

staff, (1) An organizational unit that serves in an auxiliary and facilitative role in relation to line and operating units or executives. Typical functions of staff are data gathering, preparing plans, giving advice and recommendations, and rendering specialized service. Types of staff are personal or assistant to, general, specialized, and service. (2) Those persons who occupy a staff role in organizations.

staff, specialized, A type of staff that provides specialized or technical information, counsel, and service to operating units of the organization. Generally serves the entire organization and often deals in specialized staff work usually possess specialized or technical training and skills.

staff, service, A type of staff that provides essential services to the organization and its members. These services are primarily of a physical, maintenance, protection, or welfare nature.

status, Relative ranking of a person or group in terms of duties, rights, privileges or prestige. Relative social rank.

supervision, The function of leading, coordinating, and directing the work of others to accomplish designated objectives. Those people who perform this function, usually applied to those occupying positions of leadership in the lower strata of an organization such as foremen, general foremen, and office supervisors.

supervisor, acting (or manager), A temporary managerial position that may or may not become permanent.

supervisor, working, A non-supervisory employee who devotes a portion of his time to assigning work to other employees, job instruction, clearing troubles in the work, and checking progress of the work. *Syn:* group leader.

system, A system is characterized by: (a) a set of components of subsystems linked by information channels, (b) engagement in coordinated, goal-directed activity, (c) information flow as the basis for control, (d) a set of subgoals associated with the individual subsystems or components, (e) an external environment which influences the system. A system is said to be an open system if it reacts to its environment and is a closed system if it does not. It is an adaptive system if it reacts to environmental changes in a way that is favorable toward achieving the system goals.

systems 1, 2, 3, 4, These are four patterns or models of management and organizational climate developed by Rensis Likert. System 1 is designated as exploitive-authoritative, system 2 is benevolent-authoritative (paternalistic), system 3 is consultative, and system 4 is participative. The climate of a given organization is measured by a questionnaire covering such elements as leadership, communication, interaction-influence, and control processes.

systems analysis, A method of problem-solving that encompasses the identification, study, and evaluation of interdependent parts and their attributes, functioning as an ongoing process, and constituting an organic whole. For organization theory this involves the analysis of the functioning and integration of systems of people and physical entities.

T-Group, An unstructured learning group in a face-to-face setting under the general guidance of a trainer who utilizes a relatively non-directive style. The T-Group constitutes the heart of sensitivity or laboratory training. (See sensitivity training.)

task, A duty, an elementary component of a job or an operation to be performed. Every job normally consists of a number of tasks or basic work requirements.

task force, A temporary unit consisting of resources drawn from different parts of the organization and assigned to execute a specific mission. *Syn:* Project Team.

Theory X, A set of assumptions about the nature of people in organizations that was first formulated by Douglas McGregor. Those holding Theory X views believe that man is basically lazy, lacks ambition, and is indifferent to the needs of the organization. Therefore, the task of management is to employ pressure, coercion, and tight control in order to achieve organizational objectives.

Theory Y, A set of assumptions about the nature of people in organizations that was first formulated by Douglas McGregor. Those holding Theory Y views believe that people, under proper conditions, seek responsibility and like to work. They have the capacity to exercise imagination and initiative in solving organizational problems. Therefore, management should emphasize self-direction and self-control and the integration of individual and organizational goals.

transactional analysis (TA), A method designed to help the manager analyze and understand the behavior of others as well as his own. TA focuses on such areas as (a) the structure of the personality (personality is made up of three different ego states — Parent, Adult, and Child), (b) the way in which people interact (interactions between people involve a transaction between their ego states, with transactions classified as compelmentary, crossed, or ulterior), (c) the way in which people structure their time (the six basic ways of structuring time are withdrawal, rituals, pastimes, games, activities, and authenticity), and (d) the roles that people learn to play in life (life scripts).

unity of command, The concept that each subordinate in an organization should report to and be subject to the authority of only one superior. Often expressed as a principle of organization.

value added, See productivity.

Production Planning and Control

10

Z94.10

Enormous progress has been made in the field of production planning and control over the recent past. Within the last ten years, there has been a revolution in this field. Today, we are the possessors of an organized and coherent body of knowledge, most of which simply didn't exist ten years ago, and which is based on valid, widely-recognized principles consistent with the realities of life in a manufacturing organization.

In such a rapidly developing field, terminology can be a problem; it tends to lag the state of the art. Now, however, given the existence of a body of knowledge which has coalesced, we believe that standardized terminology is a practical matter. We hope that this publication assists materially in such a development, which would provide further impetus to the maturing process already well underway in the field of production planning and control.

A good case in point regarding the terminological difficulties in this field is the use of the term "MRP." Your chairman has felt for some time that the most serious terminology problem in the entire field centers around the use of the term "MRP," and that the imprecision surrounding it represents the greatest obstacle to achieving a standardized terminology. What we've been calling material requirements planning first evolved in the 1960's, but it didn't carry that label then. We called it "time-phased requirements planning" or "time series planning" or things like that. Somewhere along the way, however, "material requirements planning" became widely used and quickly became "MRP" in the verbal shorthand that we routinely employ in a technically complex world.

Initially we viewed MRP (material requirements planning) as merely a better technique for releasing orders. We thought of it as nothing more than an inventory ordering technique, albeit a superior one. It wasn't long, however, before we realized that material requirements planning was much more than that. Due to its precision and its time-phasing, we found that MRP had the ability to evaluate orders already released and to detect when the due date of an order was out of phase with when it would be needed. MRP could tell us when to reschedule open orders. This was a major breakthrough, and MRP had become a scheduling tool — much more important to us than merely an inventory ordering technique.

The next major development was when we recognized that we really needed to get feedback from the plant and purchasing to be assured that the plans were in fact being executed. Capacity requirements planning techniques had become well accepted by this time and we found ourselves talking about "closed-loop MRP," which specifies a great deal more than material requirements planning: capacity planning and control, plant and vendor scheduling and feedback, master production scheduling, etc. The evolution of closed-loop MRP was a development of enormous significance. For the first time we had a complete system with which we could effectively plan and monitor factory and vendor operations.

Closed-loop MRP also paved the way for the next major development in the field: manufacturing resource planning. Manufacturing resource planning is all of closed-loop MRP and more: it includes direct tie-ins to financial and business planning; it is capable of reporting, at any level of detail, in dollars for financial control as well as units for operational control; and it contains a simulation capability to test alternative plans and approaches. A handful of companies today are successfully utilizing manufacturing resource planning, and are reaping enormous benefits. It works, and in time it will become the standard for the management of a manufacturing company. Recognizing its origins, we refer to it as MRP II. Within this dictionary, therefore, are definitions for MRP (material requirements planning), closed-loop MRP, and manufacturing resource planning (MRP II).

Subcommittee on Production Planning and Inventory Control Z94.10

Chairman: Thomas F. Wallace, President
T. F. Wallace & Company
Cincinnati, Ohio

Individuals who have directly assisted the Chairman in the development of this terminology are as follows:

Jim Burlingame
Twin Disc

Jack Durben
Miles Laboratories

Romeyn Everdell
Rath & Strong

Walt Goddard
Oliver Wight, Inc.

Doug Hoelscher
Tennant Company

Luther Jenkins
Rohm & Hass

Bill O'Keefe
Martin Marietta
Data Systems

Jay Paul
Xerox

John Schorr
Independent Consultant

R. C. "Bing" Sherrill
Moog, Inc.

Dr. Tom Vollmann
Indiana University

Bill Wassweiler
MRM, Inc.

Ollie Wight
Oliver Wight, Inc.

Acknowledgements

Thanks are owed to the following organizations who provided various glossaries and other information to assist in the development of this dictionary:

Computer Aided Mfg. — International, Inc.
Arlington, Texas

Dow Corning Corporation
Midland, Michigan

IBM Corporation, General Systems Division
Atlanta, Georgia

Martin Marietta Data Systems
Baltimore, Maryland

In addition, the following publications proved to be of significant help to the Chairman and contributors:

Manufacturing Control – The Last Frontier for Profits, George W. Plossl, Reston Publishing Company, 1973.

Master Production Scheduling – Principles and Practice, William L. Berry, Thomas E. Vollmann and D. Clay Whybark, APICS, 1979.

Material Requirements Planning, Joseph Orlicky, McGraw-Hill Book Company, 1975.

Production and Inventory Control – Principles and Techniques, G. W. Plossl and O. W. Wight, Prentice-Hall, Inc., 1967.

Production and Inventory Control Handbook, James H. Green, APICS, 1970.

Production and Inventory Management in the Computer Age, Oliver W. Wight, Cahners Books International, 1974.

ABC classification, Classification of the items in an inventory in decreasing order of annual dollar volume. This array is then split into three classes, called A, B, and C. Class A contains the items with the highest annual dollar volume and receives the most attention. The medium Class B receives less attention, and Class C, which contains the low-dollar volume items, is controlled routinely. The ABC principle is, that effort saved through relaxed controls on low-value items will be applied to reduce inventories of high-value items. *Syn:* distribution by value.

ABC inventory control, An inventory control approach based on the ABC classification.

accessory, A choice or feature offered to customer for customizing the end product. In many companies, this term means that the choice does not have to be specified prior to shipment but could in fact be added at a later date. In other companies, however, this choice must be made prior to shipment. (See attachment, option.)

action message, An output of an MRP system that identifies the need for and the type of action to be taken to correct a current or a potential material coverage problem. Examples of action messages are "Release Order," "Reschedule In," "Reschedule Out," "Cancel," etc. *Syn:* exception message.

aggregate inventory, The sum of the inventory levels for individual items. For example, the aggregate finished goods inventory would be made up of one half the sum of all the lot sizes plus the sum of all of the safety stocks plus any anticipation inventory plus transportation inventory.

allocation, (1) In an MRP system, an allocated item is one for which a picking order has been released to the stock room but not yet sent out of the stock room. It is an "uncashed" stock room requisition. (2) A process used to distribute material in short supply. (See reservation.)

anticipated delay report, A regular report, normally issued by both Manufacturing and Purchasing to the material planning function, regarding jobs or purchase orders which will not be completed on time, why not, and when they will be completed. This is one essential ingredient of a closed-loop MRP system.

anticipation inventories, Additional inventory above basic pipeline stock to cover projected trends of increasing sales, planned sales promotion programs, seasonal fluctuations, plant shutdowns and vacations. (See base series.)

assemble-to-order product, See make-to-order product.

attachment, A choice or feature offered to customers for customizing the end product. In many companies, this term means that the choice, although not mandatory, must be selected prior to the final assembly schedule. In other companies, however, the choice need not be made at that time. (See accessory, option.)

available to promise, The uncommitted portion of a company's inventory or planned production. This figure is frequently calculated from the master production schedule and is maintained as a tool for order promising.

backlog, All of the customer orders booked, i.e. received but not yet shipped. Sometimes referred to as "open orders" or the "order board."

backorder, An unfilled customer order or commitment. It is an immediate (or past due) demand against an item whose inventory is insufficient to satisfy the demand. (See stock-out.)

backward scheduling, A scheduling technique where the schedule is computed starting with the due date for the order and working backward to determine the required start date. This can generate negative times, thereby identifying where time must be made up. (See forward scheduling.)

base series, A standard series of demand-over-time observations used in forecasting seasonal items. This series of factors is usually based upon the relative level of demand during the corresponding period of previous years. The average value of the base series over a twelve-month period will be 1.0. The base series is superimposed upon the average trend in demand for the item, for forecasting purposes. *Syn:* base index.

bill of capacity, See product load profile.

bill of labor, See product load profile.

bill of material, A listing of all the sub-assemblies, parts and raw materials that go into a parent assembly showing the quantity of each required to make an assembly. There are a variety of formats of bill of material, including single level bill of material, indented bill of material, modular (planning) bill of material, transient bill of material, matrix bill of material, costed bill of material, etc.

bill of material processor, Refers to the computer applications supplied by many manufacturers for maintaining, updating, and retrieving bill of material information on direct access files.

bill of material structuring, The process of organizing bills of material to perform specific functions. (See bill of material, planning bill, transient bill.)

bill of resources, See product load profile.

bin reserve system, See two bin system.

book inventory, An accounting definition of inventory units or value obtained from perpetual inventory records rather than by actual count.

bottleneck, A facility, function, department, etc., that impedes production — for example, a machine or work center where jobs arrive at a faster rate than they leave.

bucketed system, An MRP system in which all time-phased data is displayed in accumulated time periods, or "buckets." If the period of accumulation would be one week, then the system would be said to have weekly buckets. (See bucketless system, time buckets, horizontal display, vertical display.)

bucketless system, An MRP system in which all time-phased data is received, stored, processed and reported by specific dates and not in weekly (or larger) time buckets. (See bucketed system, time buckets, horizontal display, vertical display.)

business plan, A statement of income projections, costs and profits usually accompanied by budgets and a projected balance sheet as well as a cash flow (source and application of funds) statement. It is usually stated in terms of dollars only. The business plan and the production plan, although frequently stated in different terms, should be in agreement with each other. (See manufacturing resource planning.)

closed-loop MRP, A system built around MRP and also including the additional planning functions of production planning, master production scheduling, and capacity requirements planning. Further, once the planning phase is complete and the plans have been accepted as realistic and attainable, the execution functions come into play. These include the shop floor control functions of input-output measurement, detailed scheduling and dispatching, plus anticipated delay reports from both the shop and vendors, purchasing follow-up and control, etc. The term "closed-loop" implies that not only is each of these elements included in the overall system but also that there is feedback from the execution functions so that the planning can be kept valid at all times. See diagram. (See manufacturing resource planning.)

common parts bill (of material), A type of planning bill which groups all common components for a product or family of products into one bill of material. (See planning bill, modular bill, super bill.)

component, An inclusive term used to identify a raw material, ingredient, part or subassembly that goes into a higher level assembly, compound or other item. May also include packaging materials for finished items.

configuration control, The function of ensuring that the product being built and shipped corresponds to the product ordered and designed. This means that the correct features, customer options and engineering changes have been incorporated.

control center, In a centralized dispatching operation, the place at which the dispatching is done. (See centralized dispatching.)

costed bill of material, A form of bill of material which, besides providing the normal information such as components, quantity per, effectivity data, etc. also extends the quantity per of every component in the bill by the cost of the components.

CPIM, Certified in Production and Inventory Management. A certification awarded by APICS.

critical ratio, A dispatching rule which calculates a priority index number by dividing the time to due date remaining by the expected elapsed time to finish the job.

$$\frac{\text{Time Remaining}}{\text{Work Remaining}} = \frac{30}{40} = .75$$

Typically ratios of less than 1.0 are behind, ratios greater than 1.0 are ahead, and a ratio of 1.0 is on schedule. (See dispatching rules.)

CRP, Abbreviation for Capacity Requirements Planning.

cumulative lead time, The longest length of time involved to accomplish the activity in question. For any item planned through MRP it is found by reviewing each bill of material path below the item, and whichever path adds up to the greatest number defines cumulative *material* lead time. *Syn:* aggregate lead time, stacked lead time, composite lead time, critical path lead time.

customer service, Delivery of product to the customer at the time which the customer or corporate policy specifies. (See percent of fill.)

customer service ratio, A measure of delivery performance usually in the form of a percentage. In a make-to-stock company this percentage usually represents the number of items or dollars on the customer's order for a specific time period. In a make-to-order company it is usually some comparison of the number of jobs that were supposed to be shipped in that time period. (See percent of fill, stockout percentage.)

cycle counting, A physical inventory-taking where inventory is counted on a periodic schedule rather than once a year. For example, a cycle inventory count may be taken when an item reaches its reorder point, when new stock is received, or on a regular basis usually more frequently for high-value fast-moving items and less frequently for low-value or slow-moving items. Most effective cycle counting systems require the counting of a certain number of items every work day. (See ABC classification.)

decentralized dispatching, The organization of the dispatching function into individual departmental dispatchers. (See centralized dispatching.)

delivery cycle, The actual time from the receipt of the customer order to the time of the shipment of the product.

delivery policy, The company's goal for the time to ship product after receipt of a customer's order. The policy is sometimes stated as "our quoted delivery time."

demand, A need for a particular product or component. The demand could come from any number of sources, i.e. customer order, forecast, interplant, branch warehouse, service part, or to manufacturing the next higher level. At the finished goods level, "demand data" is usually different from "sales data" because demand does not necessarily result in sales, i.e., if there is no stock there will be no sale. (See dependent demand, independent demand.)

demand management, The function of recognizing and managing all of the demands for products to insure that the master scheduler is aware of them. It encompasses the activities of forecasting, order entry, order promising, branch warehouse requirements, interplant orders, and service parts requirements. (See master production schedule.)

dependent demand, Demand is considered dependent when it is directly related to or derived from the demand for other items or end products. Such demands are therefore calculated, and need not and should not be forecast. A given inventory item may have both dependent and independent demand at any given time. (See independent demand.)

direct-deduct inventory transaction processing, A method of doing bookkeeping which decreases the book (computer) inventory of an item as material is issued from stock, and increases the book inventory as material is received into stock. The key concept here is that the book record is updated coincident with the movement of material out of or into stock. As a result, the book record is a representation of what is physically in stock. (See pre-deduct inventory transaction processing, post-deduct inventory transaction processing.)

dispatching, The selecting and sequencing of available jobs to be run at individual work stations and the assignment of these jobs to workers. (See centralized dispatching, decentralized dispatching, expediting, shop planning, closed-loop MRP.)

dispatching rule, The logic used to assign priorities to jobs at a work center. (See critical ratio.)

distribution requirements planning (DRP), The function of determining the needs to replenish inventory at branch warehouses. Frequently a time-phased order point approach is used where the planned orders at the branch warehouse level are "exploded" via MRP logic to become gross requirements on the supplying source. In the case of multi-level distribution networks, this explosion process can continue down through the various levels of master warehouse, factory warehouse, etc., and become input to the master production schedule. Demand on the supplying source(s) is recognized as dependent, and standard MRP logic applies. (See time-phased order point, physical distribution, pull distribution system.)

distribution resource planning, The extension of distribution requirements planning into the planning of the key resources contained in a distribution system: warehouse space, manpower, money, trucks and freight cars, etc. (See distribution requirements planning.)

due date, The date at which purchased material or production on order is due to be available for use.

economic order quantity (EOQ), A type of fixed order quantity which determines the amount of product to be purchased or manufactured at one time in order to minimize the total cost involved, including the ordering costs (set-up of machines, writing orders, checking receipts, etc.) and carrying costs (cost of capital invested, insurance, taxes, space, obsolescence and spoilage). The economic order quantity may be calculated from the equation:

$$EOQ = \sqrt{(2AS)/rv}$$

where EOQ is the quantity to be ordered, S is the annual sales, A is the ordering cost, r is the carrying cost, and v is the unit cost. (See carrying cost, ordering cost.)

effective date, The date on which a component or an operation is introduced or severed from a bill of material or an assembly process as the case may be. The effective dates are used in the explosion process to create demands for the correct material or assembly labor. Normally, bill of material systems provide for an effective "Start Date" and "Stop Date," signifying the start or stop of a particular relationship. Control may also be by serial number rather than date.

effectivity, See effective date.

EOQ, Abbreviation for Economic Order Quantity.

expediting, The "rushing" or "chasing" of production orders which are needed in less than the normal lead time. (See dispatching.)

explosion, An extension of a bill of material into the total of each of the components required to manufacture a given quantity of upper-level assemblies or sub-assemblies.

exponential smoothing, A type of weighted average forecasting technique in which past observations are geometrically discounted according to their age. The heaviest weight is assigned to the most recent datum. The smoothing is termed "exponential" because data points are weighted in accordance with an exponential function of their age. The technique makes use of a smoothing constant to apply to the difference between the most recent forecast and the critical sales datum, which avoids the necessity of carrying historical sales data. The approach can be used for data which exhibit no trend or seasonal patterns or for data with either (or both) trend and seasonality.

extrinsic forecast, A forecast based on a correlated leading indicator such as estimating furniture sales based on housing starts. Extrinsic forecasts tend to be more useful for large aggregations such as total company sales, than for individual product sales. (See intrinsic forecasts.)

feature, See accessory, attachment, option.

final assembly schedule (FAS), Also referred to as the "finishing schedule" as it may include other operations than simply the final operation. It is a schedule of end items either to replenish finished goods inventory or to finish the product for a make-to-order product. For make-to-order products, it is prepared after receipt of a customer order, is constrained by the availability of material and capacity, and it schedules the operations required to complete the product from the level where it is stocked (or master scheduled) to the end item level. *Syn:* blending schedule, pack-out schedule.

finished products inventories, Inventories on which all manufacturing operations, including final test, have been completed. These may be either finished parts, like renewal parts or finished products which have been authorized for transfer to the finished stock account. These products are now available for shipment to the customer either as end items or repair parts.

finishing lead time, The time that is necessary to finish manufacturing a product after receipt of a customer order. The time allowed for the FAS. The finishing lead time should be equal to or less than the company's goal for shipping its product after receipt of a customer order.

finite loading, Conceptually the term means putting no more work into a factory than the factory can be expected to execute. The specific term usually refers to a computer technique that involves automatic shop priority revision in order to level load operation by operation. (See infinite loading.)

firm planned order, A planned order that can be frozen in quantity and time. The computer is not allowed to automatically change it; this is the responsibility of the planner in charge of the item that is being planned. This technique can aid planners working with MRP systems to respond to material and capacity problems by firming up selected planned orders. Additionally, firm planned orders are the normal method of stating the master production schedule.

fixed order quantity, A lot sizing technique in MRP that will always cause a planned order to be generated for a predetermined fixed quantity (or multiples thereof if net requirements for the period exceed the fixed order quantity). (See economic order quantity, lot-for-lot, period order quantity.)

floor stocks, Stocks of inexpensive production parts held in the factory from which production workers can draw without requisitions.

fluctuation inventories, Inventories that are carried as a cushion to protect against forecast error. (See safety stock.)

forecast, A forecast is the extrapolation of the past into the future. It is an objective computation involving data as opposed to a prediction which is a subjective estimate incorporating management's anticipation of changes and new factors influencing demand. The implied objective is to minimize the forecast error.

forecast error, The difference between actual demand and forecast demand, typically stated as an absolute value.

forecast horizon, The period of time into the future for which a forecast is prepared.

forecast interval, The length of time over which the forecast is computed.

forecast period, The time unit for which forecasts are prepared, such as monthly, weekly, or quarterly. *Syn:* forecast interval.

forward scheduling, A scheduling technique where the scheduler proceeds from a known start date and computes the completion date for an order usually proceeding from the first operation to the last. (See backward scheduling.)

FPO, Abbreviation for Firm Planned Order.

Gantt chart, The earliest and best known type of control chart especially designed to show graphically the relationship between planned performance and actual performance. Named after its originator, H. L. Gantt, scientific management pioneer. Used especially for machine loading, where one horizontal line is used to represent capacity and another to represent load against that capacity or for following job progress where one horizontal line represents the production schedule and another parallel line represents the actual progress of the job against the schedule in time.

gross requirements, The total of independent and dependent demand for a part or an assembly prior to the netting of on hand inventory and scheduled receipts.

group classification code, A part or material classification technique which provides for designation of characteristics by successively lower order groups of code. Classification may denote, for example, function, type of material, size, shape, etc. (See group technology.)

group technology, An engineering and manufacturing philosophy which identifies the "sameness" of parts, equipment or processes. It provides for rapid retrieval of existing designs and anticipates a cellular type production equipment layout.

hedge, (1) In master production scheduling, a quantity of stock used to protect against uncertainty in demand. The hedge is similar to safety stock, except that a hedge has the dimension of timing as well as amount. A typical example of a "product mix" hedge might be for a company that makes a box with either gold or silver handles (all else same). If 40% of the boxes usually took gold handles and 60% silver, a master production schedule for 200 would explode into 200 common boxes, 80 gold handles, and 120 silver handles. The mix hedge might be to carry 50 extra gold and 50 extra silver handles. If at some point in time 180 of the 200 boxes are sold, uncertainty only exists as to the remaining 20; any silver or gold handles above 20 can be rescheduled until the next uncertainty occurs. Another hedge is called the "volume" hedge, which would be for the common boxes in this example. This hedge is typically carried by the master production scheduler just beyond some time fence in the future such that rolling the hedge over the time fence will result in planned orders for major expense items to now be in the action time bucket. (2) In purchasing, any purchase or sale transaction having as its purpose the elimination of the negative aspects of price fluctuations.

horizontal display, A method of displaying output from an MRP system where requirements, scheduled receipts, projected balance, etc. are displayed horizontally, i.e. across the page. Horizontal displays are difficult to use in conjunction with bucketless systems. (See vertical display, bucketless system.)

indented bill of material, A form of multi-level bill of material. It exhibits the highest level sub-assemblies closest to the left side margin and all the components going into these sub-assemblies are shown indented to the right of the margin, all subsequent levels of components are indented farther to the right. If a component is used in more than one sub-assembly within a given product structure, it will appear more than once, under every sub-assembly in which it is used.

independent demand, Demand for a finished good or a component is considered independent when such demand is unrelated to the demand for other items. Demand for finished goods, parts required for destructive testing and service parts requirements are some examples of independent demand. (See dependent demand.)

infinite loading, Showing the work behind work centers in the time periods required regardless of the capacity available to perform this work. The term infinite loading is considered to be obsolete today, although the specific computer programs used to do infinite loading can now be used to perform the technique called capacity planning. Infinite loading was a gross misnomer to start with, implying that a load could be put into a factory regardless of its availability to perform. The poor terminology obscured the fact that it is necessary to generate capacity requirements and compare these with available capacity before trying to adjust requirements to capacity. (See capacity requirements planning, finite loading.)

input, (1) Work entering a production facility. (2) Data to be processed on a computer.

input/output control, A technique for capacity control where actual output from a work center is compared with the planned output developed by CRP. The input is also monitored to see if it corresponds with plans so that work centers will not be expected to generate output when jobs are not available to work on. (See capacity control, closed-loop MRP.)

interplant demand, Material to be shipped to another plant or division within the corporation. Although it is not a customer order, it is usually handled by the master production scheduling system in a similar manner. (See demand management.)

intransit lead time, The time lag between the date of shipment (at supplier shipping point) and the date of receipt (at the customer's dock). Normally customer's orders specify the date by which goods should be at his dock. Consequently this date should be offset by intransit lead time for establishing a ship date for the supplier.

intrinsic forecast, A forecast made based on past history, such as a forecast made from a moving average. (See extrinsic forecast.)

inventory, Items which are in a stock point or work-in-process and which serve to decouple successive operations in the process of manufacturing a product and distributing it to the consumer. Inventories may consist of finished goods ready for sale; they may be parts or intermediate items; they may be work-in-process; or they may be raw materials.

inventory control, The activities and techniques of maintaining the stock of items at desired levels, whether they be raw materials, work-in-process, or finished products. (See inventory management.)

inventory management, The branch of business management concerned with the planning and control of inventories. (See inventory control.)

inventory policy, A definite statement of the philosophy of management on inventories.

inventory shrinkage, Losses resulting from scrap, deterioration, pilferage, etc.

inventory turnover, The number of times that an inventory "turns over," or cycles during the year. One way to compute inventory turnover is to divide the average inventory level into the annual cost of sales. For example, if average inventory were three million dollars and cost of sales were twenty-one million dollars, the inventory would be considered to "turn" several times per year.

inventory valuation, The value of the inventory at either its cost or its market value. Because inventory value can change with time, some recognition must be taken of the age distribution of inventory. Therefore, the cost value of inventory, under accounting practice, is usually computed on a first-in-first-out (FIFO), last-in-first-out (LIFO) basis, or a standard cost system to establish the cost of goods sold.

inventory write-off, A deduction of inventory dollars from the financial statement because the inventory is no longer saleable or because of shrinkage, i.e., the value of the physical inventory is less than its book value.

kitting, The process of removing components of an assembly from the stockroom and sending them to the assembly floor as a kit of parts. This action may take place automatically whenever a full set of parts is available and/or it may be done only upon authorization by a designated person. (See picking.)

lead time, A span of time required to perform an activity. In a production and inventory control context, the activity in question is normally the procurement of materials and/or products either from an outside supplier or from one's own manufacturing facility. The individual components of any given lead time, queue time, move or transportation time, receiving and inspection time. (See manufacturing lead time, purchasing lead time.)

lead time offset, A term used in MRP where a planned order receipt in a one time period will require the release of that order in some earlier time period based on the lead time for the item. The difference between the due date and the release date is the lead time offset.

level, Every part or assembly in a product structure is assigned a level code signifying the relative level in which that part or assembly is used within that product structure. Normally the end items are assigned level "0" and the components/sub-assemblies going into it level "1" and so on. MRP explosion process starts from level "0" and proceeds downwards one level at a time.

limiting operation, In a series of operations with no alternative routings, the capacity of the total system can be no greater than the operation with the least capacity. As long as this limiting condition exists, the total system can be effectively scheduled by simply scheduling the limiting operation. *Syn:* bottleneck.

line balancing, An assembly line process can be divided into elemental tasks, each with a specified time requirement per unit of product and a sequence relationship with the other tasks. Line balancing is the assignment of these tasks to work stations so as to minimize the number of work stations and to minimize the total amount of unassigned time at all stations. Line balancing can also mean a technique for determining the product mix that can be run down an assembly line to provide a fairly consistent flow of work through that assembly line at the planned line rate. For example, if an automotive assembly line happened to be scheduled one day with nothing but convertibles, some workers would be standing idle while others would not be able to keep pace with the line.

load, This is the amount of scheduled work ahead of a manufacturing facility, usually expressed in terms of hours or work units of production.

load leveling, Spreading orders out in time or rescheduling operations so that the amount of work to be done in the time periods tends to be distributed evenly. (See finite loading.)

load profile, A display of future capacity requirements based on planned and released orders over a given span of time. *Syn:* load projection.

logistics, In an industrial context, this term refers to the art and science of obtaining and distributing material and product. In a military sense (where it has greater usage), its meaning can also include the transportation of personnel.

long-range resource planning, A planning activity for long-term capacity decisions, based on the production plan and perhaps on even more gross data (e.g., sales per year) beyond the time horizon for the production plan. This activity is to plan long term capacity needs out to the time period necessary to acquire gross capacity additions such as a major factory expansion.

lot-for-lot, A lot sizing technique in MRP which generates planned orders in quantities equal to the net requirements in each period. *Syn:* discrete order quantity.

lot size, The amount of a particular item that is ordered from the plant or a vendor. *Syn:* order quantity.

lot size inventory, Inventories which are maintained whenever quantity price discounts, shipping costs, or set-up costs, etc. make it more economical to purchase or produce in larger lots than are needed for immediate purposes.

low level code, Identifies the lowest level in any bill of material at which a particular component may appear. Net requirements for a given component are not calculated until all the gross requirements have been calculated down to that level. Low level codes are normally calculated and maintained automatically by the computer software. (See level.)

machine loading, The accumulation by work station(s), machine, or machine group of the hours generated from the scheduling of operations for released orders by time period. Machine loading differs from capacity planning in that it does not use the planned orders from MRP but operates solely from scheduled receipts. As such, it has very limited usefulness. (See capacity requirements planning.)

make-to-order products, The end item is finished after receipt of a customer order. Frequently, long lead time components are planned prior to the order arriving in order to reduce the delivery time to the customer. Where options or other subassemblies are stocked prior to customer orders arriving, the term "assemble-to-order" is frequently used.

make-to-stock product, The end item is shipped from finished goods, "off the shelf," and therefore is finished prior to a customer order arriving.

manufacturing lead time, The total time required to manufacture an item. Included here are order, preparation time, queue time, set-up time, run time, move time, inspection and put-away time.

manufacturing order, A document or group of documents conveying authority for the manufacture of specified parts or products in specified quantities.

manufacturing resource planning, A method for the effective planning of all the resources of a manufacturing company. Ideally, it addresses operational planning in units, financial planning in dollars, and has a simulation capability to answer "what if" questions. It is made up of a variety of functions, each linked together: Business Planning, Production Planning, Master Production Scheduling, Material Requirements Planning, Capacity Requirements Planning and the execution systems for capacity and priority. Outputs from these systems would be integrated with financial reports such as the business plan, purchase commitment report, shipping budget, inventory projections in dollars, etc. Manufacturing resource planning is a direct outgrowth and extension of MRP. Often referred to as MRP II. (See closed-loop MRP.)

market hedge, See hedge.

master production schedule (MPS), For selected items, it is a statement of what the company expects to manufacture. It is the anticipated build schedule for those selected items assigned to the master scheduler. The master scheduler maintains this schedule and, in turn, it becomes a set of planning numbers which "drives" MRP. It represents what the company plans to produce expressed in specific configurations, quantities, and dates. The MPS should not be confused with a sales forecast which represents a statement of demand. The master production schedule must take forecast plus other important considerations (backlog, availability of material, availability of capacity, management policy and goals, etc.) into account prior to determining the best manufacturing strategy. (See closed-loop MRP.) *Syn:* master schedule.

master schedule, See master production schedule.

master schedule item, A part number selected to be planned by the master scheduler. The item would be deemed critical in terms of its impact on lower level components and/or resources such as skilled labor, key machines, dollars, etc. Therefore, the master scheduler, not the computer would maintain the plan for these items. A master schedule item may be an end item, a component, a pseudo number or a planning bill of material.

master scheduler, The job title of the person who manages the master production schedule. This person should be the best scheduler available as the consequences of the planning here has a great impact on material and capacity planning. Ideally, the person would have substantial product and shop knowledge.

materials management, A term to describe the grouping of management functions related to the complete cycle of material flow, from the purchase and internal control of production materials to the planning and control of work-in-process to the warehousing, shipping and distribution of the finished product. Differs from materials control in that the latter term traditionally, is limited to the internal control of production materials.

matrix bill of material, A chart made up from the bills of material for a number of products in the same or similar families. It is arranged in a matrix with parts in columns and assemblies in rows (or vice versa) so that requirements for common components can be summarized conveniently.

minimum order quantity, An order quantity modifier, applied after the lot size has been calculated, that increases the order quantity to a pre-established minimum.

modular bill (of material), A type of planning bill which is arranged in product modules or options. Often used in companies where the product has many optional features, e.g., automobiles. (See planning bill, common parts bill, super bill, option.)

move ticket, A document used in dispatching to authorize and/or record movement of a job from one work center to another. It may also be used to report other information such as the active quantity or the material storage location.

MPS, Abbreviation for Master Production Schedule.

MRP, Material Requirements Planning. A system which uses bills of material, inventory and open order data, and master production schedule information to calculate requirements for materials. It makes recommendations to release replenishment orders for material. Further, since it is time-phased, it makes recommendations to reschedule open orders when due dates and need dates are not in phase. Originally seen as merely a better way to order inventory, today it is thought of primarily as a scheduling technique, i.e., a method for establishing and maintaining valid due dates on orders. (See closed-loop MRP, manufacturing resource planning.)

MRP II, See manufacturing resource planning.

multi-level bill of material, A multi-level bill shows all the components that are directly used in an assembly together with the quantity required of every component. If a component is a sub-assembly, all the components of the sub-assembly will also be exhibited in the multi-level bill.

net change MRP, An approach via which the material requirements plan is continually retained in the computer. Whenever there is a change in requirements, open order or inventory status, or engineering usage, a partial explosion is made only for those parts affected by the change. Net changes systems may be continuous and totally transaction oriented, or done in a periodic (often daily) batch. (See regeneration MRP, requirements alteration.)

net requirements, In MRP, the net requirements for a part or an assembly are derived as a result of netting gross requirements against inventory on hand and the scheduled receipts. Net requirements, lot sized and offset for lead time, become planned orders.

netting, The process of calculating net requirements.

on hand, The balance shown in perpetual inventory records as being physically present at the stocking locations.

on order, The stock on order is the quantity represented by the total of all outstanding replenishment orders. The on order balance increases when a new order is released, and it decreases when material is received to fill an order, or when an order is cancelled. (See on hand, open order.)

open order, (1) An active manufacturing order or purchase order. (2) An unfilled customer order. *Syn:* scheduled receipt.

option, A choice or feature offered to customers for customizing the end product. In many companies, the term "option" means a mandatory choice — the customer must select from one of the available choices. For example, in ordering a new car, the customer must specify an engine but need not necessarily select an air conditioner. (See accessory, attachment.)

order entry, The process of accepting and translating what a customer wants into terms used by the manufacturer. This can be as simple as creating shipping documents for a finished goods product line, to a more complicated series of activities including engineering effort for make-to-order products.

order multiple, An order quantity modifier applied after the lot size has been calculated, that increments the order quantity to a predetermined multiple.

order point, The inventory level such that if the total stock on hand plus on order falls to or below the order point, action is taken to replenish the stock. The order point is normally calculated as: forecasted usage during the replenishment lead time plus safety stock. (See time-phased order point.) *Syn:* reorder point, trigger level.

order promising, The process of making a delivery commitment, i.e., answering the question "when can you ship?" For make-to-order products this usually involves a check of uncommitted material and availability of capacity. *Syn:* order dating, customer order promising.

order quantity modifiers, Order quantities are calculated based upon one of the lot sizing rules. However, it may be necessary to adjust the calculated lot size due to some special considerations. These adjustments are called order quantity modifiers. (See minimum order quantity, order multiples.)

ordering cost, In calculating economic order quantities, refer to the costs which increase as the number of orders placed increases. Includes costs related to the clerical work of preparing, issuing, following and receiving orders, the physical handling of goods, receiving orders, inspections, and machine set-up costs, if the order is being manufactured. (See carrying costs.) *Syn:* acquisition cost.

output, (1) Work being completed by a production facility. (2) The result of a computer program.

overlapped schedule, The "overlapping" of successive operations, whereby the completed portion of a job lot at one work center is processed at one or more succeeding work centers before the pieces left behind are finished at the preceding work center(s). *Syn:* lap-phasing, telescoping.

Pareto's Law, A concept developed by Vilfredo Pareto, an Italian economist, that simply says that a small percentage of a group account for the largest fraction of the effort, value, etc. For example, twenty percent of the inventory items comprise eighty percent of the inventory value. (See ABC classification.) *Syn:* 80/20 rule.

part number, A number which serves to uniquely identify a component, product, or raw material. *Syn:* stock code, product coding.

pegging, In MRP, pegging displays for a given item the details of the sources of its gross requirements and/or allocations. Pegging can be thought of as "live" where-used information.

percent of fill, A measure of the effectiveness with which the inventory management system responds to actual demand. The percent of customer orders filled off the shelf can be measured in either units or dollars. (See stock out percentage.)

period order quantity, A lot sizing technique under which the lot size will be equal to the net requirements for a given number of periods (e.g., weeks) into the future. (See fixed order quantity, lot-for-lot.) *Syn:* days' supply, weeks' supply.

perpetual inventory, Usually used to describe an inventory record-keeping system where each transaction in and out is recorded and a new balance is computed. (See physical inventory.)

phantom bill of material, See transient bill of material.

physical distribution, The combination of activities associated with the movement of material, usually finished products, from the manufacturer to the customer. In many cases, this movement is made through one or more levels of field warehouses. (See distribution requirements planning.)

physical inventory, The determination of inventory quantity by actual count. Physical inventories can be taken on a continuous, periodic, or annual basis. (See cycle counting.)

picking, The process of withdrawing from stock the components to make the products, or the finished goods to be shipped to a customer. (See kitting.)

picking list, A document which is used by operating personnel to pick manufacturing or shipping orders.

planned issue receipt, An issue or receipt transaction which updates the on hand balance and the related allocation or open order.

planned order, A suggested order quantity and due date created by MRP processing, when it encounters net requirements. Planned orders are created by the computer;

exist only within the computer; and may be changed or deleted by the computer during subsequent MRP processing if conditions change. Planned orders at one level will be exploded into gross requirements for components at the next lower level. Planned orders also serve as input to capacity requirements planning, along with released orders, to show the total capacity requirements in future time periods.

planning bill (of material), An artificial grouping of items, in bill of material format, used to facilitate master scheduling and/or material planning. (See common parts bill, option, super bill.)

planning horizon, In an MRP system, the planning horizon is the span of time from the current to some future date for which material plans are generated. This must cover at least the cumulative purchasing and manufacturing lead time, and usually is quite a bit longer to facilitate MRP II.

post-deduct inventory transaction processing, A method of doing inventory bookkeeping where the book (computer) inventory of components is reduced only after completion of activity on their upper-level parent or assembly. This approach has the disadvantage of a built-in differential between the book record and what is physically in stock. (See direct-deduct inventory transaction processing, pre-deduct inventory transaction processing.)

pre-deduct inventory transaction processing, A method of doing bookkeeping where the book (computer) inventory of components is reduced prior to issue, at the time a scheduled receipt for their parent or assembly is created. This approach has the disadvantage of a built-in differential between the book record and what is physically in stock. (See direct-deduct inventory transaction processing, post-deduct inventory transaction processing.)

prediction, An intuitive estimate of demand taking into account changes and new factors influencing the market; as opposed to a forecast, which is an objective projection of the past into the future. (See forecast.)

priority, In a general sense, refers to the relative importance of jobs, i.e. which jobs should be worked on and when. It is a separate concept from capacity. (See capacity, scheduling.)

product load profile, A statement of the key resources required to manufacture one unit of a selected item. Often used to predict the impact of the item scheduled in the master production schedule on these resources. *Syn:* bill of labor, bill of resources, resource profile.

product structure, The way components go into a product during its manufacture. A typical product structure would show, for example, raw material being converted into fabricated components, components being put together to make subassemblies, subassemblies going into assemblies, etc.

production control, The function of directing or regulating the movement of goods through the entire manufacturing cycle from the requisitioning of raw materials to the delivery of the finished product. (See inventory control.)

production plan, The agreed upon strategy that comes from the production planning function. (See production planning.)

production planning, The function of setting the overall level of manufacturing output. It's prime purpose is to establish production rates that will achieve management's objective in terms of raising or lowering inventories or backlogs, while usually attempting to keep the production force relatively stable. The production plan is usually stated in broad terms (e.g., product groupings, families of products). It must extend through a planning horizon sufficient to plan the labor, equipment, facilities, material, and finances required to accomplish the production plan. Various units of measure are used by different companies to express the plan such as standard hours, tonnage, labor operators, units, pieces, dollars, etc. As this plan affects all company functions, it is normally prepared with information from marketing, manufacturing, engineering, finance, materials, etc. In turn, the production plan becomes management's authorization for the Master Scheduler to convert into a more detailed plan. (See business plan, closed-loop MRP.) *Syn:* production program.

production rates, The quantity of production usually expressed in units, hours, or some other broad measure, expressed by a period of time, i.e. per hour, per shift, day, week, etc. (See production planning.) *Syn:* production levels.

production schedule, A plan which authorizes the factory to manufacture a certain quantity of a specific item. Usually initiated by the production planning department.

projected available balance, In MRP, the inventory balance projected out into the future. It is the running sum of on hand inventory minus requirements plus scheduled receipts.

projection, Estimation based on past data. *Syn:* extrapolation.

pseudo bill (of material), See planning bill.

pull distribution system, A system for replenishing field warehouse inventories wherein replenishment decisions are made at the field warehouse itself, not at the central supply warehouse or plant. (See push distribution system.)

purchasing lead time, The total lead time required to obtain a purchased item. Included here are procurement lead time, vendor lead time, transportation lead time, receiving, inspection and put away time.

push distribution system, A system for replenishing field warehouse inventories wherein replenishment decision-making is centralized, usually at the manufacturing site or central supply facility. (See distribution requirements planning, pull distribution system.)

queue, A waiting line. In manufacturing, the jobs at a given work center waiting to be processed. As queues increase, so do average queue time and work-in-process inventory.

queue time, The amount of time a job waits at a work center before set-up or work is performed on the job. Queue time is one element of total manufacturing lead time. Increases in queue time result in direct increases to manufacturing lead time.

regeneration MRP, An approach where the master production schedule is totally re-exploded down through all bills of material, at least once per week to maintain valid priorities. New requirements and planned orders are completely "re-generated" at that time. (See net change MRP, requirements alteration.)

reorder point, See order point.

reorder quantity, In a fixed order system of inventory control, the fixed quantity which should be ordered each time the available stock (on hand plus on order) falls below the order point. However, in a variable reorder quantity system the amount ordered from time period to time period will vary. (See economic order quantity, lot size.) *Syn:* replenishment order quantity.

repair parts demand, See service parts demand.

replenishment lead time, The total period of time that elapses from the moment it is determined that a product is to be reordered until the product is back on the shelf available for use.

requirements alteration, Processing a revised master production schedule through MRP in order to review the impact of the changes. Not to be confused with net change, which, in addition to changes to the MPS, also processes changes to inventory balances, bill of material, etc. through MRP. (See net change MRP, regeneration MRP.) *Syn:* alteration planning.

rescheduling, The process of changing order or operation due dates, usually as a result of their being out of phase with when they are needed.

rescheduling assumption, A fundamental piece of MRP logic which assumes that existing open orders can be rescheduled into nearer time periods far more easily than new orders can be released and received. As a result, planned order receipts are not created until all scheduled receipts have been applied to cover gross requirements.

reservation, The process of designating stock for a specific customer order. (See allocation.)

resource planning, See long-range resource planning.

resource profile, See product load profile.

resource requirements planning, The process of converting the production plan and/or the master production schedule into the impact on key resources, such as man hours, machine hours, storage, standard cost dollars, shipping dollars, inventory levels, etc. Product load profiles or bills of resources could be used to accomplish this. The purpose of this is to evaluate the plan prior to attempting to implement it. Sometimes referred to as a rough-cut check on capacity. Capacity requirements planning is a detailed review of capacity requirements. (See closed-loop MRP.) *Syn:* rough-cut capacity planning, rough-cut resource planning.

rework lead time, The time required to rework a lot of material in house or at a supplier.

rough-cut resource planning, See resource requirements planning.

run-time, The standard hours allowed to produce one item on the operation. The actual time taken to produce one piece may vary from the standard but the latter is used for loading purposes and is adjusted to actual by dividing by the appropriate work center efficiency factor.

safety capacity, The planning or reserving for excess manpower and equipment above known requirements for unexpected demand. This reserve capacity is in lieu of safety stock. *Syn:* reserved capacity procurement.

safety stock, (1) In general, a quantity of stock planned to be in inventory to protect against fluctuations in demand and/or supply. (2) The average amount of stock on hand when a replenishment quantity is received. (3) In the context of Master Production Scheduling, safety stock can refer to additional inventory and/or capacity planned as protection primarily against forecast errors and/or short term changes in the backlog. This investment is often under the control of the master scheduler in terms of where it should be planned. Sometimes referred to as "Overplanning" or a "Market Hedge." (See hedge.)

safety time, In an MRP system, material can be ordered to arrive ahead of the requirement date. The difference between the requirement date and the planned in-stock date is safety time. (See safety stock.)

schedule, A listing of jobs to be processed through a work center, department, or plant and their respective start dates as well as other related information.

scheduled receipts, Within MRP, open production orders and open purchase orders are considered as "scheduled receipts" on their due date and will be treated as part of available inventory during the netting process for the time period in question. Scheduled receipt dates and/or quantities are not normally altered automatically by the MRP system. Further, scheduled receipts are not exploded into requirements for components as MRP logic assumes that all components required for the manufacture of the item in question have been either allocated or issued to the shop floor. (See planned order, firm planned order.)

scheduling, Establishing the timing for performing a task. There are various levels of scheduling within a manufacturing company. The master production schedule establishes the overall logistics plan for supplying material to support production and sales. Material requirements are generated, and scheduled due dates are established for this material to support the master production schedule. Shop orders may be broken down into more detailed schedules for each operation and desired completion (or start) dates for each of these operations established to show when they must be completed in order to get the shop order completed on time. (See master production schedule, dispatching, priority.)

scheduling rules, Basic rules that are spelled out ahead of time so that they can be used consistently in a scheduling system. Scheduling rules usually specify the amount of calendar time to allow for a move and for queue, how load will be calculated, etc.

service parts, Parts used for the repair and/or maintenance of an assembled product. Typically they are ordered and shipped at a date later than the shipment of the product itself.

service parts demand, The need for a component to be sold by itself, as opposed to being used in production to make a higher level product. *Syn:* repair parts demand.

set-up time, Time required to adjust a machine or line and attach the proper tooling to make a particular product. *Syn:* start up time.

shop floor control, A system for utilizing data from the shop floor as well as data processing files to maintain and communicate status information on shop orders and work centers. The major subfunctions of shop floor control are: (1) Assigning priority to each shop order. (2) Maintaining WORK IN PROCESS quantity information for MRP.

(3) Conveying shop order status information to the office. (4) Providing actual output data for capacity control purposes. (See closed-loop MRP.)

shop packet, A manufacturing order that travels with the job and includes a group of documents like the routings, time tickets, etc. Usually many of the documents are in the form of punched cards.

shop planning, The coordination of material handling, material availability, the set-up and tooling availability so that a job can be done on a particular machine. Shop planning is often part of the dispatching function and the term shop planning is sometimes used interchangeably with dispatching although dispatching does not have to necessarily include shop planning. For example, the selection of jobs might be handled by the centralized dispatching function while the actual "shop planning" might be done by the foreman or one of his representatives. (See dispatching, closed-loop MRP.)

single-level bill of material, A single-level bill shows only those components that are directly used in an upper-level item. It does not show any relationships more than one level down.

single-level where used, Single-level where used for a component lists each assembly in which that component is directly used and in what quantity. This information is usually made available through the technique known as "implosion."

SKU, See stockkeeping unit.

slack time, The difference in calendar time between the scheduled due date for a job and the estimated completion date. If a job is to be completed ahead of schedule, it is said to have slack time; if it is likely to be completed behind schedule, it is said to have negative slack time. Slack time can be used to calculate job priorities using methods such as the critical ratio. In the Critical Path Method, total slack is the amount of time a job may be delayed in starting without necessarily delaying the project completion time. Free slack is the amount of time a job may be delayed in starting without delaying the start of any other job in the project. (See critical ratio.)

stockkeeping unit (SKU), A stockkeeping unit represents an item at a particular location. For example, if product A is stocked at several locations, each combination of product A and a stocking location is a different SKU.

stockout, The lack of materials or components which are needed to be on hand in stock. (See backorder.)

stockout percentage, A measure of the effectiveness with which the inventory management system responds to actual demand. The stockout percent can be a measurement of total stockouts to total line item orders, or of line items incurring stockouts during a period to total line items in the system. (See customer service ratio.)

stock status, A periodic report showing the inventory on hand and usually showing the inventory on order and some sales history for the products that are covered in the stock status report.

sub-assembly, An assembly which is used at a higher level to make up another assembly. For example, an automobile carburetor is a sub-assembly going into the automobile engine which is an assembly. (See component.) *Syn:* intermediate.

summarized bill of material, A form of multi-level bill of material, which lists all the parts and their quantities required in a given product structure. Unlike the Indented bill of material, it does not list the levels of assembly and lists a component only once for the total quantity used.

super bill (of material), A type of planning bill, located at the top level in the structure, which ties together various modular bills (and possibly a common parts bill) to define an entire product or product family. The "quantity per" relationship of super bill to modules represents the forecasted percentage popularity of each module. The master scheduled quantities of the super bill explode to create requirements for the modules which also are master scheduled. (See planning bill, modular bill, common parts bill.)

time bucket, In an MRP system, this refers to the number of days summarized into one columnar display. A weekly time bucket would contain all of the relevant planning data for an entire week. Weekly time buckets are considered to be the largest possible (at least in the near and medium term) to permit effective MRP. (See bucketless system.)

time fence, A policy or guideline established to note where various restrictions or changes in operating procedures take place. For example, changes to the master production schedule can be accomplished easily beyond the cumulative lead time whereas changes inside the cumulative lead time becomes increasingly more difficult to a point where changes should be resisted. Time fences can be used to define these points.

time phased order point (TPOP), MRP for independent demand items. Gross requirements come from a forecast, not via explosion. This technique can be used to plan warehouse inventories as well as planning for service (repair) parts since MRP logic can readily handle items with dependent demand, independent demand or a combination of both. (See distribution requirements planning.)

TPOP, Abbreviation for Time Phased Order Point.

transient bill of material, A bill of material coding and structuring technique used primarily for transient (non-stocked) subassemblies. For the transient subassembly item, lead time is set to zero and lot-sizing is lot-for-lot. This permits MRP logic to drive requirements straight through the transient item to its components, but retains its ability to net against any occasional inventories of the subassembly. This technique also facilitates the use of common bills of material for engineering and manufacturing. *Syn:* phantom bill of material, blow through.

transit time, A standard, arbitrary, allowance that is given on any given order for the physical movement of items from one operation to the next.

transportation inventory, Inventories that exist because material must be moved. For example, if it takes two weeks to replenish a branch warehouse on the other side of the country, inventory equivalent to approximately two weeks of sales will normally be in transit and therefore this will be extra inventory.

traveling purchase requisition, A purchase requisition designed for repetitive use. After a purchase order has been prepared for the goods requisitioned, the form is returned to the originator who holds it until a repurchase of the goods is required. The name is derived from the repetitive travel between the originating and purchasing departments. *Syn:* traveling requisition.

traveling requisition, See traveling purchase requisition.

two bin system, A type of fixed order system in which inventory is carried in two bins. A replenishment quantity is ordered when the first bin is empty. When the material is received, the serve bin is refilled and the excess is put into the working bin. This term is also used loosely to describe any fixed order system even when physical "bins" do not exist.

two-level MPS, A master scheduling approach wherein a super bill (of material) is master scheduled along with selected key options, features and attachments.

unit of measure, The unit in which quantitative data regarding an item is expressed. Examples are each, pounds, gallons, feet, etc.

unplanned issue/receipt, An issue or receipt transaction which updates the quantity on hand but for which no order or allocation exists in the data base.

vendor lead time, The time that normally elapses between the time an order is placed with a supplier and his shipment of the material.

vendor scheduler, An individual whose main responsibility is insuring vendor performance to the schedule. The position usually consists of taking the MRP output reports, communicating directly with the vendor in terms of ordering and rescheduling the required purchased material, and assuming the responsibility for delivery on the promised date. This includes communicating in advance with the master scheduler when parts will not arrive on time to support the schedule. By using vendor scheduler approach, the buyer is then freed from day-to-day order placement and expediting and thus has the time to do cost reduction, negotiation, vendor selection, alternate sourcing, etc.

vertical display, A method of displaying or printing output from an MRP system where requirements, scheduled receipts, projected balance, etc. are displayed vertically, i.e., down the page. Vertical displays are often used in conjunction with bucketless systems. (See horizontal display, bucketless system.)

wait time, See queue time.

warehouse demand, The need for an item in order to replenish a branch warehouse. *Syn:* branch warehouse demand.

what if analysis, The process of evaluating alternate strategies. Answering the consequences of changes to forecasts, manufacturing plans, inventory levels, etc. Some companies have the capability of submitting various plans as a "trial fit" in order to find the best one. *Syn:* simulation.

work center, A specific production facility, consisting of one or more people and/or machines, which can be considered as one unit for purposes of capacity requirements planning and detailed scheduling.

work in process, Product in various stages of completion throughout the plant including raw material that has been released for initial processing and completely processed material awaiting final inspection and acceptance as finished product or shipment to a customer. Many accounting systems also include semi-finished stock and components in this category. (See movement inventory.) *Syn:* in-process inventory.

Applied Psychology

Z94.11

11

APPLIED PSYCHOLOGY is the term used by the AMERICAN INSTITUTE OF INDUSTRIAL ENGINEERS in the classification of INDUSTRIAL ENGINEERING in the AIIE RESEARCH ABSTRACTS, and was accepted by the Z94 COMMITTEE in one of its earliest sessions (1967) in a breakdown of the field.

The ABSTRACTS included under this classification are psychological, sociological, and human aspects, wage incentives, personnel and industrial relations, and biomechanics. As explained in the INTRODUCTION, Distribution and Marketing comprised an important aspect in the practice of industrial engineers, yet had been omitted. Hence, the COMMITTEE added a separate section in the 1972 Edition (viz., Z94.10: APPLIED PSYCHOLOGY), and a revision appears here in Section 11 of this Z94.0 TERMINOLOGY. On the other hand, it was felt that BIOMECHANICS (= BIOLOGY + MECHANICS) really did not belong in an APPLIED PSYCHOLOGY treatment, and was extricated. It appeared as Z94.1: BIOMECHANICS - 1972, and appears, in its revision here in Z94.0, as Section 1. Also Distribution and Marketing (originally Z94.4 but now Section 4) was also removed from the original APPLIED PSYCHOLOGY category.

Although there were many psychological and sociological aspects that were properly retained under APPLIED PSYCHOLOGY, it was apparent that, with the passing of time, many would develop and mature into disciplines of their own. Among these are HUMAN FACTORS, WAGE AND SALARY ADMINISTRATION, ANTHROPOMETRY, SAFETY, and OCCUPATIONAL MEDICINE, which now appear, in this present Z94.0 Edition, as Sections 13, 14, 15, 16, and 17, respectively.

Meanwhile it was considered proper to retain the APPLIED PSYCHOLOGY heading, since it would surely be a source of additional developments. For this reason, the terms, appearing in the 1972 Edition (Z94.11) have, for the most part been retained. Some additions and deletions were made, however, in this chapter.

The 1972 TERMINOLOGY in the field was essentially written by Mr. Irving A. Delloff (Motivation Programs), Z94 Secretary, and Dr. William J. Jaffe (Emeritus Distinguished Professor, N.J. Institute of Technology), Z94 Chairman. This chapter was revised under the editorship of Dr. Jaffe, but for committee membership purposes, Mr. Bertram Gottlieb (Gottlieb Associates) was listed as this Sub-committee's Chairman.

ability, Demonstrable knowledge or skill. Ability includes aptitude and achievement.

absenteeism, Term is most commonly applied to unjustified failure of workers to report to work when scheduled. In determining *rate of absenteeism*, however, it may apply to all absences whether for justified (e.g., sickness) or unjustified reasons.

accident and sickness benefits, Regular payments to workers who lose time from work due to off-the-job disabilities through accident or sickness. Usually insured and part of a private group health and insurance plan financed in whole or in part by the employer. (See health and insurance plan, temporary disability insurance, workmen's compensation, sick benefits, sick leave.)

accidental death and dismemberment benefits, An extra lump-sum payment made under many group life insurance plans for loss of life, limb, or sight as a direct result of an accident. Coverage is usually for both occupational and nonoccupational accidents, but may be limited to the latter. (See life insurance plan, workmen's compensation.)

achievement, Acquired ability to accomplish something of value or importance in relation to standards.

across-the-board increase, A general wage increase affecting simultaneously all or most of the employees within a plant, company, or industry. Such an increase may be granted in uniform percentage or cents-per-hour terms. In the former case, the absolute amount of increase will differ among employees in accordance with their original rate levels.

active employees, Employees at work, as distinguished from retired or laid-off employees.

advance notice, In general, an announcement of an intention to carry out a certain action, given to an affected or interested party in sufficient time to prepare for it, as in informing a union of planned changes in production methods or plant shut-down, notifying a worker that he will be laid-off on a certain date, and notifying management of the union's intention to terminate of modify a collective bargaining agreement on its expiration date. (See pay-in-lieu-of notice.)

advance on wages, Refers to any practice by which employees are entitled to draw wages or salaries in advance of actual work performance.

agency shop, Provision in a collective bargaining agreement which requires that all employees in the bargaining unit who do not join the union pay a fixed amount monthly, usually the equivalent of union dues, as a condition of employment to help defray the union's expenses in acting as bargaining agent. Under some arrangements, the payments are allocated to the union's welfare fund or to a recognized charity. May operate in conjunction with a modified union shop. (See union shop.)

agreement, Written contract between an employer (or an association of employers) and a union (or unions), usually for a definite term, defining conditions of employment (wages, hours, vacations, holidays, overtime payments, working conditions, etc.), rights of workers and union, and procedures to be followed in settling disputes of handling issues that arise during the life of the contract. (See collective bargaining agreement, union contract.)

allowed time, Under incentive wage systems, the total time allowed or set as standard to complete a task. Also relates to the amount of time permitted a worker for care of tools, for rest period, or for other reasons. This time is added to operating time in computing the standard time allowed for a particular operation as the basis for establishing piece rates or production bonuses. The standard definition is: a normal time value increased by an appropriate allowance. (See Work Measurement and Methods.)

annual earnings, The total amount of compensation received for services by a worker during the year, including wages, salaries, and bonuses. The total annual earnings of a worker may be the result of work performed for a single employer or a number of employers in a given year.

annual improvement (productivity) factor, As introduced in the 1948 agreement between General Motors Corp. and the United Automobile Workers, and since retained, this was the term for wage increases granted automatically each contract year, in addition to cost-of-living adjustments. The provision was prefaced with the following words which set it apart from ordinary deferred wage increases: "The annual improvement factor provided herein recognizes that a continuing improvement in the standard of living of employees depends upon technological progress. It further recognizes the principal that to produce more with the same amount of human effort is a sound economic and social objective." (See deferred wage increase.)

annual wage or employment guarantee, An arrangement under which any employer guarantees his workers a minimum amount of wages or employment during the year.

annuity, A fixed payment to a specific person at stated intervals either for a definite number of years or for life. Usually associated in industrial relation with a pension plan. (See pension plan.)

anti-racketeering law, Federal law making it a felony to obstruct, delay, or affect interstate commerce by robbery or extortion, e.g., Hobbs Act (1934).

anti-strikebreaker law, Federal law prohibiting the interstate transportation of strikebreakers, e.g., Byrens Act (1936).

anxiety, A state of apprehension or uneasiness related to fear. The object of anxiety is ordinarily less specific than the object of fear.

apprentice, A person, usually young, who enters into agreement to learn a skilled trade and to achieve a journeyman status through supervised training and experience, usually for a specified period of time. Practical training is supplemented by related technical off-the-job instruction.

apprentice rate, The schedule of rates applicable to workers being given formal apprenticeship training for a skilled job, in accordance with set standards. The rate schedule is usually established in such a manner as to permit the gradual achievement of the minimum journeyman rate.

aptitude, The capacity to learn. Aptitude tests are designed to predict the outcome of training, hence to predict future ability on the basis of present ability.

arbitration (voluntary, compulsory, advisory), Method of settling labor-management disputes through recourse to an impartial third party, whose decision is usually final and binding. Arbitration is *voluntary* when both parties agree to submit disputed issues to arbitration, and *compulsory* if required by law. (A court order to carry through a voluntary arbitration agreement is not generally considered as compulsory arbitration.) *Advisory arbitration:* As provided in federal government agreements, arbitration without a final and binding award.

arbitrator, An impartial third party to whom disputing parties submit their differences for decision (award). An *ad hoc* arbitrator is one selected to act in a specific case or a limited group of cases. A *permanent arbitrator* is one selected to serve for the life of the contract or a stipulated term. (See impartial chairman.)

arbker, Part-time employees in garment industry.

assessment, A charge levied by a union on each member for a purpose not covered by the regular dues. Assessments may be either one-time or periodic charges.

association agreement, An agreement negotiated and signed by an employers' association, on behalf of its members, with a union or unions. (See multiemployer bargaining.)

attendance bonus, Payment or another type of reward (e.g., a day off) for employees whose record of daily reporting for work, without absences, meets certain standards of excellence.

attention, The focusing of perception leading to heightened awareness of a limited range of stimuli.

attitude, An orientation toward or away from some object, concept, or situation; a readiness to respond in a predetermined manner to the object, concept, or situation.

attrition arrangement, A process of relying upon voluntary quits, deaths, and retirements to reduce a company's labor force overtime instead of resorting to dismissal of workers.

authorization card, A statement signed by the worker authorizing a union to act as his representative in dealings with management, or authorizing the company to deduct union dues from his pay. (See card check, checkoff.)

automatic progression, A policy by which rates of pay workers in jobs with established rate ranges are increased automatically and at set time intervals. The width of the rate ranges and the number of steps within each range may vary among occupations, establishments, and industries, but under all fully automatic plans increments are received at specified time intervals until the minimum rate for the job is reached. Some plans combine automatic progression up to a specific point (for example, the midpoint) within the range with discretionary increases usually based on some type of merit review, up to the top of the range. Also refers to the automatic movement from a trainee rate to a job classification single rate or to the minimum of a job classification rate range.

automatic retirement, See compulsory retirement.

automation, As used by engineers, applied to several types of technical developments, including (a) a continuous flow production process which integrates various mechanisms to produce a finished item with relatively few or no worker operations, usually through electronic control; including robots; (b) self-regulating machines (feedback) that can perform highly precise operations in sequence; and (c) electronic computing machines. In common usage, however, the term is often used in reference to any type of advanced mechanization or as a synonym for technological change; more specifically, it is often associated with cybernetics.

average hourly earnings exclusive of overtime payments, In general, average hourly earnings from which the effect of premium payments for overtime work has been eliminated. Also, a measure of average hourly earnings published by the Bureau of Labor Statistics in which gross average hourly earnings in manufacturing are adjusted statistically to eliminate the influence of premium overtime payments at time and one half the regular rate of pay after a fixed number (presently, 40) hours of work a week. The adjustment does not compensate for other forms of overtime payment nor for other types of premium pay. (See average straight-time hourly earnings, gross average hourly earnings.)

average straight-time hourly earnings, Average wages earned per hour excluding premium overtime payments and shift differentials. Commissions, production bonuses, and cost-of-living bonuses are included, but nonproduction bonuses (such as Christmas, profit-sharing, attendance, and service), tips and allowances for room or board or other payments in kind are excluded. This concept is used by the Bureau of Labor Statistics in virtually all of its occupational wage rate studies and by numerous private organizations engaged in wage survey work. This definition, however, is not universally accepted. There appears to be general agreement on the elimination of overtime premium payments; but differences are found as to the treatment of shift differentials and the monetary value of some other wage and related practices. (See average hourly earnings exclusive of overtime payments, gross average hourly earnings.)

back pay, Delayed payment of part of the wages for a particular period of time arising from arbitration awards, grievance procedure regarding particular rates, errors in computation of pay or current legal interpretation of wage legislation. (See retroactive pay.)

back-to-work movement, Return of some or all striking workers to their jobs before the strike is ended.

bargaining, The process by which persons or groups with partly conflicting and partly harmonious interests try to agree on a procedure for dividing available resources. Bargaining is likely to occur when each person controls resources desired by the other and a range of agreements can be made that will benefit both persons more than no agreement.

bargaining agent, Union designated by an appropriate government agency, such as the National Labor Relations Board, or recognized voluntarily by the employer, as the exclusive representative of all employees in the bargaining unit for purposes of collective bargaining.

bargaining rights, Legally recognized right of unions to represent workers in dealings with employers.

bargaining unit, Group of employees in a craft, department, plant, firm or industry recognized by the employer or group of employers, or designated by an authorized agency such as the National Labor Relations Board, as appropriate for representation by a union for purposes of collective bargaining.

base rate, The amount of pay for a unit of time; e.g., hour, day, week, month, or year, exclusive of premium pay for overtime or other premium payments.

beginner rate, Learner rate.

behavioral sciences, The sciences concerned in one way or another with the behavior of man and lower organisms, especially social anthropology, psychology, and sociology, but including some aspects of biology, economics, political science, history, philosophy, and other fields of study.

benchmarks, A standard of measurement with enough characteristics common to individual units of a population to facilitate economical comparison of attributes for selected unit from a sample which, therefore, need not be directly measured. (See Work Measurement and Methods. See peg points.)

bereavement pay, See funeral leave pay.

blue-collar workers, Term for manual workers, usually those employed in production, maintenance, and related occupations, and paid by the hour or on an incentive basis. (See white-collar workers.)

board of inquiry, A board appointed by the President of the United States, under the Labor Management Relations Act, to examine and report on the facts and positions of the parties in a "national emergency" dispute. The term is often used for any board set up by a public agency to investigate a labor dispute. (See fact-finding board.)

bogus, A term used in the printing industry to designate typesetting work which is not needed for printing but which is required by the collective bargaining agreement. Other terms for unused work are "reproduction," "reset," and "dead horse."

bonus, A broad term which refers to any payment above regular or base rates.

book member, See union member.

bootleg wages, The wages above those at the prevailing rate or the union scale which an employer may pay in a tight labor market to hold or attract employees. May also refer to wages at rates below the prevailing or union rate which an employee may accept in order to obtain employment. (See kick-back.)

boycott, Efforts by a union, usually in collaboration with other unions, to discourage the purchase, handling, or use of products of an employer with whom the union is in dispute. When such action is extended to another company doing business with the employer involved in the dispute, it is termed a *secondary boycott.* (See hot-cargo clause.)

brainstorming, A problem-solving conference technique. Participants announce ideas as fast as possible and no one may criticize or evaluate them while the meeting is in session. The objective of brainstorming is to make public the maximum number of possible problem solutions, useful as well as useless ones.

break time, See rest period.

bridge benefits, See survivors' benefits.

bug, See union label.

bumping, Practice that allows a senior employee (in seniority ranking or length of service) to displace a junior employee in another job or department during a layoff or reduction in force. (See seniority, rolling.)

Bureau of Labor Statistics (BLS), U.S. Department of Labor.

Bureau of National Affairs, Inc. (BNA), A commercial non-governmental organization engaged in providing various types of reports and services dealing with industrial relations and labor affairs.

business agent, Generally a full-time paid employee or official of a local union whose duties include day-to-day dealing with employers and workers, adjustment of grievances, enforcement of agreements, and similar activities.

business unionism ("bread-and-butter" unionism), Union policy that places primary emphasis on securing higher wages and better working conditions for its members through collective bargaining rather than through political action or radical reform of society. The term has been widely used to characterize the objectives of the trade union movement in the United States.

bylaws, Generally, provisions supplementing charters or constitutions of unions or other organizations, setting forth the rules for the organization.

call-back pay, The pay (usually at premium rates) received by a worker called back to duty after completing his regular assignment.

call-in pay, The amount of pay guaranteed to a worker who is called to work on a day on which he otherwise would not have reported, and finds no work available or is not given a full or half shift's employment. Call-in pay may be higher than the amount of reporting pay, and may be provided for at premium rates on specified premium days, such as Saturdays, Sundays, and holidays. (See reporting pay.)

card check, Procedure whereby signed union authorization cards are checked against a list of workers in a prospective bargaining unit to determine if the union has majority status. The employer may recognize the union on the basis of this check without the necessity of a formal election. Often conducted by an outside party, e.g., a respected member of the community. (See authorization card.)

casual workers, Workers who have no steady employer, but who shift from employer to employer. Also used in longshoring to refer to workers not regularly attached to a particular work group. Sometimes applied to temporary employees.

catastrophe insurance, See major medical expense benefit.

central labor council, An AFL-CIO organization formed by association of local unions in a community or other geographical area, to further union interests and activities. Also known as City Central Body.

certification, Formal designation by a government agency, such as the National Labor Relations Board, of the union selected by the majority of the employees in a supervised election to act as exclusive bargaining agent for all employees in the bargaining unit.

changeover time, The time required to modify or replace an existing facility or workplace, usually including both tear-down time for the existing condition and setup for the new condition. (See Work Measurement and Methods.)

chapter, See local union.

charter, Written authorization to establish a subordinate or affiliated body of a union.

check-off, The practice whereby the employer, by agreement with the union, regularly withholds from the wages of his union workers assessments and dues, and transmits these funds to the union. Under the Labor Management Relations Act of 1947, the employer must receive from each employee a written assignment which shall not be irrevocable for a period of more than 1 year, or beyond the termination rate of the agreement, whichever occurs sooner.

Civil Rights Act, Under title VII of this federal act (1964) employers, unions, and employment agencies are required to treat all persons equally, regardless of race, color, religion, sex, or national origin, in all phases of employment, including hiring, promotion, compensation, firing, apprenticeship, job assignments, and training. An Equal Employment Opportunity Commission was created to assist in carrying out this section of the act.

classification act employees, Federal government employees, typically professional, administrative, technical and clerical employees, whose salary rates and certain other conditions of employment are determined by the Congress. (See wage board employees.)

cleanup time, Paid time allowed to workers to clean their workplaces or tools or to wash up before leaving the plant at the close of the workday or for lunch. Also known as Washup time. (See clothes changing time.)

closed shop, Form of union security provided in an agreement which binds the employer to hire and retain only union members in good standing. The key distinction between a closed shop and a union shop lies in the hiring restriction, a restriction prohibited by the Labor Management Relations Act, 1947. Legal closed shops may be found outside the scope of this act (which applies to employers and employees in industries affecting interstate commerce) and outside of states with "right-to-work" laws.

closed union, A union which bars new members or makes membership acquisition difficult (e.g., by very high initiation fees) in order to protect job opportunities for its present members, or for other reasons. Some unions accept only sons of present members. (See open union.)

clothing allowance, An allowance granted by an employer to those of his employees who are required to buy special clothing such as uniforms and safety garments, in connection with the performance of their work.

clothes changing time, Time allotted within the paid workday for changing from street wear to working clothes, or from working clothes to street wear, or both. (See cleanup time.)

codes of ethical practices, Rules adopted by the AFL-CIO in 1956-57, setting standards of behavior for unions and their officers.

coffee break, See rest period.

collective bargaining, Method whereby representatives of the employees (the union) and employer determine the conditions of employment through direct negotiation, normally resulting in a written contract setting forth the wages, hours, and other conditions to be observed for a stipulated period (e.g., two years). The term is also applied to union-management dealings during the term of the agreement. (See agreement.)

commission earnings, Compensation to sales personnel based on a percentage of value of sales. Commission earnings may be in addition to a guaranteed salary or may constitute total pay. Sales personnel on straight commission usually have a fixed drawing account which is balanced against actually realized commission earnings at specific periods. (See drawing account.)

common labor, General term used to designate unskilled workers performing heavy labor. In specific plants, may refer to unskilled workers not assigned to a particular job. The latter use is probably now the more frequent one.

common labor rate, The hourly rate paid to adults for physical or manual labor of a general character and simple nature, requiring no special training or skill and requiring little or no previous experience.

community wage survey, A general term used to describe a survey designed to reveal the structure and level of wages within a particular geographic area for a given industry or, more typically for broad categories of industry.

company union, Term used to designate a labor organization that is organized, financed, or dominated by the employer and is thus suspected of being an agent of the company rather than the workers. This practice is prohibited under the Labor Management Relations Act, 1947. The term also survives as a derogatory charge leveled against a union suspected of being ineffectual. (See single-company union.)

companywide bargaining, See multiplant bargaining.

comparable rate, A rate paid for work agreed or determined to be comparable within a plant or within an area. Comparisons of this type may be limited to virtually identical occupations within a specific industry or be broadened to include occupations with similar characteristics in various industries. Such comparisons are used in wage negotiations and wage determinations.

compensable injury, A work injury for which compensation indemnity benefits are payable to the injured worker or his beneficiary under workmen's compensation laws.

compensation, See earnings.

competitive wage, In economic theory, the wage within a given labor market required to balance the demand for and supply of labor of a particular type. More popularly the wage level a company must maintain to compete with other firms in the same labor market for particular types of labor. Also used in the sense of the wage level that is required by a company to maintain a competitive price position with other firms in the same industry.

compulsory arbitration, See arbitration.

compulsory retirement, Involuntary separation from employment in a company upon reaching a specified age. In precise pension terms, a distinction is usually made between compulsory and automatic retirement. The age of compulsory retirement is that point at which worker loses the right to decide whether he should retire or continue on his job. The age of automatic retirement is the age beyond which no employee may continue to work under the terms of the pension plan. In other words, an employee may work beyond the compulsory retirement age if the employer consents, but automatic retirement rules out the option on both sides. (See automatic retirement.)

conciliation, See mediation.

conflict, The simultaneous presence of opposing or mutually exclusive impulses, desires, or tendencies.

conformity, A general term referring to adherence to a group *norm* concerning *beliefs, values, attitudes,* or *behavior.* Such adherence may reflect a variety of underlying psychological processes.

consultation, An obligation on the part of management to consult the union on particular issues (e.g., contracting-out) in advance of taking action is frequently provided by agreements. What consultation actually means in each situation is what the parties want it to mean. In general, the process of consultation lies between notification to the union, which may amount simply to providing information, and negotiation, which implies agreement on the part of the union before the action can be taken.

consumer price index (CPI), A government index, issued monthly by the Bureau of Labor Statistics, which measures the average change in prices of goods and services purchased by urban wage-earner and clerical-worker families. The CPI measures price change only, describing shifts in the purchasing power of the consumer's dollar. It is often incorrectly referred to as a "cost-of-living index." (See cost-of-living adjustment, escalator clause.)

continuous bargaining committees (interim committees), Committees established by management and union in a collective bargaining relationship to keep the agreement under constant review, and to discuss possible contract changes, long in advance of the contract expiration date. May provide for third-party participation. (See human relations committees, crisis bargaining.)

continuous operations, Necessary plant operations (powerhouse, maintenance, protection, etc.) that must continue to function on a 24-hour, 7-day basis. (See continuous process. Also known as round-the-clock operations.)

continuous process, A process which, once begun, must continue without interruption for a long period, making the use of multiple shifts necessary, often applied to a service (e.g., electric utility) needed continuously.

contract, See agreement.

contract bar, A denial of the request for a representation election, based on the existence of an agreement. Such an election will not be conducted by the National Labor Relations Board if there is in effect a written agreement which is binding upon the parties, has not been in effect for more than a "reasonable" time, and its terms are consistent with the National Labor Relations Act.

contract wage payment, An arrangement whereby the worker contracts to perform a specific job for a predetermined amount of compensation.

contributory pension plan, A pension plan for the benefit of the employee under which the cost is shared by both the employer and the employee. (See noncontributory pension plan.)

cooling-off period, A period of time which must elapse before a strike or lockout can begin or be resumed, by agreement or by law. The term derives from the hope that the tensions of unsuccessful negotiation will subside in time and that a work stoppage will be averted. (See national emergency dispute.)

corridor, Deductible.

cost-of-living allowance, Regular cents-per-hour or percentage payments made to workers through the operation of escalator clauses or other types of cost-of-living allowance not incorporated into base rates. Also known as float.

cost-of-living adjustment (COLA), An adjustment of wages or salaries in accordance with changes in the cost of living as measured by an appropriate index of the retail prices of goods and services that enter into the consumption of low- or moderate-income families. Limitations as to the extent of wage adjustment are sometimes provided for. (See escalator clause.)

cost-of-living index, This term is often used, incorrectly, to designate the Bureau of Labor Statistics Consumer Price Index. (See consumer price index.)

Council of Economic Advisers, This small federal government agency, established under the terms of the Employment Act of 1946, advises the President on economic developments, appraises government economic growth and stability and assists in the preparation of the President's annual economic report to the Congress. (See guideposts.)

craft, Usually, a skilled occupation requiring a thorough knowledge of processes involved in the work, the exercise of considerable independent judgement, usually a high degree of manual dexterity, and, in some instances, extensive responsibility for valuable product or equipment.

craft union, Term applied to a labor organization which limits membership to workers having a particular craft or skill or working at closely related trades. In practice, many so-called craft unions also enroll members outside the craft field; some come to resemble industrial unions in all major respects. The traditional distinction between craft and industrial unions has been substantially blurred. (See industrial union.)

craft unit, A bargaining unit composed solely of workers having a recognized skill; for example, electricians, machinists or plumbers.

credited service, Years of employment counted for retirement, severance pay, seniority, etc. The definition of a credited year of service varies among companies and plans.

critical incidents, Those events which are particularly conducive to success or failure in performing any task.

daily rate, For a worker hired on a daily rate basis, the rate of pay is normally expressed as a rate for a standard number of work hours per day. E.g., the hourly rate, the quotation of a daily rate normally excludes premiums that may be paid for late-shift work or overtime hours, as well as bonuses for special conditions of work or for other reasons unrelated directly to production.

daywork, Usually refers to work for which pay is based on time as contrasted to pay based on output. (See Work Measurement and Methods.)

deadheading pay, A special payment to a transportation worker who is required to report for work at a point far removed from his home terminal. "Deadheading" refers to the extra time consumed in traveling to and from the place of work.

dead time, See delay time, down time.

dead work, A term used in mining, referring to nonproductive work, including the removal of rock, debris, and other waste matter from the product mined.

death benefit, Payment, usually a lump sum, provided to a worker's beneficiary, in the event of his death. May be provided by a pension plan or another type of employer-sponsored welfare plan, or by a union to its members. Commonly called life insurance. (See life insurance plan.)

decertification, Withdrawal by a government agency, such as the National Labor Relations Board, of a union's official recognition as exclusive bargaining representative.

decision making, The response to a need or stimulus by means of acquiring and organizing information to yield alternate courses of action; and selecting one course of action from the alternatives. (See Organization Planning and Theory.)

deductible, In major medical insurance plans, term is applied to that portion of covered hospital and medical charges which the insured worker must pay before the plan's benefits begin. If the worker also has basic health insurance protection, the worker's liability is often referred to as the corridor because it lies between two insured areas. Also known as corridor. (See major medical expense benefit.)

deferred wage increase, General term for negotiated wage changes which are not to become effective until some specified date in the future, e.g., a year later. (See annual improvement [productivity] factor.)

delay time, See down time.

demotion, See downgrading.

differential piece rates, Plan under which piece rates vary at different levels of output.

direct labor, (1) Work which is readily chargeable to or identifiable with a specific product. (2) Work performed on a product or service that advances the product or service towards its ultimate specifications or objectives. (See Work Measurement and Methods.)

disability, Any injury or illness, temporary or permanent, which prevents a worker from carrying on his usual occupation. (See permanent and total disability.)

disability retirement, Retirement because of physical inability to perform the job. (See permanent and total disability.)

disaffiliation, Withdrawal of a local union from membership in a national or international union, or of a national or international union from a federation.

discharge, Dismissal of a worker from his employment. Term implies discipline for unsatisfactory performance, and is thus usually limited to dismissals for cause relating to the individual, e.g., insubordination, absenteeism, inefficiency.

discrimination, Term applied to a prejudice against or unequal treatment of workers in hiring, employment, pay, or conditions of work, because of race, national origin, creed, color, sex, religious beliefs, age, union membership of activity, or any other characteristic not related to ability or job performance.

dismissal compensation, A specific payment, in addition to regular wages, which is given an employee upon permanent termination of employment through no fault of his own. The dismissal compensation, or wage, may be paid in one sum at the time employment is terminated, or it may be spread over a period of time. The amount of the payment to any one employee is usually related to the amount of his regular wages and his length of service.

dispute, Any disagreement between union and management which required resolution in one way or another; e.g., inability to agree on contract terms, an unsettled grievance, etc. (See labor dispute.)

doubleheading pay, Refers to the extra compensation given the railroad engineer where very steep grades require the use of more than one engine for some distance. The specific rules regarding payment under these conditions are found in the various union agreements.

double time, Penalty or premium rate (e.g., for overtime work, for work on Sundays and holidays) amounting to twice the employee's regular rate of pay for each hour worked.

downgrading: downgrade, (1) The lowering of a particular job in scope, authority, responsibility, degree of difficulty, etc., with a possible reduction in wage or salary. (2) Dilution of skills required for the task. (See Work Measurement and Methods.)

down time: downtime, A period of time during which an operation is halted due to a lack of material, and machinery breakdown, or the like. (See dead time, Work Measurement and Methods.)

drawing account, A weekly allowance given to sales personnel working on a straight commission basis in retail and wholesale trade. The drawing account is balanced against total commission earnings at specified periods, in some cases 6 months or more. The anticipated earning power of sales personnel is usually the determining factor in establishing the amount of the drawing account. (See commission earnings.)

dual pay, A system of wage payment used by railroads under which employees are paid on a mileage or hourly basis. A standard mileage is defined as a basic day, usually 8 hours, for the purpose of determining the daily rate. Wages are computed on the number of hours or miles, whichever yields the greater compensation to the employee.

dual unionism, A charge (usually a punishable offense) leveled at a union member or officer who seeks or accepts membership or position in a rival union, or otherwise attempts to undermine a union by helping its rival.

earnings, The total remuneration of a worker or group of workers for services rendered, including wages, overtime pay, bonuses, commissions, etc.

Economic Opportunity Act, Especially a 1964 act "to mobilize the human and financial resources of the Nation to combat poverty in the United States." An important part of this act, the work-training program (administered by the U.S. Department of Labor), is directed to encouraging young unemployed persons (age 16-21, inclusive) to stay in school or obtain job experience that would prepare them for meaningful work careers.

economic strikes, Union-authorized strikes to bring about changes in wages, hours, or working conditions, usually associated with contract negotiations.

efficiency, work, The ratio of a standard performance time to an actual performance time, usually expressed as a percentage. (See Work Measurement and Methods.)

emergency boards, Under the Railway Labor Act, the President, upon notification by the National Mediation Board that a contract negotiation dispute threatens seriously to interrupt interstate commerce, may appoint an emergency board to investigate and report within a specified time period. After the board reports, no change may be made in the conditions underlying the dispute, except by agreement of the parties. This is the last formal step in the act's procedures regarding contract disputes.

employee, General term for an employed wage earner or salaried worker. Used interchangeably with "worker" in the context of a work situation, but a "worker" is not an "employee" when he is no longer on the payroll.

employee benefit plan, See welfare plan.

employer, General term for any individual, corporation, or other operating group, which hires workers (employees). The terms "employer" and "management" are often used interchangeably when there is no intent to draw a distinction between owners and managers.

employers' association, Voluntary membership organization of employers established to deal with problems common to the group. It may be formed specifically to handle industrial relations and to negotiate with a union or unions. (See association agreements.)

entrance rate, The hourly rate which a worker received upon being hired into an establishment. In the case of a skilled worker, the rate may be slightly less than, or equivalent to, the minimum job rate; it may also be identical with the probationary rate. The entrance rate for unskilled workers may be synonymous with the minimum plant rate, and may be increased after designated time intervals. (See minimum plant rate, probationary rate.)

equal pay for equal work, The payment of equal compensation to all employees within an establishment or other unit performing the same kind and amount of work, regardless of race, sex, or other characteristics of the individual workers.

escalator clause, A provision in a union agreement allowing for the adjustment of wages in accordance with specific changes in the cost of living as measured by an appropriate index, or in the price of materials used in production, or in accordance with some other agreed-upon criterion, such as production index or price of product. (See cost-of-living adjustment.)

escape clause, General term signifying release from an obligation. One example is found in maintenance-of-membership arrangements which give union members an "escape-period" during which they may resign from membership in the union without forfeiting their jobs.

exclusive bargaining rights, The right and obligation of a union designated as majority representative to bargain collectively for all employees, including nonmembers, in the bargaining unit.

executive board, Constitutional union administrative body composed of elected officials and other elected or appointed members generally responsible for overseeing union operations and carry out union policies.

expected earning level, See target.

expense account, An account of expenses paid or incurred by an employee in connection with the performance of his services, usually covering such items as transportation, meals, and lodging while in a travel status and away from home. These expenses are reimbursable to the employee; payment is usually made after the expense account has been audited.

experience rating, Process of basing tax rates or insurance premiums on the employer's own record — as in workmen's compensation, unemployment insurance, and commercially insured health and insurance programs — so that he may benefit from a good record.

expiration date, Formal termination date established in a collective bargaining agreement, or the earliest date at which the contract may be terminated.

extended leave plan, In general, a plan allowing a worker to take extended, unpaid leave without loss of job or seniority, such leave being usually granted for a limited number of specific reasons.

extended vacation plan, A plan providing extra-long paid vacations to qualified (long service) workers at regular intervals supplementing an annual paid vacation plan.

fact-finding board, A group of individuals appointed under government authority to investigate, assemble, and report the facts in a labor dispute, sometimes with authority to make recommendations for settlement. (See board of inquiry.)

fair employment practice laws, These laws forbid discrimination in hiring, promotion, discharge, or conditions of employment on the basis of race, creed, color, national origin, and in some cases, sex and age. Such laws exist at federal, state, and local levels. (See civil rights act.)

Fair Labor Standards Act, The 1938 federal law which prohibited oppressive child labor and established a minimum hourly wage and premium overtime pay for hours in excess of a specific level (now time and one-half after 40 hours per week) for all workers engaged in, or producing goods for, interstate commerce. The minimum wage and the coverage of the act have been modified several times since enactment. Also known as FLSA; wage-hour law.

"favored nations" clause, An agreement provision indicating that one party to the agreement (employer or union) should have the opportunity to share in more favorable terms negotiated by the other party with another employer or union.

featherbedding, A derogatory term applied to a practice, working rule, or agreement provision which limits output or requires employment of excess workers and thereby creates or preserves soft or unnecessary jobs; or to a charge or fee levied by a union upon a company for services which are not performed, or not to be performed. Also known as make work.

Federal Mediation and Conciliation Service (FMCS), An independent U.S. Government agency which provides mediators to assist the parties involved in negotiations, or in a labor dispute, in reaching a settlement; provides lists of suitable arbitrators on request; and engages in various types of "preventive mediation." Mediation services are also provided by several state agencies.

federation, Association of unions formed to promote common interests. The major federation is the AFL-CIO (formed by the merger of two federations).

finish-go-home basis of pay, A practice under which employees are permitted to go home after completing a specific work assignment generally considered a standard day's work. An example of the "finish-go-home" basis of pay may be found in nonferrous smelters. Charging the furnace in less than 8 hours permits more time for the actual smelting process, and workers are encouraged to complete charging in less than 8 hours but are paid for a full shift. Similar standards for a day's work are set up in unloading of railroad cars, or other shipping facilities, in shake-out work in foundries, and in other industries. Essentially, this may be considered an incentive method of pay; it generally involves making costly facilities available for use as quickly as possible.

fink, See strikebreaker.

fixed shift, The term applied to the type of shift on which a group of workers maintains the same schedule of hours week after week rather than rotating time-of-day assignments periodically with other groups. (See rotating shift, shift, split shift, swing shift.)

flagged rate, See red circle rate.

flat rate, A system of remuneration used especailly in auto repair shops. The flat rate refers to the labor charge made for a repair job, and is usually based on the standard time specified in an official automobile repair manual. The auto mechanic receives a percentage of the total labor cost, which is computed at a rate which allows a margin of profit for the employer. If the actual time spent on a job greater than the standard allowance, the actual time is the determining factor in computing the labor cost.

float, See cost-of-living allowance.

floor under wages, See minimum wage.

foiler, Derogatory term applied to employees who share in benefits without making contributions.

form agreement, Uniform agreement signed by individual members of an employers' association and often by employers in the same line of work but outside the association. (See standard agreement.)

free riders, A derogatory term applied by unions to nonmembers who, because of being in the bargaining unit, share in benefits resulting from union activities without paying dues.

fringe benefit, A benefit supplemental to wages received by workers, at a cost to employers. Among these benefits, commonly designated as "fringe" are paid holidays, paid vacations, pensions, and insurance benefits (life, accident, health, hospitalization, and medical.)

full crew law or rule, Generally used to refer to laws or regulations of several states which require a minimum number of workers having specified skills for each railway train, e.g., engineer, firemen, conductors, brakemen, and flagmen.

full-time earnings, Earnings received for working a regular schedule of hours over a stated period of time. Full-time earnings may be defined in terms of a day, week, month, or other period.

full-time worker rate, A rate paid to a full-time worker or regular worker, as distinguished from that paid to a part-time, or temporary worker.

fund, In general, money and investments set aside in a separate account, usually administered by trustees, to take care of the payment of pensions, supplemental unemployment benefits, strike benefits, etc. Also used as verb, i.e., to set up a (fund); to set aside adequate reserves. Also known as trust fund.

funeral leave pay, Pay to a worker, usually for a limited period, for time lost because of the death and funeral of a member of his immediate family. Also known as bereavement pay.

garnishment of wages, The practice of legally attaching the wages of a debtor and collecting the debt directly via his employer.

general strike, See strike.

general wage changes, Wage adjustments which affect large numbers of workers in a similar manner at the same time.

going rate, See prevailing rate.

gonveal, Illegal return of a portion of his earnings by a worker to an employer or supervisor as a tribute for retaining his job.

goon, Slang term for a person hired by either management or union during a labor dispute to make trouble and intimidate the opposition by violence or the threat of violence.

graveyard shift, The shift beginning around midnight and ending at dawn. See shift.

grievance, Any complaint or expressed dissatisfaction by an employee in connection with his job, pay, or other aspects of his employment. Whether it is formally recognized and handled as a "grievance" depends on the scope of the grievance procedure.

grievance procedure, Typically a formal plan, specified in the agreement, which provides a channel for the adjustment of grievances through discussions at progressively higher levels of authority in company and union, usually culminating in arbitration if necessary. Formal plans may also be found in nonunion companies, with the important difference that there is no union to represent the workers.

grievance steps, The regular steps in a grievance procedure by which a grievance dispute moves from one level or authority in the company and union, to the next higher level. The steps are usually specified in the agreement.

gross average hourly earnings, As used by the Bureau of Labor Statistics, a measure of hourly wages obtained by dividing total compensation prior to payroll deduction for taxes, social security, or other purposes for a given payroll period by man-hours worked plus hours paid for sick leave, holidays and vacations, total compensation includes premium payments for overtime and late-shift work, as well as recurrent production and nonproduction bonuses. Computed monthly on an industry basis for comprehensive groups of manufacturing and nonmanufacturing industries by the Bureau of Labor Statistics in its employment and payroll reporting program. (See average straight-time hourly earnings, average hourly earnings exclusive of overtime payments.)

gross average weekly earnings, A measure of weekly wages typically obtained by dividing total compensation prior to payroll deductions for taxes, social security payments, and other purposes for a given weekly payroll period (or, by the use of a conversion factor for a period of longer duration) by total employment or, alternatively, by multiplying average hourly earnings by average weekly hours. Computed monthly on an industry basis for comprehensive groups of manufacturing and nonmanufacturing industries by the Bureau of Labor Statistics in its employment and payroll reporting program.

group, An aggregation of persons becomes a group when all see their actions as interrelated and their fates to some extent interdependent. The members of a group usually expect that their continued affiliation will provide a means of obtaining desired goals.

group annuity plan, See pension plan.

group incentive plan, Payment of incentive earnings based on the output of a group of workers (team, gang, department, etc.) rather than the output of the individual worker.

guaranteed rate, The rate of pay guaranteed to an incentive worker. In the case of a production bonus worker, this rate may be equivalent to, or higher than, the base rate. When earnings at incentive are lower than the guaranteed rates, allowances are made to bring earnings up to the guaranteed levels. (See base rate.)

guaranteed time, Rate of hourly or weekly pay guaranteed to a worker under an incentive system. May differ from base rate. The term is sometimes used for weekly wage or employment guarantees (e.g., a guarantee of 36 hours work for employees called to work on the first day of the workweek).

guaranteed wage plan, An arrangement, written or unwritten, by which an employer guarantees or assures to some or all of his employees, in advance, a definite period of employment, or a specific amount of wages. (See annual wage or employment guarantee, wage advance plan.)

guarantee on trial rate, A minimum guaranteed rate that remains in effect during the time that trial runs are made on new work and a trial or temporary piece rate is in effect. The level of these guaranteed rates is usually higher than plant minimum job or base rates, and is related to past earnings of the individual or the group of workers affected. In order to provide an inducement to the worker to return on an incentive basis as soon as possible, the trial rate is frequently set at 95 or 98 percent of the average earnings of the worker for a given number of weeks prior to the trial run. (See temporary rate.)

guideposts, Standards by which unions and business leaders, and the general public, can appraise particular wage and price decisions. First formally set forth in the President's Economic Report of January, 1962, and restated in subsequent reports. In the January 1964 report, the general guidepost for wages was set forth as follows: ". . . in a particular firm or industry, the appropriate noninflationary standard for annual percentage increases in total employee compensation per man-hour (not just in straight-time, hourly rates) is the annual increase in national trend output per man-hour." Adherence to this guidepost is not compulsory. Also known as wage-price guidelines.

halo effect, The tendency for trait ratings to show a higher intercorrelation than would be revealed by more objective measurement. Typically this tendency reveals itself in the attribution of favorable characteristics to a person who is already favorably evaluated and unfavorable characteristics to a person who is already unfavorably evaluated.

handicapped workers, Workers whose capacities or earning abilities are impaired by physical or mental disability.

handicapped worker rate, A lower rate of pay for a worker whose efficiency is impaired because of physical or mental handicaps. Under the Federal Fair Labor Standards Act, rates below the legal minimum wage may be established for handicapped workers in accordance with regulations issued pursuant to section 14 of the act. (See substandard rate.)

hazard pay, Extra payments to workers in dangerous occupations or while engaged in work where the chances of injury are greater than normal. (See high time.)

health and insurance plan, A program of providing financial protection to the worker and his family against death, illness, accidents, and other risks, in which the costs are borne in whole or in part by the employer. One or more of the following major benefits may be provided for the worker and, frequently, his dependents: life insurance, accidental death and dismemberment benefits, accident and sickness benefits, hospitalization benefits, surgical and medical benefits, and major medical expense benefits (see entries for each time). Although sometimes self-insured in whole or in part, these programs usually consist of insurance purchased from Blue Cross, Blue Shield, and commercial insurance companies. Some plans provide for health centers or clinics. (See welfare plan.)

health center, Usually a clinic administered by a union, or by trustees representing employers and unions, where members and their families may receive medical examinations and treatment free or at a nominal charge.

high time, Extra pay for workers engaged in a job high above ground, and, thus, dangerous or uncomfortable, as in construction. Sometimes also applied to work below ground level with extra dangers or discomforts for the worker. (See hazard pay.)

hiring hall, An office maintained by a union, or jointly by employers and union, for referring workers to jobs or for the actual hiring operation. Common in maritime industry.

hiring rate, See entrance rate.

historical wage differential, See wage differential.

holiday pay, Pay to workers, typically at regular rates, for holidays not worked. For work on such days, payments are often provided at premium rates.

homework, Production of industrial goods by workers in their homes from materials supplied by the employer.

horizontal union, A union which includes only workers in a single craft or skill, or closely related skills, such as carpenters, electricians, etc., usually cutting across industry lines. Use of term is declining. (See craft union.)

hospitalization benefits, Plan that provides workers, and in many cases their dependents, with hospital room and board (e.g., semi-private room) or cash allowances toward the cost of such care for a specified number of days plus the full cost of specified services. Usually part of a more inclusive health and insurance program. Often referred to as Blue Cross. (See health and insurance plan.)

hot-cargo clause, An agreement provision stipulating that employees covered by the agreement cannot be required to handle or use goods shipped from, or bound to, an employer who is involved in a strike with a union. (See struck work.)

hourly rate, Typically, the rate of pay expressed in terms of cents-per-hour, usually thought of as applying to manual or other workers remunerated on a time basis. Hourly rates are normally basic rates; i.e., exclusive of extra payments for shift work and overtime, and exclusive of production or nonproduction bonus payments. However, the term "hourly rate" is sometimes interpreted to mean "earned rate per hour" under incentive methods of wage payment.

human relations, Term applied to a broad area of managerial effort and research dealing with the social and psychological relations among people at work. Bringing into play the insights and techniques of several of the social sciences, it concentrates upon workers and supervisors as individuals, all with personal problems and personality differences, and all with both psychic and material satisfactions to be fulfilled. Its practical application at the work level includes improving personal relationships, reducing friction, improving organization, and thereby enhancing efficiency.

impartial chairman, An arbitrator employed jointly by a union and employer, usually on a long-term basis, to serve as the impartial party on a tripartied arbitration board and to decide all disputes or specific kinds of disputes arising during the life of the contract. The functions of an impartial chairman often expand with experience and the growing confidence of the parties, and he alone may constitute the arbitration board in practice. (See umpire.)

improvement factor, A term used in an agreement negotiated by the United Automobile Workers (C.I.O.) with the General Motors Corporation in May, 1948, describing an annual increase in wages of a stipulated amount during the life of the agreement. The improvement factor is designed to enable the wage earner to share in the benefits resulting from increased productivity in the economy. Cost-of-living adjustments were also provided for in this contract.

incentive rate, The term "incentive rate" may apply to a piece rate, a rate of pay per unit for production above a predetermined minimum standard of output, a ratio of management-labor sharing of labor cost savings resulting from the operation of an incentive system, etc. The incentive plan may contain as an integral part of its operation, one or more kinds of special rates which also influence the worker's pay: a guaranteed rate, a base rate, a down-time rate, special rates for try-out on experimental work, etc.

incentive wage system, General term for methods of wage payment which relate earnings of workers to their actual production, individually or as a group. (See group incentive plan, piecework.)

independent union, Term applied to local, national, and international unions not affiliated with the AFL-CIO. Unaffiliated union is perhaps the better term, but it is used less frequently.

individual rate, In many establishments, there is no formal wage structure (either job rates or rate ranges), and the rates paid are known as individual rates. These rates may be based in a loose way upon the job being done, or may be related to the training, ability, skill, and bargaining power of the individual worker. The term "individual rate" is also used to indicate the rate actually received by the individual worker, as distinguished from the job rate shown in the rate structure maintained by an employer.

industrial relations, General term covering matters of mutual concern to employers and employees; the relationships, formal and informal, between employer and employees or their representatives; government actions and law bearing upon these relationships; an area of specialization in a company; a field of study whose scope is suggested by this glossary.

industrial union, A union that represents all or most of the production, maintenance, and related workers, both skilled and unskilled, in an industry or company. May also include office, sales, and technical employees of the same companies. Also known as vertical union. (See craft union.)

industry-wide bargaining, Negotiations between an employers' group and a union resulting in an agreement covering an entire "industry" or a substantial part thereof, e.g., all class I railroads. Ordinarily, the term "industry" in this connection is not consistent with standard statistical definitions, nor does "industry-wide" necessarily imply nationwide coverage. It is usually safe to assume that in whatever way "industry" is used, it does not include every establishment.

inequity, Term for a disadvantageous condition, real or alleged, generally applied to wage rates or benefits, usually the former, that are substantially lower than those prevailing elsewhere in the plant, company, locality, or industry, for the same or comparable work.

initiation fee, Payment to the union required of a worker when he joins, usually as set forth in the unions' constitution.

injunction, Court order restraining one or more persons, corporations, or unions from performing some act which the court believes would result in irreparable injury to property or other rights. Also known as labor injunction.

intercity differential, Differences in prevailing wage levels among a group of cities; usually such differences are measured by rates for comparable occupations and industries from city to city, but more general measures are sometimes employed. In particular negotiations, historical relationship in rates between various cities may be of prime importance. In such instances, the rank of a particular city with respect to other cities may be of more significance than the percentage relationships among the cities over a period of years.

job classification, An arrangement of jobs in an establishment or industry into a series of categories, each of which is based on progressively higher requirements in terms of skill, experience, training, and similar considerations. Essentially, this process results in a rough grouping of occupations where distinctions between jobs are clear and sharp. Usually job descriptions are used as a basis for classification. (See job evaluation, labor grade.)

job cycle, The period it takes to perform a job operation before having to start it all over again. A short cycle is often considered an important factor in lessening job satisfaction.

job description, A written statement listing the requirements of a particular job or occupation.

job enlargement, A technique suggested for reducing the number of frustration-producing situations inherent in many jobs. The basic notion is that jobs should be designed to use the full capacity of an individual. That is, he should have the opportunity to achieve the maximum in terms of need fulfillment.

job evaluation, The evaluation or rating of jobs to determine their position in a job hierarchy. The evaluation may be achieved through the assignment of points or the use of some other systematic rating method for essential job requirements such as skill, experience and responsibility. Job evaluation is widely used in the establishment of wage rate structures and in the elimination of wage inequalities. It is always applied to jobs rather than the qualities of individuals in the jobs. (See job classification, labor grade.)

job posting, Listing of available jobs, usually on a bulletin board, so that employees may bid for promotion or transfer.

job rate, See minimum job rate, standard rate.

joint and survivor option, A pension plan provision under which the pensioner may elect to receive a reduced benefit with a guarantee that, if he dies while his beneficiary is living, payments of a predetermined proportion of the reduced benefit will be continued to his beneficiary for life. Also known as survivor's option.

joint bargaining, Usually used to indicate two or more unions joining forces in negotiating an agreement with a single employer.

joint board (or council), Delegate body composed of representatives of locals of a single national or international union in a particular area, working together to further the interests of the union. When more than one union is involved, the term "trade council" may be used.

joint lockout, See lockout.

joint rate setting, The process of establishing rates jointly by representatives of management and labor. The extent of labor participation in the actual process of rate setting varies from industry to industry, and from establishment to establishment. In some of the apparel industries, the unions have achieved a relatively high degree of participation through formal joint organizational machinery. At the other extreme are situations where management alone sets rates, but labor through its grievance machinery has the right to "protest" specific rates, and to adjust them in conference with management if the grievance is found to have validity.

journeyman, A fully qualified craftsman, generally having mastered his trade by serving an apprenticeship.

journeyman rate, The rate of pay for a journeyman or a fully qualified worker in a skilled trade or craft, who has completed an apprenticeship or equivalent training. Typically, such rate is a minimum rate for the trade in a particular area or a union scale; some journeymen, however, may receive rates above or below the union scale.

jurisdiction, Among unions, the right or claim to represent workers within specified occupations, industries, or geographical boundaries.

jurisdictional dispute, Conflict between two or more unions over the organization of a particular establishment or whether a certain type of work should be performed by members of one union or another. *Jurisdictional strike* — a work stoppage resulting from a jurisdictional dispute.

jury-duty pay, Pay or allowance granted to an employee for working time spent on jury duty, usually in addition to fees paid by the court.

just cause, Good or fair reasons for discipline. This term is commonly used in agreement provisions safeguarding workers from unjustified discharge or a lesser punishment.

kick-back, A practice by which an employer or his representative arranges with his workers for a return of a part of their wages, established by union contract or by law, as a condition of employment. A federal Anti-kick-back Law was enacted in 1934, prohibiting kick-backs by workers employed on public construction work or on any work financed wholly or in part by federal funds. (See bootleg wages.)

labor, Dictionaries and economics textbooks provide full scope to the varied meanings and uses of this term, but the chances are that readers will most often see it used as a synonym for organized labor, the AFL-CIO, a particular union, or any union.

labor area, See labor market area.

labor dispute, See dispute.

labor force, In census terms, all persons age 14 or over, employed or unemployed (but looking for work). *Total labor force* includes members of the Armed Forces; *Civilian labor force* excludes them. Term is often used to designate total employment of a particular company or industry. Also known as work force.

labor grade, One of a series of rate steps (single rate or rate ranges) in the wage rate structure of an establishment. Labor grades are an outcome typically of some form of job evaluation in which various occupational classifications are rated on the basis of such labor requirements as skill, experience, training, working conditions, etc. The occupations are then grouped into a limited number of steps or grades so that occupations of approximately equal "value" or "worth" fall into the same grade. (See job classification, job evaluation.)

labor injunction, See injunction.

Labor Management Relations Act, Refers especially to the TAFT-HARTLEY ACT (1947) a federal law, amending the National Labor Relations Act (Wagner Act), 1935, which, among other changes, defined and made illegal a number of unfair labor practices by unions. It preserved the guarantee of the right of workers to organize and bargain collectively with their employers, or to refrain from such activities, and retained the definition of unfair labor practices as applied to employers. The act does not apply to employees in a business or industry where a labor dispute would not affect interstate commerce. Other major exclusions are: employees subject to Railway Labor Act, agricultural workers, government employees, nonprofit hospitals, domestic servants, and supervisors. Amended by Labor-Management Reporting and Disclosure Act of 1959. (See National Labor Relations Act, National Labor Relations Board, unfair labor practices, Section 14 (b); Labor Management Relations Act, 1947.)

Labor-Management Reporting and Disclosure Act of 1959, (LANDRUM-GRIFFIN ACT) this federal law was designed "to eliminate or prevent improper practices on the part of labor organizations, employers," etc. Its seven titles include a bill of rights to protect members in their relations with unions; regulations of trusteeships; standards for elections; and fiduciary responsibility of union officers. The Labor Management Relations Act, 1947, was amended in certain respects by this act. Among other changes, hot-cargo clauses in contracts were forbidden, except for apparel and construction industries. Restrictions were placed on secondary boycotts and picketing.

labor market area, General term often used to mean a geographical area from which workers may be recruited, surrounding a concentration of establishments. Usually a metropolitan area, consisting of a central city and its suburbs. Also a labor area.

labor mobility, General term to describe the extent to which workers can, are willing, or do move from job to job, employer to employer, or place to place to find employment of work of their choice.

labor movement, General term usually applied to organized labor — its growth, structure, and activities — but may sometimes encompass all concerted economic, political and social activities of workers.

labor turnover (turnover), Movement of workers into and out of employment in a company or industry through hiring, layoffs, recall, quits, etc. *Labor turnover rates* are usually expressed as the number of accessions and separations during a given period per 100 employees.

Landrum-Griffin Act, See Labor Management Reporting and Disclosure Act of 1959.

layoff (reduction in force), Involuntary separation from employment for a temporary or indefinite period, without prejudice, that is, resulting from no fault of the workers. Although "layoff" usually implies eventual recall, or at least an intent to recall workers to their jobs, the term is occasionally used for separations plainly signifying permanent loss of jobs, as in plant shutdowns.

layoff allowance, See severance pay.

learner, Generally, a beginner learning a job for which extensive technical training or experience is not required. (See apprentice.)

learner's certificate, Certificates issued by the U.S. Department of Labor, under provisions of the Fair Labor Standards Act of 1938, which permit employers to pay rates below the statutory minimum to learners, messengers, apprentices, and handicapped workers so as not to curtail opportunities for their employment.

learner rate, The rate or, more frequently, the schedule of rates applicable to workers inexperienced in the job for which they are employed, during their period of training. The schedule of rates is usually established in such a manner as to permit the gradual achievement of the minimum job rate as the learner develops competence on the job. E.g., under the Fair Labor Standards Act, an employer may be permitted to employ learners in a specific plant at a wage lower than the legal minimum, whenever employment of learners at such lower rate is believed necessary to prevent curtailment of employment opportunities. Hearings are held by the administrator to determine under what limitations as to wages, time, number, proportion, and length of service, special certificates authorizing the employment of learning at subminimum rates may be issued to an employer for certain occupations in his plant.

learning curve, A graph plotting the course of learning, in which the vertical axis plots a measure of proficiency while the horizontal axis represents some measure of practice or experience.

leave of absence, Generally, excused time (unpaid) away from work, usually for a week or more, without loss of job or seniority.

legally required benefits, Term applied to employee-benefit programs to which employers must contribute or insurance that they must purchase for employees according to law. Includes social security, unemployment compensation, workmen's compensation, and, to a lesser extent, State Temporary disability insurance, and the special programs for railroad workers.

level income option, See social security adjustment option.

level of aspiration, A goal that an individual sets as something he expects to achieve, or strives to achieve. Reaching the goal is interpreted by him as success, falling short as failure.

life insurance plan, Group term insurance coverage for employees, paid for in whole or in part by the employer, providing a lump-sum payment to a worker's beneficiary in the event of his death. (See health and insurance plan, death benefit.)

living document, This term, as used by unions, expresses the belief that the terms of an agreement, particularly a long-term agreement, should be subject to review and renegotiation by the parties if conditions change or unforeseen events come about, despite the absence of a reopening clause.

local union, Labor organization comprising the members of a union within a particular area or establishment, which has been chartered by, and is affiliated with, a national or international union. Also known as local; chapter; lodge.

lockout, A temporary withholding of work or denial of employment to a group of workers by an employer during a labor dispute in order to compel a settlement at or close to the employer's terms. A *joint lockout* is such an action undertaken at the same time by a group of employers. Technically the distinction between a strike and lockout turns on which party actually initiates the stoppage. One, however, can develop into the other. (See work stoppage, joint lockout.)

long-term contract, Generally, a collective bargaining agreement with a duration of 2 or 3 years or longer, as distinguished from a 1-year agreement.

loose rate, See runaway rate.

maintenance-of-membership clause, An arrangement provided for in a collective bargaining agreement whereby employees who are members of the union at the time the agreement is negotiated, or who voluntarily join the union subsequently, must maintain their membership for the duration of the agreement, or, possibly, a shorter period, as a condition of continued employment. (See union security.)

major medical expense benefit, Plan designed to insure workers against the heavy medical expenses resulting from catastrophic or prolonged illness or injury. If the benefit supplements the benefit payable by a basic health insurance plan (hospital, medical, or surgical), it is called a "supplementary" plan; otherwise, it is called a "comprehensive" plan. (See catastrophe insurance, deductible, health and insurance plan.)

make-up pay, Allowances given by employers to piece workers to make up differences between actual piecework earnings and earnings at guaranteed rates or statutory minimum rates. At times, the term is also associated with the practice of permitting employees to earn a full week's wages by making up for lost time.

make work, See featherbedding.

management, Term applied to the employer and his representatives, or to corporation executives who are responsible for the administration and direction of an enterprise. (See employer.)

management prerogatives, As used in union-management relationships, this term is applied to rights reserved to management, which may be expressly noted as such in a collective bargaining agreement, usually including the right to schedule production, to determine the process of manufacture, to maintain order and efficiency, to hire, etc.

man-days of strike idleness, A key measure of strike activity regularly compiled by the Bureau of Labor Statistics, reflecting working time lost because of strikes and lockouts. The figures on strike idleness do not include secondary idleness — that is, the effects of a work stoppage on other establishments or industries whose employees may be made idle as a result of material or service shortages.

manning table, A listing of the positions, or number of workers, to be used in the operation of a particular machine or process.

manpower, General term used to designate all persons able and willing to work, that is, the available labor force. The term is sometimes used in reference to a more limited group, e.g., company, industry, community, workers with particular skills, etc.

manual workers, See blue-collar workers.

master agreement, A single or uniform collective bargaining agreement covering a number of plants of a single employer or the members of an employers' association. (See multiplant bargaining, multiemployer bargaining.)

maternity benefits, Term applied to health and insurance plan benefits payable to women workers absent because of pregnancy and childbirth, and for hospital, surgical, and medical benefits related thereto, and to workers' wives where such dependents' benefits are provided. In recent years a "parallel" term (paternity benefits) has arisen.

mealtime, See paid lunch period.

mediation, An attempt by a third party to help in negotiations or in the settlement of a dispute between employer and union through suggestion, advice, or other ways of stimulating agreement, short of dictating its provisions (a characteristic of arbitration). Most of the mediation in the United States is undertaken through federal and state mediation agencies. *Mediator* — Term used to designate person who undertakes mediation of a dispute. *Conciliator* — In practice synonymous with mediation; the term lives on mainly in the name of the chief mediation agency. (See conciliation, federal mediation and conciliation service.)

medical benefits, Plans which provide workers, and in many cases their dependents, with specified medical care (other than that connected with surgery) or a cash allowance toward the cost of doctors' visits. Generally part of a health and insurance program. (See health and insurance plan, health center.)

member in good standing, See union member.

merit increase, An increase in the wage rate of an individual worker on the basis of performance or service. This is widely used as a method of advancing workers within established rate ranges, sometimes in conjunction with a provision for automatic increases over part of the range. Merit increases may be administered informally at the discretion of the employer, or provision may exist for the periodic review of the performance of employees for granting of merit increases. (See wage review.)

merit rating, A formalized system of appraising employee performance according to an established group of factors. Usually the appraisal is annual or semiannual and is by immediate supervision. Factors frequently considered include quality of work, quantity of work, reliability, adaptability, initiative, attitude, etc. (See Work Measurement and Methods.)

migratory workers, Persons whose principal income is earned from temporary employment (usually in farming) and who, in the course of a year, move one or more times, often through several states.

military leave, Excused leave of absence for military service, reserve training, National Guard duty, etc. Time lost may be paid for by the employer in whole or in part.

minimum rate, There are several kinds of minimum rates, those that are applicable to specific jobs and those that are applicable to entire establishments. Normally, those that are applicable to specific jobs are called minimum job rates and those applicable to entire establishments are called minimum plant rates. In addition, there are several varieties of guaranteed minimum rates usually applicable to individual jobs under wage incentive systems.

minimum job rate, The minimum rate of pay for workers on a given job. The minimum rate may be either a single rate or the minimum of a rate range. Union rates or union scales are usually minimum job rates. Normally, entrance rates, probationary rates, or learner rates fall below the minimum job rates.

minimum plant rate, Normally the minimum rate of pay for experienced workers in the lowest-paid job in the establishment. The term may, however, mean different things in plants with differently organized wage structures. In some plants, the term refers to the rate for the lowest-paid production job, although lower rates may exist for such jobs as common labor or janitor. In some plants the so-called minimum rate may actually be a hiring or probationary rate. (See entrance rate.)

minimum wage, Rates of wages, established legally or through collective bargaining, below which workers cannot be employed. The Fair Labor Standards Act establishes the legal minimum wage to be paid to workers engaged in interstate commerce, unless such workers are covered by state laws which provide for higher minimum wages. Minimum rates are also established by collective bargaining and are applicable to individual plants, or to groupings of plants within an area or within an industry.

modified union shop, See union shop.

money-purchase plan, See pension plan.

monitorship, Supervision or surveillance of a union by an outside party, usually for a limited time, imposed by order of a court or parent union organization.

Monthly Labor Review, A monthly magazine devoted to general economic and labor matters, issued by the Bureau of Labor Statistics, United States Department of Labor.

moonlighting, Term applied to the simultaneous holding of more than one paid employment by a worker, e.g., a full time and a second supplementary job with another employer, or self-employment.

moving allowance, See relocation allowance.

multiemployer bargaining, Collective bargaining between a union or unions and a group of employers, usually represented by an employer association, resulting in a uniform or master agreement.

multiplant bargaining, Collective bargaining between a company and the union or unions representing workers in more than one of its plants, usually resulting in a master agreement. If all or most plants are involved, the term "companywide" is appropriately used. (See companywide bargaining.)

nafets, Unemployed steamfitter; used interchangeably with "folled." Both are terms of derision; origin coal industry.

national emergency dispute, Term used in the Labor Management Relations (Taft-Hartley) Act to designate an actual or threatened strike or lockout which may imperil the national health or safety. If such a possibility exists in the opinion of the President, he may appoint a board of inquiry to investigate the issues in the dispute. Upon receiving a report from the board, the President may direct the Attorney General to petition the appropriate district court for a time-specified injunction, during which the board of inquiry and the National Labor Relations Board have certain functions to perform. If no settlement is reached during this cooling-off period, the parties are free to resume their dispute, and the President may recommend appropriate action to the Congress.

National Labor Relations Act (Wagner Act of 1935), Basic federal act guaranteeing workers the right to organize and bargain collectively through representatives of their own choosing. Also, defined "unfair labor practices" as regards employers.Amended by the Labor Management Relations Act, 1947 and the Labor-Management Reporting and Disclosure Act of 1959.

National Labor Relations Board (NLRB), Agency created by the National Labor Relations Act, 1935, and continued through subsequent amendments, whose functions are to define appropriate bargaining units, to hold elections to determine whether a majority of workers want to be represented by a specific union or no union, to certify unions to represent employees, to interpret and apply the act's provisions prohibiting certain employer and union unfair practices, and otherwise to administer the provisions of the act. (See Labor Management Relations Act, 1947.)

National Mediation Board, Agency established by the Railway Labor Act, 1926, to provide aid in settling disputes between railway and airline companies and unions over union representation, negotiation of changes in agreements, and interpretation of agreements reached through mediation.

National Railroad Adjustment Board, Federal agency established in 1934 which functions as a board of arbitration, handing down final and binding decisions on disputes arising out of grievances, or the application and interpretation of agreements, in the railroad industry (airline industry not covered). Board is composed of 36 members, 18 of whom represent and are paid by the carriers and 18 by national railway labor organizations.

national union, Ordinarily, a union composed of a number of affiliated local unions. The Bureau of Labor Statistics, in its union directory, defines a national union as one with agreements with different employers in more than one state, or an affiliate of the AFL-CIO, or a national organization of government employees. (See international union.)

negotiation, See collective bargaining.

night shift, See shift.

noncontributory pension plan, A pension plan for the benefit of the employee under which the entire cost is borne by the employer. (See contributory pension plan.)

nonproduction bonus, An extra payment to employees which depends on factors other than the output of the individual worker, such as profit-sharing, safety, attendance, and Christmas bonuses. (See bonus plan, production bonus.)

no-raiding agreement, See raiding.

normal retirement, See retirement.

normal retirement age, In technical terms, the earliest age at which a worker under a pension plan may retire of his own accord and receive the full amount of benefits to which he is entitled under the normal benefit formula of the plan. In most plans, and under social security, this age is 65 years. Historically, it is the age "set" by Prince von Bismarck.

no-strike, no-lockout clause, Provision in a collective bargaining agreement through which the union agrees not to strike and the employer agrees not to lockout employees for the duration of the contract. These pledges may be hedged by certain qualifications, e.g., the union may strike if the employer violates the agreement.

occupational rate, Rates (single or ranges) that are designated for particular occupations in an establishment, area, or industry. Generally, these rates are formal rates, and are paid to any worker who is qualified to perform the work of the occupation.

occupational wage relationship, The relationship of wage rates among occupations representative of a range of duties, skills, and responsibilities. Relationships may be analyzed within an individual plant, a community or region, or on an industry basis.

off the books, Refers especially to work performed but not recorded officially in order to escape scrutiny especially by financial, union, safety, tax and/or governmental committees, and/or agencies.

Old-Age Survivors and Disability Insurance Benefits (OASDI), Retirement income and survivors' and disability payments available to eligible workers covered by federal social security legislation.

open-end agreement, Collective bargaining agreement with no definite termination date, usually subject to reopening for negotiations or to termination at any time upon proper notice by either party.

open shop, Term commonly applied to an establishment with a policy of not recognizing or dealing with a labor union. Term may sometimes be applied to an organized establishment where union membership is not a condition of employment. (See union security.)

open union, A union which will admit any qualified person to membership, usually upon payment of reasonable initiation fees. (See closed union.)

organizer, Employee of a union or federation (usually paid but sometimes a volunteer) whose duties include recruiting new members for the union, assisting in forming unions in nonunion companies, assisting in campaigns for recognition, etc. (See union organizer.)

outlaw strike, See strike.

out-of-line rate, See red circle rate, runaway rate.

out-of-work benefits, Usually, payments made by a union to unemployed members.

output per man-hour, Productivity.

overtime, Work performed in excess of basic workday or workweek, as defined by law, collective bargaining agreement, or company policy. Sometimes applied to work performed on Saturdays, Sundays, and holidays at premium rates.

overtime premium pay, Payment of wages at premium rate for time worked beyond the regular hours of employment established by union agreement, employer or industry practice, or law. In the United States, payment is typically made at one and a half times the regular rate of pay. Higher premium rates are found to a limited extent. (See premium rate.)

pace setter, A worker who is better than average on a particular job, and whose production is used by the employer as a standard for measuring the amount of work which can be done in a given period of time.

package settlement, A term used to describe a combination of benefits received by workers as a result of collective bargaining. A package may include wage increases and other benefits of monetary value, such as insurance, paid holidays, paid vacations, and sick leave. The term generally implies that during the bargaining process, the parties agreed that a specific amount of increase was to be applied partly to rates of pay, and partly to the financing of the related benefits.

pact, See agreement.

paid absence allowance, Payment for lost working time available to workers for various types of leave not otherwise compensated for, e.g., excused personal leave.

paid holidays, Holidays are days of special religious, cultural, social, or patriotic significance on which work or business ordinarily ceases. Paid holidays are those, established by agreement or by company policy, for which workers receive their full daily pay without working. Half-day holidays are also common. (See holiday premium pay, unpaid holidays.)

paid lunch period, Time allowed for eating lunch (or the mid-shift meal on late shifts), commonly 20-30 minutes, counted as part of the paid workday. Usually practiced where employees cannot leave their workplaces for meals (e.g., coal mining). Agreements sometimes also require company to furnish meals when workers remain in the plant for overtime work.

paid vacations, Excused leave of absence of a week or more, with full pay, granted to workers annually for purposes of rest and recreation. Paid vacations are provided in private industry by collective bargaining agreements or company policy, not by law. Vacations are frequently graduated by length of service. (See extended vacation plan.)

paper locals, Local unions which exist only "on paper" (charter) with no actual membership.

part-time employee, Worker employed on a temporary or regular basis for a workweek substantially shorter than the scheduled week for full-time employees.

part-time worker rate, A rate paid to a part-time temporary, or contingent worker, as distinguished from that paid to a regular or full-time worker. Part-time rates may be equal to, or lower or higher than, regular or full-time rates. During periods of ample labor supply, part-time rates are usually lower, but may become equivalent or higher when the labor market is tight because of keen competition for such help. Retail trade establishments and restaurants are among the industries dependent on part-time or temporary help to carry on their normal functions.

past practice, Existing practices in the plant or company, sanctioned by use and acceptance, that are not specifically included in the collective bargaining agreement, except, perhaps, by reference to their continuance.

past service, Under a pension plan, years of employment or credited service prior to the establishment of the plan or a change in the plan's benefits.

pattern bargaining, Term applied to follow-the-leader negotiating practices in an industry.

pay-as-you-go, See unfunded plan.

pay-in-lieu-of notice, Where employers are required to provide advance notice of layoff and fail to do so, agreements often require the employer to pay workers for the full notice period as a penalty.

payment by result, Refers to any method of wage payment where the amount of the wage depends upon the amount of output. The term applies to straight piecework or other types of incentive systems. The production to which wages are related may be the output of an individual worker or the output of a group of workers.

payment certain guarantee, See period certain option.

payroll deduction, A deduction from an employee's gross earnings made by his employer for social security, unemployment insurance, federal income tax, local government payroll tax, union dues, special union assessments, group insurance premiums, etc.

payroll period, The established frequency with which workers are paid in a particular industry, regardless of the time to which the rate applies.

payroll tax, Taxes levied by the government and paid by employers, employees, or both, creating funds from which employees receive retirement, unemployment, or other benefits. Also may refer to employer contributions, based on fixed percentages of total payroll, to union or other private health and welfare and vacation funds, and to payroll taxes levied by cities.

peg point, An occupational rate for a key unskilled, semiskilled, or skilled job, establishing differentials within the wage structure. The term first used by the National War Labor Board in its decisions on wages in the cotton textile industry in 1945 (See 21 W. L. B. 882), and thereafter applied to the wage structure through collective bargaining.

penalty rate, An extra rate which is paid for hazardous jobs, late-shift work, or for overtime. (See premium rate.)

pension plan, Any plan whose primary purpose is to provide regular payments for a determined period of time, usually life, to employees upon retirement. Additional benefits are often provided. The term *private pension plans* is often used to distinguish voluntary plans from the social security system. If the employee shares in the cost, the plan is *contributory;* if the cost is borne entirely by the employer, the plan is *noncontributory.*

group annuity plan — Pension plan underwritten and administered by an insurance company.

trusteed plan — Noninsured; contributions deposited with bank, trust company, or board of trustees, who administers the program.

deposit administration — Insurance company assumes role of trustee, as above.

money purchase plan — Fixed contributions to the worker's account, his pension thus determined by the amount contributed. (See retirement plan, group annuity plan, annuity.)

per capita tax, Regular payments made on the basis of a paid-up membership count by a local union to its national organization, or by a national union to a federation, to finance the activities of the parent organization. Amount usually set by union constitution.

period certain option, Provision in a pension plan under which the pensioner may elect to receive a reduced benefit for life, on the condition that if he dies before receiving a specified number of payments the balance is continued to his beneficiary. A guarantee of a specified number of payments may be a standard plan provision, in which case it is called a *payment certain guarantee.*

permanent and total disability, Inability of a worker to perform his job, or any job, owing to physical or mental impairment which is expected to be of long-continued and indefinite duration. The existence of the impairment must be certified by a physician, under the Social Security Act and most private pension plans, in order to qualify for benefits. Mental disabilities may be excluded by some pension plans.

permanent piece rate, A rate established for a piece-work job calculated to yield an appropriate level of earnings and based, generally, on experience with trial rates for the job assignment; such rates are expected to persist until basic conditions change.

permit card, Card issued by a union to a nonmember which permits him to accept or retain employment on a temporary basis in a union shop or on a union job. Also known as work permit.

perquisite, Relates to the furnishing by employers of food, lodging and other payments in kind to workers in addition to monetary compensation, thus, waitresses are generally allowed a certain number of meals, depending upon the length of the shift; board and lodging are usually supplied to workers in lumber camps and in some cases to farm labor. Often called Perks.

personal leave, Excused leave for reasons important to the individual worker, but not otherwise provided for, e.g., getting married.

personalized rate, See red circle rate.

picketing, Patrolling near employer's place of business by union members (pickets) to publicize the exitence of a labor dispute, persuade workers to join the union or the strike, discourage customers from buying or using employer's goods or service, etc.

organizational picketing — Picketing carried on by the union for the purpose of persuading employees to join the union or authorize the union to represent them.

recognitional picketing — Picketing to compel the employer to recognize the union as the exclusive bargaining agent for his employees.

informational picketing – Picketing directed toward advising the public that an employer does not employ members of, or have a contract with, a union.

piece rate, Under an incentive wage system, the predetermined amount paid to a worker for each unit of output. Rates may be based on individual or group output. (See incentive rate.)

piece scale, See price list.

piecework, Method of wage payment based on the number of units produced, or any work for which piece rates are paid.

pink collar worker, Term used to identify the great influx of women to the industrial work force. Clearly, a "pink collar" can be "white" or "blue." (See blue collar worker, also white collar worker — but the term frequently identifies jobs at which women "traditionally" (sic) worked.)

p.m., An incentive payment to sales personnel in retail trade to push and sell items on which the margin of profit is large, to dispose of slow moving items, or the clear out old stock. Also referred to as premium money, push money.

portal-to-portal pay, In mining, the computation of hours worked and paid for so as to include travel time between the mine entry and the place of work at the start and completion of the work shift. Also applies to payments made by other industries for time spent on company premises in getting to and from working places.

posting, See job posting.

power, A person's power in a relationship is his capacity to influence others in that relationship. This is determined by the range of outcomes through which he can move his partner, or by his "outcome control." Power may also derive from "cue control."

preferential hiring, Agreed-upon arrangement whereby the employer gives preference in hiring to union members, to applicants with previous training and experience in the industry, to workers displaced from another plant or from another part of a particular plant, or by order of the National Labor Relations Board to employees found to be discriminatorily discharged.

premium money, See p.m.

premium overtime pay, See overtime pay.

premium rate or pay, An extra rate paid for overtime, work on late shifts, holiday and Sunday work, or for work in particularly dangerous or unpleasant occupations. The term is also used in reference to extra rates paid to employees, usually because of exceptional ability or skill in the occupation. (See overtime premium pay, penalty rate, skill differential.)

prevailing rate, Typically, the predominant or more common rate paid to a group of workers, usually with reference to specific occupations in an industry or labor market area. In actual application, the term "prevailing rate" is used in a variety of ways. Some of the variations arise from differences in the concept of the geographic unit or industry that is pertinent to a particular situation. For example, where comparable occupations are found in an area in numerous establishments, the geographic unit may be as narrow as, for example, the metropolitan area. If labor has to be induced to migrate to an area where the prevailing rate is to be set, and no local labor is available, the geographic unit upon which determination is made may be a locality far removed from the site under consideration; at times, additional pay may be provided to induce labor to move. Another set of variations in the application of the term "prevailing rate" arises from differences in industry limits used in an area, for example, are in effect rates found in that portion of industry that employs such workers, whether or not the limits are specified in the determination. Other occupations, on the other hand, may be found in all industries in an area. Another source of variation in the use of prevailing rate concept arises from the use of quantitative or statistical descriptions of prevailing rates, some of these terms are rather loosely conceived as, for example, "going rate," when reference is made to the rate received by a substantial number of workers, possibly the modal rate. In some situations, notably locality wage surveys, measures of prevailing rates relate to the arithmetic mean, or to the median.

In view of these variations, the use of the term "prevailing rate" requires specific mention of the area, occupation, industry, rate, and type of quantitative measure involved to have definite meaning.

prevailing wage law, Federal act requiring the payment of prevailing wage rates in the locality on construction, alteration, or repair of public works performed under contract with the federal government, e.g., DAVIS-BACON Act (1931) amended in 1964 to include certain payments for fringe benefits as part of the prevailing rate.

preventive mediation, A function of the Federal Mediation and Conciliation Service involving the development of procedures by union and management designed to anticipate and to study potential problems. This may take the form of early entry into labor disputes before a strike threatens.

price list, A listing of piece prices or rates to be paid by a company or a group of companies making similar products. In unionized establishments, price lists are established typically upon agreement between the union and the employer.

probationary period, Usually a stipulated period of time during which a newly hired employee is on trial prior to establishing seniority or otherwise becoming a regular employee. Sometimes used in relation to discipline, e.g., a period during which a regular employee, guilty of misbehavior, is on trial.

probationary employee – A worker in a probationary period. Where informal probation is the practice, a worker who has not yet attained the status of regular employee may be called a *temporary employee.* (See regular employee.)

probationary rate, The rate of pay for worker during the period of his employment on a new job or in a new plant. The probationary rate is usually lower than the minimum rate for the job. (See entrance rate.)

production bonus, A bonus payment directly related to the output of an individual worker or a group of workers. Usually paid for production in excess of a quota or for the completion of a job in less than standard time. The bonus may be a flat amount paid for all production above standard, or it may increase in various proportions as production increases. (See bonus, nonproduction bonus.)

production standards, Usually, expected output of a worker or group of workers, consistent with the quality of workmanship, efficiency of operations, and the reasonable working capacities of normal operators.

production workers, Usually, employees directly involved in manufacturing or operational processes, as distinguished from supervisory, sales, executive, and office employees. The term "production and related workers" as used in federal government statistics is usually specifically defined for survey purposes.

productivity, (1) The ratio of output to total inputs. (2) The ratio of acutal production to standard production — applicable to either an individual worker or group of workers. (See Work Measurement and Methods.)

productivity factor, See annual improvement factor.

profit-sharing plan, Any procedure under which an employer pays his employees, in addition to their regular pay, a share of the profits of the business as a whole.

deferred profit-sharing plan – Share of profits set aside in a fund to be distributed at some later date, usually when the employee retires (a form of retirement plan).

progression system, Wage progression.

protest price, In some industries, notably pottery and women's dresses, piece rates on new work are determined on the basis of previously developed time elements. A worker may not be able to earn an appropriate amount under such estimated time allowances and piece rates. If he does not earn enough, he enters a protest, but continues to work at these rates until a review is made and new rates are set. Any adjustment in rates is usually retroactive to the time of protest, or to the time the worker was started on the new work. (See temporary work.)

push money, See p.m.

pyramiding, Double payment of overtime rates for overtime work which may result from paying both daily and weekly overtime rates for same hours of work; sometimes applied to any premium added to another premium rate.

quickie strike, See strike.

quit, Voluntary termination of employment initiated by the employee, as distinguished from dismissal or layoff which are involuntary.

raiding (no-raiding agreement), Term applied to a union's attempt to enroll members belonging to another union or already covered by a collective bargaining agreement negotiated by another union, with the intent to usurp the union's bargaining relationship.

no-raiding agreement – Written pledge signed by two unions or more to abstain from raiding, applicable only to signatory unions.

rank and file, Members of an organization, exclusive of officers and employees.

rate cutting, Term applied to a reduction by management of established incentive or time wage rates in the absence of comparable changes in job content, or any actions by companies in reducing wages.

rate range, A range of rates for the same job, with the specific rates of individual workers within the range determined by merit, length of service, or a combination of various concepts of merit and length of service. Rate ranges may be set up with various degrees of formality and more or less rigid rules respecting the position within the range at which new workers are hired and the rules concerning their automatic or nonautomatic advancement to the maximum rate. The range may be expressed as a spread from a set minimum to a set maximum rate.

rate setting, (1) The establishment of pay per unit for incentive work. (2) The establishment of a standard time. (See Work Measurement and Methods.)

ratification, Formal approval of a newly negotiated agreement by vote of the union members affected.

real wages, Real wages are represented by the goods and services typically consumed by workers that can be purchased with money wages; i.e., real wages are an expression of the purchasing power of money wages. Over periods of time, changes in real wages are obtained by dividing indexes of money wages by an appropriate index of consumers' prices.

recall, Process of bringing laid-off employees back to work, usually based on the same principles that governed order of layoff in inverse order (e.g., last worker laid off is first to be recalled). In union affairs, recall is a procedure for removing (disciplining) an officer by means of a membership vote.

recognition, See union recognition.

red circle rate, Rate of pay higher than the contractual, or formally estbalished, rate for a job. The special rate is usually attached to the incumbent worker, not to the job as such. This procedure is commonly used to protect long-service workers from a decline in earnings through no fault of their own. Also known as out-of-line rate, personalized rate, flagged rate.

reduction in force, layoff.

referendum, Process by which all members of a union vote, usually as individuals, for the election of officers, changes in union constitution, etc., as distinguished from decision-making through delegates assembled in convention.

regional differential, Differences in wage levels among several broad geographic subdivisions especially in the United States.

regular employee, Usually, a full-time employee who has fulfilled formal or informal probationary requirements, as distinguished from seasonal, part-time, probationary, and temporary employees.

regular rate, The rate of pay received by a worker for all hours of work performed at straight-time rates. Also refers to the rate of pay at which a worker is predominantly engaged when he is subject to assignments at varying rates.

rehire, As a verb — to reemploy a worker previously separated. As a noun — a former employee returned to his job as a new employee.

relief time, Time during which a worker is permitted to leave his workplace, usually for personal needs, with his place being taken by a substitute when necessary. (See rest period, spellout.)

relocation allowance, Payment to employees of all or part of their expenses in moving to a new location, or a fixed allowance to be used for this purpose. (See moving allowance.)

reopening clause, Clause in a collective bargaining agreement stating the time or the circumstances under which negotiations can be requested, prior to the expiration of the contract. Reopenings are usually restricted to wage issues and, perhaps, other specified economic issues, not to the contract as a whole. (See wage reopening.)

reporting pay, The amount of pay guaranteed to a worker who reports for work at the usual hour, without notification to the contrary, and finds no work available or is not given a full shift's employment. Typically, pay for a minimum number of hours at regular rate is provided for in union agreements. (See call-in pay.)

representation election (election), Election conducted to determine by a majority vote of the employees in an appropriate unit, which, if any, union is desired as their exclusive representative. These elections are usually conducted by the National Labor Relations Board or by state labor relations agencies. (See bargaining unit.)

rest period, Brief interruption in the workday, usually of 5 to 15 minutes duration, during which the worker rests, smokes, or takes refreshments without loss of pay. (See coffee break, break time, relief time.)

retirement, Withdrawal from working life or from a particular employment because of age, disability, etc., with an income. *Normal retirement* is retirement for age, presently at age 65. *Early retirement* is retirement prior to the normal retirement age. *Disability retirement* is retirement prior to the normal retirement age because of poor health or injury disabling the worker. *Special early retirement* — extra early retirement benefits provided under specified circumstances, e.g., involuntary separation. (See pension plan, social security act.)

retirement plan, See pension plan.

retraining, Development of new skills for workers through a definite program, so that they are able to qualify for new or different work.

retroactive pay, Delayed payment of part of the wages for a particular period, resulting from a retroactive application of wage increases arising from wage negotiations. (See back pay.)

right-to-work law, Term applied to state legislation which prohibits any contractual requirement that a worker join a union in order to get or keep a job, thus banning provision in agreements requiring employees to become and remain union members (otherwise permissible under Labor Management Relations Act).

role-playing, A method for teaching principles affecting interpersonal relations by having subjects assume various roles during interaction.

rolling, See bumping.

rotating shift, The system of rotating the crews where two or more shifts are worked in an establishment. This system is designed to distribute day and night work on an equal basis among the various workers. In some industries, where 7-day operations are common, the work schedules may be arranged so that workers are given different days off in each week. (See fixed shift, shift, split shift, swing shift.)

round-the-clock operations, See continuous operations.

royalty, In relation to wages, the payment to union health and welfare funds, such as those benefiting members of the United Mine Workers and the American Federation of Musicians, although the term is not the official designation for such payments. In these cases, the application of the term stems, at least in part, from the fact that employer contributions are based on tons of coal mined and number of musical records produced. For some types of professional workers, such as musicians, singers, and writers, payment for work is frequently based on a percent of sales of the final product (book, article, or song). Such payments are referred to as royalties.

runaway rate, A standard (piece or incentive) rate which is considered "loose." A standard which yields very high earnings when compared with earnings for a task of similar difficulty.

runaway shop, Term used by unions to characterize a business establishment which moves to evade a union or State labor laws, or to reap a competitive advantage from low wage standards in another area, dismissing all or most of its regular employees in the process.

runoff election, A second election conducted after the first produces no winner according to the rules. If more than two contenders were in the first contest, the runoff may be limited to the two highest. (See representation election.)

salary and commission, See commission earnings.

salary rate, For workers hired on a weekly, monthly, or annual basis, the rate of pay is normally expressed in terms of dollars per week, month, or year. Workers employed on a monthly or annual salary basis may actually be paid monthly, semi-monthly, or more frequently. Usually, the length of the work week is specified and a policy is established for compensation in the event that longer or shorter hours than a full week are worked.

savings plan, See thrift plan.

scab, See strikebreaker.

scale, See union rate.

seasonal employment, employment during part of the year only, arising out of the seasonal character of an industry. Agricultural, cannery, construction, and lumber workers are examples of workers subject to seasonal employment.

secondary boycott, See boycott.

Section 14 (b), Labor Management Relations Act, 1947, This section of the Taft-Hartley Act provides the opening through which states may enact "right-to-work" laws. It reads as follows: "Nothing in this act shall be construed as authorizing the execution or application of agreements requiring membership in a labor organization as a condition of employment in any State or Territory in which such execution or application is prohibited by State or Territorial law." (See right-to-work law.)

seniority, Term used to designate an employee's status relative to other employees, as in determining order of promotion, layoff, vacations, etc. *Straight seniority* — seniority acquired solely through length of service. *Qualified seniority* — other factors such as ability considered with length of service. *Departmental or unit seniority* — seniority applicable in a particular section of a plant rather than in the entire establishment. *Plantwide or Companywide seniority* — seniority applicable throughout the plant or company. *Seniority list* — individual workers ranked in order of seniority. (See superseniority.)

sensitivity training, A group process that is designed to create an opportunity for individuals to expose their behavior and to give as well as receive feedback. Members frankly say how they react to certain behaviors so that each gains an awareness of the way his behavior affects others.

service fee, Fee required by unions to be paid by nonmembers applying for employment in union hiring halls, as a condition of referral to employment.

severance pay, Monetary allowance paid by employer to displaced employees, generally upon permanent termination of employment with no chance of recall, but often upon indefinite layoff with recall rights intact. Plans usually graduate payments by length of service. Also known as dismissal pay or allowance, termination pay, separation pay, layoff allowance.

shapeup, System of hiring work gangs from groups of workers assembled to seek employment. Used in longshore work in some ports, and in the hiring of migratory farm workers.

share-the-work, See work sharing.

shestil, Term used to describe an additional initiation fee required by some craft unions to permit part-time employees to accept full-time employment. This practice has been outlawed by the Taft-Hartley Act.

shift, A term applied to a work period where two or more groups of workers are employed at different hours during the operating time of an establishment; e.g., an establishment may operate two shifts of 8 hours each or 16 hours a day. In some industries, the term "trick" or "tour" is used instead of "shift." (See fixed shift, rotating shift, split shift, swing shift.)

shift differential, Added compensation to workers who are employed on a work schedule other than the regular daytime schedule. Shift differentials may be paid in a number of ways: (1) a fixed amount per hour above the rate paid on the regular day shift; (2) a percentage over earnings at the regular day shift rates; (3) shorter hours with full daily pay, or; (4) both shorter hours and additional monetary compensation above full daily pay.

shop committee, Group of workers selected by fellow employees, usually union members, to represent them in their dealings with management. Special types are: grievance committee, negotiating committee.

shop rules, Either regulations established by an employer dealing with day-to-day conduct in the plant — operations, safety, hygiene, records, etc. — or working rules set forth in collective bargaining agreements and in some union constitutions. Also known as working rules.

shop steward, A local union's representative in a plant or department selected by union members (or sometimes appointed by the union) to carry out union duties, adjust grievances, collect dues, and solicit new members. Usually a fellow employee. (See union steward.)

short workweek benefit, As part of a supplemental unemployment benefit plan, payment to worker for the difference between a specified level of weekly hours and the hours actually worked or paid for.

sick benefits, See accident and sickness benefits.

sick leave, Period of time during which a worker may be absent without loss of job or seniority if unable to work because of illness or accident. A *paid sick leave* provides for full or partial pay for such absence, usually up to a stipulated maximum. Sick leave plans differ from accident and sickness benefits, principally in that the former cover shorter periods of absence and are uninsured. (See accident and sickness benefits.)

single company union, An independent or unaffiliated union of employees of one company, usually with no formal ties to any other labor organization.

single rate, A rate which is the same for all workers on the same job or in the same job classification, and under which the individual worker on a job receives the same rate during the entire time that he is holding the job. The single rate usually is paid to experienced workers in jobs requiring varying degrees of skill. Learners or apprentices may be paid according to rate schedules which start below the single rate and permit the worker to achieve the full job rate over a period of time. In the less skilled jobs, the rates for beginners and experienced workers may be identical because the period of time necessary to become familiar with all phases of the work is relatively short. Individual workers may occasionally be paid above or below the single rate for special reasons, but such payments are regarded as exceptions to the usual rule. The definition of a "job" or "classification" may be very narrow or very broad, and the single rate may therefore be applicable in as few as one or two workers doing identical jobs, or as many as several thousands performing a number of essentially different jobs which are nevertheless regarded as meriting the same rate of pay.

sitdown strike, See strike.

skill differential, Differences in wage rates paid to workers engaged in occupations requiring varying levels of skill in work performance. May also refer to differentials in rates of workers in the same occupation, higher rates being paid to those who usually perform the more complex tasks. (See premium rate.)

slowdown, See strike.

Social Security Act, 1935, Federal law establishing a national social insurance program. The law provides for: old-age, survivors' and disability benefits (an all-federal program); public assistance to the aged, the blind, and to needy families; and unemployment insurance (both federal-state programs). The coverage and other provisions have been modified several times since enactment.

social security adjustment option, Pension plan provision under which a worker eligible for an early retirement benefit may elect to get a larger plan benefit than is actually due up to the time his social security benefit is payable, and a small benefit thereafter, so that a level income is maintained throughout retirement. (See level income option.)

social security offset, Under some pension plans, the amount of social security benefits to which a retiring worker is entitled that is to be deducted from the private plan benefit, as computed, to obtain the actual benefit payable. The offset or deduction may be all or part of the social security benefit.

special permit rate, A rate paid to a union worker who comes from another city and is employed under a special permit because of local labor shortages. The rate received is the same as that paid to a permanent worker in the area. In the unionized brewery industry, this term refers to the rates paid to special workers who are temporarily employed during the peak summer period. These rates are usually lower than those received by "regular" union workers.

speedup, Workers' term for conditions which force them to increase effort or production within a given time without a compensating increase in earnings. (See stretchout.)

spellout, See relief time.

spendable earnings, In general, the monthly earnings of worker less various amounts deducted for taxes and other purposes from payrolls; hence, "spendable earnings" may be identified broadly with "take-home" earnings. Term also used in the sense of the earnings available for private spending or saving, but this usage would include certain types of deductions (e.g., union dues) as spendable earnings. "Net spendable average weekly earnings" is a series developed by the Bureau of Labor Statistics in which Federal Social Security and Income Taxes are deducted from gross average weekly earnings for workers with specified number of dependents. (See take-home pay.)

split shift, The daily working time that is not continuous, but split into two or more working periods. "Split shifts" are usually found in industries such as local transportation, which is affected by peak or rush periods at various times of the day. (See fixed shift, rotating shift, shift, swing shift.)

spread-the-work, See work sharing.

standard agreement, Collective bargaining agreement prepared by a national or international union for use by, or guidance of, its local unions, designed to produce standardization of practices within the union's bargaining relationships. (See form agreement.)

standard rate, A basic rate of pay established for an occupation in a plant, industry, or community through collective bargaining, company regulation, or by law. May also refer to established rates for services rendered in a community in connection with maintenance and repair of automobiles, appliances, buildings, etc.

starting rate, See entrance rate.

step rates, Fixed levels between the minimum rates for an occupation in wage progression system. (See wage progression.)

stint, Shift.

stock option plan, Plan allowing employees or officers the privilege of purchasing company stock (shares) at a certain price at a time of their own choosing.

stock purchase plan, Plan enabling employees to purchase stock (shares) in the company, with or without employer contributions, generally under more favorable terms than are available on the open market.

straight time, Time worked at regular rate, as distinguished from overtime. (See earnings, overtime.)

stretchout, Term used by workers when they are required to tend more machines or assume additional duties within a given time without a corresponding increase in earnings. (See speedup.)

strike (wildcat, outlaw, quickie, slowdown, sympathy, sitdown, general), Temporary stoppage of work by a group of employees (not necessarily members of a union) to express a grievance, enforce a demand for changes in the conditions of employment, obtain recognition, or resolve a dispute with management.

wildcat or outlaw strike, A strike not sanctioned by union and one which violates agreement.

quickie strike, A spontaneous or unannounced strike.

slowdown, A deliberate reduction of output without an actual strike in order to force concession from employer.

sympathy strike, Strike of workers not directly involved in a dispute, but who wish to demonstrate worker solidarity or bring additional pressure upon company involved.

sitdown strike, Strike during which workers stay inside the plant or work-place, but refuse to work or allow others to do so.

general strike, Strike involving all organized workers in a community or country (rare in the United States).

walkout, Same as strike.

strike benefits, Union payments made to members who are on strike.

strikebreaker, Worker or person on hire who accepts employment or continues to work in a plant where an authorized strike is in process, filling the job of a striker and knowingly assisting in defeating the strike. (See anti-strike-breaker law, scab, fink.)

strike deadline, Time set by the union for beginning a strike if a satisfactory settlement is not reached. Typically, this is at midnight of the last day of the contract term or the start of the next day's first shift.

strike fund, Money allocated by a union or set aside in a separate account to pay strike benefits and to defray other expenses of strikes.

strike insurance, Payment by companies in an association to a fund, or for the purchase of insurance, to reimburse a struck member company for lost business.

strike notice, Formal notice of an intention to strike, presented by the union to the employer, or to the appropriate federal government agency, e.g., the Federal Mediation and Conciliation Service.

strike vote, Vote conducted among members of a union to determine whether or not a strike should be called.

struck work or goods, Goods produced or services performed by a firm while a strike of its employees is in process. (See hot-cargo clause.)

style development rate, Similar to temporary, experimental, or trial rate. The term is used in the hosiery manufacturing industry and relates to work on new styles for which no piece rates have yet been set. Generally, hourly rates are paid on such work. Usually, these hourly rates average close to the workers' previous piece-rate earnings. The style development rates are in effect for a specified time and are then replaced by new piece rates.

subminimum rate, A rate below the minimum established for an occupation, establishment, industry, or area by union agreement, law, or policy. Such rates may be paid to learners and to substandard, superannuated, probationary, or special permit workers.

subsistence allowance, A payment to a worker for expenses covering meals, lodging, and transportation while in a traveling status for his employer. Such allowances may be based on a fixed amount for meals and lodging plus other expenditures or on the actual expenses incurred for all items. There also are cases where institutional workers (e.g., nurses) receive a subsistence allowance for living outside the institution, since free room and board are incorporated into the wage structure.

substandardard rate, A rate of pay below the prevailing or standard level for a worker whose efficiency is impaired because of physical or mental handicaps. The term is also used to refer to rates below Federal or State minimum wage levels or below prevailing levels for an occupation in an industry or area. (See handicapped worker rate.)

suggestion system, Plan whereby employees' ideas that may increase efficiency or improve operations or safety are channeled to the attention of management; usually combined with a system of rewards for acceptable ideas.

superannuated rate, A rate of pay below the prevailing level for a worker above a certain age. Such rates are frequently allowed in union agreements. At times, the agreement requires the employment of a certain ratio of older workers at superannuated rates. Superannuated workers with long service are sometimes retained in an employed status because of their economic need; also their services are sometimes sought during periods of labor shortages.

superannuated workers, Term sometimes applied to employees who are unable to perform their jobs, or any job, at the normal level because of advanced age and its attendant infirmities.

superseniority, A position on the seniority list ahead of what the employee would acquire solely on the basis of length of service or other general seniority factors. Usually such favored treatment is reserved to union stewards, or other workers entitled to special consideration in connection with layoff and recall to work.

supplements to wages and salaries, As defined by the U.S. Department of Commerce for national income purposes: "Supplements to wages and salaries is the monetary compensation of employees not commonly regarded as wages and salaries. It consists of employer contributions to private pension and welfare funds, compensation for injuries, director's fees, pay of the military reserve, and a few other minor items of labor income." Term sometimes used more broadly to refer to all supplements to basic wage or salary rates.

Supplemental Unemployment Benefit plans (SUB), Introduced by agreement between Ford Motor Company and the United Automobile Workers in mid-1955 and subsequently adopted by other companies in the automobile, steel, and related industries, these plans provide regular weekly payments to laid-off workers receiving State unemployment insurance, through funds financed by the employer. Other benefits (e.g., short-workweek benefits and severance pay) were added to many plans.

supplementary benefits, See fringe benefits.

surgical benefits, Plans which provide workers, and in many cases their dependents, with specified surgical care or a cash allowance toward the cost of such care, usually in accordance with a schedule of surgeon's fees. Often referred to as Blue Shield. Generally part of a health and insurance program. (See health and insurance plan.)

survivors' benefits (transition benefits, bridge benefits, widow's allowance), Payments to dependents of employees who die prior to retirement, financed in whole or in part by the employer. May be in the form of payments for a fixed period (e.g., 24 months) supplementing regular life insurance benefits, a benefit for life out of a pension program, a lump-sum payment, etc.

survivor's option, See joint and survivor option.

suspension, Form of disciplinary action of a temporary nature, as in removing a worker from his job for a stipulated time with the consequent loss of pay as punishment, or in removing a union official from office until his affairs are checked or put into order.

sweatshop, Term of contempt applied to an establishment employing workers for long hours at low wages under unfavorable working conditions.

sweetheart agreement, A derogatory term for a union contract exceptionally favorable to a particular employer, in comparison with other contracts, implying less favorable conditions of employment than could be obtained under a legitimate collective bargaining relationship.

swing shift, An extra shift of workers required in establishments where continuous or seven-day operations are scheduled, to provide the other crews with days off. The "swing crew" usually rotates among all of the other shifts. Also, refers to the practice of one of three rotating shifts staying on the job through two shift periods, thus "swinging" the shifts into their new assignments. (See fixed shift, rotating shift, split shift.)

sympathy strike, See strike.

Taft-Hartley Act, See Labor Management Relations Act.

take-home pay, Typically, earnings for a pay-roll period, less required deductions. (See spendable earnings.)

target, In piece rate systems, a rate is set with the objective of making it possible for a worker to earn, on the average, 10, 15, or some other percentage above the base rate. The expected earnings to which the piece rate is geared is referred to as a target.

technological unemployment, Displacement of workers caused by, or attributed to the installation of labor-saving methods or machinery.

temporary disability insurance, Provision presently enacted into law in four states (California, New Jersey, New York, and Rhode Island) providing payments for a limited period of time to workers suffering loss of wages due to sickness or disability incurred off the job. In all except the Rhode Island plan, employers may substitute privately insured coverage meeting certain standards. (See workmen's compensation, accident and sickness benefits.)

temporary employee, See probationary period.

temporary rate, A rate set tentatively on new work. When new work is started under piece rates in some industries and it is not known whether or not the initial rates can be properly set for the tasks involved, temporary rates are established. These rates are later revised and are made permanent when found to be satisfactory. Sometimes they are called "experimental" or "trial" rates. These are alternate designations in various industries. (See guarantee on trial rate, protest price.)

termination pay or allowances, See severance pay.

thrift plan, An arrangement under which payroll deductions are made, with the worker's consent, for investment and saving, to which the employer contributes. The accumulated amounts (contributions plus earnings on the contributions) become available to the worker, usually after certain conditions are met. (See saving plan.)

time study, A work measurement technique consisting of a careful time measurement of the task with a time measuring instrument, adjusted from any observed variance from normal effort of pace and to allow adequate time for unavoidable or machine delays, rest to overcome fatigue, and personal needs. (See Work Measurement and Methods. See also motion analysis and motion economy in above-noted Section.)

time and one-half, Premium rate consisting of one and one-half times the employee's regular rate.

tip, A gratuity given by a customer or patron in recognition of satisfactory personal service or through custom. Tips are considered as compensation by the Bureau of Internal Revenue, thus constituting taxable income. A substantial proportion of the earnings of some categories of workers in hotels, restaurants, steamships, and barber and beauty shops is realized from tips.

tonnage rate, Pay for a unit of work applicable to incentive workers, and common in such industries as coal mining and basic iron and steel, where output for important categories of workers can be measured on a tonnage basis.

tool allowance, Allowance to an employee, paid by employer, as reimbursement for the cost of tools and their upkeep, where the employee furnishes his own tools or is responsible for their maintenance.

tool maintenance time, See allowed time.

tour of duty, See shift.

trade council, See joint board.

trade union, See union.

trainee, The term "trainee" applies to workers who receive formal training for occupations requiring a limited degree of skill. The training may include some classroom work. A Trainee differs from a learner in that a learner does not receive formal training but learns his job through actual performance, under supervision.

transfer card, Card issued by a local union to a member in good standing, certifying his eligibility to join another local of the same union in a different location.

transition benefits, See survivors' benefits.

travel time, The time spent traveling to and from and a designated point and place of work. Such travel includes portal-to-portal in mining, deadheading on railroads, and out-of-town work performed by building tradesmen, mechanics, musicians, etc.

trick, See shift.

trustee, A person, bank, or trust company who administers and takes responsibility for a trust fund, or a person who is a member of a board of trustees. (See fund.)

trusteeship, In union affairs, the taking of direct control of a local union by the national or international union, generally to correct mismanagement or illegal practices on the part of local officers, to prevent secession, or to strengthen the local. Control of the local is returned to members or officers after the cause for trusteeship has been corrected. (See monitorship.)

trust fund, See fund.

tuition payment plan, Plan providing for payment by the employer of a part or all of the costs of job-related training courses undertaken by employees.

turn, See shift.

turnover, See labor turnover.

umpire, See impartial chairman.

unaffiliated union, See independent union.

unemployment insurance (unemployment compensation), Joint federal-state program, established in 1935 under the Social Security Act and subject to the standards set forth in the Federal Unemployment Tax Act, under which state-administered funds obtained through payroll taxes provide payments to the eligible unemployed persons for specified periods of time. Levels of benefits and tax rates are established by each state. Generally, excluded groups include among others, railroad workers (covered by Railroad Unemployment Insurance Act), agricultural workers, State and municipal employees and workers in nonprofit institutions.The federal part of the program is administered by the United States Department of Labor.

unfair labor practice, Action by either an employer or union which violates the provisions of national or state labor relations acts, such as refusal to bargain in good faith.

unfair labor practice strike — a strike caused, at least in part, by an employer's unfair labor practice.

unfair list, Union list of employers designated as unfair to organized labor.

unfunded plan, Plan (e.g., pension plan) under which benefits are paid, like wages, directly from an employer's general assets, often as a payroll item, as distinguished from a fund irretrievably segregated from the general assets of the firm and separately administered. Some pension plans are funded in part and unfunded with regard to certain benefits. (See pay-as-you go.)

union, Any organization in which workers participate as members, which exists for the purpose of dealing with employers concerning grievances, wages, hours, and conditions of employment. Unions are voluntary organizations and need no license from the government to operate. Unions may incorporate if they wish. Also known as labor union, labor organization. (See trade union.)

union agreement, See agreement.

union contract, See agreement.

union convention, An assembly of delegates meeting periodically to act on union problems, elect officers, and determine policies. The convention is typically the chief governing body of the union in constitutional terms.

union dues, Fee paid periodically, usually monthly, by members of a union, typically as a condition of continued membership. Each union sets its own dues requirement. (See checkoff.)

union label (bug), Tag, imprint, or design attached to an article as evidence that it was produced by union labor.

union leave, Paid or unpaid, but excused, leave for union representatives, shop stewards, etc., to attend to union business, e.g., participating in union conventions.

union-management cooperation, Voluntary joint participation of union and management in solving problems such as production and safety, or in engaging in certain outside activities, such as community or charitable work. The term is usually reserved to joint actions outside of the process of collective bargaining itself.

union member, A union member may de defined in broad terms as a worker who has met the union's qualifications for membership, has joined the union, and has maintained his membership rights. Each union usually determines its own qualifications. In general, *dues-paying members* are those who pay dues to the union on a regular basis. *Members in good standing* include dues-paying members and members exempted for various reasons (unemployed, on strike, ill, etc.) but carried on the union rolls as full-fledged members. *Book members* are those listed on the union rolls, dues-paying or not.

union organizer, See organizer.

union rate, An hourly rate, usually a single rate for an occupation or trade, established by agreement reached through collective bargaining. A union rate or scale is usually the minimum rate that may be paid to qualified persons in the job; there are usually no restrictions prohibiting the employer from paying higher rates.

union recognition, Employer acceptance of a union as the representative of his employees, the first step in the establishment of a collective bargaining relationship.

union scale, See union rate.

union security, Protection of a union's status by a provision in the collective bargaining agreement establishing a closed shop, union shop, agency shop, or maintenance-of-membership arrangement. In the absence of such provisions, employees in the bargaining unit are free to join or support the union at will, and, thus; in union reasoning, are susceptible to pressures to refrain or to the inducement of a "free ride."

union shop, Provision in a collective bargaining agreement that requires all employees to become members of the union within a specified time after hiring (typically 30 days), or after a new provision is negotiated, and to remain members of the union as a condition of continued employment.

modified union shop — variations on the union shop. Certain employees may be exempted, e.g., those already employed at the time the provision was negotiated who had, up until then, not joined the union.

union steward, See shop steward.

unlicensed personnel, Seamen not required to have an official license, such as deckhands, stewards, firemen, etc., as distinguished from licensed masters, mates, and engineers.

unpaid holidays, Holidays observed by an establishment only to the extent of providing premium pay for work on that day. Paid time off not provided. (See paid holidays.)

upgrading, The process of a more rapid than usual advancement of workers to jobs having greater skill requirements and commanding higher rates of pay. This measure is largely resorted to when experienced help is difficult to secure, as was the case during World War II, and when new work of long or permanent duration is undertaken requiring higher degrees of skill on the part of the labor force. (See downgrading.)

urban wage rate index, Series maintained by the Bureau of Labor Statistics, beginning in 1943, to measure the movement of wage rates in urban areas in manufacturing, major manufacturing industry groups, and selected non-manufacturing industries.

U.S. Department of Labor, The department was established by Act of Congress in 1913 to "foster, promote, and develop the welfare of the wage earners of the United States, to improve their working conditions, and to advance their opportunities for profitable employment." The Department also has important functions in the field of international labor affairs. The major bureaus and offices of the Department include: Bureau of Apprenticeship and Training, Bureau of Employment Security, Bureau of Labor Statistics, Bureau of Labor Standards, Wage and Hour and Public Contracts Divisions, Women's Bureau, Bureau of International Labor Affairs, Office of Manpower, Automation and Training, Office of the Solicitor, Neighborhood Youth Corps, Labor-Management Services Administration, and Manpower Administration. The Secretary of Labor heads the Department.

vacation pay, Payment for a period of time received by workers for vacation purposes. The time period frequently varies with length of service. During busy times or in a tight labor market, workers may be given the option of accepting vacation pay in lieu of time off.

vertical union, See industrial union.

vesting (vested rights), As commonly used in connection with pension plans, a guarantee to a worker of his equity in the plan, based on contributions by the employer on his behalf, should his employment terminate before he becomes eligible for retirement. The worker usually must meet specified minimum age and service requirements for qualification. The vested worker receives the pension he has earned when he reaches retirement age, wherever he is then employed.

wage advance plan, Advancing of wages in work-weeks of short duration under plans obligating employers to maintain weekly wages up to a specified minimum level. Wages must be repaid during later weeks in which regular or longer hours worked. No repayment is required unless the employer provides sufficient work to enable the advance to be repaid. (See annual wage of employment guarantee, guaranteed wage plan.)

wage arbitration, The referral of wage disputes between employers and unions to an arbitrator or Board of Arbitration. The arbitrator's award or decision is customarily binding upon both parties. Arbitration is usually voluntary, both parties having agreed to refer the dispute to a third party for a decision. (See wage mediation.)

wage assignment, A voluntary transfer by a worker of some of his earned wages or commissions to another party or parties. Such assignments may be used for payment of purchased goods and debts, purchase of savings bonds, and payment of union dues and assessments.

wage board employees, Federal government employees, typically manual workers, whose rates of pay are determined on the basis of prevailing rates for comparable work in the area, as distinguished from classification act employees.

wage determination, The process of establishing wage rates and wage structures through collective bargaining, arbitration, individual employer determination, etc., the process may involve comparisons with rates paid by other firms, the use of job evaluation, or other techniques. The term is also applied to findings, orders, or decisions of wage regulatory bodies such as minimum wage boards.

wage differential, Differences in wages among occupations, industries, or areas. Historical wage differentials, to which frequent reference is made, are those which have existed over long periods of time.

wage drift, Term generally used to describe the differential change in average earnings levels over time as measured against negotiated changes. The difference between the level of actual earnings, which is influenced by many factors, and the level at which earnings would be if formal general wage changes alone are taken into account, is thus likened to an upward drift.

wage escalation, See escalator clause.

wage-hour law, See Fair Labor Standards Act.

wage inequality, An unadjusted disparity between rates of workers whose duties and responsibilities are similar or identical. Wage inequalities can be considered either on an intraplant or interplant basis. The elimination of wage inequalities is often accomplished through job review or the adoption of job evaluation plans. (See wage inequity.)

wage inequity, An unadjusted relationship between the wage rates of workers or of job classifications. (See wage inequality.)

wage leadership, The influence exercised by the wage settlements reached by a large firm or group of firms on other settlements in an industry or labor market. "Follow-the-leader" wage adjustments appear to be particularly significant in some industries. May also relate to a policy adopted by a firm maintaining a position of wage leadership in an industry or area.

wage level, The level of wages received by workers in an occupation, establishment, industry, or area. Wage levels are generally indicated by average rates.

wage mediation, The entrance of a disinterested third party into a wage dispute in an effort to effect a settlement. Unlike arbitration, the mediator merely makes recommendations and assists the disputant parties in reaching a settlement. This is the principal function of the Federal Mediation and Conciliation Service. (See wage arbitration.)

wage policy, A formalized practice of an establishment or industry to elements of wages, such as wage rate scales, shift differentials, overtime provisions, nonproduction bonuses, automatic increments, paid holidays, paid vacations, pensions, and insurance benefits. In a broader sense, criteria for wage adjustments are stated in terms of objectives (e.g., stabilization, rising standard of living, etc.) or in terms of prevailing economic conditions.

wage rate, The monetary compensation for a given unit of time or effort by which a worker's pay is calculated. There are several kinds of wage rates, related to the system of wage payment used in an establishment. The principal kinds are hourly rates, daily rates, weekly rates, monthly rates, annual rates, and various kinds of incentive rates.

wage rate bracket, In the administration of wartime wage stabilization policy by the National War Labor Board, the term referred to a range of "sound and tested going rates" for an occupation in a labor market area. The minimum of the range or bracket, the most important point in actual wage administration, was frequently set at the first substantial cluster of rates in a wage distribution. The minimum of the bracket was the point up to which the War Labor Board would permit adjustments in interplant inequity cases.

wage reopening, A provision or clause in a union agreement permitting the question of wages to be reopened for negotiation before the expiration of the agreement.

wage review, A periodic review of the performance of workers to determine or select those who deserve merit increases, or advancement to higher paying jobs. (See merit increase.)

wages and salaries, As defined by the U.S. Department of Commerce for national income purposes: "Wage and salaries consist of the monetary remuneration of employees commonly regarded as wages and salaries, inclusive of executives' compensation, commissions, tips, and bonuses, and of payments in kind which represent income to the recipients." More generally, this term refers to remuneration to individuals for productive effort.

wage and salary administration, The managing and supervision of the wage structure of an employer. It involves the application of wage and salary adjustments, according to established policies, and the analysis of data such as cost of living, prices, wage and salary surveys, which have a direct bearing on the wage structure and are used in wage negotiations. May also involve the establishment of new rates through job evaluation, job analysis, and time studies.

wage and salary receipts, As defined by the U.S. Department of Commerce for national income purposes: "Wage and salary receipts are equal to wages and salaries less employee contributions for social insurance, except that retroactive wages are counted when paid rather than when earned."

wage structure, The sum total of the various elements and considerations that characterize a specific rate schedule in an establishment, industry, area, or country as a whole. Typical of such elements are: (1) relationship between rates of occupations of different skills; (2) relationship between rates of pay for men, women, and workers of different races and color in the same occupations; (3) provision for extra pay for late shift work, overtime, hazardous, unpleasant, or unhealthful work; (4) interarea and interregional variations in rates of pay; (5) methods of pay; (6) provisions for lunch and rest periods; and (7) supplementary benefits, such as vacations, insurance, sick leave, and holiday provisions.

wage survey, A general term used to describe a wage study based on the collection, tabulation, and analysis of original data. Wage surveys are of many types, and the kinds of data collected depend upon the uses to which the surveys are put.

waiting time, See dead time, down time.

welfare plan, An arrangement which provides various health and other employee interest benefits — including death benefits. The term is used also in connection with various employee benefit plans.

white-collar worker, Usually associated with the description of an employee on the office, clerical, sales, semi-technical and/or professional staffs. More specifically, the term is used to describe an employee not "on production work." Usually, it is regarded as the antonymn of blue-collar worker (q.v.).

work sharing, Spreading available work among all employees in a group, in order to prevent or reduce the extent of a layoff.

work standard, An established measurement base, usually stated as time per unit of production or per task, that a qualified, trained, and experienced employee should take to perform the specific task, when working at an average pace, with allowances for personal needs and normal delays associated with the task.

work stoppage, Any temporary halt to work. It may be initiated either by the employees or by the employer. The term has been adopted by the Bureau of Labor Statistics to replace the terms strike and lockout.

workmen's compensation, A system of insurance, financed by the employer, which provides financial aid to compensate an employee for damages or injuries incurred in the course of employment.

Work Measurement and Methods

12

Z94.12

This aspect of industrial engineering — WORK MEASUREMENT AND METHODS — is historically *the* basic field. Taylor's original interest was an attempt at answering the fundamental question: "WHAT IS A FAIR DAY'S WORK?" (See INTRODUCTION, *Industrial Engineering Field.*) Although a good part of Taylor's work was associated with stopwatch studies, it remained for the Gilbreths to search for THE ONE BEST WAY. For many years there was a schism, then finally a combination of Motion and Time Study, which dominated the IE field. It was from these pioneers that the field was expanded into the psychology of the worker as well as the physiology of her/his performance, into the measurement of the task as well as the methods of performing it.

The TERMINOLOGY in this section contains several hundreds of terms of interest to the worker as well as the manager, and should prove of value not only in measuring performance but also in paying for it. Although the terms in this section represent a revision of the 1972 Z94.12 edition, it would not be amiss to point out that Dr. Williams' earliest contributions appear in the INDUSTRIAL ENGINEERING TERMINOLOGY, which appeared in the JOURNAL OF INDUSTRIAL ENGINEERING, (AIIE, November - December, 1965).

The present subcommittee consisted of the following:

W. Bruce Ballantyne
Westinghouse Electric Corporation

Franklin H. Bayha, P.E.
Professional Engineer

Mitchell Fein, P.E.
Mitchell Fein Inc.

Bertram Gottlieb, P.E.
Gottlieb and Associates

Dr. Charles M. Overby, P.E.
Ohio University

Dr. Robert L. Williams, Chairman
Professor and Chairman, Industrial and Systems Engineering, Ohio University

abnormal reading, See abnormal time.

abnormal time, A time value which is outside of statistical or policy variance limits. *Syn:* abnormal reading.

accumulative timing, A multiple (usually three) stopwatch technique for time study in which a mechanical linkage pressed at successive cycle breakpoints instantaneously stops, starts, and resets the individual watches so that, respectively, they may be: read for recording the latest element time, timing the element currently being observed, and ready to time the next element.

activity sampling, See work sampling.

actual hours, See actual time.

actual time, The unadjusted time for the accomplishment of a defined task or task element as obtained by a timing device. *Syn:* observed time.

allowance, (1) Work measurement: a time value or percentage of time by which the normal time is increased, or the amount of non-productive time applied, to compensate for justifiable causes or policy requirements which necessitate performance time not directly measured for each element or task. Usually includes: irregular elements, incentive opportunity on machine controlled time, minor unavoidable delays, rest time to overcome fatigue, and time for personal needs. (2) Dimensional: the minimum clearance or maximum interference distance between two interpenetrated objects. (See training allowance.)

allowed hours, See standard hour, allowed time.

allowed time, A normal time value increased by an appropriate allowance(s). (See standard time.)

assignable cause, A source of variation in a process which can be isolated, especially when its significantly larger magnitude or different origin readily distinguishes it from random causes of variation.

auxiliary process time, The time required for essential supplementary process operations which assure the continuity and completion of the principal process operations. Such auxiliary operations as deburring, straightening, cleaning, and finishing usually result in relatively minor changes in the appearance of physical characteristics of the workpiece in comparison with the effects due to such principal operations as cutting, forming, welding, casting, and assembly.

available machine time, The portion of a time cycle during which a machine could be performing useful work.

available process time, The portion of a time cycle during which a process agent or system could be acting usefully on the product.

average cycle time, (1) The sum of observed or actual work times, excluding abnormal times, divided by the number of such cycle observations. (2) The sum of the average element times. (See average element time.)

average element time, The sum of a series of observed or actual element times, excluding abnormal times, divided by the number of such element observations. (See average cycle time.)

average time, See average cycle time, average element time.

avoidable delay, A time delay not allowed in standard time calculations because it is unnecessary and is due to factors under worker control and responsibility.

balance, (1) The act of distributing the work elements between the two hands performing an operation or between the different operations in a process to achieve essentially equal performance times among them. (2) The state of approximately equal working time distribution among the various components of an operation or process, e.g., the stations on an assembly line.

balanced motion pattern, (1) The sequence of concurrent arm and hand movements over symmetrical paths that produce approximately equal momentum between the arms in directions which facilitate muscular equilibrium. (2) A series of movements with both hands involving negligible delay or idle time for either hand while the other is working.

balancing delay, (1) The idle time of one hand in an operation due to imperfect balancing. (2) The idle time of one or more operations in a series due to imperfect balancing. (See balance.) *Syn:* balance delay.

ballistic movement, A motion of a body extremity with relatively simple muscle action that is rapid, smooth, and flowing from start to finish. Such action results when a protagonistic muscle group causes motion in the intended direction to attain a peak force and velocity, which retards to zero as an antagonistic muscle group changes the direction or causes motion to cease. This sequential muscle action contrasts with that producing nonballistic motion, in which the protagonistic and antagonistic muscle groups act concurrently to precisely control the force and velocity of movement. Historically, the term is associated with the similarity of the motion plot to projectile trajectory plots.

bank, See float.

basic division of work, See therblig.

basic element, See elemental motion.

basic motion, A human motion closely related to primary physiological and/or bio-mechanical performance capabilities of the body or its members (e.g., a therblig or other standard motion defined within a predetermined time system). Compare: elemental motion.

benchmark, A standard of measurement with enough characteristics common to the individual units of a population to facilitate economical comparison of attributes for units selected from a sample. Benchmarks may be used for job evaluation, performance rating, establishing operational standards, standard data development, cost estimating, and other purposes. (See benchmark job, key job.)

benchmark job, A job with enough characteristics common to other jobs, judged acceptable as a gauge for those other jobs without their direct measurement for time standards, job evaluations, or other purposes. (See benchmark.)

breakpoint, A point in a work cycle readily distinguished by sight and/or sound which is selected as the boundary between two elements for time recording or element definition in motion study. *Syn:* reading point; endpoint.

building block, An approach used in standard data development of creating fixed groups or modules of work elements which may be added together to obtain time values for elements and entire operations.

camera study, See micromotion study, memomotion study.

changeover time, The time required to modify or replace an existing facility or workplace, usually including both teardown time for the existing condition and setup of the new condition. (See setup, teardown.)

check study, A partial or complete review of a job or operation to evaluate the validity of a standard time.

chronocyclegraph technique, A modification of the cyclegraph technique to permit the computation of motion velocities and accelerations from the spacing of light signals on the film whose shapes indicate the direction of movement, produced by pulsing the lights on and off at regular time intervals. (See cyclegraph technique.)

chronological study, A detailed study and recording of a sequence of events in the order of occurrences. (See production study.)

coding, (1) Translation of a data processing machine program from descriptive, symbolic, or diagram form into machine language (code) or into an explicit symbolic language that may be translated directly into machine language by means of an assembly program or compiler. (2) Referring numbers to a convenient origin and/or scale for ease of computation. (3) Assigning a numerical and/or alphabetical symbol or group of symbols to a class or variable to achieve consistent identification, location, or interpretation. A desirable property of such symbology is the mnemonic, or memory-jogging, characteristic to promote efficient association of the code meaning.

combined motions, Two or more nonconsecutive elemental motions performed during the same time interval by the same body member.

combined work, (1) The production of a person working with one or more machines in which the output is controlled by the operator. In calculating a standard, the machine portion of the work cycle is not taken into account. (2) The total accomplishment of a group of workers.

consistency, (1) The absence of noticeable or significant variation in behavioral or numerical data as, for example, in the work pace or method used by a worker. (2) Uniformity or agreement, within stated limits, between repetitive occurrences of an event or a numerical value.

constant element, A job or task element which occurs without significant variation in work content and/or performance time. May be used to describe elements within a given operation or elements common to different operations.

continuous method, See continuous timing.

continuous reading, See continuous timing.

continuous timing, A stopwatch technique in which the watch runs continuously throughout the study and readings are made of the cumulative time at the end of each element. Individual element times are then obtained by subtraction. *Syn:* continuous method, continuous reading.

control system, A system that has as its primary function the collection and analysis of feedback from a given set of functions for the purpose of controlling the functions. Control may be implemented by monitoring and/or systematically modifying parameters or policies used in those functions, or by preparing control reports that initiate useful action with respect to significant deviations and exceptions.

coverage, actual, The number of jobs, the number of personnel, or the total hours which are covered by standards during the reporting period. Commonly expressed as a percentage of the total number of jobs, personnel, or hours. *Syn:* standard coverage.

coverage, potential, An estimate of the number of jobs, the number of personnel, or the total hours which can be covered by time standards during the reporting period. Commonly expressed as a percentage of the total number of jobs, personnel, or hours.

coverage, standard, See actual coverage.

cycle, The complete sequence of activities, operations, and machine or process times required to complete one segment, unit, or batch of work. (See motion cycle, work cycle.)

cycle timing, Timing a complete work cycle as a single time value rather than timing the individual elements of the cycle. (See differential timing.)

cyclegram, Similar to cyclegraph except that the film being exposed is moved slowly to one side so that retraced motions are shown side by side on the film rather than superimposed on earlier motions.

cyclegraph, The pattern developed by the cyclegraph technique. (See cyclegraph technique.)

cyclegraph technique, The use of small lights on the hands or other body members to indicate their motion patterns. The lights are recorded by a still camera in a darkened room with an exposure time equal to at least one motion cycle.

cyclic element, An element of an operation or process that occurs in every cycle of the operation or process.

cyclic timing, See cycle timing.

daywork, Work for which pay is based on time as contrasted to pay based on performance.

decimal-hour stopwatch, A timing device with two hands whose movement may be started, stopped, or reset to zero by depressing control buttons on the perimeter of the watch case. A small dial is calibrated in hundredths of an hour and a large dial is calibrated in ten-thousandths of an hour. Thus, the time interval may be read in decimal hours to four decimal places.

decimal-minute stopwatch, A timing device with two hands whose movement may be started, stopped, or reset to zero by depressing control buttons on the perimeter of the watch case. A small dial is calibrated in whole minutes and a large dial is calibrated in hundredths of a minute. Thus, the time interval may be read in decimal minutes to two decimal places.

delay, A pause or interruption in the scheduled work activity of the human, machine, or product flow. (See avoidable delay, unavoidable delay, inherent delay.) *Syn:* interruption, stoppage.

delay allowance, (1) A time increment to allow for contingencies and minor delays beyond the control of the operator. May be included in a time standard as a percentage or as nonproductive time. (2) A separate credit (in time or money) to compensate the operator on incentive for a specific instance of delay not covered by the piece rate or standard. (See unavoidable delay allowance.)

delay time, A period during which an employee is idle due to breakdown of equipment, lack of tools or materials, or any other factor beyond personal control. (See downtime, idletime.)

diagnostic study, A brief investigation or cursory methods study of an operation, process, group, or individual to discover causes of operational difficulties or problems for which more detailed remedial studies may be feasible. An appropriate work measurement technique may be used to evaluate alternatives or to locate major areas requiring improvement. *Syn:* survey.

differential timing, The time study technique used to obtain the time value of an element of extremely short duration. It consists of: (1) obtaining cycle values, first including and then excluding the element for which the time is required, and obtaining the required element time by subtraction; (2) timing the element by combining it with preceding and/or following elements in successive cycles and then obtaining the time for the short element by subtraction.

direct labor, (1) Work which is readily chargeable to or identifiable with a specific product. (2) Work performed on a product or service that advances the product or service towards its completion or objectives.

direct labor standard, A standard time set on a direct labor operation. (See direct labor.)

discontinuous timing, See repetitive timing.

division of labor, The separation of jobs or tasks into less complex jobs or tasks usually to allow use of workers possessing less skill than that required by the overall job or task, or to make use of special skills. *Syn:* division of work.

downgrade, (1) The lowering of a particular job in scope, authority, responsibility, degree of difficulty, etc., with a possible reduction in wage or salary. (2) Dilution of skills required for the task. (See division of labor.)

downtime, A period of time during which an operation is halted due to the lack of materials, a machinery breakdown, or the like. (See downtime, Applied Mathematics, Applied Psychology, Production Planning and Control.)

drop delivery, (1) The movement of a component or object to some location where it is merely let go rather than being placed where it is to go. (2) Provisions in a workplace for disposal of objects by dropping.

earned hours, The time in standard hours credited to a worker or a group of workers as a result of their completion of a given task or group of tasks; usually calculated by summing the multiproduct of applicable standard times and the completed work units.

effectiveness, (1) The ratio of earned hours to actual hours spent on prescribed tasks. When earned hours equal actual hours, the effectiveness equals 100%. (2) The ratio of standard, estimated, or budgeted performance to actual performance expressed as a percentage. (3) The performance or output received from an approach or program. Ideally it is a quantitative measure which can be used to evaluate the level of performance in relation to some standard, set of criteria, or end objective. (See efficiency, labor.)

efficiency, labor, (1) The ratio of standard performance time to actual performance time, usually expressed as a percentage. (2) The ratio of actual performance numbers (e.g., the number of pieces) to standard performance numbers, usually expressed as a percentage. (See productivity.)

effort, The apparent physical and mental exertion exhibited by the worker while performing a segment of work.

effort controlled cycle, See manually controlled work.

effort rating, See performance rating.

elapsed time, (1) The actual time taken by a producing worker or machine to complete a task, an operation, or an element of an operation. (2) The total time interval from the beginning to the end of a time study. (See actual time.)

element, A subdivision of the work cycle composed of one or a sequence of several basic motions and/or machine or process activities which is distinct, describable, and measurable. (See manual element, machine-controlled time.)

element breakdown, (1) The separation of a work cycle into two or more elements. (2) A listing of work elements with individual descriptions and/or calculations for each.

element time, The time to perform a given element. May refer to the observed (raw), average, selected, normal, or standard time.

elemental motion, Individual manual motions or simple motion combinations used to describe the sensory-motor activity in an operation. Generally refers to the more basic and elementary therbligs. An attempt often is made to define these precisely with associated time values. Typical elemental motions are: reach, move, assemble, pre-position, turn.

elemental standard data, (1) Standard data. (2) A time value of an individual element in a standard data system.

endpoint, See breakpoint.

engineered performance standard, See standard time.

engineered standard, See standard time.

engineered time standard, See standard time.

ergonomics, The study of work tasks with emphasis on reducing to a practical minimum the physiological cost of doing the work. (See work design, methods engineering, motion analysis, motion economy.)

ergonometrics, See ergonomics, work measurement.

estimated time, An element or operation time that has been predicted on the basis of such information as may be available without detailed study.

excess work allowance, An allowance to compensate for work required to be performed on an operation or job in addition to that specified in the standard method. Sometimes applied as a separate grant of time or as a grant of money in a piece-work system. (See allowance.)

expected attainment, See fair day's work.

expected work pace, (1) The work pace necessary for an operator to maintain in order to achieve a specified level of earnings under an incentive system. (2) The work pace required to meet non-incentive production standards.

external element, See external work.

external work, Any element of an operation which must be performed by the operator while the machine or process is not in operation and which results in a loss of potential machine or process operating time. The term "external" implies that the element occurs outside the machine or process cycle.

fair day's work, The amount of work which is expected daily from an employee. May be established solely by management or through mutual agreement. May or may not be established through the use of various measurement techniques. (See normal performance.) *Syn:* expected attainment.

fatigue, A psychological and physiological process that reduces the performance capacity and motivation of humans subjected to excessive or repeated work stresses.

fatigue allowance, Time included in the production standard to allow for rest to overcome the effects of fatigue. May be applied either as a percentage of the leveled, normal, or adjusted time or as a stated number of nonproductive minutes per hour. (See allowance, standard time.)

film analysis, A systematic, detailed analysis of work from a motion picture film. Usually related to micromotion or memomotion study. (See micromotion study, memomotion study.)

film analysis chart, For recording a film analysis. Generally records each successive elemental motion, element, or operation, the beginning and ending clock time (if a clock is included in the picture) or frame number, and its descriptive symbol. (See simo chart, therblig chart.) *Syn:* film analysis record.

film analysis record, See film analysis chart.

films, rating, Motion picture films containing a consistent or random sequence of work scenes being done at varying performance levels, used to train work measurement analysts in identifying different performance levels. May also be used to attempt to standardize the concept of normal performance, such as in card dealing, walking, or typical shop operations.

first piece time, The allowed time to produce the first piece in an order of several pieces. Intended to compensate for delays resulting from unfamiliarity with the work method, or for extra first piece work such as setup.

fixture, A device used to position and hold materials which are being worked upon or assembled.

float, (1) The amount of material in a system or process, at a given point in time, that is not being directly employed or worked upon. (2) The total cushion or slack in a network planning system. *Syn:* bank.

flow analysis, Detailed examination of the progressive travel, either of personnel or material, from place to place and/or from operation to operation.

flowchart, Tabular material, standardized symbols, and explanations depicting the predetermined route of either personnel or material, from place to place and/or from operation to operation in the manufacturing or processing sequence of events.

flow diagram, A representation of the location of activities or operations and the flow of materials between activities on a pictorial layout of a process. Usually used with a flow process chart.

flow line, (1) The direction taken either by personnel or material as they progress through the manufacturing or processing sequence of events. (2) The path along which personnel or material travel in progressing through the plant. *Syn:* line of flow.

flow path, The route taken and/or space occupied by the personnel, material, subassembly, or assembly as these progress through the manufacturing process.

flow process chart, A graphic, symbolic representation of the work performed or to be performed on a product as it passes through some or all of the stages of a process. Typically, the information included in the chart is quantity, distance moved, type of work done (by symbol with explanation), and equipment used. Work times may also be included. Flow process chart symbols generally used are:

ASME* Standard Symbol ○

operation: a subdivision of a process that changes or modifies a part, material or product, and is done essentially at one workplace location.

A specialized application for paperwork uses two standard symbols:

◎ creation of a record or set of papers

addition of information to a record or set of papers

➡ *transportation (move):* change in location of a person, part, material, or product from one workplace to another.

□ *inspection:* comparison of observed quality or quantity of product with a quality or quantity standard.

▽ *storage:* keeping a product, material, or part protected against unauthorized removal.

D *delay:* an event which occurs when an object or person waits for the next planned action.

combined activity: adjustment during testing, e.g., would combine the separate operation and inspection symbols.

*See ASME Standard 101

Syn: flow chart, production process chart, product analysis chart.

foreign element, An element with a random, usually unpredictable, frequency of occurrence, not part of a normal method.

form process chart, A graphic, symbolic representation of the process flow of paperwork forms. Similar to a flow process chart except that the item of interest is one or more forms. A form process chart may show organizations, operations, movements, temporary and controlled storages, inspection or verification, disposal of all forms charted, as well as the source and type of information transmitted between forms. Flow process chart symbols may be adapted to reflect the form processing activity. *Syn:* information process analysis chart, functional forms analysis, forms analysis chart.

frame counter, A mechanical counter which can be used to determine the number of frames that have passed a predetermined point in a motion picture. The frame counter may be attached to any device for showing or viewing motion pictures.

frequency, (1) The number of times a specified value occurs within a sample of several measurements of the same dimension or characteristics on several similar items. (2) In work measurement, the number of times an element occurs during an operation cycle.

fumble, An unintentional human activity referred to as a sensory-motor error that may or may not be avoidable depending upon the working environment or the skill of the operator.

gang chart, See multiple activity process chart.

Gantt chart, A graphic representation on a time scale of the current relationship between actual and planned performance. (See Production Planning and Control.)

Gilbreth basic element, See therblig.

gravity feed, The principle of using the force of gravity to convey materials from one location to another by having them slide or roll downward and laterally.

hand time, The time required to perform a manual element. (See manual element.) *Syn:* manual time.

handling time, The time required to move parts or materials to or from an operation or work area.

high task, Performance of an average experienced operator working at an efficient pace, over an eight-hour day under incentive conditions, without undue or cumulative fatigue. Often stated as a percentage above normal performance. (See normal effort, low task.) *Syn:* incentive pace.

historical time, See standard time, statistical.

human engineering, See human factors engineering.

human factors engineering, A merging of those branches of engineering and the behavioral sciences which concern themselves principally with the human component in the design and operation of human-machine systems. Based on a fundamental knowledge and study of human physical and mental abilities and emotional characteristics.

idle machine time, See machine idle time.

idle time, Time during which a worker is not working. (See avoidable delays, unavoidable delays, waiting time.)

incentive pace, See high task.

incentive operators, (1) Employees whose pay is determined all or in part by the quantity and/or quality of output. (2) Employees working under a wage incentive plan.

incidental element, See irregular element.

indirect labor, Labor which does not add to the value of a product but which must be performed to support its manufacture. May not be readily identifiable with a specific product or service. (See indirect labor standard.)

indirect labor standard, (1) An established standard of time for labor performed while rendering services necessary to production, the cost of which cannot be assessed against any part, product, or group of parts or products accurately or without undue effort and expense. (2) A standard time for indirect labor. (See indirect labor, standard time.)

industrial engineer, A person qualified to practice industrial engineering as defined by the Institute of Industrial Engineers.

industrial engineering, Concerned with the design, improvement, and installation of integrated systems of people, materials, equipment, and energy. It draws upon specialized knowledge and skills in the mathematical, physical, and social sciences together with the principles and methods of engineering analysis and design, to specify, predict, and evaluate the results to be obtained from such systems. (Official definition of the Institute of Industrial Engineers.)

inherent delay, See delay time.

instruction card or sheet, A written description provided to a worker which states the standard method or standard practice to be used.

interference allowance, An allowance to compensate for interference time. (See interference time.)

interference time, Idle machine time resulting from the inability of a machine operator, when assigned to two or more semi-automatic machines, to serve one or more of them when they require service. (See machine interference.)

intermittent element, See irregular element, regular element.

internal element, See internal work.

internal work, Manual work performed by an operator while the machine or process is operating automatically. *Syn:* fill up work, inside work.

irregular element, An element which occurs randomly and can be statistically determined.

jig, (1) A mechanical device used to guide a cutting tool along a predetermined path when in contact with the material or workpiece supported in the device. (2) A device used to hold parts in position (e.g., welding jig, airframe jig).

job, (1) The combination of tasks, duties, and responsibilities assigned to an employee and usually considered as a normal or regular assignment. (2) The contents of a work order.

job analysis, Determination of the requirements of a job through detailed observation and evaluation of the work performed, facilities required, conditions of work, and the qualifications required of a worker. *Syn:* job study.

job breakdown, The systematic division of an operation into elements, or the results of such an analysis. *Syn:* operation breakdown.

job characteristic, See job factor.

job design, See work design.

job factor, (1) An element characteristic of a job which provides a basis for selecting and training workers and establishing the wage range for the job. Such characteristics include mental and physical requirements, responsibilities, hazards, and other working conditions. (2) A predetermined element of a job evaluation plan against which jobs are compared. *Syn:* job characteristic.

job skill, The manual and mental proficiency required to perform a given task. *Syn:* skill.

job standardization, The procedure of specifying a standard practice or a standard method for a job.

job study, See job analysis.

key job, A job that is considered representative of similar jobs in the same plant, company, industry, or labor market and hence may be used for comparing the descriptions of other jobs with the key job for job evaluation and job classification purposes. May also be used as an aid in establishing wages for other jobs.

kymograph, An electronic time study device used to measure extremely short work time intervals. Consists of a system of transducers (principally micro-switches and photoelectric cells) that are activated by an operator performing a job, and a tape puller that records the impulses as a function of time.

labor cost, That part of a firm's total costs attributable to wages, salaries, supplementary benefits, and other employment costs.

labor saving ratio, The labor cost saved by an improved method divided by the unit labor cost of the original method.

labor standard, See direct labor standard, indirect labor standard.

learner's allowance, See training allowance.

learning curve, A plot of productive output or unit work times of an individual or group as a function of time or output per unit time; used to predict the learning rate in starting up a new job or project. A learning curve is usually exponential and flattens out with time. (See start-up curve, progress curve.) *Syn:* learner's curve.

leveled element time, See normal element time.

leveled time, See normal time.

leveling, See performance rating.

line of flow, See flow line.

loose standard, A standard time greater than that required by a qualified worker with normal skill and effort, following a prescribed method and utilizing allowances for delays, personal needs, and rest.

lost time, See waiting time.

low task, A term used to indicate that performance rating or production standards are based on daywork levels as contrasted to high task or incentive work performance. Sometimes taken to mean a level of performance below the level expected under measured daywork conditions.

machine assignment, (1) The equipment assigned to an operator in the performance of a job. (2) Equipment designed to perform jobs as in production scheduling.

machine attention time, Time during which a machine operator must observe the machine's functioning and be available for immediate servicing, while not actually operating or servicing the machine. *Syn:* service time.

machine capability, Qualitative and quantitative measures of acceptable output from a given piece of power equipment.

machine-controlled time, The time portion of an operation cycle required by a machine to complete the machine portion of the work cycle. The operator does not control this portion of the cycle time, whether or not attending the machine. *Syn:* independent machine, machine-controlled time allowance, allowance for machine-controlled time.

machine element, See machine-controlled time.

machine hour, A unit for measuring the availability or utilization of machines. It is equivalent to one machine working for 60 minutes, two machines working for 30 minutes, or an equivalent combination of machines and working time.

machine idle time, (1) Time during which a machine is idle during a work cycle awaiting the completion of manual work. (2) Interference time.

machine interference, The occurrence of conflicting demands for service by two or more units of equipment.

machine load, (1) The planned usage of a unit of equipment during a specified interval of time. (2) The percentage of maximum load at which the machine is actually used.

machine pacing, Machine or mechanical control over the rate at which the work progresses, as opposed to pacing by the worker(s). (See machine-controlled time.)

machine time, See machine-controlled time.

machine time allowance, See machine-controlled time, allowance.

macroelement, An element of a work cycle long enough to permit observation and timing by a stopwatch. (See microelement.)

maintenance, Preventive and/or correctional activities to insure that facilities and equipment are functionally capable of expected operation. As a result of these activities, equipment should be in good operating condition (clean, free from hazards, etc.) within specified limitations such as those imposed by age and prior use.

man-hour, A unit of measure representing one person working for one hour. The combination of "n" people working for "h" hours produces nh man-hours. Frequent qualifications to the definition include: (1) designation of work effort as normal effort; (2) designation of time spent as actual hours. (See man-minute.)

man-machine chart, See multiple activity process chart.

man-minute, A unit for measuring work. It is equivalent to one person working at normal pace for one minute, two people working at normal pace for thirty seconds, or an equivalent combination of people working at normal pace for a period of time. (See man-hour.)

man-process chart, A graphic, symbolic representation of the work steps or activities performed or to be performed by a person. Typically, the information included on the chart is the distance the person moves and type of work done (by symbol with description). Equipment used and work times may also be included.

manual element, A distinct, describable, and measurable subdivision of a work cycle or operation performed by hand or with the use of tools, and one that is not controlled by process or machine.

manual time, The time required to perform a manual element. (See manual element.) *Syn:* hand time.

manually-controlled work, A work cycle consisting completely of manual elements or where the manual time controls the pace at which the work progresses. *Syn:* effort-controlled cycle.

marstochron, An electric motor driven paper-tape puller used to record motion or work element times. An observer visually detects the endpoints of successive motions or elements and presses one or both of two keys that record these endpoints as successive marks along a time base on the tape. *Syn:* chronograph, marstograph.

maximum working area, That portion of the working area that is easily accessible to the hands of an operator, with arms fully extended, who is in the normal working position with trunk erect and stationary.

mean time, See average time.

measured daywork, (1) Work performed for a set hourly nonincentive wage where performance is compared to established production standards (most frequent use). (2) An incentive plan wherein the hourly wage is adjusted up or down and is guaranteed for a fixed future period (usually a quarter) according to the average performance in the prior period (infrequently used).

measured work, A term used to describe work, operations, cycles, etc., on which a standard has been set using time study or another standard setting technique.

mechanization, The act or process of using power-driven machinery to perform specific operations or functions usually with the intent of improving productivity and/or quality of the work performed.

median time, That time which is greater than or equal to half of the observed times, excluding abnormal times. It is also less than or equal to the other half of the observed times.

memomotion study, A work measurement and methods analysis technique using a motion picture camera that records events at less than normal camera speed, e.g., 50, 60, or 100 frames per minute. Used for the analysis of long events, group activities, or processes that do not move rapidly. *Syn:* camera study, time-lapse photography.

mental work, Work done principally by the mind: logical decision-making, such as sorting, classifying, or inspecting (monitoring); recalling (memory); calculation, such as performing mathematical or verbal operations and inductive policy or hypothesis formulation. The complexity may vary from elementary mental reactions to highly involved judgments based on a large number of variable factors.

merit rating, A formalized system of appraising employee performance according to an established group of factors. Usually the appraisal is annual or semi-annual and is by immediate supervision. Factors frequently considered include quality of work, quantity of work, reliability, adaptability, initiative, and attitude. (See Production Planning and Control and Wage and Salary Administration.) *Syn:* performance appraisal, employee rating, personnel rating.

method, (1) The procedure or sequence of motions by workers and/or machines used to accomplish a given operation or work task. (2) The sequence of operations and/or processes used to produce a given product or accomplish a given job. (3) A specific combination of layout and working conditions; materials, equipment, and tools; and motion patterns involved in accomplishing a given operation or task.

methods analysis, That part of methods engineering normally involving an examination and analysis of an operation or a work cycle broken down into its constituent parts for the purpose of improvement, elimination of unnecessary steps, and/or establishing and recording in detail a proposed method of performance.

methods engineering, That aspect of industrial engineering concerned with the analysis and design of work methods and systems, including technological selection of operations or processes, specification of equipment type and location, design of manual and worker-machine tasks. May include the design of controls to insure proper levels of output, inventory, quality, and cost. (See work design, motion analysis, motion economy, methods analysis.)

methods study, A systematic examination of existing methods with the purpose of developing new or improved methods, tooling, or procedures.

microchronometer, A large-faced electric clock with rapidly moving hands used in micromotion studies (within the camera's view) to indicate the passage of time. The clock usually measures to the nearest wink, or 0.0005 minutes. *Syn:* wink counter.

microelement, An element of work too short in time to allow it to be observed with the unaided eye. (See elemental motion.)

micromotion study, A work measurement or methods analysis technique using a motion picture camera to record events at normal (960 frames per minute) or faster than normal camera speed. Used for the analysis of short, highly detailed movements that are too rapid for satisfactory visual observation. The camera may be driven so as to act as a timing device for the measurement of motions or elements, or there may be a timing device such as a microchronometer in the camera's field of view. *Syn:* camera study.

minimum time, The shortest actual time recorded during a time study for each element of work.

modal time, The actual time value for an element or operation that occurs more often than any other time value.

motion analysis, The study of the basic divisions of work involved in the performance of a given operation for the purpose of eliminating all useless motions and arranging the remaining motions in the best sequence for performing the operation. (See principles of motion economy.)

motion cycle, The complete sequence of motions and activities required to do one unit of work or to perform an operation once. (See cycle.)

motion economy, See principles of motion economy, motion, analysis.

motion study, See motion analysis.

multiple activity operation chart, See multiple activity process chart.

multiple activity process chart, A chart of the coordinated synchronous or simultaneous activities of a work system of one or more machines and/or one or more workers. Each machine and/or worker is shown in a separate, parallel column indicating their activities as related to the rest of the work system. Examples: multiworker process chart, Gantt chart, multiworker-machine process chart, worker-machine process chart, worker-multimachine process chart. *Syn:* multiple activity operation chart; multiple activity chart.

multiple watch timing, See accumulative timing.

nonballistic motions, See ballistic movement.

noncyclic element, An element of an operation or process that does not occur every cycle of the operation or process, but its frequency of occurrence in the operation or process is specified by the method.

nonrepetitive, (1) Generally an operation or process that is performed for only one or a few cycles before it has to be changed significantly to adapt to new requirements. (2) Odd-job production. (3) An operation that does not have a predictable order of elements. (4) An occasional and/or varying element, operation, or job.

normal effort, The effort required in manual work to produce normal performance. (See normal performance.)

normal element time, The selected (average, modal, or other) element time adjusted by performance rating to obtain the time required by an average qualified worker to perform a single element of an operation while working at a normal pace.

normal pace, The manual pace required to produce normal performance. (See normal performance.)

normal performance, (1) The work output of a qualified employee which is considered acceptable in relation to standards and/or pay levels, which result from agreement, with or without measurement, by management or between management and the workers or their representatives. (2) An acceptable amount of work produced by a qualified employee following a prescribed method under standard conditions with an effort that does not incur cumulative fatigue from day to day. (See fair day's work.)

normal task, See normal performance.

normal time, The time required by a qualified worker to perform a task at a normal pace to complete an element, cycle, or operation using a prescribed method. (See normal performance.) *Syn:* base time, leveled time.

normal working area, (1) The area at the workplace which is bounded by the arc drawn by the worker's fingertips moving in the horizontal plane, with the elbow as a pivot, when the worker is standing or seated in the normal working position with the arm close to the body hanging in a stationary position; the section where the right and left hands overlap in front of the worker constitutes the normal working area for the two hands. (2) In a vertical plane, the space on the surface of the imaginary sphere which would be generated by rotating about the worker's body as an axis, the arc traced by the worker's fingertips of the right or left hand when the forearm is moved vertically about the elbow as a pivot. (3) The space within reach of a worker's fingertips as they develop arcs of revolutions, the elbows acting as a pivot when the worker is standing or sitting in the normal working position with the upper arm hanging from the shoulder close to the body in a stationary position.

numbering system, (1) A plan for the assignment of numeric keys to items or cases included within a given classification. (2) Means of identifying individual or group arrangements.

numerical control, A system of controlling a unit, usually a machine tool, whereby either a binary or decimal digit system is programmed to carry out machining operations through electronic circuits and related activating mechanisms.

objective rating, See performance rating.

observation, (1) In time study, the act of noting and recording the elapsed time taken by a worker performing an operation or an element of an operation. (2) In motion study, the act of noting and recording the motions used by a worker to perform an operation or an element of an operation. (3) In work sampling, the act of noting and recording what a worker is doing or what is happening in an operation at a specific instant.

observed time, See actual time.

occurrence (frequency), (1) The number of times an event takes place, usually in a specific time period. (2) The number of times an element occurs per cycle.

one best way, The concept that for every job there is an optimal work method that can be developed and specified. A concept originated by Frank and Lillian Gilbreth.

operation, (1) A job or task, consisting of one or more work elements, usually done essentially in one location. (2) The performance of any planned work or method associated with an individual, machine, process, department, or inspection. (3) One or more elements which involve one of the following: the intentional changing of an object in any of its physical or chemical characteristics; the assembly or disassembly of parts or objects; the preparation of an object for another operation, transportation, inspection, or storage; planning, calculating, or the giving or receiving of information.

operations analysis, A study of an operation or series of operations involving people, equipment, and processes for the purpose of investigating the effectiveness of specific operations or groups so that improvements can be developed which will raise productivity, reduce costs, improve quality, reduce accident hazards, and attain other desired objectives.

operation breakdown, See job breakdown.

operation chart, See right- and left-hand chart.

operation process chart, A graphic, symbolic representation of the act of producing a product or providing a service, showing operations and inspections performed or to be performed with their sequential relationships and materials used. Operation and inspection time required and location may be included.

operation time chart, See operator process chart.

operator process chart, An operation process chart describing the work done by one operator. (See right- and left-hand chart.) *Syn:* operation time chart.

operator productivity, The ratio of standard time or other performance standard to the actual time or other performance measure for the same task. When this ratio is equal to 1.00 (100%) the operator is meeting standard output. (See performance index.) *Syn:* operator performance.

outlier, In a group of data, a value which is so far removed from the rest of the distribution that its presence cannot be reasonably explained by the random combination of chance causes. (See abnormal time.)

outside work, See external work.

pace rating, See performance rating.

Pareto's Law, Sometimes called the law of the trivial many and the critical few. A principle which states that, in most activities, a small fraction (commonly estimated at 20%) of the total activity creates the major portion (commonly estimated at 80%) of the work, cost, profit, or other measure of importance. *Syn:* rule of 80-20 (q.v.).

performance evaluation, A critical and objective appraisal of performance measurement data and related information to obtain an accurate picture of the overall status of a specific area or persons to ascertain exceptional accomplishments, identify shortcomings and their causative factors, and develop meaningful recommendations.

performance index, The ratio of a performance standard established for a certain quantity of work to the performance actually achieved. When this ratio is equal to 1.00 (100%) the worker or group is meeting standard performance. (See operator productivity.)

performance indicator, A significant quantitative measure of performance which provides the best perspective of total management effort being applied in an area.

performance measurement, The assessment of accomplishments in terms of historical or objective standards or criteria. (See performance evaluation.)

performance rating, (1) The process whereby an analyst evaluates observed operator performance in terms of a concept of normal performance. (2) The performance rating factor. *Syn:* leveling, pace rating, effort rating, objective rating.

performance rating factor, The number (usually a percentage) representing the performance rating.

performance rating scale, A numerical scale of performance which may or may not include defined benchmarks. For example, normal performance might be expressed as 100% or 60 minutes per hour. The 100% scale is the most common scale used.

performance ratio, See performance index.

performance sampling, A technique for determining the performance rating factor to be applied to an operator or a group of operators determined by short randomly spaced observations of the performance.

performance standard, A criterion or benchmark to which actual performance is compared.

personal allowance, An allowance to provide time for the personal needs of the worker during the workday. (See allowance.) *Syn:* personal time.

personal time, See personal allowance.

predetermined motion-time system, See predetermined time system.

predetermined time system, An organized body of information, procedures, techniques, and motion times employed in the study and evaluation of manual work elements. The system is expressed in terms of the motions used, their general and specific nature, the conditons under which they occur, and their previously determined performance times. *Syn:* predetermined motion time system.

predetermined time, See predetermined time system.

principles of motion economy, A general listing of common sense steps and procedures to simplify and improve the effectiveness of manual work.

process, (1) A planned series of actions or operations (e.g., mechanical, electrical, chemical, inspection, test) which advances a material or procedure from one stage of completion to another. (2) A planned and controlled treatment that subjects materials or procedures to the influence of one or more types of energy (e.g., human, mechanical, electrical, chemical, thermal) for the time required to bring about the desired reactions or results.

process chart, A graphic, symbolic representation of the specific steps in a processing activity. (See flow process chart, operation process chart, man process chart, flow chart, multiple activity process chart, operator process chart.)

process chart symbols, Graphical symbols or signs used on process charts to depict the type of events that occur during a process. (See flow process chart.)

process design, The act of prescribing the production process to produce a product as designed. This may include specifying the equipment, tools, fixtures, machines, and the like required; the methods to be used; the personnel necessary; and the estimated or allowed times. (See methods analysis, process.)

process engineer, An individual qualified by education, training, and/or experience to prescribe efficient production processes to safely produce a product as designed and who specializes in this work. This work includes specifying all the equipment, tools, fixtures, human job elements, and the like that are to be used and, often, the estimated cost of producing the product by the prescribed process. (See process, process design.)

process planning, A procedure for determining the operations or actions necessary to transform material from one state to another.

process sheet, A sketch, diagram or listing of the operations in the sequential order necessary to accomplish the desired result (such as transforming material from one state to another).

process time, (1) Time required to complete the machine or process-controlled portion of a work cycle. (2) Time required to complete an entire process.

processing, The carrying out of a production process. (See process.)

production standard, See standard time.

production study, (1) A detailed analysis of a job, operation, process, or group of activities using the techniques of methods engineering and work measurement with the objective of improvement. (2) An extended time study to determine delay allowances or verify other major variables — sometimes called an eight-hour study.

productive labor, See direct labor.

productive time, Time in which effective work is done in an operation or process, as opposed to nonproductive or idle time.

productivity, (1) The ratio of output to total inputs. (2) The ratio of actual production to standard production, applicable to either an individual worker or a group of workers.

productivity index, See productivity.

progress chart, A graphical representation of the status or extent of completion of work in process. (See Gantt chart.)

progress curve, A plot of work accomplished or productive output versus time. May be accompanied by a plot of expected or planned output for comparison purposes.

qualified operator, A worker who, by virtue of training, skill, and experience, is able to perform a task within acceptable quality and time limits.

random element, See foreign element.

random sample, A sample selected in such a way that each element of the population being sampled has an equal chance of being selected.

rate, (1) Hourly wage rate. (2) To evaluate the observed performance of a task in comparison with some concept of normal performance. (3) The quantity of output produced per unit of time. (4) The quantity of output produced expressed as a percent of either capacity or normal output. (5) Piece rate. (See performance rating.)

rate change, (1) An upward or downward adjustment of a production standard, generally made because of a revision in product design, quality requirements, production methods, materials, or conditions. (2) An upward or downward adjustment in wages paid per unit of time or unit of output.

rate cutting, The arbitrary reduction of a standard time or incentive pay rate. Not considered good practice.

rate setting, (1) The establishment of pay per unit for incentive work. (2) The establishment of a standard time. *Syn:* rate determination.

rated average element time, See normal element time.

rating, See performance rating.

ratio-delay study, See work sampling.

raw time, See actual time.

reading point, See breakpoint.

regular element, An element of an operation or process that occurs either every cycle of the operation or process or occurs frequently and in a fixed pattern with the cycles of that operation or process as, for example, once every third cycle or four cycles out of five.

relaxation allowance, See fatigue allowance, personal allowance.

repetitive element, See regular element.

repetitive timing, A stopwatch technique where a time value is read and recorded at each breakpoint and the watch is instantaneously reset to zero to begin timing the next element. *Syn:* snapback timing.

rest allowance, See fatigue allowance, personal allowance.

restricted element, See restricted work.

restricted work, Manual or human-machine work for which the pace or speed of work is not completely under the control of the worker. (See machine-controlled time.)

rework, (1) The process of correcting a defect or deficiency in a product or part. (2) Units of product requiring correction.

right- and left-hand chart, A chart on which the motions made by one hand in relation to those made by the other hand are recorded, using standard process chart symbols or basic therblig abbreviations or symbols. (See operator process chart.)

rule of 80-20, See Pareto's Law.

runout time, Time required by machine tools after cutting time is completed before the tool and material are completely free of interference so that the next sequence of operations can proceed.

select(ed) element time, See select(ed) time.

select(ed) time, The time which is chosen by simple observation or by statistical means as being representative of the actual time values (prior to applying a performance rating factor) obtained from the observation of an element or operation. (See average cycle time, average element time.)

sequencing, Specifying the order of performance of tasks so that available production facilities are utilized in an optimal manner.

setup, Preparation of a workplace or a machine for a specific work method, activity, or process. Includes installation of all necessary hand tools, jigs, fixtures, and other tools or equipment in the location and condition for proper performance of the work.

simo chart, (simultaneous motion chart), A chart for displaying two-handed work with motion symbols plotted vertically against time. The therblig or motion abbreviation and a brief description are shown for each activity. In addition, individual time values and body member detail may be shown. (See right- and left-hand chart.)

simplified practice, (1) The practices or operations resulting from a work simplification or methods study. (2) A description of the work method of a job, specified in somewhat less detail than in a standard practice.

simultaneous motions, Two or more nonconsecutive elemental motions performed during the same time interval by different body members.

skill, See job skill.

snapback timing, See repetitive timing.

speed rating, See performance rating.

standard, (1) An established norm for the measure of quantity, weight, extent, value, quality, or time. (2) Standard time.

standard allowance, An allowance calculated, arbitrarily set, or negotiated to provide in advance for specified conditions. (See allowance.)

standard data, A structured collection of normal time values for work elements codified in tabular or graphic form. The data are used as a basis for determining time standards on work similar to that from which the data were collected without making additional time studies. (See synthetic data.)

standard element time, A standard time for an individual work element. (See standard time.)

standard hour, The quantity of output required of an operator to meet exactly the production quota for one hour. The production quota is normally based on a standard time. Also used to refer to an hour of less than 60 minutes when allowances are expressed as nonproductive minutes. (See allowed hours.)

standard output, The reciprocal of standard time expressed in appropriate units (e.g., dozens of units per hour, tons per day, or hundreds of barrels per week).

standard performance, The performance of a person or group achieving standard output.

standard practice, A description of a work method wherein all of the significant variables of the method have been specified in detail. Usually follows a specified format. (See method.) *Syn:* standard method, written standard practice.

standard system, The common name for a codified set of time-motion data, often covering both general and proprietary data sets, considered the usual practice for a given plant or location and thus regarded as authoritative. (See predetermined time system.)

standard time, A unit time value for the accomplishment of a work task as determined by the proper application of appropriate work measurement techniques by qualified personnel. Generally established by applying appropriate allowances to normal time. Standard time and normal time are identical when nonproductive time is granted in lieu of allowances. (See normal performance.) *Syn:* direct labor standard, engineered performance standard, engineered standard, output standard, production standard, time standard.

standard time, statistical, A standard time developed from statistical analysis of past performance time data. *Syn:* historical time.

standard time data, See standard data.

standards audit, A work measurement study or sequence of studies intended to test the correctness of existing standard times and methods. By means of periodic sampling of work times, an attempt is made to detect significant changes.

standby, A category of time in which the worker is not actively engaged in producing a unit of output but is available to take appropriate action when needed. Standby is recognized only when it is essential to the task and when no other work can be done during the standby period. (See delay.)

standby time, The time expended in standby status, e.g., the time spent by workers in awaiting equipment, labor crews, or work assignment; or due to failure of utilities, inclement weather, and other similar occurrences.

startup curve, A learning curve applied to a job or process to adjust for work times longer than standard, or average, as a result of the introduction of a new job or new worker(s). (See learning curve.)

static work, Work performed by the hands or arms where no significant motion occurs, e.g., holding.

statistical time, See standard time, statistical.

stopwatch, A portable timing device that can be started or stopped at will by the user to register continuous and/or elapsed time. (See decimal-hour stopwatch, decimal-minute stopwatch.)

subtracted time, The difference between successive stopwatch readings when using a continuous timing technique. Usually represents the time for one element.

synchronization allowance, See interference allowance.

synthetic data, (1) Work measurement time values not obtained from direct measurement of the work to which they are applied. Generally represent values for task elements that are sufficiently basic as to occur in several jobs, obtained from measuring task elements in similar jobs or from predetermined time systems. (2) Any production data not measured directly from but applicable to a given situation. (See standard data, predetermined time systems.)

synthetic time standard, A standard time determined from synthetic data.

task, See job.

teardown, All work items required between the end of one operation or job and the start of setup for the next operation or job, both jobs requiring the same machinery or facilities.

temporary rate, (1) An output rate based on a temporary standard. (2) Wage incentive pay rate based on a temporary standard.

temporary standard, An approximate standard time intended to apply for a limited time to account for some unusual job condition or while awaiting restudy of the task to which it applies.

therblig, A short manual work segment used to described the sensory-motor activities or other basic elements of an operation. Developed by Frank and Lillian Gilbreth, therbligs form a basic language for methods description and, in modified form, for elemental motion time data. The original seventeen are: search, select, grasp, transport empty, transport loaded, hold, release load, position, pre-position, inspect, assemble, disassemble, use, unavoidable delay, avoidable delay, plan, rest for overcoming fatigue. *Syn:* Gilbreth basic element, basic division of accomplishment, fundamental motion, basic motion, basic element.

therblig chart, An operation chart with the suboperations broken down into individual motions, and all motions designated with their appropriate therblig symbols. *Syn:* right- and left-hand chart, simo chart.

thruput, The quantity of acceptable product processed through a manufacturing line or machine in a specified time (such as an eight-hour shift.)

tight rate, See tight standard.

tight standard, A standard time less than that required by a qualified workman with normal skill and effort following a prescribed method and including allowances for delays, personal needs, and rest. *Syn:* tight rate.

time allowance, See allowance.

time formula, A formula for determining the normal time or standard time of a task as a function of one or more variables in the task. Included are coefficients for the variables so that insertion of the variable values allows direct time computation.

time standard, See standard time.

time study, A work measurement technique consisting of careful time measurement of the task with a time measuring instrument, adjusted for any observed variance from normal effort or pace and to allow adequate time for such items as foreign elements, unavoidable or machine delays, rest to overcome fatigue, and personal needs. Learning or progress effects may also be considered. If the task is of sufficient length, it is normally broken down into short, relatively homogenous work elements, each of which is treated separately as well as in combination with the rest.

time study observation sheet, A form for the systematic, detailed recording of element time values, and irregular occurrences observed during a time study. Generally space is also provided for entering other pertinent information and for computation of standard times from the data. *Syn:* time study computation sheet, time study form.

tolerance, (1) A permissible variation in a characteristic of a product or process, usually shown on a drawing or specification. (2) In work measurement, the permissible variation of a time value for an operation or other work unit.

tolerance limits, (1) The upper and lower extreme values permitted by the tolerance. (2) In work measurement, the limits between which a specified operation time value or other work unit will be expected to vary.

training allowance, (1) An allowance to compensate for an untrained worker's learning effect. (2) An adjustment to an incentive pay piece rate for the same reason. (3) An adjustment to earnings of an individual or group for the same reason. Usually involves adjusting actual earnings under an incentive plan to some guaranteed percentage, or to prior average earnings. (See allowance.) *Syn:* learner's allowance.

travel chart, A table giving distances travelled between points in a manufacturing facility. Values may be adjusted to reflect weight, value, or some other factor depending on circumstances.

travel time, Time required to move material, equipment, personnel, or information from one work or storage area to another.

unavoidable delay, A delay which is outside the control or responsibility of the worker.

unavoidable delay allowance, An allowance intended to provide time for expected unavoidable delays in a task. (See allowance, unavoidable delay.)

unrestricted element, An operation element that is completely under the control of the worker. *Syn:* manual element.

unrestricted job, A job that is completely under the control of the worker.

value analysis, Review of product costs to evaluate contribution to product value. May include phases of work design, methods engineering, and motion economy to reduce manufacturing costs. May include phases of work simplification and brainstorming to evaluate use of alternate materials, components, or work specifications.

value engineering, See value analysis.

variable element, (1) An element whose normal time varies significantly from cycle to cycle as a function of one or more job variables. (2) An element common to two different jobs and whose time varies due to differences between the jobs.

waiting time, The time a unit waits for service or a worker waits for parts. (See standby.) *Syn:* idle time.

wink, One division on the microchronometer equal to 1/2,000 (.0005) minute.

wink counter, See michrochronometer.

work cycle, (1) A pattern or sequence of tasks, operations, and/or processes. (2) A pattern of manual motions, elements, operations, and/or activities that is repeated without significant variation each time a unit of work is completed. (See motion cycle.)

work design, The design of work systems. System components include people, machines, materials, sequence, and the appropriate working facilities. The process technology and the human characteristics are considered. Individual areas of study may include analysis and simplification of manual motion components; design of jigs, fixtures, and tooling; human-machine analysis and design; or the analysis of gang or crew work. *Syn:* ergonomics, job design, methods engineering, methods study, motion study, operation analysis, work simplification, motion economy.

work measurement, A generic term used to refer to the setting of a time standard by a recognized industrial engineering technique, such as time study, standard data, work sampling, or predetermined motion time systems. *Syn:* ergonometrics.

work sampling, An application of random sampling techniques to the study of work activities so that the proportions of time devoted to different elements of work can be estimated with a given degree of statistical validity.

work simplification, A management philosophy of planned improvement using any or all of the tools and techniques of industrial engineering in an atmosphere of creative participation which enables employees to achieve individual goals through the achievement of organizational goals. (See work design.)

work station, See workplace.

work study, The techniques of methods study and work measurement employed to ensure the best possible use of human and material resources in carrying out a specific activity.

work task, A specific quantity of work, set of duties or responsibilities, or job function assigned to one or more persons.

working area, That portion of the workplace within which an operator moves about in the normal job performance.

working conditions, The condition of the physical environment within which people work. The environment includes the presence and amount of illumination, heat, air movement and pollution, radiation, cleanliness, spaciousness, and safety. It may also include the conditions of the social environment, including type and intensity of supervision, emotional impact of the nature of the job, and opportunities for interaction with peers.

workplace layout, The manner in which all of the items necessary to perform a work task, as specified by the standard method, are arranged.

written standard practice, See standard practice.

Human Factors

13

Z94.13

This terminology is for anyone directly or indirectly involved in Human Factors work. For Human Factors practitioners there are specialty areas, and for those not fully engaged in the field there are many basic terms. Most physical quantities are not included since they are easily available in common reference works.

The Human Factors Subcommittee consulted military, industrial, governmental and academic sources to provide a balanced representation of the field of Human Factors. The authors welcome comments regarding achievement of the aim, comprehensiveness, clarity and accuracy.

The Subcommittee on Human Factors included:

CHAIRMAN:
Brian Crist, Eastman Kodak Company

MEMBERS:
Dr. Mary-Carol Cahill, Fordham University

Dr. Edwin Cohen, Singer Company

Mr. Terrence W. Faulkner, Eastman Kodak Company

Dr. Alan G. Hundt, SAI Comsystems Corporation

Dr. John V. Fechter, Honeywell, Inc.

Dr. Mark Ominsky, IBM

PREVIOUS CHAIRMAN:
Dr. Richard Lucas (formerly of Eli Lilly & Company), Eastman Kodak Company

Assisted by Colleagues of the chairman in the Human Factors Section of Eastman Kodak Company

BIBLIOGRAPHY

Following are the references consulted to find existing definitions of terms and terms or concepts for which definitions were developed. Because of the numerous modifications of definitions, the sources are not cited individually in the glossary.

Air Force Systems Command. Design Handbook, Series 1-0, General; AFSC DH 1-1, General Index and Reference. (3rd ed.) December, 1973.

American National Standards Institute, *Psychoacoustical Terminology,* ANSI S3.20-1973.

Anastasi, A. *Fields of Applied Psychology.* New York, McGraw-Hill, 1964.

Army Missile Command. Military Standard: *Human Engineering Design Criteria for Military Systems, Equipment, and Facilities*. (MIL-STD-1472A) Redstone Arsenal, Alabama. 1970.

Bard, P. *Medical Physiology.* (10th ed.) St. Louis, Mosby, 1956.

Bartels, H., Dejours, P., Kellogg, R. H., & Mead, J. Glossary on Respiration and Gas Exchange. *Journal of Applied Physiology,* April, 1973, *34*(4).

Bilodeau, E. A. *Principles of Skill Acquisition*. Academic Press, New York, 1969.

Blesser, A. *Systems Approach to Bio-medicine*, McGraw-Hill, 1969.

Chapanis, A. *Man-Machine Engineering*. Wadsworth Publishing Co., Belmont, CA 1965.

Chapanis, A. *Research Techniques in Human Factors Engineering*. Baltimore, Johns Hopkins, 1959.

Chapanis, A., Garner, W. R., & Morgan, C. T. *Applied Experimental Psychology: Human Factors in Engineering Design*. New York, Wiley, 1949.

Damon, A., Stoudt, H. W., & McFarland, R. A. *The Human Body in Equipment Design*. Cambridge, Mass., Harvard University Press, 1966.

DeGreene, K. B. *Systems Psychology.* New York, McGraw-Hill, 1970.

Department of Defense: *Definitions of Effectiveness Terms of Reliability, Maintainability, Human Factors and Safety.* MIL-STD-721B, 25 August 1966.

Fogel, L. J. *Biotechnology: Concepts and Applications*, Prentice-Hall, Inc., Englewood Cliffs, N.J., 1963.

Gagne, R. M., & Fleishman, E. A. *Psychology and Human Performance*. New York, Holt, Rinehart & Winston, 1959.

Gagne, R. M. (Ed.) *Psychological Principles in System Development*, Holt, Rinehart and Winston, New York, 1962.

Kubokawa, C. *Databook for Human Factors Engineers*. Man Factors, Inc., 1969.

McCormick, E. J. *Human Factors Engineering*. (3rd ed.) New York, McGraw-Hill, 1970.

Meister, D., & Rabideau, G. F. *Human Factors Evaluation in System Development*. New York, Wiley-Interscience, 1965.

Morgan, C. T., A. Chapanis, J. S. Cook, III, and M. W. Lund, (Ed.) *Human Engineering Guide to Equipment Design*. U.S. DOD Joint Services Steering Committee, McGraw-Hill, New York 1963.

Murrell, K. F. H. *Human Performance in Industry.* (2nd ed.) New York. Reinhold, 1969.

Parsons, H. M. *Man-Machine System Experiments*, The Johns Hopkins Press, Baltimore, 1972.

Post Office Department. *The Post Office Challenge to Industry.* Section VI. *Contractors Guide to Terminology,* 1967.

Shurtleff, D. *Design Problems in Visual Displays.* ESD-TR-66-299, 1966.

Stevens, S. S. (Ed.) *Handbook of Experimental Psychology.* New York, Wiley, 1951.

Tiffin, J., & McCormick, E. J. *Industrial Psychology.* (5th ed.) Englewood Cliffs, N.J., Prentice-Hall, 1965.

VanCott, H. P. and Kinkade, R. G. *Human Engineering Guide to Equipment Design*. Washington, D.C.: American Institutes for Research, 1972. For sale by Superintendent of Documents, U.S. Gov. Printing Office.

Woodson, W. E., & Conover, D. W. *Human Engineering Guide for Equipment Designers*. (2nd ed.) Berkeley, University of California Press, 1970.

Woodworth, R. S., & Schlosberg, H. *Experimental Psychology.* (Rev. ed.) New York, Holt, 1954.

abduction, Movement of a limb away from midline axis of body.

absolute pitch, Refers to the ability of a person to identify the pitch of a pure tone without any external reference.

absolute threshold, That stimulus value which marks the transition between no response and response on the part of an observer; occasionally used analogously to describe a qualitative change in the nature of the response, as the "threshold of feeling" between sound and pain at intense levels of acoustic stimulation; or the "pricking pain threshold" between perceived warmth and burn. Values obtained for the threshold depend on the statistical properties of the psychophysical method used. (See lower threshold.)

accessibility score, A rating of the passage space available for a person to reach his work station based on the width of the passageway. The rating is multiplied by the number of people affected and averaged across people.

accident, An event that is not planned or anticipated. The event is preceded by preventable act(s) and/or unsafe condition(s). The usual result is personal injury, property damage, work stoppage or interference, or any combination.

accommodation, The adjustment of a sense organ to receive an impression distinctly. In vision, the ability of the eye to focus on objects at varying distances effected by the gradual thickening and increasing curvature of the crystalline lens as objects become closer.

acoustic scattering, The irregular reflection, refraction, or diffraction of a sound in many directions.

acoustic (or auditory) stimulus, The waveform of air pressure at the eardrum.

acoustic trauma, Injury to the ear caused by an intense auditory stimulus resulting in a certain degree of temporary or permanent hearing loss.

activity analysis, A structural listing of the operations, behaviors, or tasks that make up a job or work. Data gathered typically includes the sequence, frequency, and importance of activities and the time devoted to each, as well as their interrelationships. Various procedures and formats are used, according to the purpose of the analysis. (See activity sampling.)

activity sampling, Performing an activity analysis by observing only portions or segments of the ongoing work process, with observation periods being scheduled in advance according to a sampling rationale. (See activity analysis.)

adaptation, sensory, The maintenance of sensory effectiveness under changing stimulation (positive adaptation) or the reduction of sensory responsiveness with continued stimulation (negative adaptation); for example, maintenance of visual acuity under reduced illumination, or reduced responsiveness to an unpleasant odor.

adduction, Movement of a limb toward midline axis of body.

advisory light, A signal indicating safe or normal performance, operation of essential equipment, or to alert an operator for routine action purposes.

afterimage, A visual image or other sense impression that persists after the stimulus is no longer operative. Visual after-images are complex perceptions that can vary in hue, brightness, saturation, shape, pattern, texture, focus, latency, duration and developmental sequence.

anechoic room, A room in which the boundaries absorb almost all incident sound resulting in virtual free field conditions.

animated-panel display, Diagrammatic representation of components of a system hardware and/or simple semi-functioning models of such components.

anoxia, Lack of oxygen. Extreme condition of hypoxia. Anoxia usually results in permanent damage.

anthropometry, Classically, the study of people in terms of physical dimensions. (1) The measurement of static and dynamic features or characteristics of the human body. Typically included are linear and arc dimensions, mass and volume. (2) The techniques used to quantitatively express the form and dimensions of the body. (See Anthropometry.)

apparent motion, An illusion of motion brought about by certain patterns of non-moving stimuli. (See autokinetic effect, phi-phenomenon.)

articulation index, A numerically calculated measure of the intelligibility of speech in the presence of background noise. It is generally based on the speech/noise ratio for each of twenty frequency bands. This ratio is weighted, multiplied by .05, and summed over the twenty bands to yield the AI which ranges from 0.0 (voice communication nearly impossible) to 1.0 (excellent communication).

artificial horizon, A pictorial display used in aircraft to provide information as to the orientation of the plane with respect to the horizontal, indicating to the pilot whether he is flying straight and level.

audiogravic illusion, Errors in auditory localization which accompany an error in the perception of body position.

audiogyral illusion, Errors in auditory localization following rapid rotation of the body in the absence of visual cues.

audiometer, Instrument used to assess human auditory sensitivity as a function of frequency. Tones of varying frequencies and intensities are presented through earphones to a subject who indicates the point at which a given tone just becomes, or ceases to be, audible.

auditory after-effect, A phenomenon lasting several seconds in which familiar sounds are modulated following listening to rapid, high-intensity pulses for about one minute.

auditory fatigue, A temporary increase in auditory threshold due to a previous auditory stimulus.

auditory flutter, A wavering auditory sensation resulting from the periodic interruption of a continuous sound at a sufficiently slow rate.

auditory fusion, A phenomenon in which a series of short duration primary sounds with successive arrival times at the ear(s) produce the sensation of a single secondary sound.

auditory lateralization, The determination by a subject that the apparent direction of a sound is either left or right of the frontal-medial plane of the head.

auditory localization, The determination by a subject of the apparent direction and/or distance of a sound.

autokinetic effect, An illusion of the apparent movement of a stationary point light source presented in a dark stimulus field in the absence of a visual frame of reference.

availability, measure of, The ratio of the total time a system is capable of performing its function to the total time that there is a requirement for its operation.

"average man" concept, An oversimplified method of describing the characteristics of a varied population based on mean measurements. When more than one measurement is used the "average man" disappears.

backlash, Tendency for control system response to be reversed when a control movement is stopped.

beats, Periodic variations in sound resulting from hearing two equally intense simple tones of slightly different frequencies. As the difference between the frequencies is increased, a train of subjective experiences is produced, beginning with waxing and waning and progressing to throbbing (beats), then changing to rattling or roughness. That stage lasts until the two frequencies are separated enough that the subjective interaction disappears and two distinct tones are heard.

beginning spurt, A transient burst of activity at the beginning of the work period.

binocular parallax, The lack of correspondence between the images which fall on each retina due to the interpupillary distance between the two eyes; a cue to depth. (See parallax.)

biomechanics, Application of mechanical principles and methods of analysis to the structure and movements of the body and body parts. Frequently-studied aspects include the range, strength, endurance, and speed of movements and responses to such physical forces as acceleration and vibration. (See Biomechanics.)

bionics, A study of living systems or organisms with the intention of deriving technological knowledge applicable to the design and performance of artificial systems.

bit, An information processing unit. (See Data Processing and Systems Design.)

blind-positioning reactions, Movements that require the localization of objects in space without the aid of visual cues.

BTPS conditions, The temperature and pressure of gas in the lungs at body temperature (37°C) and saturated with water (47 torr or 6.266 134 K Pa) at ambient pressure. (See STPD conditions.)

central hearing loss, Hearing impairment following damage in the auditory pathways or the auditory areas of the brain.

channel capacity, By analogy to engineering usage, the concept of a theoretical upper limit on the amount of information that human beings can receive and process, usually within a given time. The term is most frequently applied to the sensory modalities and/or the central nervous system.

choice reaction time, Amount of time required to make different responses to different stimuli. The time increment above what is needed for a response to a single stimulus is sometimes taken as a measure of the time to process the cognitive choice. (See reaction time.)

clo unit, A measure of the thermal insulation provided by clothing.

compatibility, How consistently the spatial movement or conceptual relationships of stimuli and responses meet human expectations. With reference to controls and displays, the naturalness of the movement of a control compared to what is displayed to the operator.

compensatory display, In tracking tasks, a display on which either the input signal (target) or the output signal (cursor) is fixed and the other moves; the operator's task in manipulating the controls is to eliminate or minimize system error by superimposing the cursor onto the target.

concept trainers, Aids used when the concepts to be learned are too complex to be absorbed in a few trials from verbal descriptions, and when the principles to be used in task performance can best be simulated by physical objects and real actions such as miniature representation of control components, and check points of a complex electronic system.

conductive hearing loss, Hearing impairment due to interference with the transmission of sound waves through the external or middle ear.

cones, Visual receptor cells found most densely toward the center of the retina. They are responsible for sharp vision and color discrimination, but cease functioning at low light intensities.

consonance, The phenomenon in which tones presented simultaneously produce a blended or pleasant sensation.

contingency analysis, The part of a task analysis performed to identify non-routine situations with which a system may have to deal, in order to determine any special human performance required by such events.

contrast, The brightness relationships of two non-specular surfaces adjacent to each other when the illumination of both the objects and the immediate surroundings are the same.

control, A device, usually mechanical, electrical, or electronic, which directs the action of some mechanism or produces some change in the operation of a process or system.

control-display ratio, The ratio of movement of a control device to the movement of a display indicator.

control-force curve, A function describing the relationship between the amount of force applied by the operator and the resulting displacement of a control.

convergence, Coordinated directing of both eyes at the same point so as to obtain single vision by bringing the image of the object to the fovea of each eye.

coriolis effect, The difference in effective G force at different distances from the rotation axes in a short-axis rotating environment. When limbs or head move across a G-force gradient, misperceptions of body orientation (tilting, rotation) or of limb position may be experienced, with concomitant physiological effects such as nausea or vertigo.

critical flicker frequency, The minimum number of alternations per second at which a regularly intermittent light ceases to appear as flickering and fuses to yield a perception of steady light. Also called critical fusion frequency or flicker-fusion frequency.

critical incident technique, An approach to job analysis which calls for factual description of specific events or behavior. The technique is also applicable to other situations, such as accident investigation in which reports of errors or near-accidents may be collected.

cursor, A moving element on display. On an instrument, it is used to show position discrepancy or error in reference to another "target" element. On a CRT display, it is used to show row and column positions or to indicate where the next activity will occur.

cybernetics, Study of communication and control functions common to both living and engineered systems with the emphasis on attempting to understand organisms through making analogies to machines. ("I made up the word . . . from the Greek . . . 'Steersman' . . . from the same root as 'governor' and refers to control and communication in the animal and the machine." — Norbert Wiener.)

dark adaptation, A process of neuro-chemical changes whereby the visual system becomes much more sensitive to light.

dazzle, Extreme brightness of direct or reflected light that causes difficulty in seeing.

dead band, Loci of control positions, usually around the center position of a knob or lever, for which there is no control output.

dead man control, Device for shutting off equipment or stopping a vehicle in the absence of the kind of force exerted by a conscious operator.

decibel, A logarithmic scale expressing quantities of sound, or electrical power. Measurement is relative to a specified reference level.

decision time, Delay attributable to cognitive processing of input data.

Delphi method, A procedure for obtaining the consensus of experts in areas such as long-range forecasts by successive iterations of questioning interspersed with feedback to each expert on the others' opinions and supporting reasons.

depth perception, Visual discrimination along the dimension of distance from an observer, using monocular cues (e.g., interposition, accommodation, size) and/or binocular cues (convergence, stereopsis).

describing function, Mathematical model of human performance, especially in tracking tasks such as aircraft piloting.

detectability, Quality of a signal, display, or stimulus that affects the probability of its presence being perceived.

dichotic, Pertaining to differential stimulation of the two ears.

diotic, Pertaining to the stimulation of the two ears by identical stimuli.

difference limen, Difference threshold — the amount by which two supraliminal stimuli must differ if this difference is to be perceived in an exactly specified portion (usually 50 or 70%) of trials. (See limen.)

display, The presentation of output data from a device or system in a form designed for human perception through vision, audition, or other sensory modality.

dynamic visual acuity, The measure of visual acuity used when a target moves with respect to an observer. (See visual acuity.)

dynamometer, A device for measuring the strength of a muscular contraction; typically a hand-grip device with a scale showing squeeze force.

empty field myopia, The normal accommodation of the eye for near, rather than far, vision when viewing a homogeneous field such as an empty sky.

energy expenditure, The amount of energy used during activity or rest. Usually expressed in calories per unit time.

engineering psychology, One of several terms used to define approximately the same discipline. Other terms are human factors, human factors engineering, and ergonomics. Ergonomics is used predominantly outside the U.S.A.; the others predominantly within. The aim of the discipline is the evaluation and design of facilities, environments, jobs, training methods and equipment to match the capabilities of users and workers. Exact definition of these terms of the scope or work is improbable. (See ergonomics.)

ergometer, A device for measuring the energy output of human muscular work; for example, a stationary bicycle.

ergonomics, One of several terms used to define approximately the same discipline. Other terms are engineering psychology, human engineering, human factors, and human factors engineering. Ergonomics is used predominantly outside the U.S.A.; the others predominantly within. The aim of the discipline is the evaluation and design of facilities, environments, jobs, training methods and equipment to match the capabilities of users and workers. Exact definition of these terms or the scope of work is improbable.

error, constant, Deviation from true value or established standard due to factors remaining unchanged during the series of measurements.

error, variable, Deviation from true value or established standard due to random factors that affect each measurement separately.

face validity, Relevance or appropriateness of a measure.

fail safe, Characteristic of system that avoids injury to humans and/or catastrophic damage when an element malfunctions.

fatigue, Decreased ability to perform and/or feelings of tiredness following effort.

fault localization, Systematic procedure for identifying malfunctioning element of a system.

fault tree analysis, A logical approach to identify the areas in a system that are most critical to uninterrupted and safe operation.

feedback, Data on the performance of a system, equipment, group, or individual for use in modifying performance.

fidelity, The degree to which a system's output duplicates the input.

field of view, The solid angle within which an optical sensor (such as the human eye) provides useful data.

flash blindness, Temporary loss of vision for low-luminance objects following brief exposures to intense light.

flight trainer, A ground-based device containing a representation of an aircraft cockpit, used for pilot training. It need not have the fidelity of a flight simulator.

flight simulator, A ground-based device containing a representation of an aircraft cockpit with a computer driving functional displays and controls. It may be used for training, for research, or for engineering development.

flybar (flying by auditory reference), A system providing turn, bank, and air speed information via auditory signal, rather than with conventional flight instruments.

g-tolerance, A tolerance in a person or animal, or in a piece of equipment, to an acceleration of a particular value and direction.

gain, (1) An increase in signal power in transmission from one point to another. Gain is usually expressed in decibels and is widely used to denote transducer gain. (2) An increase or amplification. (3) More generally, the sensitivity or degree of control provided by any manual adjustment.

ganged controls, Controls which are grouped or "stacked" on a single axis so as to reduce crowding on panels of limited dimensions. For example, two control knobs attached to concentric shafts.

glare, A sensation produced by a luminance in the visual field that is sufficiently greater than the luminance to which the eyes are adapted. Glare may cause annoyance, discomfort, disability, distraction or a reduction in visibility.

go, no-go display, A visual display which provides only two alternate choices of information (e.g., on-off, start-stop, etc.).

grayout, A temporary condition in which vision is hazy, restricted, or otherwise impaired, owing to insufficient oxygen.

ground, The unfocused surroundings and interstices of a figure or object, perceived as lying beyond and not belonging to the figure or object, e.g., the background in a painting. Figure and ground are sometimes reversible, as when an interwoven black-white pattern may appear either as a white figure on a black background, or vice versa.

heart rate, A physiological measure that is used as an estimate of job stress, work load, or environmental stress. Usually expressed in beats per minute.

heat stress, Additional physiological load induced by working in a hot environment. The effects of heat stress include increased heart rate and increased body temperature, and may also include increased sweat rate and feelings of fatigue. Severe heat stress can result in heat exhaustion or heat stress.

heat stress index, Any of several methods that attempt to combine some or all of the interactive effects of temperature, humidity, air velocity, work load, and clothing into a single measure predictive of physiological effect.

heuristic program, A set of instructions that imitates the behavior of human operations (i.e., response modification based on previous, current and anticipated conditions which are not pre-planned).

hue, The attribute of color determined primarily by the wavelength of light entering the eye. Spectral hues range from red through orange, yellow, green, and blue to violet.

human engineering, One of several terms used to define approximately the same discipline. Other terms are human factors, human factors engineering, and ergonomics. Ergonomics is used predominantly outside the U.S.A.; the others predominantly within. The aim of the discipline is the evaluation and design of facilities, environments, jobs, training methods and equipment to match the capabilities of users and workers. Exact definition of these terms or the scope of work is improbable. (See ergonomics.)

human factors, One of several terms used to define approximately the same discipline. Other terms are human engineering, human factors engineering, and ergonomics. Ergonomics is used predominantly outside the U.S.A.; the others predominantly within. The aim of the discipline is the evaluation and design of facilities, environments, jobs, training methods and equipment to match the capabilities of users and workers. Exact definition of these terms or the scope of work is improbable. (See ergonomics.)

human operator, A person who participates in some aspect of operation or support of a system and its associated equipment and facilities. (Generally refers to one who operates equipment as opposed to one who maintains the equipment.)

human transfer function, The mathematical description of a human operator's outputs (e.g. control movement) in a tracking task as a function of his inputs (display indications), described in terms of a linear differential equation.

hypoxia, An insufficiency of the oxygen supply to living tissues relative to their existing requirement, such as may occur with reduced atmospheric pressure at high altitudes. The central nervous system and heart are most vulnerable: marked cardiovascular and behavioral symptoms may develop over a period of minutes, and include disorientation, dizziness, dimness of vision, and breathlessness, progressing to unconsciousness and collapse ("blackout").

impulse, (1) The product of a force and the time during which the force is applied. (2) Psychology; human response based more on emotion than cognition.

impulse noise, Noise generated in discrete energy bursts, which has a characteristic wave shape of its own.

incremental transfer effectiveness, A measure of effectiveness that compares successive increments of time spent in one training task with successive increments saved in subsequent training.

indicator, An instrument or device for displaying information; e.g. over-limit, location, or speed. Indicators may be mechanical (symbolic or pictorial), electrical (warning lights), or electronic (cathode-ray tube or screen).

information, In information theory, a purely quantitative property of an ensemble of items that enables categorization or classification of some or all of them. The amount of information in an ensemble (symbolized by H) is measured by the average number of operations (statements, decisions, tests, etc.) needed to effect categorization of the items.

information theory, An interdisciplinary study dealing with the transmission of messages or signals, or the communication of information. It draws upon communications theory (which includes much from physics and engineering), linguistics, psychology, and sociology.

ingress, Access for entering an area, such as an operating or passenger station within a vehicle or work area.

injury, Generally any physical harm or damage to a person. Section 1 defines a disabling injury as one which prevents a person from performing his regular job for a full day beyond the day of the accident which produced the injury. In OSHA usage, an injury may or may not involve loss of time from work.

input, (1) The path through which information is applied to any device. (2) The means for supplying information to a machine. (3) Information or energy entering into a system. (4) The quantity to be measured, or otherwise operated upon, which is received by an instrument. Also called input signal.

integrated controller, A control device which combines more than one aspect of an operation (e.g., control of steering, acceleration and braking in a single joystick).

integrated display, A visual display which combines several related parameters into a single function format.

interface, (1) A common boundary between two parts of a system, whether material or non-material. (2) The physical boundary between an operator and the equipment used, e.g., control, display, seat, etc.

international system of units, See SI Units.

job analysis, A study based upon measurement, observation, and interviews to determine and identify the duties, tasks, and functions involved, together with the skills, knowledge, and responsibilities required. (See Applied Psychology and Wage and Salary Administration.)

job enlargement, The addition of duties to a job. It may include vertical job enlargement which means adding the planning, controlling, and decision-making responsibilities or it may include horizontal enlargement which means adding a wider span of operations to the job. Job enlargement, sometimes called job enrichment, is often considered to add variety, interest, and significance for the worker. (See Organization Planning and Theory.)

just noticeable difference, The least amount of a stimulus which, added to or subtracted from a standard stimulus, produces a detectably different experience. Also called just perceptible difference, least noticeable difference, minimal change.

kinesthesis, The sense that yields knowledge of the movements of the body or of its members; awareness of movement due to the mechanical stimulation of special receptors in the muscular tissue, tendons, and joints.

knowledge of results, Feedback of information about performance.

landolt ring, A ring with a small gap at one point, used to test visual acuity by having observer report orientation of the gap.

lateral, Of or pertaining to the side or lateral axis; directed, moving or located along, or parallel to, the lateral axis.

learning, Any change in behavior that occurs as a result of training, practice, or experience.

learning, trial-and-error, The process in which a person or animal, having no already established response to the requirements of the task, responds at first to only the tasks general features with a wide variety of acts, then gradually eliminates the responses that prove unsatisfactory and repeats with increasing frequency those that prove satisfactory.

learning, whole vs. part, In "whole learning," all of the material to be learned is worked through in successive repetitions from first to last; in "part learning" (or "piecemeal learning"), the material is broken into smaller segments to be learned separately and then combined into the whole.

life support, An area of human factors which focuses on health promotion, biomedical aspects of safety, protection, sustenance, escape, survival, and recovery of personnel.

light adaptation, A process of neuro-chemical changes whereby the visual system becomes less sensitive to light.

limen, Threshold; a psychophysical concept denoting the lowest detectable intensity of any sensory stimulus.

link analysis, An analysis of the visual, auditory, and tactual links between man and machine or between one man and another involved in an operation. Primary objectives are determination of the importance of links, frequency of their use, and their adequacy.

load, The demand made by inputs on performance.

load, sensory, The number and variety of stimuli to which responses must be made. For example, if the operator must discriminate among several different classes of visual stimuli, the load on the visual system is greater than if discrimination is of only one type or of limited types.

localization, auditory, The capability of an observer to identify the position of a sound source with reference to himself.

loudness, The attribute of auditory sensation by which sounds may be ordered on a scale extending from soft to loud. The unit of loudness is the sone.

loudness contour, A graph of sound pressure level versus frequency, showing the sensitivity of the ear.

loudness level, The (judged) loudness level of a sound, in phons, is numerically equal to the median sound pressure level, in decibels relative to 20 micronewtons per square meter, of a 1000 hertz reference tone presented to subjects facing the sound source and judged by the subjects in a number of trials to be as loud as the sound under test.

lumen, A unit of luminous flux equal to the luminous flux radiated into a unit solid angle (steradian; sr) from a point source having a luminous intensity of 1 candela (cd). SI symbol: lm; formula: cd • sr.

man-machine system, A system in which the functions of the man and the machine are interrelated and necessary for the operation of the system.

manual, Operated by people rather than automatically.

masking, Masking is the amount by which the threshold of audibility of a sound is raised by the presence of another (masking) sound. The unit customarily used is the decibel.

medial plane, The medial plane is a vertical plane that divides the body into left and right. It is perpendicular to the frontal plane.

memory, (1) Recall and recognition of anything previously learned or experienced. (2) The component of a computer, control system, or the like, designed to provide ready access to data or instructions previously recorded.

metabolic reserves, The energy source stored in chemical form, such as carbohydrates, that can be efficiently mobilized and utilized by the body, particularly for muscular activity and work beyond the normal level of activity of an individual.

metabolism, The sum of all the physical and chemical processes by which living organized substance is produced and maintained; also the transformation by which energy is made available for the uses of organisms.

method of adjustment, A psychophysical method used primarily to determine thresholds; in this procedure the subject varies some dimension of a stimulus until that stimulus appears equal to or just noticeably different from a reference stimulus.

method of constant stimuli, A psychophysical method used primarily to determine thresholds; in this procedure a number of stimuli, ranging from rarely to almost always perceivable (or rarely to almost always perceivably different from some reference stimulus), are presented one at a time. The subject responds to each presentation: "yes-no," "same-different," "greater than-equal to-less than," etc.

method of equal sense distances, A psychophysical method used primarily to scale sensations; in this procedure the subject adjusts a set of stimuli until the elements of the set appear to be equidistant along some dimension.

method of limits, A psychophysical method used primarily to determine thresholds; in this procedure some dimension of a stimulus, or of the difference between two stimuli, is varied incrementally until the subject changes his response.

method of magnitude estimation, A psychophysical method used primarily to scale sensations; in this procedure the subject assigns to a set of stimuli numbers which are proportional to some subjective dimension of the stimuli.

method of magnitude production, A psychophysical method used primarily to scale sensations; in this procedure the subject adjusts a stimulus along some dimension until the magnitude of the stimulus appears equal to some specified magnitude.

method of paired comparisons, A psychophysical method used primarily to scale responses; in this procedure stimuli are presented in pairs to a subject who compares them along some dimension.

method of rank order, A psychophysical method used primarily to scale sensations; in this procedure stimuli are presented to a subject and he orders them along some dimension.

method of ratio production, A psychophysical method used primarily to scale sensations; in this procedure the subject adjusts a stimulus along some dimension until that stimulus appears to be a specific fraction or multiple of a reference stimulus.

method of single stimuli, A psychophysical method used primarily to scale sensations; in this procedure stimuli are presented singly to a subject who rates them along some dimension.

mets, A functional energy expenditure classification, using multiples of the resting metabolic value. Thus, 2 mets is two times resting metabolism.

minimum separable acuity, Smallest space between two lines that can be discriminated as a gap. It is measured in terms of the angle subtended by the gap, measured at the eye.

monitor, To observe, listen in on, keep track of, or exercise surveillance over by any appropriate means, as, to monitor radio signals; to monitor the flight of a rocket by radar; to monitor a landing approach.

motion parallax, The apparent difference in rate of movement of two objects actually moving at the same velocity but at different distances from the observer.

motion and time study, A systematic study of work systems, which, according to Ralph Barnes, have the purposes of: (1) developing a preferred system and method (usually one with lowest cost); (2) standardizing this system and method; (3) determining the time required by a qualified and properly trained person working at a normal pace to perform a specific task or operation; (4) assisting and training a worker in the preferred method. Motion study (or methods design), finding the preferred method of doing work. Time study (or work measurement), determining standard time for performing a specific task. Taylor used the term time study almost indiscriminately — including what the Gilbreths called motion study — much to their chagrin, especially when he took time studies without first studying the "one best way." (See Work Methods and Measurement.)

motor skill, The ability to achieve the more or less complicated adjustments of hands, fingers, legs, feet, or other parts of the body in an integrated, smoothly flowing sequence resulting in the performance of some act.

negative transfer, A mental process occurring when past learning interferes with new learning. This can occur when the stimuli in the new situation are similar to the stimuli in the old situation, but different responses are required. It can also occur when the stimuli in a new situation are different from the stimuli in the old situation, but the responses are the same.

neuromuscular, Pertaining jointly to nerves and muscles, as neuromuscular junction.

night vision, Vision which occurs in faint light, or after dark adaptation. Sometimes called twilight or scotopic vision. Hues and saturations cannot be distinguished.

noise, (1) Unwanted signal(s) that typically interfere with wanted signals. Noise may be audible (voice communications), or visible (radar, hard copy). (2) Unwanted energy (or the voltage produced), usually of random character, present in a transmission system due to any cause.

noise criterion (NC) curves, Any of several versions (SC, NC, NCA, PNC) of criteria used for rating the acceptability of continuous indoor noise.

nystagmus, An involuntary oscillation of the eyeballs, especially occurring as a result of eye fixations and stimulations of the inner ear during rotation of the body.

open loop, A control system to which there is no self-correcting action, operating without feedback, or with only partial feedback.

operations research, A mathematical modeling procedure aimed at discovering how, under the necessary conditions, requirements, and constraints, a system can best accomplish what it is designed or expected to do. Beginning with a clear statement of objectives, it involves identifying the relevant variables, expressing them in quantitative terms, measuring their fluctuations in given situations, and discovering how they interact. The system may be a machine, a social organization, or a combination of man and machine. (See Applied Mathematics.)

operator inputs, In a man-machine system, information received (sensed) by the operator from instructions or displays or directly from the environment.

operator outputs, In a man-machine system, the action taken by the operator; the result of a decision based on the input, e.g., manipulation of controls and verbal communication.

operator overload, Condition in which an operator is required to perform more functions than he is able to handle effectively within given time limitations, such as identifying signals, processing information, making decisions, etc.

optimum-location principle (equipment design), The principle of arrangement such that each display and control should be placed in its optimum location in terms of some criterion of use (convenience, accuracy, speed, force to be applied, etc.).

oxygen consumption, Rate of oxygen use of organisms, tissues, or cells.

parallax, The difference in the apparent direction or position of an object when viewed from different points. (See binocular parallax.)

past task simulation, That type of simulation which gives the subject an opportunity to learn selected aspects of the total task.

perception, (1) Perception is the process by which sensations are interpreted. (2) The awareness of external objects, qualities, or relations, which ensues directly upon sensory processes.

perceptual-motor task, A task involving an overt movement response to a nonverbal stimulus situation; the response is determined by the organization of sensory cues within the individual.

perceptual skill, Relative ability to detect and interpret information received through the sensory channels; contrast with motor skill.

performance decrement, A decrease in human proficiency often attributable to operator overload, stress, or fatigue, and characterized by increasing errors, misjudgments, omission of task elements, reduced intensity of effort, etc.

performance measures, Objective and subjective measures developed to evaluate personnel effectiveness; objective measures include productivity, job knowledge tests, job performance samples, proficiency tests and checklists; subjective techniques include peer ratings, supervisory ratings, and self-ratings.

peripheral vision, "Side" vision, primarily sensed by the rod cells. May produce an orienting reflex whereby the head turns and the visual object is focused on the fovea for optimal perception.

personnel subsystem (PS), The segment of system management which provides the human performance necessary to operate, maintain, support, and control the system in its operational environment.

phon, The unit of loudness level of sound, numerically equal to the sound pressure level in decibels, relative to 20 μ N/m^2 $\times$ (20 μ Pa) of a simple 1000 Hz tone judged by listeners to be equivalent in loudness. Compare sone.

population stereotype, A commonly anticipated behavioral sequence, or "the way most people expect something to be done." For example, when a control is rotated clockwise, it is expected that the value of the setting will increase. A design where the value decreases with clockwise rotation would be a violation of the population stereotype.

photopic adaptation, The decreased visual sensitivity to light, sometimes manifested by decreased perceived brightness of a fixed stimulus.

photopic vision, Vision associated with levels of illumination (luminance) about 30 cd/m^2 or higher, characterized by the ability to distinguish colors and small details. Also called foveal vision.

position, tracking, A tracking task in which movement of the operator's control is associated with a direct displacement or movement of the tracking indicator; a linear relationship is implied between system error and control movement.

positioning movement, primary, The first movement an operator makes in positioning a control; movement which carries the body member approximately to the point of aim; also called "gross adjustment" or "slewing."

positioning movement, secondary corrective, That part of the positioning movement made to bring the body member into exact relation with the point of aim; also called "fine adjustment."

predictor display, Shows the operator some aspect of what will happen to the system in the future. Such displays may be pursuit or compensatory, symbolic or pictorial; and they may predict the vehicle output, the system input, or both. The distinguishing feature is that they provide advanced information that allows the operator to anticipate future requirements for control movements.

proactive inhibition, The state or process hypothesized to account for the lessened ease of learning of the later members of a series following learning of an earlier member.

proprioceptor, Any receptor sensitive to the position and movement of the body and its members, including (1) receptors in the vestibule of the inner ear and in the semicircular canals sensitive to the orientation of the body in space, and to bodily rotation, and (2) receptors in the muscles, tendons, and joints giving rise to kinesthetic sensations.

psychometrics, The measurement of psychological processes through application of mathematical and statistical techniques.

psychomotor ability, Of or pertaining to muscular action ensuing directly from a mental process, as in the coordinated manipulation of aircraft or spacecraft controls.

psychomotor task, A task which involves precise coordination of a sensory or ideational process and a motor activity; e.g., a tracking task.

psychophysical methods, Standardized procedures for presenting stimulus material to a subject for judging and for recording the results. Originally developed for determining functional relations between physical stimuli and correlated sensory responses, but now used more widely.

psychophysical quantity, A physical measurement, as a threshold; dependent on human attributes or perception.

psychophysics, A branch of psychology that deals with the quantitative relationship between physical and psychological events.

Purkinje effect, The response of the human eye which makes it less sensitive to lights of longer wavelengths under conditions of decreased illumination, e.g., red appears darker at night than blue having the same brightness under photopic conditions.

pursuit tracking, A task in which the subject is required to keep a response marker in line with a moving target.

quantitative display, A display which provides numerical values as opposed to one in which only qualitative information is provided.

quickening, A modification of a closed-loop system which reduces the need for the operator to perform analog differentiations or to sense and utilize derivative information separately. Quickening provides an operator with immediate feedback about tracking actions.

radiale height (elbow height), Vertical distance from the floor to radiale, the depression at the elbow between the bones of the upper arm (humerus) and forearm (radius); subject standing erect, arms hanging naturally at sides. (See Anthropometry.)

rate tracking, A task in which the operator moves a control in such a way as to keep an indicator in line with a moving object, thus measuring the objects speed of motion; involves an exponential relationship between system error and control movement. (See position tracking, tracking.)

reaction time (RT), The interval between application of a stimulus and the beginning of the subject's response (sometimes measured to response completion).

reaction time, complex, The time required to react when a discrimination needs to be made; e.g., a response is made to one, but not to others, of two or more expected signals (discrimination reaction time), or a different response is specified for each kind of stimulus (choice reaction time).

reaction time, simple, The time required to make a predetermined response as quickly as possible to prearranged signal; a combination of sensing time and response time.

real time, The absence of delay, except for the time required for the transmission by electromagnetic energy, between the occurrence of an event or the transmission of data, and the knowledge of the event, or reception of the data at some other location.

receiver-operating-characteristic, A receiver-operating-characteristic is a graphical summary of the performance of a detector. Detection probability is plotted on the ordinate and false alarm probability is plotted on the abscissa. These are conditional probabilities, that is, the probability that the condition (signal present or signal absent) is true. A family of non-intersecting curves is often plotted with either constant detectability index or constant signal-to-noise ratio as the parameter. Each curve shows how detection probability and false-alarm probability vary monotonically as a function of the decision criteria (operating point) of the detector.

receptor, A sensory nerve ending or end-organ in a living organism that is sensitive to physical or chemical stimuli.

recognition, The psychological process in which an observer so interprets stimuli received from an object that he forms a correct conclusion as to the nature of that object.

red-green blindness, A common form of partial color blindness, or dichromatism, in which red and green stimuli are confused because they are seen as various saturations and brightnesses of yellow, blue, or gray.

redout, The condition occurring under negative g in which objects appear to have red coloration due to uncertain causes, possibly venous congestion of engorged eyelids.

redundancy, The existence of more than one means for accomplishing a given function. Each means of accomplishing the function need not necessarily be identical. (Active) — that redundancy wherein all redundant items are operating simultaneously rather than being switched on when needed. (Sometimes referred to as parallel redundancy.) (Standby) — the redundancy wherein the alternative means of performing the function is inoperative until needed and is switched on upon failure of the primary means of performing the function.

refractory period, A brief period following stimulation of a nerve or muscle during which it is unresponsive to a second stimulus; may be absolute (no response) or relative (response only if the stimulus is very strong.)

reinforcement, The action of a symbolic or concrete reward upon a response so that the response is strengthened (positive reinforcement); or, the action of symbolic or physical punishment on a response so that the response is weakened or replaced by an escape or avoidance response (negative reinforcement).

reliability, The probability that a system will perform a required function under specified conditions for a specified period of time or at a given point in time.

reliability (psychological test), The complex property of a series of observations or measuring instruments that makes possible the obtaining of similar results upon repetition.

remote handling, Process by which manipulative skills are transferred from the proximity of the human operator to a more distant area by means of mechanically or electronically linked remote-control systems.

remote indicating, Of an instrument, displaying indications at a point remote from its sensing element, often by electrical or electronic means.

resolving power, The capability of the eyes to see two objects viewed simultaneously as two distinct objects; the capability to perceive as distinct two objects, in close proximity, casting images on the retina.

respiratory quotient, Quotient of the volume of CO_2 produced, divided by the volume of O_2 consumed by an organism, an organ, or a tissue during a given period of time.

response, (1) Physiological — the muscular contraction, glandular secretion, or other activity of an organism which results from stimulation. (2) Psychological — a behavioral action, most commonly motor or verbal, that occurs following an external or internal stimulus.

response time, The time required for a human operator to make a given response after a stimulus is perceived. Under simple conditions (e.g., pushing a button), it may be a few hundredths of a second.

retinal disparity, The difference which exists between the images formed in the right and left eyes when a solid object is viewed binocularly.

retinal field, The extended mosaic of the rod and cone receptor elements of the retina, which forms something of an anatomical correlate of the stimulus field.

retinal illuminance, The illuminance of the retina, the usual units being the troland and the lux.

retinal rivalry, Alternation of sensations first from one eye and then from the other, when the two eyes are simultaneously stimulated by different colors or figures. Also called binocular rivalry. Contrast with binocular fusion, in which the two impressions are fused into a single impression.

rhodopsin, A substance found in the rods of the dark-adapted eye, which bleaches rapidly on exposure to light, and is believed to be the substance underlying scotopic or twilight vision.

root mean square pressure of fundamental speech sounds, A method of relating the measurements of individual sounds as uttered, by making the intervals of time over which the squared pressures are averaged correspond directly to the intervals during which the specific vowels and consonants are spoken. Values are expressed in decibels relative to the r.m.s. pressure of the weakest speech sound (the initial consonant, "th," of "thin").

rotation (zero g), The controlled movement of a free-floating orbital worker to achieve a desirable work position after translation from one location in space to another; a positioning maneuver required in order to perform maintenance or other duties inside or outside an orbiting space vehicle.

saccadic movements, Sudden movement of the eyes from one fixation point to another.

safety, Freedom from those conditions which can cause injury or death to personnel and damage to or loss of equipment or property.

saturation, Extent to which a chromatic color differs from a gray of the same brightness, measured on an arbitrary scale from 0% to 100% (where 0% is gray).

scintillation, (1) Generic terms for rapid variations in apparent position, brightness, or color of a distant luminous object viewed through the atmosphere. (2) A flash of light produced in a phosphor by an ionizing event. (3) On a radar display, a rapid apparent displacement of the target from its mean position. Also called target glint or wander. This includes but is not limited to shift of effective reflection point on the target.

scotoma, A blind or partially blind area in the visual field.

scotopic adaptation, Like dark adaptation, but with more explicit reference to the part played by the rod system of the retina.

scotopic vision, Vision which occurs in faint light, or after dark adaptation. Sometimes called twilight or night vision. Hues and saturations cannot be distinguished.

seat reference point (SRP), The point at which the midlines of the seat and backrest intersect.

secchi disk, A white disk which, when submerged to varying depths, aids in determining the color and depth of light penetration in the sea.

sensation, Subjective response or any experience aroused by stimulation of a sense organ.

sensing time, The time required for a human operator to become aware of a signal.

sensor, The nerve endings or sense organs which receive information from the environment, from the organism, or from both.

sequence-of-use principle (equipment design), The principle of arrangement that controls and displays should be so positioned that those used in sequence would be physically arranged in order of operation.

service test model, A model used to determine the characteristics, capabilities, and limitations of a piece of equipment or a complete system under either simulated or actual service operational conditions.

shade, Any color darker, i.e., of lower lightness, than median gray.

shake table, A test device for determining the effects of vibration on human or animal subjects or on equipment.

shape coding, Varying the shape of controls to make them distinctive. Shape coding is effective both visually and tactually, that is, the difference can be both seen and felt. The shape of a control should suggest its purpose, and the shape should be distinguishable not only with the naked hand, but also with gloves.

short-term memory, The storage of recently received inputs for a period of seconds or minutes.

shoulder-elbow distance, From top of acromion to tip of elbow, measured when subject is sitting erect, with upper arm vertical, forearm horizontal.

SI Units, Abbreviation of Le Systeme International d'Unites (International System of Units).

side tone, The signal from the talker's microphone usually returned to the talker via his earphones. This feedback signal is called the side tone. The talker's speech can be manipulated by varying the side tone.

signal detection theory, A psychophysical model which views human perception as a process in which decisions based on uncertain sensory information are made under conditions of risk.

simulation, A set of test conditions designed to duplicate field operating and usage environments.

simulator, Any machine or apparatus that simulates a desired condition or set of conditions, such as a flight simulator.

sitting height, The vertical distance from the sitting surface to the top of the head. The subject sits erect, looking straight ahead, with knees at right angles.

skinfold measurement for estimating body fat, Permits reasonably close estimation of body fat. Calipers are used to exert a pressure of 10 g/mm^2 over an area of 20 to 40 mm^2. Measurements are made over the right triceps muscle and below the inferior angle of the right scapula.

snow-blindness, A temporary abnormality of the color sense, in which all objects are tinged with red. Caused by long-continued exposure to very bright light, as in Arctic exploration, on glaciers, in telescopic observation of the sun, watching welding operations, etc.

somatotyping, A nearly obsolete means of numerical classification of various human body types. Endomorphy: soft, round form with loose, flabby tissue. Mesomorphy: massive, solid form with cubical head, heavy muscles. Ectomorphy: slender limbs and body, with slight head.

sone, Unit of loudness; a simple tone of 1000 Hz 40 dB above a listener's threshold produces a loudness of one sone.

space myopia (empty-field myopia), A disturbance of the visual accommodative mechanism due to insufficiently sharp detail, making it difficult to detect small objects in the relatively unstructured visual environment of space.

spectrum colors, The series of saturated colors normally evoked by photopic stimulation of the retina with radiant energy of continuously differing single wavelengths through the visible range.

specular reflection, Reflection in which the reflected radiation is not diffused; reflection as from a mirror. Also called regular reflection, simple reflection.

speech articulation index, The method of estimating the intelligibility of speech by measuring differences (in decibels) between speech and noise levels at different frequency bands.

speech interference level (SIL), Used as a gross basis for comparing the relative effectiveness of speech in noise. It is usually the simple numerical average of the decibel level of noise in three octave bands, namely, those with centers at 500, 1000, and 2000 Hz.

spin table, A round platform on which human or animal subjects can be placed in various positions and rapidly rotated in order to simulate and study the effects of prolonged tumbling at high rates. Complex types of tumbling can be simulated by mounting the spin table on the arm of a centrifuge.

squatting height, The height from the top of the subject's head to the floor when the subject balances on toes, body erect.

squeeze, Squeeze in diving is due to the effect of increasing external pressure upon the ears and sinuses, the face plate or the swim suit, uncompensated by an equal increase in pressure from within.

stabilization (zero g), A condition required during translation and after arrival of a free-floating orbital worker at a desired position inside or outside an orbiting space vehicle. A mechanism is necessary to stabilize the worker against accelerations that may cause visual disorientation, labyrinthine reactions, and movement away from the work area.

standard observer, A hypothetical observer with a visual response mechanism possessing the colorimetric properties defined by the 1931 ICI tables of the distribution coefficients, x, y, z, and the trichromatic coefficients, x, y, z, of the equal energy spectrum. The y coefficients of the equal energy spectrum are the relative luminosity values defining the standard observer for photometry.

standing knee height, Vertical distance from top of kneecap to floor when subject is standing.

stature, The vertical distance from the floor to the top of the head. The subject stands erect and looks straight ahead.

sternum height, Vertical distance from lower tip of sternum to floor when subject is standing.

stimulus, Energy, external or internal, which excites a receptor.

stimulus generalization, An aspect of learning characterized by the spreading of a response learned in association with one stimulus to other stimuli which are similar but not identical to the first.

STPD conditions, Standard temperature and pressure, dry. These are conditions of a volume of gas at 0°C, at 760 torr (101.324 72kPa) without water vapor. A STPD volume of a given gas contains a known number of moles of that gas. (See BTPS conditions.)

stress, (1) The force per unit area on a body that tends to produce a deformation. (2) The effect of a physiological, psychological, or mental load on a biological organism which may produce fatigue and degrade proficiency.

strophospheres, Define the region in a person's fixed workplace that is common to the hand motions made with various hand manipulations.

subject, A member of a specified population from whom an experimenter obtains response data for specific stimulus variables. The data obtained are often generalized to that population.

subjectively expected utility (SEU), Risks will be taken only if the pay-off is judged to be sufficiently high. If the value to the subject of a "dollar to be gained" is multiplied by the subject's estimate of the probability of success, one obtains the SEU for the given situation.

surround brightness, The brightness of the area immediately adjacent to an area of visual work.

system, A composite of equipment, skills, and techniques (including all related facilities, equipment, material, services, and personnel) that is capable of performing and/or supporting an operational role.

system analysis, Identification of the dynamic relationships between elements of a system.

system engineering, The study and planning of a system so that the relationships of various parts of the system are fully established before designs are committed, or to look for improvements in existing systems.

tabular display, Display of alphanumeric and other symbols in a row and column format.

tachistoscope, An experimental device for controlling the exposure of visual stimulus material, e.g., symbols.

tactile, Of or relating to the sense of touch. Tactual.

target, (1) Any object, point, etc., toward which something is directed. (2) An object which reflects a sufficient amount of a radiated signal to produce an echo signal on detection equipment.

target acquisition, The process of optically, manually, mechanically, or electronically orienting a tracking system in direction and range to lock on a target.

target discrimination, (1) Perceiving a desired signal within a background of noise. (2) Distinguishing multiple signals.

task, A group of related job elements performed within a work cycle and directed toward a goal; a composite of the discriminations, decisions, and motor activities required of an individual in accomplishing a unit of work. A set of operationally similar tasks constitutes a duty.

task analysis, An analytical process employed to determine the specific behaviors required of human components in a man-machine system. It involves determining, on a time base, the detailed performance required of man and machine, the nature and extent of their interactions, and the effects of environmental conditions and malfunctions. Within each task, behavioral steps are isolated in terms of the perceptions, decision, memory storage, and motor outputs required, as well as the errors which may be expected. The data are used to establish equipment design criteria, personnel and training requirements, etc.

task coding, A numerical and/or alphabetical index sometimes used during task analysis to indicate in relative terms: (1) how demanding a task is with respect to perceptual, judgmental, and motor skills (Skill Coding); (2) how critical the correct performance of the task is to system performance or mission success (Task Criticality); (3) whether personnel are already doing the task or whether it is new (Task Newness).

task element, A basic unit of behavior, made up of the smallest logically definable set of perceptions, decisions, and responses which the human being is required to perform in completing a task; e.g., a single action identified by a specific signal on a specified indicator or display, a specific control actuation, and a feedback signal or response adequacy.

task equivalence, The condition existing when two tasks have sufficient elements in common to permit an individual to transfer from one to the other at the same level of skill without additional training or experience.

task, continuous, Those tasks which involve a continuously changing response to a continuously changing stimulus. The operator compares the desired output with the error signal which is fed back and makes continuous adjustments accordingly; e.g., maintaining a given vehicle attitude by manual control movements in response to instrument indications.

tasks, procedural, Those tasks which involve stepwise, discontinuous, all-or-none responses to discrete perceptual cues. The operator observes a displayed condition and performs a single response, or observes a series of conditions and performs a series of separate nonrepeated responses; e.g., turning on deicing equipment when icing conditions are indicated.

threshold, absolute visual, The minimum intensity of light that can be seen by the human eye after complete dark adaptation, generally considered to be about 1/1,000,000,000 of a lambert in the periphery. Theoretically, as few as six quanta of light reaching the retina have a good probability of yielding a visual sensation.

threshold contrast, The smallest difference between luminances (or brightnesses) that is perceptible to the human eye under specified conditions of adaptation luminance and target visual angle. Also called contrast threshold, liminal contrast. Compare threshold illuminance. Psychophysically, the existence of a threshold contrast is merely a special case of the general rule that for every sensory process there is a corresponding lowest detectable intensity of stimulus, i.e., a limen.

threshold, difference, The minimum difference between two stimuli that can be perceived as different; statistically, that difference which is perceptible 50% of the time.

threshold (limen), The statistically determined point at which a stimulus is just barely adequate to elicit a specified organismic response (absolute threshold) or at which it differs enough from another stimulus to elicit a different response (difference or differential threshold).

threshold, masked differential, The least amount of change of a stimulus necessary to elicit a response when the stimulus is obscured (masked) by another stimulus presented simultaneously; applies to brightness, loudness, etc.

threshold, terminal, A maximum stimulus that will produce a given type of sensory experience or elicit a given kind of response.

threshold of audibility, For a specified signal, the minimum effective sound pressure level of the signal that is capable of evoking an auditory sensation in a specified fraction of the trials. The characteristics of the signal, the manner in which it is presented to the listener, and the point at which the sound pressure level is measured must be specified. Also called threshold of detectability.

threshold of detectability, Threshold of audibility.

threshold of discomfort, In acoustics, for a specified signal, the minimum effective sound pressure level of that signal which, in a specified fraction of the trials, will stimulate the ear to a point at which the sensation of feeling becomes uncomfortable. The term applies similarly for other senses.

threshold of pain, A specified stimulus signal, the minimum level of that stimulus which, in a specified fraction of trials, will stimulate a sense organ to a point at which the discomfort gives way to definite pain that is distinct from mere nonnoxious feeling of discomfort.

threshold shift, A threshold shift is a change in the threshold of audibility for an ear at a specified frequency. The amount of threshold shift is customarily expressed in decibels.

through-put capacity, The output of thoroughly processed items from a machine.

timbre, That attribute of auditory sensation by which a listener discriminates between two sounds of similar loudness and pitch, but of different tonal quality.

time and link analysis, An analysis of the number and duration of activities assigned to each potential element of a system. It provides an overview of what the human component is expected to do at each point in an operation, an outline of the process of information exchange among men and between men and equipment, and data useful in the determination of crew size for each major component of the system.

time-line analysis, An analytical technique for the derivation of human performance requirements which provides an indication (by means of a time-line chart) of the functional and temporal relationships among tasks, as well as the behavior and task loadings for any given combination of tasks.

time-sharing, The division of an operator's perceptual, decision-making, or response time among activities or tasks which must be performed at or about the same time. For example, flight control movements must be time-shared with engine control adjustments, communications, etc.; in general, tasks which must be time-shared in the operational situation should also be time-shared during training.

tint, Any color lighter, i.e., of higher lightness, than median gray. May imply weak saturation as well as relatively high lightness.

tonal gap, A tonal gap is a frequency band of low auditory sensitivity above and below which the sensitivity is better.

tonal island, A tonal island is a frequency region bounded above and below by tonal gaps.

tonality (tonal), Tonality is the phenomenon in which tones spaced an octave apart sound more alike than those spaced along any other interval.

tone, (1) Physics — a sound wave capable of exciting an auditory sensation having pitch. (2) Psychology — a sound sensation having pitch.

tracking, The process by which an operator continually attempts to minimize some measure of the difference between a desired and an actual output of a system; an activity that involves following a target moving in two- to three-dimensional space with reference to the observer or equipment.

training, The totality of instructions, planned circumstances, and directed activity by which personnel acquire and/or strengthen new concepts, knowledges, skills, strength habits, or attitudes which will enable them to perform assigned duties with maximum reliability, efficiency, uniformity, safety, and economy.

transfer of learning, The phenomena through which information acquired in one situation is applied or "transferred" to some other situation, most usually an operational one.

transillumination, (1) Indirect illumination used on console panels and utilizing edge- and back-lighting techniques on clear, fluorescent, or sandwich-type plastic materials. (2) Transmitting light through material to enhance seeing quality deviations.

transmittance, The ratio of luminous flux transmitted through a medium to the incident flux, expressed as a percentage. "Selective transmittance" refers to the transmission of particular wavelengths of light through a transparent or translucent medium, such as a red filter.

tremor, Shaking, trembling, or oscillation which may result from natural impairment or from excitement, emotion, cold, or an effort to maintain a part of the body in a fixed position. It can be measured by the distance or the number of departures from a fixed path or position in a given time. Fatigue or efforts to prevent tremor may increase it; strong support of the body or part of the body affected, or the application of friction, tend to decrease the trembling.

trial, A trial is a sequence of events during an experimental session in which a subject (or subjects) is presented a stimulus set and the resulting response set is recorded.

trianomaly, Rare type of trichromatism in which an abnormally large portion of blue stimulus is required in a blue-green mixture to match a given cyan.

trichromatic theory, A color theory based upon the facts that all hues may be derived from the mixture of two or more of three primaries.

trichromatism, Form of vision yielding colors which require in general three independently adjustable primaries (such as red, green, and blue) for their duplication by stimulus mixture.

tritanopia, Form of dichromatism in which reddish blue and greenish yellow stimuli are confused. Tritanopia is a common result of retinal disease, but in rare cases may be inherited. Sometimes called blue blindness.

trunk, Human body torso.

turn-around time, That element of maintenance time needed to service or check out an item.

ultrasonic, In acoustics, of or pertaining to frequencies above those that affect the human ear, i.e., more than 20,000 Hz.

unburdening, A method of easing the human task in tracking by using computers to perform mathematical integrations; or, the mechanization of a system to reduce physical effort on the part of the operator.

uptime, That element of active time during which a system element is either alert, reacting, or performing.

validity (test), The degree to which a test measures what it was designed to measure, estimated by means of a coefficient of correlation between test scores and a criterion measure, such as acutal on-the-job performance; empirical validity.

value, (1) The dimension of the Munsell system of color which corresponds most closely to lightness. (2) Numerical quantity. (3) Worth, as in value engineering.

verbal mediation, The technique of verbalizing instructions or directions so as to emphasize cue-response relationships in learning or performing a task; used especially during the early stages of learning.

vertigo, Sensation of dizziness or whirling, in which objects, though stationary, appear to move in various directions, and the person finds it difficult to maintain an erect posture. It may result from overstimulation of the sense organs, from changes in the blood supply of the brain, or from disease. "Aviator's vertigo," caused by conflict between visual and gravitational cues, results in a disturbance in the pilot's orientation with respect to the earth.

viewing angle, The angle formed by a line from the eye to the viewed surface.

vigilance, An activity or type of research involving continuous watch and the ability of individuals to detect signals of varying frequencies during these watches.

virtual image, An image that cannot be shown on a surface but is visible, as in a mirror.

visibility, The capacity of radiant energy, within a certain range of wavelengths, to excite a visual receptor process and thereby evoke the phenomenon of brightness.

vision, foveal, Visual sensations or perceptions due to stimulation of the fovea centralis, or center of the retina.

vision, peripheral, Visual sensations or perceptions due to stimulation of the outlying portions of the retina.

vision, persistence of, The tendency of visual excitation to outlast the stimulus, or more generally the tendency of changes in visual sensory response to lag behind changes in the stimulus.

visual acuity, The ability to perceive detail at various distances. More specifically, it can be expressed as the reciprocal of the visual angle, in minutes of arc, that is subtended by the smallest detail that can be visually discriminated.

visual adaptation, Adjustive change in visual sensitivity due to continued visual stimulation. Three recognized types are: (1) scotopic or dark adaptation, (2) photopic or light adaptation, and (3) chromatic or color adaptation.

visual angle, The angle subtended by an object in the visual field at the nodal point of the eye. Determines the size of the image on the retina. Objects of different sizes or distances have the same-sized image on the retina if they subtend the same angle.

visual field, That part of space that can be seen when head and eyes are motionless, (or) the totality of visual stimuli which act upon the unmoving eye at a given moment.

visual photometry, A subjective approach to photometry, wherein the human eye is used as the sensing element; to be distinguished from photoelectric photometry.

visual range, The distance, under daylight conditions, at which the apparent contrast between a specified type of target and its background becomes just equal to the threshold contrast of an observer; to be distinguished from the night visual range. Also called daytime visual range.

visual space, This term, like visual field, refers to the extended world as perceived by means of the eyes but is commonly used in a more generic and abstract way in discussions of the perception of distance and length, of depth or distance away from the retina, and of form or figure in two and three dimensions.

volume (tonal), Volume is that attribute of auditory sensation in terms of which sounds may be ordered on a scale extending from spatially small to spatially large.

waist depth, The horizontal distance between the back and abdomen at the level of the greatest lateral indentation of the waist (if this is not apparent, at the level at which the belt is worn).

warm-up time, Time measured from the application of power to an operable system to the instant when the system is capable of functioning in its intended manner.

watchkeeping, That task in which the operator monitors some condition such as the status of automatic equipment.

weak color, A color of low saturation.

Weber-Fechner Law, An approximate psychophysical law relating degree of response or sensation of a sense organ and the intensity of the stimulus.

Weber's Law, A law that states that for each sense parameter the ratio between the just noticeable difference and the stimulus intensity remains constant.

weightlessness, (1) A condition in which no acceleration, whether of gravity or other force, can be detected by an observer within the system in question. (2) A condition in which gravitational and other external forces acting on a body produce no effect, either internal or external on the body. (3) The absence of gravitational pull on a body or object. (4) Condition in free fall or in space beyond the earth's gravitational field; may be simulated by air-bearing devices or during parabolic flight of high-performance aircraft. (5) A condition that is characterized by the absence of gravity. For clarity, it has been proposed that the term zero g or null g be used to refer to the physical state of an object or body and that the term weightlessness be used to refer to the physiological and psychological experience of living organisms under zero g conditions.

white noise, (1) Randomly fluctuating noise. (2) The noise that is heard when many sound waves of different lengths are combined so that they reinforce or cancel one another in a non-uniform fashion. (3) A noise for which the spectrum density is substantially independent of frequency range. (4) A sound or electromagnetic wave whose spectrum is continuous and uniform as a function of frequency.

whiteout, An atmospheric and surface condition in the arctic in which no object casts a shadow, the horizon being indiscernible, and only very dark objects being seen.

windchill, A commonly used scale for expressing the severity of cold environments, indicates the effects of wind velocity and temperature combined.

work, (1) Expression for human effort measured in physical units, or specific output results. (2) General description of task.

work cycle, The total series of actions and events which characterizes or describes a work assignment or a single operation.

work pace, The rate at which a task or activity is done. Work may be paced externally, as by machine or mission rates, or self-paced by the worker.

work physiology, The study of bodily responses to physical and mental effort. Includes measures of stress on the cardiovascular, respiratory and musculoskeletal systems in particular.

work rest cycle, The ratio of work time to non-work time. The larger the ratio the more opportunity there is for fatigue.

work space layout, A design of a work area or work station to include provisions for seating, physical movement of human operators, operational maintenance, and other factors permitting adequate person-to-person contact and man-machine interaction.

work study, Analysis of work methods, techniques, and procedures.

workspace, The physical area in which an individual performs some duty or task.

yaw, (1) The rotational or oscillatory movement of an aircraft, rocket, or the like about a vertical axis. The amount of this movement is the angle of yaw. (2) To cause to rotate about a vertical axis.

yaw axis, A vertical axis through an aircraft, rocket, or similar body, about which the body yaws. Also called a yawing axis.

yawing moment, A moment that tends to rotate an aircraft, etc., about a vertical axis.

yes-no experiment, An experiment in which each trial contains a single observation interval and the subject judges whether or not a stimulus was presented during that interval.

Wage and Salary Administration

14

Z94.14

At its organization meeting in 1966, the main Z94 INDUSTRIAL ENGINEERING TERMINOLOGY Committee, in an attempt to take advantage of the prevailing expertise in the various fields of the cooperating societies, used, as a basis for working subcommittees — and, consequently, their resulting terminologies — the breakdown then in use by the AIIE RESEARCH ABSTRACTS. However, the Committee felt that some changes were necessary in its own division of its task. Thus, DISTRIBUTION AND MARKETING was added. BIOMECHANICS was the result of separating this discipline from the APPLIED PSYCHOLOGY category, because it was strongly felt that this field was essentially made up of "Biology PLUS Mechanics" and ought not be a part of a psychological treatment. When the time came for the present revision of the entire Z94 field, the committee felt that WAGE AND SALARY ADMINISTRATION was also deserving of a place of its own, and it is, therefore, being removed from the APPLIED PSYCHOLOGY classification and expanded. (See Introduction, Makeup of the Z94 Standard.)

In few fields is the need for standardized terms so urgent as in labor relations, especially because of the often opposing and sometimes antagonistic positions assumed by the concerned parties. Thus, although the term *right to work laws* has become recognized as standard, union officials are more likely to refer to *union busting laws*. Again, *givebacks* is a term that received particular prominence during the 1980 New York Transit negotiations, and, despite the union's counterpart, *takeaways,* the NEW YORK TIMES (see MAGAZINE, Sunday, April 27, 1980, Section 6, Page 18) used the former term in its definition ("nonwage concessions involving reductions in benefits and increases in productivity").

Surely there is a great need for understanding the "meanings" behind the terms — but there must be clear connotation of the term if prejudice is to be avoided. Otherwise, negotiations will be complicated by such public relations and journalese observations as "union making demands" and "management insisting on concessions." It is interesting to note that often some terms survive even though the precise historical details may have eventually assumed more sophisticated methods: e.g., sabotage no longer involves a shoe in the machinery. Again, consider a term now obsolete: greenhands. A century ago it referred to an attempt to reduce the "number of masters" and an increase in the number of "green apprentices." In the 1980 New York Transit strike, the term was not used, but one of the matters discussed was the "increase of part-time help" (not necessarily "green").

Even if the field of WAGE AND SALARY ADMINISTRATION had not taken on the increased importance and diversification it had assumed — especially with the growth of labor unions, the expansion of governmental regulations, and the further development of the field of personnel relations — worker payment and relations deserved individualized attention, for they have always been an essential part of the development of Industrial Engineering. As a matter of fact, modern wage incentive plans were basic in Taylor's work at Midvale Steel just about one hundred years ago. Taylor's search for what constituted a "fair day's work" really began with the recognition of the "evil" that came from "systematic soldiering." As his biographer, Copley, noted, he felt that "paying all men, on similar work, at the same rate tended to allow the poorest worker to establish the amount of work output." As Taylor expressed it, "Men will not do an extraordinary day's work for an ordinary day's pay." Thus began a series of wage incentive plans mostly identified by the names of such pioneers as Taylor, Gantt, Towne, Halsey, et alia. This trend continued until Lytle (MANAGEMENT ENGINEERING, May, 1922) developed a generalized mathematical treatment for the study and development of incentive plans.

Job evaluation similarly presented a series of plans, each identified by a personal, company, and/or association name: e.g., Benge, Westinghouse, NEMA (National Electric Manufacturers Association), etc. Even merit rating followed this pattern.

Lytle quotes J. C. Hopwood:

> There is no difference between office and plant with respect to the principles involved in labor

relations and no difference between wages and salaries as compensation to individuals employed.

Hence the study and application expanded from the plant into the office. Incidentally, on the basis of work by Bishop about a third of a century ago, Lytle discussed the question of rates for women workers in his treatment of "Equal Pay for Equal Work" — a matter that only now is receiving the attention it deserves. Essentially because of the expansion and importance of governmental regulations, there is an increasing need for standardized terms in this important aspect of Industrial Engineering.

Many of the terms in this section come from the expertise and involvement in the practical aspects of the field of Wage and Salary Administration of the members of this Subcommittee as well as from such sources as *Encyclopedia of Professional Management* (Lester R. Bittel, Editor, New York, McGraw Hill Book Co., Inc., 1978), *Industrial Engineering Handbook* (Third Edition, H. B. Maynard, Editor, New York, McGraw Hill Book Co., Inc., 1971), *Industrial Engineering Terminology* (New York, American Society of Mechanical Engineers, 1955).

Special mention is due Dr. William J. Jaffe for assistance in coordinating and editing this section as he did with all sections of this TERMINOLOGY and compiling the overall indexes.

This Subcommittee was comprised of the following who contributed their efforts to complete this section:

MEMBERS:

Mr. Harvey Gittler, P.E., Vice President, Gilford Laboratories

Dr. William J. Jaffe, P.E., Emeritus Distinguished Professor, N.J. Institute of Technology and Chairman 1972 IE TERMINOLOGY project and present TERMINOLOGY project.

CHAIRMAN:

Prof. James L. Rigassio, P.E., Chairman, Industrial and Management Engineering, Department, N.J. Institute of Technology

Note: In formulating this section, the original Z94 (1972) TERMINOLOGY was, among other sources, used. Hence, the reader might want to refer to that publication, but should also refer to the analagous sections in this present TERMINOLOGY.

ability, Demonstrable knowledge or skill. A criterion or job factor of a job evaluation plan.

ability to pay, Concept sometimes expressed in collective bargaining relating to the economic base upon which the cost of wages or benefits is to be borne. Uses the effects of wage levels on costs to an organization rather than as income to employees and, therefore, helps determine whether or not an organization can afford (ability) to maintain a specific wage level.

accrual of benefits, The process, method or formula normally based on various factors such as length of service, hours worked, or level of responsibility which is used to determine either when a benefit comes into existence as a legally enforceable claim or to increase in number, value, or percentage the benefit(s).

across-the-board increase, A general wage increase applicable to a group or class of employees.

activity plateau, A level of performance established informally, resulting from past practice, or by design such as under the application of a wage incentive plan.

affirmative action, Plan of recruitment taken to effect a balanced or non-discriminatory work force representing minority groups.

allowance, A pay or work time adjustment to compensate an employee for job fatigue, unavoidable delays, personal needs and rest.

allowance, fatigue, Time provided during a normal work period for recovery from physical or mental exertion. (See fatigue.)

allowance, personal time, Time provided during a normal work period for personal needs.

allowance, rest, Time provided during a normal work period for recovery from fatigue.

allowance, unavoidable delay, A delay outside the control or responsibility of a worker.

allowed time, The basic time established for the performance of a task increased by appropriate allowance.

aptitude, Ability to develop requisite performance skills.

aptitude test, An actual or simulated trial to determine qualification or fitness to perform a task or job. A method to determine aptitude.

area differential, Additional compensation paid for similar jobs performed in one geographical location as compared to another.

attendance bonus, Payment for punctual and consistent presence at place of work.

authority pay, The promotion of an employee or the granting of greater authority to an employee without any financial compensation.

automatic increases, Additional pay resulting from predetermined policy. (See automatic progression.)

automatic progression, Increases in job pay or classification resulting from predetermined formula or policy.

average earnings, Individual or group earnings over a period of time divided by the number of time or production units in the period under consideration.

average hourly earnings, See average straight time hourly earnings.

average straight time hourly earnings, Average wages earned per hour excluding premiums for overtime and shift assignment.

average time, Mathematical mean of selected work cycle or elemental times.

avoidable delay, Delay controllable by a worker and therefore not allowed in the job standard.

back pay, Compensation for work performed during a past period of time.

base pay, Minimum rate of compensation for a given task when performed by a fully qualified employee. May differ from rate paid during training or probationary periods.

base points, Term used in job analysis referring to minimum point score value for a given evaluated factor.

base rate, The amount of pay set for a given job or classification based on a period of time, exclusive of bonus, premium or differential amounts which may be added for specific reasons.

base salary, See base pay.

base wage rate, Also applied to the minimum rate of compensation of a wage structure. (See base pay.)

basic piece rate, A set payment for a unit of production exclusive of additional allowances.

beginner rate, Compensation rate during initial probationary period or while learning a new job. (See Applied Psychology.)

benchmark, A standard usually established by concensus, for use in job evaluation.

benchmark evaluation method, A job evaluation technique in which the pay rate for a job under study is established by comparison with pay rates for jobs selected as standards or benchmarks.

benchmark job, A job or task accepted as a gauge for comparison of other jobs or tasks.

benchwork job, A task performed at a table or fixture of similar configuration.

benefit, Compensation other than direct wages or salary. Usually health, disability and life insurance, social security, unemployment compensation and pension contributions paid by employer.

benefit limitations, The minimum and maximum restrictions placed on a benefit which define the parameters of the program.

bereavement pay, See funeral leave pay.

bogey, A target level of performance usually based on minimum acceptable productivity.

bonus, Extra compensation paid for specific reason. Unlike premium, amount of compensation may be non-recurring and established by a wide variety of methods.

bonus determinant, The relationship in an incentive plan between the production rate and the earning rate which establishes the earnings potential under the plan.

bonus earnings, Additional compensation over and above base earnings paid for specific reasons such as high productivity, low waste or defects, low absenteeism, employment anniversary, end of year, company profitability, etc.

bonus plan, A formula or plan used to determine bonus eligibility and payment.

bonus restriction, The limits placed on an incentive plan so that earnings come from extra operator effort rather than from technological improvements.

broad banding, Consolidation of various separate job titles with narrowly defined duties into more generally defined jobs thereby reducing specialization by job title.

cafeteria plan, A benefits program incorporating a wide variety of options designed to permit choice by individual employee to suit preferences within a fixed benefit cost entitlement.

call-in pay, Minimum compensation guarantee for being required to return to a job after completion of a regular shift of work or at times other than designated as regular.

checkoff, Deduction from earnings of employees for union dues and initiation fees.

class rate, Pay applicable to a given grouping of jobs or employees.

classification, A category or group in which a job is located on basis of pay rate, competence or training level, seniority or other job identification criteria.

classification index, A numerical value assigned to a job classification which establishes the relationship of that job classification to others.

COLA, See cost of living adjustment.

cost of living adjustment, Changes in compensation made to reflect changes in the consumer price index.

collective bargaining, A process of negotiating proposals and counter-proposals by representatives of management and labor in order to settle or reach agreement on terms and conditions of employment.

commission, Compensation for services rendered in arranging a transaction. May be a percentage of the transaction amount.

comparable wages, Equal pay for equal work.

compensation, The total of value received for the performance of a job.

constant sharing plan, A series of incentive plans in which the percent earnings of operators above a base level is either greater than or less than the percent increase in production. An incentive plan with a straight line incentive pay curve which does not pass through the (0,0) point. Most common constant sharing plans pay less than 1:1 incentive bonus; usually a 50-50 sharing plan in which the employees receive a 1% bonus for each 2% increase in production.

constant total cost plan, An incentive plan with a pay curve that keeps the total cost per piece (labor plus overhead) a constant. Rare.

constant unit labor cost plan, An incentive plan with a pay curve that keeps the labor cost per piece a constant.

consumer price index, A numerical measure of retail price changes of selected goods and services prepared by the U.S. Department of Labor which reflects retail price trends compared to a selected base year.

control (wage & salary), The administration of compensation plans in accordance with set predetermined policy.

correlation, As used in job evaluation — the relationship between total point score value of a job as determined by job analysis and pay grade.

cost-of-living, See consumer price index.

cost-of-living index, See consumer price index.

CPI, Consumer Price Index.

cycle time, The time required to perform one complete sequence of job activities.

daily rate, See day rate.

day rate, Rate of pay for day work jobs. Rate upon which compensation for working a given period of time is calculated. Differs from incentive rate.

day work, Tasks for which fixed wage rates are paid based upon time applied as contrasted to piecework for which a standard dollar or earned hour payment is made for each unit of production.

deferred bonus payment, A plan in which bonus earned in one period is paid out in subsequent periods either in a lump sum or over a period of time. Usually used with executives' bonus plans for tax advantage.

deferred compensation, Money or other financial rewards earned by an executive within a corporation but which is not paid at the time of earning.

Deferred compensation can be salary deferrals, bonus deferrals, or supplementary retirement payments. The earnings are paid at a later date usually when the individual is in a lower tax bracket.No use of the funds can be made by the recipient until the agreed-upon transfer date.

Postponed earnings designated for future income declaration under a qualified plan for postponing income tax on earnings until future years.

deferred wage increase, A postponed payment of a wage increase.

defined contribution plan, A procedure for investment of set contribution amounts for purposes of retirement, the value of which, at time of retirement, being dependent upon the market value of the investments made.

degree (job evaluation), Designation of selected factor with a range of factor criteria. Normally designated numerically such as 1,2,3,4, or 5 in range of 5.

delay, An interruption of a specified sequence of work activity.

delay allowance, Compensation for interruption of work activity.

dictionary of occupational titles, A comprehensive compilation of job descriptions and titles published by the U.S. Department of Labor.

differential piecework, A wage incentive plan using two or more piece rates. One rate is paid for production of a fixed quantity in a given time, and a higher rate(s) is paid for production beyond the initial level. (See Taylor differential piecework plan, multiple piece rate plan.)

differentials, skill, Those objectively measurable factors (skills) which are required for a position and, thereby, define that position as uniquely different from another position. The skills demonstrated by one employee as compared to those of any other employee.

differential time plan, A wage incentive plan that pays a higher base wage rate as production levels increase. A series of step bonuses for increased production.

direct labor, Work which can be closely identified with the production of a product and which, from a cost accounting standpoint, may be assessed to a particular unit of product.

direct labor budget, A projection correlated with future time periods of total direct labor required to meet forecast work or production.

dirty money, Extra pay for especially disagreeable work.

disability pay, Compensation for time away from work due to work-related injury or illness.

double back pay, Penalty reimbursement to discharged employees. Specified by Labor Reform Act of 1977 (NLRA) as penalty for discharge of employees for protected activities during a union organization campaign.

double dipping, Receiving multiple retirement pension payments from having earned retirement benefits through employment subsequent to retirement from prior job.

downgrading, See Applied Psychology.

downtime, Periods during a scheduled work shift when employees are available for work but not performing work or when required equipment is inoperable or incapable of being used productively.

draw, A designated amount against which expenses may be charged or from which payment for expenses may be withdrawn.

earned hours, Time in standard hours credited to a worker and used as the basis for calculating pay earnings.

earned rate, Used to denote pay rate for job performed such as in the case of job transfer. (See earned hours.)

earnings, Gross pay or compensation for work performed. (See Applied Psychology.)

earnings efficiency, Actual earnings, normally calculated in terms of earned hours, compared to standard or expected earnings.

educational assistance, Reimbursement of all or a share of tuition and expenses for employee's studies at a training institution not directly associated with the employee's firm.

efficiency, A measure of work performance determined by comparing actual work done with standard amount of work for this job determined as "normal."

efficiency ratio, The numerical value of actual quantity of work done divided by that quantity of work established as "normal" or standard.

effort, Physical and mental effort required to be expended to perform a given job or task. A criterion or job factor of a job evaluation plan.

effort rating, Evaluated worker performance.

elapsed time, Actual clock time expended in performing a job or task.

element time, Time to perform a given part of a job or task.

employee classification, A designation given to an employee based upon some selected basis for differentiation such as pay grade, skill or trade, length of employ, membership of a work unit, etc.

employee services, A wide range of employer provisions at no cost to the employee or made available at discounted or token prices.

employment contract, A legal agreement concerning period of employment, earnings, conditions of employment and duties of the employee and the employer.

entry level, Basic or lowest job in a line of progression in which employees entering that line of progression first accumulate seniority rights in that line.

entry rate, Compensation paid new or transferred employees during a probationary or training period.

equal pay, Equal pay for equal work as measured in terms of job skill requirements, worker effort, job responsibilities and working conditions which includes exposure to hazards.

ERISA, The Employee Retirement Income Security Act. This act was passed by Congress in 1974, legislating the manner in which pensions, pension plans, and other service-related benefits shall be administered. Compliance for ERISA legislation is the responsibility of the Department of Labor.

escalator clause, See Applied Psychology.

evaluated rate, The base pay for a job determined by an analytical procedure for comparing jobs and setting rates of pay.

executive compensation, Money and fringe benefits paid to an executive for services performed. Executive compensation may include money, deferred financial payment, car, club membership, stock options and other perquisites.

executive compensation plan, Any of a number of special pay arrangements to motivate and reward key management employees. Includes incentive bonuses, stock plans, insurance, deferred compensation and perquisites.

exempt employee, Any employee excluded from the Wage and Hour Law. Includes executive, administrative, professional, or outside salesperson who meets certain criteria covering earnings, direction of work force, exercise of discretion and independent judgment, advanced knowledge in a field of science or learning, or selling outside the employer's place of business.

expected earnings, Expected earnings level and expected earnings target. (See Applied Psychology.)

experience curve, See learner curve.

experience employee, A worker who has performed a job or task for a lengthy period of time.

factor comparison, A method of comparing and evaluating different jobs for purposes of pay determination based on selected key factors or attributes common to the jobs being compared.

factor scales, The points assigned to the various factors (skill, effort, responsibility, etc.) in a job evaluation plan which define the range of the particular factor and thereby the importance of the factor in the overall plan.

fair day's pay, The level of pay for a job for which a standard level of work performance is expected. Equated to a "fair day's work." Normally the base rate or guaranteed pay rate of a job. The "normal" or "100%" base of a wage incentive plan.

fair day's work, Standard level of work performance for the base or guaranteed pay rate of a job representing the "fair day's pay."

Fair Labor Standards Act, See FLSA.

fatigue, See Work Measurement and Methods.

fatigue allowance, See Work Measurement and Methods.

featherbedding, See Applied Psychology.

financial incentive, Monetary compensation other than base pay to encourage and reward above standard work performance.

financial incentive plan, A method or system for providing financial incentives for worker productivity.

flagged rate, See red circle rate.

flat increase, A given adjustment in compensation similarly applied to all employees within a group. May also be called an across-the-board increase.

flat rate, A fixed compensation for a job or task. (See basic piece rate.)

flextime, Work schedule arrangement in which employee is allowed flexibility within certain time constraints to choose individual or group working hours.

FLSA, Fair Labor Standards Act establishing minimum wage provisions, rules for overtime, equal pay, child labor and standards for records for employees covered by the legislation.

foreign service premium, Additional compensation made available to employees assigned to positions outside the continental United States. Such premiums may be a percentage of pay or a flat dollar allowance. Separate allowances for housing and schooling may be added. There may be supplemental allowances for hazardous duties or hardships. Such payments are intended to cover cost-of-living differences and socio-economic differences borne by an employee and family living in a foreign country.

fringe benefits, Compensation other than direct wages or salary. May refer to a wide spectrum of employer supported services ranging from those mandated by law to those strictly discretionary on the part of the employer.

front end, A compensation arrangement covering a set period of time in which the greater portion or rate of compensation occurs during the early part of the time period.

gain sharing, A financial incentive plan in which the savings from improved productivity (savings gained) are shared between the company and the employee. (See profit sharing.)

giveback, Negotiated elimination from a future labor agreement of past benefits.

going rate, Usual or prevailing compensation for a job or job classification.

going wages, Usual or prevailing earnings for a particular class of employees. (See going rate.)

grade description, See job description, position description.

grandfathering, Maintaining of benefits or other advantages for designated individuals in specific instances because of seniority or prior circumstance. (See red circle rate.)

group incentive, An incentive plan in which individual worker earnings are based upon a share of the total incentive earnings reflecting the combined performance of the entire group.

group payment, Compensation amount paid to employees comprising an identifiable work unit. May be paid as incentive earnings.

group piecework plan, A method of compensating all workers in a designated group on the basis of total group piecework earnings.

guarantee, A minimum rate of pay, normally time rate for a job, upon which compensation for work is calculated. (See minimum guarantee.)

guaranteed annual wage, Minimum assured compensation under a plan providing guaranteed annual employment.

guaranteed base rate, Minimum pay rate for specific tasks or for work performed under special circumstances.

guaranteed earnings, See guaranteed rate, guaranteed time, guaranteed wage plan and guarantee on trial rate.

guaranteed employment, A plan providing a minimum number of days employment in a work year.

guaranteed hourly rate, An assured rate of pay which, when applied to the number of hours worked, would determine an employee's earnings.

guaranteed standard, An assured rate of pay based upon quantity of work produced.

guaranteed wage rate, An assured rate of pay for a given job or classification. Usually applied per hour worked.

halo effect, A biasing effect in an evaluation resulting from undue emphasis of one or a few, characteristics or attributes.

hazard pay, Extra payment for work performed while exposed to abnormal hazards or unsafe conditions.

high task, The 100 percent or normal level of production for a standardized task. Compares to low standard of 60 to 65 percent for unstandardized task.

high time, Extra payment for work performed while exposed to hazardous height conditions.

hiring rate, Also hire-in rate. Hiring or entrance rate for new employees.

hourly rate, Pay for each hour worked at a given job.

hundred per cent incentive, A wage incentive plan which, unlike a sharing plan, permits employees to receive the entire monetary value of direct labor savings resulting from above standard performance.

idle time, Time during which a worker is not working. (See down time.)

incentive earnings, Applies to standard hours earned under incentive plan. (See incentive pay.)

incentive opportunity, Earnings potential over and above a base wage for productivity which exceeds an established standard of performance.

incentive pace, Performance of a qualified worker at a speed above that established as standard or normal for a task or job.

incentive pay, Extra pay for work performed in excess of a set quantity or "standard."

incentive performance, Worker output in excess of a set predetermined standard of productivity. (See incentive pace.)

incentive plan, A method for compensating work performed over and above an established level or standard.

incentive plan, group, See group incentive.

incentive plan, piecework, An incentive plan directly based upon the number or quantity of units produced. Earnings are calculated on earnings per piece or unit of production.

incentive plan, standard hour, An incentive plan directly based on pay hours earned above actual clock hours worked. Incentive earnings are based on excess of earned hours over clock hours.

incentive rate, Rate of pay for incentive work. Rate upon which compensation for producing a given quantity of work is based. Differs from day work.

income, Any remuneration contributed by an employer which is perceived by the employee to have utility.

income security plans, Essentially a plan financed by an employer to provide weekly payments to laid-off workers. (See SUB.)

indirect compensation, Money paid by an employer for an employee to a third party such as hospitalization, life insurance, disability insurance and other fringe benefits or benefits derived by an employee from such services provided for by the employer.

indirect labor, Work which is not readily identifiable with the production of a product and which is normally applied as a service to those employees identified as direct labor.

individual job rate structure, See individual rate.

individual retirement account (IRA), Tax-deferred savings plan entered into by an individual. The intent of an IRA is to provide tax-deferred savings until period of retirement.

individual wage determination, The method of arriving at the actual wage to be paid an individual within a wage range. Such wage rates reflect the individual's performance, wage history, position in a range, experience, and other factors.

inequity adjustment, Pay adjustment for inequity. (See Applied Psychology.)

integrated salary structure, A compensation program that brings together theories, facts, discoveries, principles, and considerations of many fields and disciplines. The development involves sound, meaningful, and effective policy recommendations across a broad spectrum of compensation matters which anticipate changing conditions. Such programs go beyond the traditional labor market/collective bargaining surveys in establishing salary structure.

interference time, See Work Measurement and Methods.

job, A term used to identify a work assignment, a task or related series of tasks, a position classification or quantity of work. May be used in connection with either wage or salary activities.

job analysis, Systematic study of a job (or position) leading to a detailed specification of qualification and performance requirements for purposes of wage (or salary) administration, employee selection or training, determination of skill transfer opportunity, comparison and consolidation of jobs or organization design.

job breakdown, A job analysis detailing various elements pertaining to skill requirements, training and experience, mental and physical demands, working conditions, hazards exposure and requirements or responsibilities of performance.

job characteristic, An attribute of a particular job related, normally, to skill, experience, responsibility, effort or working conditions factors.

job class, A designation given to jobs grouped by some criteria such as pay rate, skill level, permanency, etc. (See job classification.)

job classification, A group in which a particular job is included. May also refer to the procedure by which jobs are related by pay rate. (See classification.)

job cluster, A grouping of jobs resulting from relatively equal job evaluation scores or related within an organization and linked together by technology, organization, or custom. A wide cluster would include all factory jobs; a narrow cluster would be assembly jobs or maintenance jobs.

job comparison scale, A listing of key job factors of each job being compared based upon their quantifiable values in order to permit relative analysis of job factors.

job content, Reference to basic job elements and scope of duties involved in a given job (or position) primarily those pertaining to skills, mental and physical demands and responsibilities.

job definition, See job description; job specification.

job description, An established written summary statement of a job or position which describes its functional requirements. It may include such specifics as equipment or tools used, physical and mental skills, training, working conditions, duties, responsibilities, location and designation of supervisor.

job evaluation, The procedure of rating, comparing, and ranking various jobs according to set comparison criteria to determine the basis for job pay. (See job analysis.)

job evaluation manual, A document defining policy, procedure, evaluation standards, jobs and the resulting analysis of these jobs under a system of job evaluation.

job evaluation point method, A method of job evaluation wherein job factors are measured according to a numerical index, the sum total of these numerical values for a given job becoming the relative measure of its value (and pay rate) to other jobs.

job factor, A defined characteristic or attribute of a job common to other jobs and suitable for basing a comparison of value or relative worth between jobs.

job freezing, Deferral of permanent or non-temporary assignment to a given job for a set period of time or until the person designated for the job meets job qualification requirements.

job posting, An announcement of job openings for the purpose of soliciting bids for promotion or transfer from employees.

job ranking, A method of job evaluation wherein various jobs are set in a compensation level array based on relative appraisal of job factors.

job rate, (1) The established pay for a given job. Also applied to the performance standard upon which pay is based. (2) Pay established for a specific job. May be designated by job classification or group in which job is listed.

job rating, See job evaluation.

job scope, The range of duties activities, responsibilities and authority involved in the proper performance of a job. (See job description.)

job sharing, The redesign of a full time job resulting in two or more employees holding responsibility for a single job, dividing time, salary or wages and fringe benefits.

job specification, See job description.

job standardization, The establishing of proper work methods, materials, tools, equipment and procedural routines by which a job or task is to be performed.

journeyman, A fully qualified trade craftsman. Designation given to those successfully completing an apprentice program.

journeyman rate, Rate of pay applicable to trade jobs requiring fully qualified craftsmen or journeymen.

jury duty pay, Compensation for absence caused by jury duty.

Keough Retirement Plan, An individual retirement account designed for self-employed persons.

key employee plan, A supplementary bonus or incentive plan designed to compensate key employees, usually department heads and department supervisors as opposed to officer/executive levels, for improved or measured performance.

key job, See benchmark job.

labor area, Geographical area from which workers may be recruited. (See labor market and labor market area.)

labor cost, The cost of human resources in wages that is a part of the cost of goods or services. Usually refers to those wages which go directly into the product, but may include the indirect functions which go to support direct labor.

labor class, (labor grade, salary grade, job grade, job class), A series of rate steps (single rate/2-rate range) in the wage structure. Labor classes are usually determined by job evaluation plan or by negotiations so that different jobs or positions of the same relative worth fall into the same grouping (class) and are therefore, paid the same rate.

labor grade, A pay or rate classification in a series of related classification levels. (See Applied Psychology.)

labor market, See labor market area.

labor productivity, The rate of output of employees per unit of time. Usually measured against a standard or expected rate of output.

labor relations, In an organization it is concerned with the working relationship between the employees and the organization (often called personnel relations). In a union organization, the working relationship between the union representatives and the management representatives.

labor-saving ratio, The mathematical relationship of the unit labor cost of one method to the unit labor cost of another; usually an improved method compared to an existing method.

layoff pay, Payment received because of termination. (See severance pay.)

learner rate, Pay rates for inexperienced employees payable for a period of time while developing proficiency on a job or until a set level of performance is attained.

learner's allowance, A special allowance attached to an incentive standard to encourage new employees or employees on new operations or products to reach incentive level performance.

learning curve, Also known as experience curve, a quantitative model relating the factors of time, productivity improvement and cost, applied to prediction of future costs under conditions of production volume increases.

legally required benefit, A provision of an employee benefit plan to which employers must contribute according to legal mandate.

leveling, A procedure for rating work performance in order to establish a standard.

leveled time, See also leveled elemental time.

line of progression (LOP), A designated sequence of related job classifications through which an employee may progress for purposes of promotion or demotion in cases of layoff or work cutback.

living wage, A term used normally in collective bargaining to mean the minimum compensation under current economic conditions which is perceived by workers as necessary to sustain a particular standard of living.

loose rate, A job rate which produces a higher payment than would be paid for similar work performed under correctly applied or maintained work standards.

loose standard, See loose rate.

long term contract, A multiple-year wage contract or collective bargaining agreement as contrasted to a one-year agreement.

low task, Performance level normal for unstandardized tasks.

lump sum award, One time payment of a bonus based upon an exceptional performance.

machine attention time, See Work Measurement and Methods.

machine controlled time, The time during a work cycle in which production is entirely dependent on machine performance.

machine down time, Time during a work shift when a machine is inoperable.

machine hour, A unit of production used to measure capacity and utilization of machines based upon standard hourly performance expectations.

machine idle time, Machine down time or time when equipment is unproductive during a work cycle or work shift.

make-up pay, Earnings permitted employees to compensate for lost time or earnings below an established minimum level.

make-work, See featherbedding.

man-hour, A unit of productivity equivalent to the output of one worker performing at standard normal pace for one hour.

man-hour output, A productivity measure in which numerical quantity produced is divided by hours applied to the production of that quantity.

man-minute, A unit of productivity used in some incentive plans and equivalent to one worker performing at standard normal pace for one minute.

manning table, A tabulation of necessary positions to properly staff an operation or task force.

manual element, See Work Measurement and Methods.

market comparison study, Survey of wages paid by other companies for like or similar services.

maturity comparisons, The consideration given education, experience, age, seniority, performance, and other factors in determining an individual's wages. (See individual wage determination.)

maturity curve, The chart or curves which relate an individual's years of experience and level of responsibility to compensation within a given profession or field as opposed to compensation rates within a particular organization.

measured day work, A compromise plan combining aspects of a regular incentive plan and a no-incentive time plan.

merit budget, Allocation of funds to identified groups of employees for merit award use.

merit increase, An increase in compensation based upon a formalized system of appraising employee performance.

merit pay, Compensation over and above an employee's pay rate based upon a plan for rewarding performance meeting set criteria for excellence.

merit pool, A fund from which merit payments are allocated. May be based upon a formula for accumulation reflecting improvement in operating profit, loss reduction, productivity, or other criteria.

merit rating, A formalized system for recording and rating employee performance. An evaluated performance given an employee.

merit rating increases, See merit increase.

method, The manual procedures associated with the accomplishment of work.

methods engineer, A person who, by specialized training, analyzes and designs work methods and procedures. (Formerly, motion study engineer.)

methods engineering, Analysis and design of work methods and procedures. (Formerly identified with motion study. See motion study.)

methods study, Detailed study of work aimed at establishing the optimum or best way to perform the desired task. (Formerly identified with motion study. See motion study.)

minimum funding, The least amount of reserve investment necessary to back an employee welfare plan.

minimum guarantee, See guarantee.

minimum time, The least time wherein a task may be reasonably expected to be accomplished.

minimum wage, Lowest rate of pay guaranteed by law for a given classification of employee.

moonlighting, Practice of maintaining regular employment with more than one employer. Term refers to employee practice of taking employment in secondary job during evenings or nights. Reverse arrangement is sometimes termed "sunlighting."

motion study, Motion analysis. The study of the basic manual movements and procedures associated with the accomplishment of work. Historically associated with the Gilbreths, in contrast to time study, associated with Taylor.

motion-time analysis, Motion analysis with time consideration.

multiple piece rate plan, See also differential piece work. A wage incentive plan using two or more piece rates. One rate is paid for production of a fixed quantity in a given time, and a higher rate(s) is paid for production beyond the initial level.

multiple time plan, See also differential time plan. A wage incentive plan that pays a higher base wage rate as production levels increase. A series of step bonuses for increased production.

nonexempt employees, All employees covered by the Wage and Hour Law and not exempted from the requirements of that law.

non-financial incentive, All influences, other than financial, which tend to stimulate employees to higher levels of productivity such as promotion, training, competition with others, recognition, or praise.

non-financial incentive plan, The programs designed to administer the various non-financial incentives such as praise, recognition, training, or other socially valuable consideration.

non-financial rewards, The rewards given an employee for his/her services which have no direct financial value. Such rewards might include recognition for certain achievements by letter, postings, certificates, or announcement.

non-performance pay, Payment in recognition of certain events or occurences such as service, anniversary or generally observed holiday.

non-productive labor, Term sometimes applied to indirect labor.

normalize, To adjust observed data to a standard base.

normal operator, See Work Measurements and Methods.

normal pace, Work pace associated with normal or standard production. (See normal performance.)

normal performance, The work output of a qualified employee working at standard pace in accordance with standardized work procedures.

normal time, The time required by a qualified worker in a standardized job to produce 100 percent of a standard quantity of output.

OASDI, Old age, survivors, and disability insurance benefits.

occupational rate, Wage or salary for a job or position. (See job rate.)

occupational wage, See occupational rate.

operation, A job or task consisting of multiple work elements.

operation analysis chart, A special form (chart) used to analyze an operation and show all important factors, such as material, shipment, time, labor, etc., affecting the operation either for instructional purpose or to assist in improving the operation.

out-of-line rate, A job rate which is higher or lower than that determined as proper for the work involved based upon a systematic evaluation of a number of related jobs or in relation to a job pay regression line established through job evaluation. (See red circle rate.)

overtime, Time worked in excess of or at times other than the specified workday or workweek. Time period other than that established as the regular work shift or workday.

overtime pay, (1) Pay at higher than regular job rates for work performed at times other than the regular work shift or work day. (2) Payment at premium rate for work performed in excess of or at times other than specified workday or workweek.

pace, Rate of output or performance compared to an accepted standard. May be expressed quantitatively in terms of units per time or in terms of percent relative to standard. (See pace rating.)

pace rating, A worker's speed of performance as compared with standard or normal pace. (See performance rating.)

participation point, Established level of job performance at which employee is eligible for additional earnings or compensation.

pay grade, An assigned number (grade) which relates to a compensation range or given labor grade or labor class.

pay structure, A tabulation of wage scales for groupings of various jobs. May be applied to arrangement of wage levels or adjustments within a given job or compensation category.

percentage increase, The percent increase of wage or salary rates from the original rate as opposed to a dollar increase added to wage and salary rates.

performance, The degree with which an employee applies skill and effort to an operation or task as measured against an established standard. Standard time divided by actual time expressed as a percentage.

performance appraisal, Supervisory or peer analysis of work performance. May be made in connection with wage and salary review, promotion, transfer or employee training.

performance efficiency, A comparison of actual performance to normal.

performance rating, Observation of worker performance to determine productivity in terms standard or normal.

performance rating factor, See performance efficiency.

performance rating scale, Correlation of performance with a numerical index usually designating standard or normal performance as 100.

performance standard, See standard performance.

Perk (perquisite), Employer-paid benefits or expenses of employees extending beyond normal fringe benefits, paid compensation.

personal rate, See red circle rate.

personal time, See personal rate.

personnel relations, See labor relations.

phantom stock, Also called "shadow stock." An involved financial compensation technique in which a company buys stock for a particular employee without the use of that employee's name and at an agreed-upon time turns the proceeds or the stock itself over to the employee or an amount equal to the worth of the transactions if the company elected not to actually purchase the stock. Either way, the name of the employee is not involved in the transactions until payment is made . . . a form of deferred compensation.

phantom stock plans, The actual programs or procedures used to execute the purchase of phantom stocks.

piece rate, Established payment amount for each unit or specified quantity of work completed.

piece rate plan, A method of wage payment based on piece rates. (See piecework plan.)

piecework, Production for which wage payment is based on piecework rates.

piecework earnings, Wages received for work produced while on piece rate assignment.

piecework plan, A method of wage payment based upon a set amount of pay earned for each unit of production. (See piece rate plan.)

piecework rate, A set amount of pay for a unit of production or worker output.

piecework without base guarantee, A piece work pay arrangement wherein the amount of pay strictly reflects the amount of production or output with the extreme of no pay for no production.

point, An increment of value applied to a job factor when measured in connection with a point factor job evaluation plan.

point plan, See job evaluation point method.

point method, See job evaluation point method.

point system, See job evaluation point method.

portal-to-portal pay, (1) Payment for travel to and from a worksite from the point at which an employee enters the plant or arrives at his place of employment. (2) Payment for travel time from a point of entry to the job site and return.

position, A job or work classification. A collection of related tasks, duties, responsibilities and authority.

position class, A group designation for various positions based normally on compensation rate but which may reflect other considerations (i.e., exempt or non-exempt).

position description, See job description.

position guide, A position description usually expanded to encompass a definition of working relationships with other individuals or groups, general responsibilities and overall aspects of working procedure.

potential earnings, Projected level of compensation which may be reasonably expected from a high level of performance.

premium, (1) Payment over and above a regular job rate for various reasons such as producton performance above standard, or for working overtime or on other than regular shifts. (2) Compensation at greater than regular rate. Usually directly based on hourly rate time during which work is performed or quantity of work produced as opposed to bonus.

prevailing wage, Level of wages generally or usually paid for specific jobs or occupations in a given location. May also include payments for fringe benefits.

probationary rate, Rate of pay during initial or trial period of new employee. May apply to other employees on new assignments.

production, The generation of goods and services.

production bonus, Extra payment for increased worker output. (See incentive pay.)

productive labor, Term sometimes applied to direct labor.

productive time, Period when worker is performing assigned task. Sometimes called "up time" as opposed to "down time" when work is delayed.

productivity, The relationship of output and all associated inputs.

productivity bargaining, Negotiation of contractual obligations to increase productivity expressed in a collective bargaining agreement in exchange for increases in employee compensation or benefits. May have form of givebacks (q.v.).

productivity measurement, A gauge of productivity using real or physical volume factors used for promoting productivity and making budget and long-term projections.

profit sharing, (1) Any procedure in which employees receive, in addition to their regular pay or wages, a share of the profits of the business. (2) A disbursement of business profits or savings paid in addition to regular compensation usually based on a specific formula and in recognition of employee contributions to improved business performance.

profit sharing plan, The plan and/or procedures governing the distribution of a portion of an organization's profit to employees. The plan may be based on a single factor, such as a percentage of the profits, or on multiple factors such as profit, length of service, or position. Different amounts may be distributed among different groups of employees such as hourly, salaried non-exempt, exempt, and executive.

profit sharing trust, In a deferred profit sharing plan, the money due employees under a given formula is invested via a trust fund and the income and principle is distributed to the employees at time of separation. The investment vehicle is referred to as a profit sharing trust.

progression, Movement to a higher level of work or to a higher level of pay in same grade. (See line of progression.)

progression, automatic, See automatic increases.

promotional pay, Additional payment usually in the form of a bonus for stimulating sales or for accomplishing specific sales program objectives aimed at increasing sales performance.

purchasing power, Measure of level of compensation in terms of equivalent goods and services for which compensation may be exchanged at current market prices.

push money, An incentive payment for salespersons.

pyramiding, The application of multiple pay premiums resulting in greatly increased earnings.

quality bonus, Additional payment for maintaining or attaining specified levels of excellence of work output.

qualified operator, A worker judged to have met the minimum requirements to perform a given job.

qualified stock option, A stock option granted after December 31, 1963, which meets certain IRS codes and, allows participants to take advantage of such tax benefits as capital gains which are not allowed under other options.

quota hiring, Employment of persons in accordance with a pre-determined plan aimed at changing the composition or past characteristics of a work force or group.

ranking, See job ranking.

rate, Pay for work based on a set unit of time or productivity.

rate change, An adjustment in a production standard.

rate cutting, A reduction in a job standard not justified by compensating changes in job content.

rate erosion, A gradual development of rate incorrectness resulting from gradual changes in job conditions without compensating adjustments in the job standard.

rate range, See Applied Psychology.

rate setting, Establishing a standard time or pay for incentive work.

rating, See speed rating.

real wages, Earnings measured or correlated to actual purchasing power.

reclassification, Change in compensation grade or level of a given job or position resulting from changes in job duties, responsibilities or working conditions.

red circle rate, A job rate which is higher than justified by job evaluation which is maintained under specific circumstances for either a specified or indefinite time period.

reevaluation, The process of review of evaluated jobs for the purpose of determining changes which may lead to a new job rate.

regression line, The line of best fit among an array of job scores in a job evaluation analysis designating a trend or relationship between the jobs.

regular element, Element, a sub-division of the work cycle composed of an identifiable sequence of several basic motions.

relocation reimbursement, Payment for travel, meals, lodging and related expenses in connection with an employee's move to another location.

residency period, The time an employee must perform a given job in order to have established seniority rights to that job or to have adequately prepared for progression to an advanced job for which the lesser job is prerequisite.

restricted job, Work limited by conditions beyond the control of the worker such as machine speed or process cycle time.

retroactive pay, Past wages due for work performed during a prior time period such as may result from an adjustment in rate of compensation or settlement of a pay claim. Often called back pay.

runaway rate, A job rate which produces excessive payment. (See loose rate.)

salary, The fixed monetary compensation paid to an employee, either weekly, monthly, or yearly, for services rendered — usually based on a certain minimum number of hours per day or week. (See straight salary.)

salary administration, The methods, procedures, and policies used to control and guide various salary programs and plans.

salary compression, Occurs when new employees are hired in at pay rates (salary) close to, equal to, or even greater than the pay rates of present employees in the same or similar positions. Most prevalent during periods of high inflation.

salary curve, A curve, often linear, on a graph describing the relationship of jobs, either in points or grades, to the compensation for each of the jobs, often with job values in points or grades along the horizontal axis and dollar compensation along the vertical.

salary guide, A listing of compensation amounts for a given occupation group in order of progression based upon an application criterion for advancement such as time in position or level of job qualification.

salary differential, The difference between the rate of compensation from one position to another anywhere on the salary curve.

salary plus bonus, A type of salary compensation plan in which an employee is paid a base salary not tied to performance and a bonus over and above the base salary based upon a formula measuring performance. The bonus is usually based upon some percent of base salary.

salary plus commission, A type of salary compensation plan in which an employee is paid a base salary not tied to performance and a set dollar amount or a percent dollar amount of the value of all units sold or produced above a given level.

salary profile, A tabulation showing numbers of persons at various compensation levels in a given job, occupation, organization, geographical location or other grouping.

salary ranges, high and low designations of compensation for a given job or position. May incorporate designation of pay progression amounts from low to high applicable on basis of set formula or policy.

salary scattergram, An arrangement in graphical or matrix form of numbers of employees at each designated compensation amount or grade.

severance pay, Payment received because of permanent termination. (See layoff pay.)

sharing plan, Payment to employees of part of productivity gains over and above an established goal or standard.

shift differential, Added compensation for work performed on other than the regular day shift.

shift premium, See shift differential.

single rate, Same pay for a given group of workers.

skill, A factor of comparison in job evaluation plan. Relates to manual and mental dexterity. Developed through practice, training and experience.

sliding scale wage plan, A wage which links wages directly to profits, prices, sales, or output.

speed rating, The process of observing and evaluating worker performance by comparison with standard or normal work pace. (See performance rating.)

speedup, See Applied Psychology.

standard, An established norm of productivity defined in terms of units of ouput per set time or in standard time.

standard allowance, An established time provided for compensation for anticipated circumstances. (See allowances.)

standard allowed time, A unit time value for the accomplishment of a task inclusive of adjustments or allowances for various circumstances.

standard conditions, See Work Measurement and Methods.

standard data, A structured collection of normal time values for elements of tasks or jobs. (See standard.)

standard hour, The time earned for producing an amount of work equal to the production quota for one hour.

standard hour plan, A method or formula of compensation based on earned standard hours credited on basis of output during a given time period.

standard hours produced, The total output of a worker during a given time period expressed in terms of standard allowed time and calculated as the sum of earned time for all work produced during the period.

standard performance, Level of performance required to accomplish a standardized task or job in the standard time established.

standard rate, See standard time.

standard time, A unit time value for the accomplishment of a work task or job.

standard time data, A structured collection of time values for elements of work performed under standard conditions at normal pace. (See standard data.)

starting rate, Rate of pay during probationary period, during periods of training or for a period of time following a job transfer.

status pay, A form of a non-financial reward, such as office furnishings or office size, given to an employee by virtue of position within the organization.

step bonus, A plan of compensation under which payments are provided in increasing progression for increased productivity or accomplishment of work-related goals.

step rates, Incremental levels of compensation for a given job.

stock bonus, Stock in a company awarded to an employee by the company, usually for a particular reason such as personal performance or corporate performance, and paid for by the company.

stock option, The privilege granted to certain employees to purchase shares of the company's stock as of a given date and price at some time in the future when, it is assumed, the stock will be selling on the market at a higher price.

stock option discount, The difference between the quoted market price for a given stock and the price quoted to an employee at the time a stock option is granted.

stock option exercise, The actual purchase of shares of stock at the option price after a predetermined date.

stock option forfeit, The election by an employee not to exercise his/her rights to purchase stock under the option plan. Usually occurs when the market price falls below the option price.

stock option holding period, The time a participant in a stock option plan must hold purchased stock in order to qualify for the special tax treatments (capital gains) available. Usually refers to qualified stock option plans. Under laws governing qualified plans, participants had to hold stock for three years to take advantage of capital gains and exercise options within five years.

stock option (non-qualified), A stock option granted to employees that has fewer restrictions than a qualified plan. The option may be priced below market value, the time duration may be longer than five years, and more recently lower-priced options may be exercised before higher-priced options still outstanding. A nonqualified option is not eligible for capital gains treatment and, when exercised, becomes taxable income to the employee.

stock option plans, Plans allowing employees and/or officers the privilege of purchasing company stock at a certain price on certain dates. The formal policies and procedures that outline how stock options will be administered. The usual plan would define participation, amounts of stock, prices, time limits, and methods for exercising rights.

stock option term, The period of time in which a participant has the right to purchase stock. The maximum for qualified plans in five years.

stock purchase plan, A plan enabling employees to purchase shares of stock in a company, with or without employer contribution, generally under more favorable terms than those available on the open market. The formal policies and procedures that outline how employees may purchase stock and the method of administering the plan.

stock warrant, An option to purchase common stock at a fixed price intended to be favorable in the future. Considered a common stock equivalent. May be issued by a company to an employee or purchased by anyone in an open market.

straight commission, Compensation, usually for sales or service personnel, based on a predetermined percentage of the value of sales or services rendered.

straight salary, Compensation based on a fixed amount for a given period of time (per hour, per day, per week, per month, or per year).

straight time, Time worked at regular or base job rate.

straight time rate, Rate of pay applicable to regular time work.

SUB (income security plan), Supplemental Unemployment Benefit plan. An agreement negotiated between Ford Motor Company and the United Automobile Workers in 1955 and later extended to other companies. Such plans provide regular weekly payments to laid-off workers receiving state unemployment insurance, through funds financed by the employer. Additional pay given to laid-off employees over and above the state supplied unemployment insurance payments.

sunlighting, Secondary employment during daytime hours. (See moonlighting.)

supervisory incentive, An incentive plan designed to reward financially supervisors over and beyond their base pay for meeting predetermined objectives for productivity and costs of the operations under their supervision.

supplemental compensation, Money paid an employee over and above the agreed-upon base rate usually for exceeding predetermined targets or achieving or exceeding productivity standards. Can be either commission or bonus.

survivor protection, Compensation paid to survivors of an employee. Most commonly provided by group life insurance plans.

synthetic time standard, A time standard developed from data compiled from prior study or research rather than from study of actual job to which standard is to be applied.

take-home pay, Spendable earnings. Compensation paid less deductions. Represents net earnings.

take-home wages, See take-home pay.

task, The standard quantity of production expected under normal conditions. Usually designated as the point above which incentive pay is earned under a wage incentive plan.

task and bonus plan, An incentive plan wherein a set compensation amount is given an employee for output above a predetermined level or task.

tax shelter, For compensation purposes, usually refers to a plan for postponing declaration of earnings for tax purposes to a future period of time when tax rate for an individual is expected to be lower.

temporary rate, A standard applied for for a short period of time for special reasons or circumstances.

temporary standard, See temporary rate.

termination pay, See severence pay.

tight rate, A time standard which does not permit the usually expected earnings under a wage incentive system.

tight standard, See tight rate.

time rate, Compensation rate based on hours worked i.e. hourly wage rate. (See time wage.)

time rate earnings, Compensation based on time rate multiplied by the amount of time of the work period.

time standard, Designated time for the accomplishment of a specified task or job.

time study, A work measurement technique used to establish time standards. (See motion study.)

time wage, Compensation based on hours worked rather than quantity produced.

total compensation, The composite of all payment for work including benefits.

touch labor, Work directly applied to material or product. Refers to those workers actually in physical contact with product during manufacture.

transfer pay, Compensation for period of temporary reassignment or for period of time subsequent to permanent reassignment.

unemployment pay, Compensation to a person no longer employed.

unit of production, A measured quantity of output. May be based on variables such as physical count, weight, or accumulated standard time.

unrestricted job, A job in which the output of production is limited only by the ability of the worker to produce.

upgrading, Promotion of employees to higher paying jobs. May also be applied to upward revision of a given job such as may result from reevaluation of that job.

VEBA, Voluntary Employees' Beneficiary Association intended to provide life, sickness, accident or other benefits to members. Usually established jointly by a company and a union.

vesting, An earned guarantee of equity or share in a benefit plan, usually applied to pension plans. (See Applied Psychology.)

wage, payment for work performed. Usually applied to blue-collar work as opposed to salary applied to white-collar work.

wage arbitration, The adjudication by a neutral third party of wage disputes. (See Applied Psychology.)

wage differential, Difference in compensation of similar jobs.

wage drift, Gradual change in wages over a period of time.

wage guide, See salary guide.

wage guidelines, Standards used to measure compliance of wage increases with established policy objectives.

wage incentive, Payment for performance over a predetermined level or standard.

wage incentive payment, That portion of compensation reflecting earnings above base pay for productivity in excess of standard or normal under a wage incentive program.

wage incentive plan, A defined procedure or method for providing workers opportunity for extra compensation above base pay for productivity in excess of a set standard.

wage inequity, A disparity or unadjusted difference between compensation amounts for comparable jobs.

wage leaders, An employer or group of employees accepted as trend setters in a labor market.

wage pattern, Identifiable trends or similarities in job compensation.

wage profile, See salary profile.

wage rate, See rate.

wage and salary administration, The managing and administration of the wage and salary structure of an employer. (See Applied Psychology.)

wage and salary structure, An arrangement in an ordered manner of compensation rates reflecting relationships between jobs. May also include supplemental benefits.

wage and salary survey, The collection, tabulation and study of compensation data for various comparable jobs.

waiting time, Sometimes termed dead time or down time.

wash-up time, Also known as clean-up time. Allowance for personal needs at end of work shift.

waste bonus, Compensation for performance in reducing scrap, damage, material losses or sub-standard work.

work, The application of various skills and effort to the accomplishment of a given task or series of tasks. (Note: This is definitely *not* the physicists' concept of work, which troubled Taylor in his pioneering work.)

work content, The sum total of operations, methods, activities and procedures required to accomplish a given task or series of tasks.

work cycle, One complete sequence of operations, activities and procedures required to accomplish one unit of a given task.

Work Hours Act, Legislation applicable to federal public work or any work in which the federal government is involved and which provides an 8-hour workday, a 40-hour workweek and overtime at 1½ times basic rate.

work load, The quantity of output or production required over a given period of time.

work measurement, The process of determining the standard time for performance of a standardized task using any of a number of industrial engineering techniques. (See Work Measurement and Methods.)

working conditions, The environment under which work is performed with reference to heat or cold, light, noise, ventilation, dust, vibration, fumes, moisture and exposure to hazards and injury-producing conditions.

workmen's compensation, Also called worker's compensation. A form of insurance paid for by the employer through a private carrier or state government agency to cover compensation of an employee involved in work-related illness or injury that causes an employee to miss work or incur other costs, usually medical, related to the illness or injury.

workplace, The location at which a worker performs his job.

workplace layout, Specification or design showing positions of worker, extent of working area, contents and arrangement of workplace, location of materials and equipment.

work sampling study, A procedure for observing work activities at random intervals of time from which statistical inference may be made relative to the entire scope of work involved.

work simplification, An organized approach to improvement of worker performance involving training, implementation of improvements and follow-up.

work standard, See standard work.

work study, A procedure for observing work activities. Procedure may vary as to length of time and detail observed and may be based on sampling techniques. (See work sampling study.)

work ticket, A specification of work required. An authorization to proceed with a particular job. (See also work order.)

worktime, The time a worker or process is productively involved with a job. May also be used to designate a period of time such as a work shift.

Anthropometry

Z94.15

15

Editorial Note: Even the most casual reader will note that the data in this section are treated differently from those in the other sections: the reasons are many. First and foremost, a straight alphabetical arrangement would not be helpful. For example, why should "knuckle" follow "knee" when the reader is probably interested in more leg dimensions? Clearly a diagramatic arrangement would be more helpful and logical. However, for the sake of consistency and helpfulness, the anthropometry terms can be found in alphabetical order in the overall INDEX. Nevertheless, the reader should be alerted to the fact that when a term is indexed as being in the ANTHROPOMETRY section, the reader should refer to at least one — if not all — of the three listings in this chapter: DIMENSIONAL TERMINOLOGY, PLATE LIST, and CROSS REFERENCE LIST. Also the reader should refer to the GLOSSARY. Although these details are carefully delineated in the following INTRODUCTION, it would not be amiss to emphasize here that the numbers found in connection with the PLATES also appear in the DIMENSIONAL TERMINOLOGY. Thus, e.g., "knee-to-knee breadth, sitting, IV 30" can be found diagramatically in PLATE IV, etc. As for actual numerical dimensions, the reader should note the details in the various references in the BIBLIOGRAPHIES. Again, of special interest to engineers, designers, inventors, et alia — each of the ANTHROPOLOGY terms notes relevant applications: e.g., "knuckle height . . . carrying handles . . . suitcases . . ."

W. J. Jaffe

The ANTHROPOMETRY SUBCOMMITTEE was made up of the following:

CHAIRMAN:

Charles L. Mauro, C. L. Mauro & Associates, Inc., New York, NY 10036

MEMBERS:

Mary A. Maney, C. L. Mauro & Associates, Inc.

Carlos M. Medina, Department of Anthropology, Columbia University

Patricia A. Moore, Society of Gerontology (NYC Chapter)

Tado B. Warashina, Warashina Associates, Palo Alto, CA

INTRODUCTION

Anthropometry, the study of people in terms of their physical dimensions, has classically been performed by the physical anthropologist. More recently, in an attempt to create more efficient people-machine interfaces, engineers and designers have become increasingly aware of anthropometric terminology and dimensions.

This increased awareness on the part of practicing professionals has brought to light the need for a standard glossary of terms related to anthropometry as it is used in the fields of engineering and design. Unfortunately, most designers and engineers are not versed in medical terminology, nor are most anthropologists familiar with engineering. As a result, the majority of available references fail to communicate to both audiences. Therefore, the Z94.15 Committee was formed for the purpose of identifying those dimensions required in industry, translating their medical definitions, and compiling the resulting terms into a working glossary.

The main purpose of the standard is intended to be that of a technology reference. As such, the practicing professional should be able to consult the standard when encountering unfamiliar terms in texts or when devising a proper reference term for further definition of written or presented material.

A historical overview of the development of anthropometry is discussed in order to provide the reader with an awareness of the varying backgrounds from which anthropometry grew and also, as a result, an indication of the problems of application to the disciplines of industrial engineering and design.

A HISTORICAL OVERVIEW

The following work is intended to give the reader a reference source on the measurements most commonly taken in anthropometric research, as well as their relevance for equipment design. Since every human being that has ever lived on this planet has been physically changing every day from conception to birth to that final condition known as death, it is extremely important to have some means of quantifying the variations of these physical changes. *Anthropometry* is the formal name for the techniques used to express quantitatively the form of the human body. Thus, anthropometry is the measurement of man, whether living or dead, and consists primarily in the detailed, accurate, and consistent measurement of the dimensions of the human body (Montagu, 1960:3).

In 1920, the physical anthropologist, Ales Hrdlicka, summarized the purposes for taking anthropometric measurements. He stated that these measurements were taken for the purposes of industrial design, artistic expression of the human body, military purposes, medical, surgical and dental research and procedures, detection of bodily defects and their correction and forensic identification (Hrdlicka, 1920:8). His approach was narrowed considerably since he did admit later on in his treatise (ibid., 1920:8) that the "industrial and artistic systems are of little interest to us."

To practitioners of anthropometry and scholars in anatomy and physical anthropology, it soon became evident that standardization of measurements and measuring techniques was sorely needed and that this methodological disarray was very prejudicial to progress in the field. From 1870 to 1920, many international meetings were set up to discuss, refine, and standardize anthropometric techniques. Among the most significant of the conferences were the first and second craniometric conferences held in Munich and in Berlin in the years 1877 and 1880 respectively and the thirteenth General Congress of the German Anthropological Societies, held in 1882 at Frankfurt-on-Main.

It must be noted that the field of physical anthropology was concerned basically with anthropometric measurements until the early 1950's when the theoretical orientation of the field changed radically and interest in anthropometry was replaced by an ever-growing interest in human evolutionary processes.

Studies of movements of the body were undertaken in industry in this country during the second decade of the present century in order to improve per capita work output. During the 1920's, many management personnel realized the value of placing the work within easy reach of the workers and this led to a large amount of research geared to studying maximum and normal work area dimensions. During the 1930's and the 1940's, a large amount of military anthropological research was done on establishing effects of bodily dimensions and physiques on the design and use of such military equipment as apparel, aircraft, battleships, tanks, etc. These studies helped in synthesizing data from psychology, physiology, anthropology and medicine with engineering to form a field generally known as human engineering.

After World War II, research emphasizing the fitting of the machine to the man became more fully developed as both commercial and military outfits continued to perform research on body dimensions and work space requirements. The design of motor vehicles and mass transit facilities was reevaluated in light of this new orientation of man-machine design specifications.

Recently, it became necessary, as the technological revolution continued, to change the approaches, the types of data gathered, and the measurement instruments in anthropometric research. Dynamic anthropometry was established as it was necessary to develop a scheme of measurement taking into account three-dimensional coordinate systems. The science of biomechanics (essentially the interdisciplinary study of the mechanical nature and behavior of biological materials) is now closely related to the fields of classical anthropometry and engineering anthropometry.

With the advent of the computer it is now possible to consider the effect of more than two measurements at a time through the use of multivariate analytic routines in which one does not have to determine dependent and independent variables, and can determine the interaction between and within a large amount of variables at the same time.

TYPES OF DIMENSIONS

As mentioned in the historical overview, different types of dimensions have been defined for different applications: these include static and dynamic anthropometry. Static dimensions may be subdivided into circumferences, lengths, skinfolds, and volumetric measurements. Dynamic dimensions include link measurements and center of gravity measurements. Static and dynamic anthropometry are also referred to by the names structural and functional anthropometry, terms which more explicitly express the body and its action.

Static dimensions are taken with the particular body part held in a fixed posture. Dynamic dimensions, such as links, are likewise taken with the body part if a fixed position, however, the endpoints are defined according to functional capability of the body rather than according to bony or superficial landmarks. A body part link is the straight line connecting the axes of rotation of the joints. Body link data, unfortunately, is not based upon as extensive populations as data for static dimensions.

SKINFOLDS

The problem of the quantification of body builds (somatotyping) has been investigated for many years. One of the earliest attempts at quantification of the bodily physique was undertaken by J. Matiegka (1921) who broke down the human physique into four (4) components: gross body weight, skeletal, dermal, and muscular components. Matiegka's study was important in that he pointed to a direction of considering the human body as capable of being measured in a functionally oriented, "dynamic" fashion.

Unfortunately, Matiegka's article was long ignored in anthropometric circles until quite recently. Using more sophisticated mathematical models, Tanner et al (1959) and Tanner, Healy and Whitehouse (1959) reconfirmed with slight modifications Matiegka's original observations by performing a factor analysis of body build and finding that there are four orthogonal (independent) factors: skeletal frame size, skeletal width, muscle width, and fat width (Tanner et al, 1959:91). (Factor analysis is a multivariate analytical procedure which identifies the independent factors that explain most of the variation in a sample in decreasing order of importance.)

It must be noted that the measurement of body composition is an important advance from the grossly inappropriate height-weight tables that are currently enjoying great popularity. Taking someone's weight and comparing the resulting deviation from the corresponding "ideal weight" is almost biologically meaningless. One does not have any idea of the amount of muscle versus the amount of fat in the body. The visual system of "body typing" developed by W. C. Sheldon in 1940 is far from desirable when scrutinizing the human body in a functional frame of reference.

We shall briefly consider here the role of the skinfold as an indicator of "fatness vs. leanness," i.e., as an estimator to the individual's muscularity. The relevance of this procedure to apparel design will then become apparent. However, even though at present, the skinfolds' primary area of application would be apparel design and for biomedical and physiological studies, this does not negate their potential for use in the very near future, especially with regard to dynamic anthropometric studies.

Definition and Methodology: The procedure known as "skinfold" is used basically to determine the amount of sub-cutaneous fat. It must be noted that the thickness of the skin is relatively constant throughout the body (Montagu, 1960:85). However, the pattern of fat deposition is highly variable both within and between individuals and populations. As a result, the choice of the site for the taking of a skinfold measurement is critically important. At present, there are two (2) sites where comparable data on skinfolds can be extracted.

Upper arm skinfold: This is one of the least culturally objectionable and most readily accessible sites on the body. Individuals of both sexes in most cultures can be measured at this site with the least amount of tension and conflict. The site is located at the back of the right or left (it is important to be consistent with the choice of side) upper arm over the triceps muscle at a level approximately midway between the tip of the scapular acromial process and the elbow. The forearm is conventionally flexed at a 90° angle. Once the site is located, the subject then lowers the forearm so that the arm hangs freely as the measurement is taken by the investigator.

Scapular skinfold: The subscapular skinfold is taken with the subject standing freely. The skin is lifted most readily along a line at about a 45° angle, going medially upwards and laterally downwards. At this site, the thickness of the subcutaneous adipose tissue layer is fairly consistent.

Other sites: A variety of other sites, such as the midaxillary line, chest, abdomen, and legs have been proposed, but the variability encountered with the amount of subcutaneous adipose tissue, as well as cultural conflicts, makes these sites less desirable and as a consequence, they hardly ever are used.

Summary: In order to assure comparability of skinfold measurements, the following factors should be borne in mind:

(1) the "skin" should be lifted by firmly grasping the fold between thumb and forefinger;

(2) the width of the skin enclosed between the fingers should be minimal, but still yielding a well defined fold;

(3) the depth of the caliper placement should be the minimal distance from the crest where a true fold, with surfaces approximately parallel to each other, occurs; and

(4) the side chosen (right or left) and site locations should be consistent at all times.

Finally, note that upper arm circumference minus skin and subcutaneous adipose tissue (skinfold) is a gross measure of the muscularity of the subject.

INTENDED USE OF THE STANDARD

Only static dimensions and skinfolds are defined in this standard. The static dimensions which are included herein are those most frequently referred to in design and engineering and have a well established military data base. A few of the dimensions also have an established civilian data base. There are many dimensions not included, but these are available from the secondary bibliographic sources listed. Clearly, the standard is not in tended to be a complete listing of available anthropometric dimensions.

Although the majority of head dimensions included do not have a well established data base, their definitions have been included due to their importance in head equipment design.

Dimensions specific to safety in the design of furniture and toys for children have not been included, but appropriate references are included in the bibliography.

Each definition is cross-referenced to a drawing which illustrates the anatomical or body surface landmarks which comprise the dimension.

Sample
popliteal height, sitting (III, 25)

Plate Number Dimension Number

A Cross-Reference List of Dimension Numbers with Dimension Names has also been included since the reader may not immediately know the dimension name by inspection of the plates. In referring to the drawings, locate the number corresponding to the dimension in which you are interested; then, refer to the Cross-Reference List to locate the name of the dimension.

A Glossary of terms is included following the Cross-Reference List for explanation of anatomic terms. The reader may find that referring to the drawings in conjunction with use of the glossary explanation will clarify the term.

The drawings included represent only the adult population since bone maturation is completed by then. The focus of this reference is toward design for the adult population. A male skeleton has been used in the drawings, since, in most cases, the dimensions are similarly taken for male and for female, although the actual data bases are distinct, statistically speaking. For data taking in children's populations, some dimensions would require redefinition inasmuch as certain of the bony landmarks are not well-developed and thus are not palpable.

BIBLIOGRAPHY

If the reader needs further information on actual data bases or methods of taking anthropometric data, he should consult the annotated bibliography. The bibliography is subdivided into two sections: primary and secondary. The primary bibliography lists those sources from which the dimensions in this standard were selected. In some cases, the dimension definition has not been changed at all; changes were made only in those instances where it was necessary for clarity, i.e., by providing language understandable by the layman. The drawings, however, in this standard do include the skeleton as a reference for the dimensions whereas the figures shown in the primary sources did not provide skeletal reference.

The secondary bibliography lists sources of static and dynamic anthropometry which include a more extensive listing of dimensions as well as a data base specific to children's dimensions. Other sources included present methodological information and practical evaluations of the use of static and dynamic anthropometry.

arm reach from wall (II, 8), The horizontal distance from the wall to the tip of the longest finger, (usually the middle finger). To measure, the subject stands erect, with heels, buttocks, and shoulders (interscapular region) pressed against a wall, with the right arm and hand extended forward horizontally and maximally in the sagittal plane. To ensure that both shoulders are equally pressed against the wall, extend both arms. By convention, the right arm is then measured. The relevant application is forward arm reach which is the maximum forward distance reachable by the fingertips.

bicanthic diameter (IX, 46), The maximum distance from the most lateral point of the junction of the upper and lower eyelids of the right eye to the most lateral point of the junction of the upper and lower eyelids of the left eye. The relevant application is the greatest lateral distance required for proper design of eye protection and optical equipment, e.g., eye goggles, ophthalmic instruments.

bigonial breadth (VIII, 39), The straight-line distance between the right and left angles where the body of the mandible and the ascending ramus meet. To measure, the subject's head is tilted slightly upwards; the angle of the mandible is usually palpated both anteriorly and inferiorly from the lowest point of the ear lobes. The relevant application is the greatest distance on the outermost points of the jaw, e.g., partial face masks, chin straps.

bizygomatic breadth (VIII, 40), The maximum distance between the most laterally situated points on the zygomatic arches (cheekbones). The relevant application is the greatest distance between the outermost points of the cheekbones, e.g., helmet design.

buttock depth (II, 11), The horizontal distance between the buttocks and the abdomen at the level of the maximum protrusion of the buttocks. To measure, the subject stands erect. The relevant application is fore and aft space at buttock level while the subject is standing, e.g., turnstile design.

buttock-knee length (III, 20), The horizontal distance from the most posterior point on the buttocks to the most anterior point on the knee. To measure, the subject sits erect with each knee at a right angle, with the upper leg parallel to the floor, and feet flat on the floor. The relevant application is for establishing the distance between the seat back and objects located in front of the knees. It is a static, minimum or clearance dimension, e.g., airplane seat spacing.

buttock-leg length (III, 23), The horizontal distance from the most posterior point on the buttocks to the base of the heel. To measure, the subject sits erect on the floor with the knee fully extended and the ankle at 90°. The relevant application is clearance for the outstretched leg, e.g., wheelchair design, examination tables.

buttock-popliteal length (III, 26), The horizontal distance from the plane of the most posterior point on the buttocks to the back of the lower leg at the knee. To measure, the subject sits erect with the knees at right angles, with the upper leg parallel to the floor, and feet flat on the floor. The relevant application is for determining maximum seat depth. Too long a seat will severely discommode short-legged operators, e.g., chair depth.

chest breadth (I, 2), On males, the horizontal frontal distance across the chest at nipple level, and on females, at the level where the fourth rib meets the breastbone. To measure, the subject stands erect, breathes normally, and has the arms hanging naturally at the sides. The relevant application is for determining maximum lateral space available at chest level, e.g., clothing.

chest depth (II, 9), On males, the horizontal distance from front to back of the chest at nipple level and on females, above the breasts, at the level where the fourth rib meets the sternum or breastbone. To measure, the subject stands erect, breathing normally. The relevant application is fore and aft space available at chest level.

ear breadth (IX, 54), With the subject's head oriented in the Frankfort plane, the distance between the points where an imaginary line drawn perpendicular to the long axis of the external ear meets the most anterior point and the most posterior point of the external ear. The relevant application is the maximum breadth of the ear, e.g., earphone design.

ear height (IX, 48), With the subject's head oriented in the Frankfort plane, the distance between the highest point of the incurved rim of the external ear and the lowest point on the inferior border of the ear lobe. The relevant application is the maximum height of the ear, e.g., earphone design.

elbow-to-elbow breadth (IV, 28), The maximum horizontal distance between the lateral surfaces of the elbows. To measure, the subject sits erect with the upper arms vertical and lightly touching the sides, and the forearms extended horizontally. A relevant application is, e.g., passage-way design.

elbow height (I, 6), The vertical distance from the floor to the radiale (the depression at the elbow formed where the bones of the upper arm and forearm meet). To measure, the subject stands erect with the arms hanging naturally at the sides. The relevant application is work or rest surface height, the vertical distance between the floor and table tops, e.g., desks and workbenches used in the standing position.

elbow rest height, sitting (III, 22), The vertical distance from the sitting surface to the bottom of the right elbow. To measure, the subject sits erect with the upper arm vertical at the side, forearm at a right angle to the upper arm. The relevant application is for determining the vertical distance between the seat surface and the top surface of an elbow rest, e.g., arm rest in seating design.

extracanthic diameter (IX, 47), The distance between the lateral and the medial corners of each eye. This measurement should be taken on both the right and left eyes. The relevant application is the greatest width of the eye, e.g., binocular equipment, telescopes.

eye height, sitting (III, 16), The vertical distance from the sitting surface to the inner corner of the eye. To measure, the subject sits erect and looks straight ahead. The relevant application is seated eye level, or the vertical distance between the seat surface and the eye, e.g., location of primary functions, i.e., control panels, observation windows, cockpit design, vehicular design.

eye height, standing (II, 13), The vertical distance from the floor to the inner corner of the eye. To measure, the subject stands erect and looks straight ahead. The relevant application is eye level, or the vertical distance from the floor to the eye position which affords the best visual field to the standing subject, e.g., standing control panels for the placement of primary visual functions.

facial height, total (VIII, 44), The distance between the nasion and the lowest median point on the lower border of the mandible or jaw. The relevant application is the greatest distance between the bridge of the nose and the lowest edge of the chin, e.g., full face masks, helmets.

facial height, upper (VIII, 45), The distance in the sagittal plane between nasion and the gumline. To measure, the subject must curl up his or her upper lip so that the bottom point can be located. The relevant application is the distance between the bridge of the nose and the lowest border of the nose, e.g., partial face masks.

facial length (IX, 51), The distance between the midpoint on the forehead and the lowest median point on the lower border of the mandible. The relevant application is the greatest distance from the top point of the forehead to the outermost point of the chin, e.g., helmets, full face masks.

foot breadth (VII, 35), The maximum horizontal distance across the foot. To measure, the subject stands with weight equally distributed on both feet. The relevant application is for determining lateral space available for the foot, e.g., shoe size, brake pedal width, foot control width.

foot length (VII, 36), The maximum horizontal distance parallel to the long axis of the foot, from the back of the heel to the tip of the longest toe. To measure, the subject stands with weight equally distributed on both feet. The relevant application is for determining fore and aft space available for the foot, e.g., shoe size, brake pedal length, foot control length.

forearm-hand length (III, 17), The horizontal distance from the tip of the right elbow to the tip of the longest finger. To measure, the subject sits erect with the upper right arm vertical at the side and the forearm, hand and fingers extended horizontally. The relevant application is for determining the maximum fingertip reach from a fixed elbow point, e.g., hand control location, arm rest length.

hand breadth at metacarpal (V, 32), The maximum breadth across the hand where the fingers join the palm. To measure, the right hand is extended straight and stiff with the fingers held together. The relevant application is for determining the breadth available for palm, with fingers extended, e.g., handle widths.

hand breadth at thumb (V, 31), The maximum breadth across the palm at the first knuckle from the tip of the thumb. To measure, the right hand is extended straight and stiff with the fingers held together and with the thumb lying alongside and in the plane of the hand. The relevant application is for determining space available across the entire palm, at the level of the base of the thumb, e.g., access clearances.

hand length (V, 33), The distance from where the base of the thumb meets the wrist to the middle fingertip of the right hand extended straight on the arm. To measure, the right hand and lower arm is extended straight and stiff with the fingers held together. The relevant application is for determining maximum fingertip reach from the wrist, e.g., gloves.

hand thickness at metacarpal III (VI, 34), The maximum distance between the back and palm surfaces of the hand at the knuckle of the middle finger where it joins the palm of the right hand when the fingers are extended. To measure, the right hand is extended straight and stiff with the fingers held together. The relevant application is the space available for the flat palm. More space is required for the whole hand than the palm alone, since the muscle masses at the bases of the thumb and the little finger, (the thenar and hypothenar eminances), increase hand thickness, e.g., handles, refrigerator doors.

head breadth (VIII, 37), The maximum horizontal head breadth above the ears, at right angles to the mid-sagittal plane. The location of this dimension is highly variable. To measure, the subject is sitting erect, looking forward. The relevant application is the breadth across the head or the lateral space available for the head above the ears, e.g., headrests, earphones, helmets.

head height (VIII, 43), With the subject's head in the Frankfort plane, the distance in the midline or sagittal plane between the lowest point on the inferior border of the orbit as palpated through the skin and the top of the head. The relevant application is the greatest distance between the top of the head and the top of the jawbone under the eye, e.g., headgear.

head length (VIII, 42), The distance between the most anterior point on the forehead (between the brow ridges) and the most posterior point, on the back of the head, in the midline. To measure, the subject is sitting erect, looking forward. The relevant application is head length or the space available between the front and the back of the head, e.g., head gear.

head length, maximum (VIII, 41), The distance between the most anterior point of the nose and the most posterior point on the back of the head, in the midline. To measure, the subject sits erect looking forward. The relevant application is maximum head length, or the space available between the tip of the nose and the back of the head, e.g., face guards.

hip breadth, sitting (IV, 29), The maximum horizontal distance across the hips when seated. To measure, the subject sits erect, knees and ankles supported at right angles; knees and heels together. The relevant application is the space available across the hips, e.g., seat breadth.

hip breadth, standing (I, 3), The maximum horizontal distance across the hips. To measure, the subject stands erect, with heels together. The relevant application is the lateral space available at the level of the hips, while the person is standing, e.g., turnstile aisles.

interpupillary distance (VIII, 38), The distance between the centers of the pupils. To measure, the subject sits erect and looks straight ahead. The relevant application is binocular spacing or the separation distance between binocular eyepieces, e.g., sighting stations, stereoscopic photogrammetry, microscopy, etc.

jaw height, total (IX, 53), The distance in the midline or sagittal plane between the point at which the nasal septum merges with the upper cutaneous lip and the lowest median point on the lower border of the mandible. The relevant application is the greatest distance between the bottom of the nose and the lowest point on the chin, e.g., face masks.

knee height, sitting (III, 24), The vertical distance from the floor to the uppermost point on the knee. To measure, the subject sits erect with his knees at right angles. The relevant application is the space available above the knees, e.g., car dashboards, knee clearance in table heights.

knee-to-knee breadth, sitting (IV, 30), The maximum horizontal distance across the lateral surfaces of the knees. To measure, the subject sits erect with his knees at right angles and touching lightly. The relevant application is the space available for determining knee clearance, e.g., cock-pit clearance.

knuckle height (I, 7), The vertical distance from the floor to the largest knuckle of the middle finger, where the finger meets the palm. To measure, the subject stands erect, palm flat against the side of the thigh. The relevant application is for determining the maximum permissible vertical distance between the base of an object and the underside of an attached carrying handle, e.g., suitcases.

lower arm length (I, 4), The distance from the elbow to the wrist joint when the arm is hanging down and the palm facing inward. To measure, the subject stands erect with the arms hanging freely at the sides. The relevant application is for determining the length of the lower arm, e.g., clothing design.

maximum body breadth (I, 1), The maximum breadth across the body including the arms. To measure, the subject stands erect with arms hanging relaxed at the sides. The relevant application is for determining passage clearances, e.g., door width.

maximum body depth (II, 12), The maximum horizontal distance between the vertical lines tangent to the most anterior and posterior points on the trunk. To measure, the subject stands erect with the arms at the sides. The relevant application is as a minimum clearance dimension of the body, e.g., passage clearance.

mouth breadth, maximum (IX, 30), The distance between the two most lateral points on the mouth when the lips are joined together and the mouth is in a relaxed state. The relevant application is the greatest distance from the outermost edges of the lips across the mouth, e.g., the design of mouthpieces.

nasal breadth (IX, 49), With the subject's head tilted slightly upwards, the maximum transverse distance between the most laterally situated points on the wings of the nose. The relevant application is the greatest distance between the outer edges of the lower part of the nose, e.g., nose plugs, eye glass frames.

nasal height (IX, 52), With the subject's head tilted slightly upwards, the distance between nasion and the point where the nasal septum merges with the upper cutaneous lip. The relevant application is the greatest distance from the bridge of the nose to the underside of the nose, e.g., face masks.

popliteal height, sitting (III, 25), The vertical distance from the floor to the underside of the thigh immediately behind the knee. To measure, the subject sits erect with knees at right angles and the bottom of the thighs and the back of the knees barely touching the sitting surface. The relevant application is the measurement of the vertical distance between the floor and the highest point on the front of the sitting surface, e.g., seat height, i.e., secretarial chairs.

shoulder breadth (IV, 27), The maximum horizontal distance across the lateral surfaces of the shoulders. To measure, the subject sits erect with upper arms touching the sides and the forearms extended horizontally. The relevant application is for measuring the space available across the shoulders, e.g., shoulder pads.

shoulder elbow length (III, 18), The uppermost point on the lateral edge of the shoulder to the bottom of the elbow. The subject sits erect with his upper arm vertical at his side and the forearm making a right angle with it. The relevant application is the distance between objects located at or above the shoulder as well as at or below the elbow. It is a minimum clearance dimension. E.g., accessories and equipment hung from the shoulder.

shoulder height, sitting (III, 19), The vertical distance from the sitting surfaces to the uppermost point on the lateral edge of the shoulder with the subject sitting erect. The relevant application is for determining the vertical distance between the seat surface and objects located at or above the shoulder, e.g., seat-back design.

shoulder height, standing (I, 5), The vertical distance from the floor to the uppermost point on the lateral edge of the shoulder with the subject standing erect. The relevant application is the total distance from the floor to the shoulder, e.g., standing control position, shelf height.

sitting height (III, 15), The vertical distance from the sitting surface to the top of the head. To measure, the subject sits erect, looking straight ahead, with his knees at right angles. The relevant application is the greatest distance between the seat and the top of the head, e.g., furniture design.

stature (II, 14), The vertical distance from the floor to the top of the head. To measure, the subject stands erect and looks straight ahead. The relevant application is the maximum height of an individual, e.g., the minimum standing clearance in the workspace.

thigh clearance height, sitting (III, 21), The vertical distance from the sitting surface to the top of the thigh at its intersection with the abdomen. The subject sits erect with his knees at right angles. The relevant application is the maximum distance between the seat and the top of the thigh, e.g., table and chair height relationships.

waist depth (II, 10), The horizontal distance between the back and abdomen at the level of the greatest lateral indentation of the waist; (if this is not apparent, at the level at which the belt is worn). To measure, the subject stands erect with his abdomen relaxed. The relevant application is for determining the fore and aft space at abdominal level; clearances between the operator's back and control stick, steering wheel and work surfaces are examples.

Cross-Reference List

Plate Number	Dimension Number	Dimension Name
I	1	Maximum Body Breadth
	2	Chest Breadth
	3	Hip Breadth, Standing
	4	Lower Arm Length
	5	Shoulder Height, Standing
	6	Elbow Height
	7	Knuckle Height
II	8	Arm Reach from Wall
	9	Chest Depth
	10	Waist Depth
	11	Buttock Depth
	12	Maximum Body Depth
	13	Eye Height, Standing
	14	Stature

Place Number	Dimension Number	Dimension Name
III	15	Sitting Height
	16	Eye Height, Sitting
	17	Forearm-Hand Length
	18	Shoulder-Elbow Length
	19	Shoulder Height, Sitting
	20	Buttock-Knee Length
	21	Thigh Clearance Height, Sitting
	22	Elbow Rest Height, Sitting
	23	Buttock-Leg Length
	24	Knee Height, Sitting
	25	Popliteal Height, Sitting
	26	Buttock-Popliteal Length
IV	27	Shoulder Breadth
	28	Elbow-to-Elbow Breadth
	29	Hip Breadth, Sitting
	30	Knee-to-Knee Breadth, Sitting
V	31	Hand Breadth at Thumb
	32	Hand Breadth at Metacarpal
	33	Hand Length
VI	34	Hand Thickness at Metacarpal III

Plate Number	Dimension Number	Dimension Name
VII	35	Foot Breadth
	36	Foot Length
VIII	37	Head Breadth
	38	Interpupillary Distance
	39	Bigonial Breadth
	40	Bizygomatic Breadth
	41	Head Length, Maximum
	42	Head Length
	43	Head Height
	44	Facial Height, Total
	45	Facial Height, Upper
IX	46	Bicanthic Diameter
	47	Extracanthic Diameter
	48	Ear Height
	49	Nasal Breadth
	50	Mouth Breadth, Maximum
	51	Facial Length
	52	Nasal Height
	53	Jaw Height, Total
	54	Ear Breadth

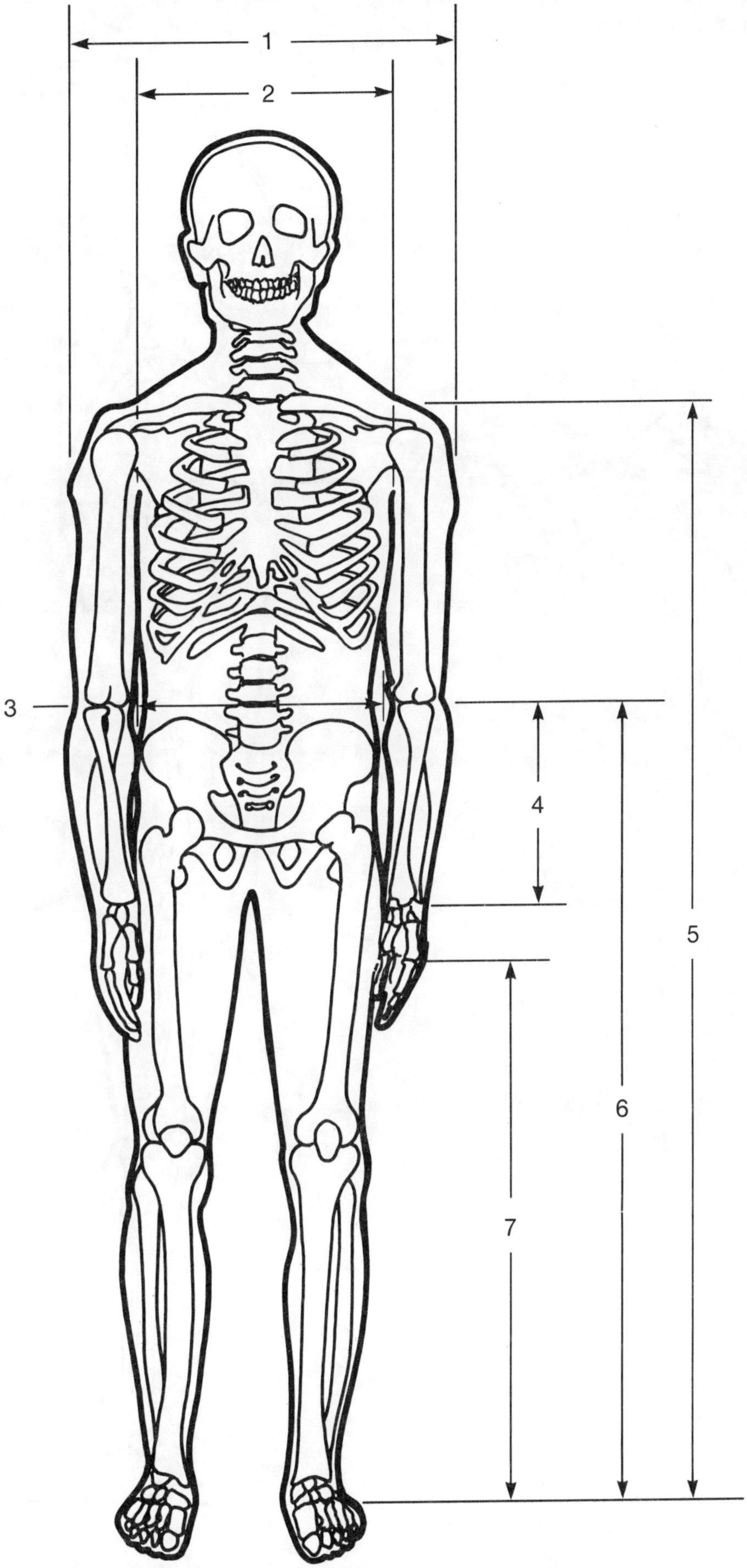

PLATE NO. I

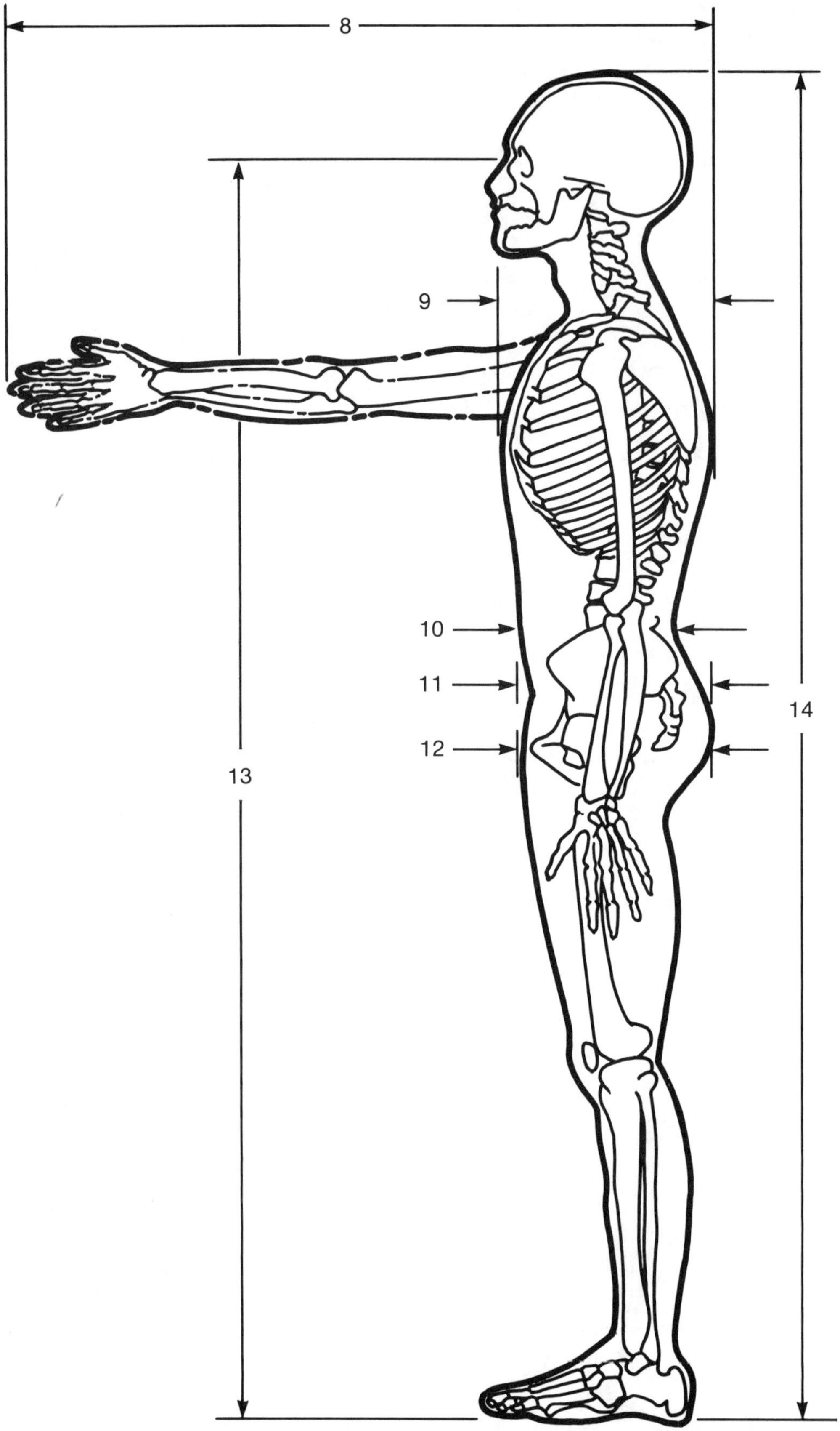

PLATE NO. II

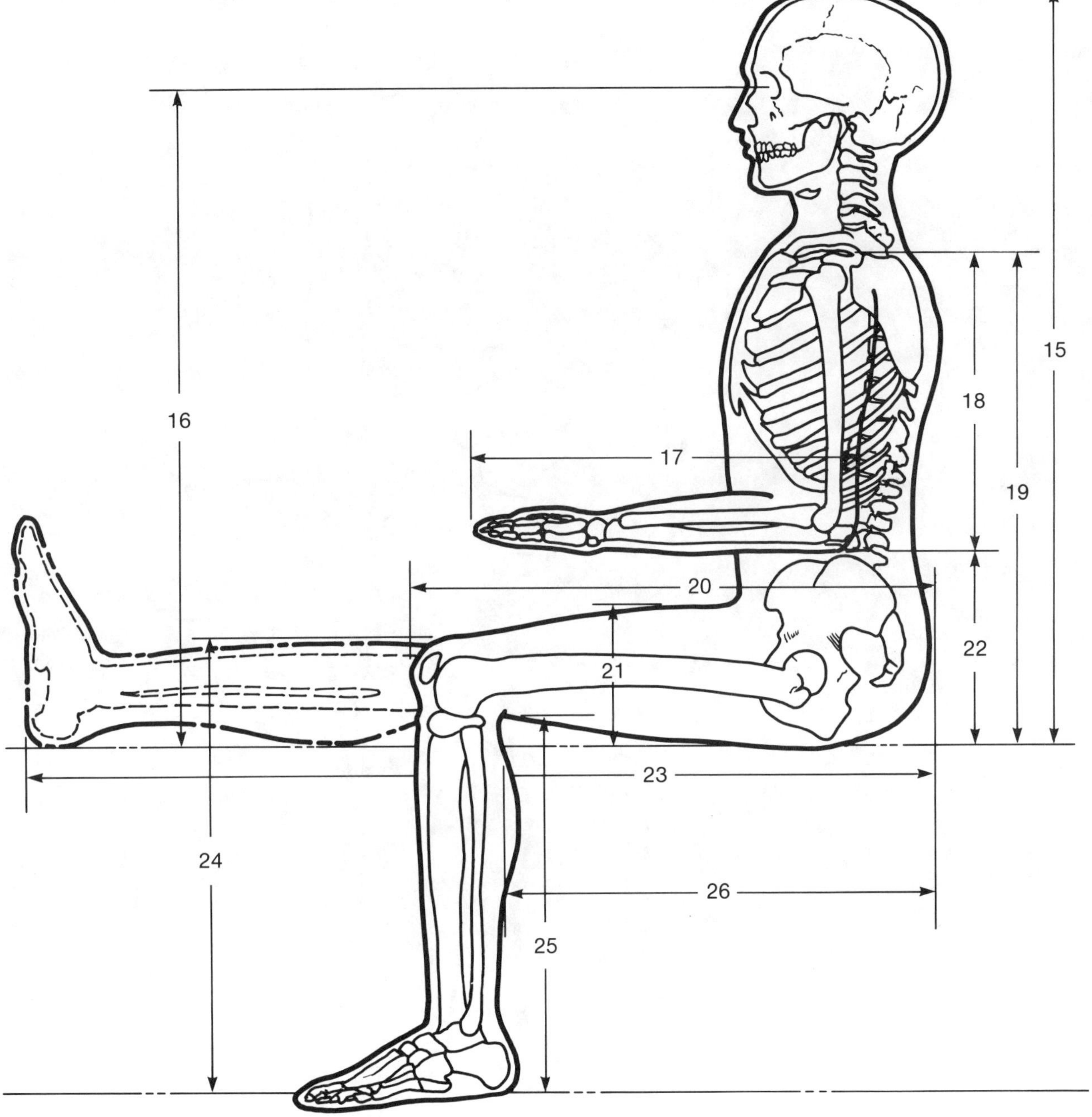

PLATE NO. III

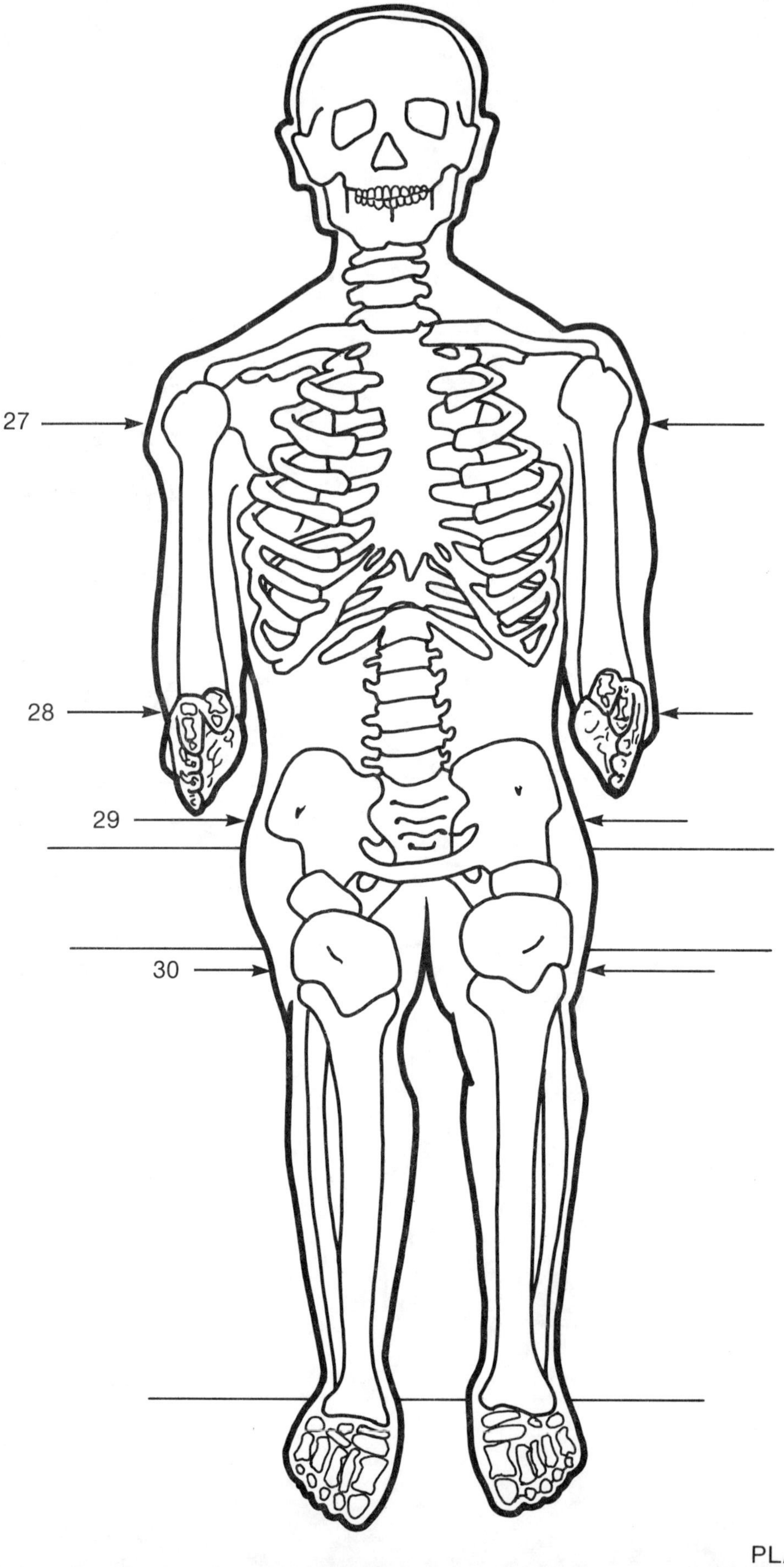

PLATE NO. IV

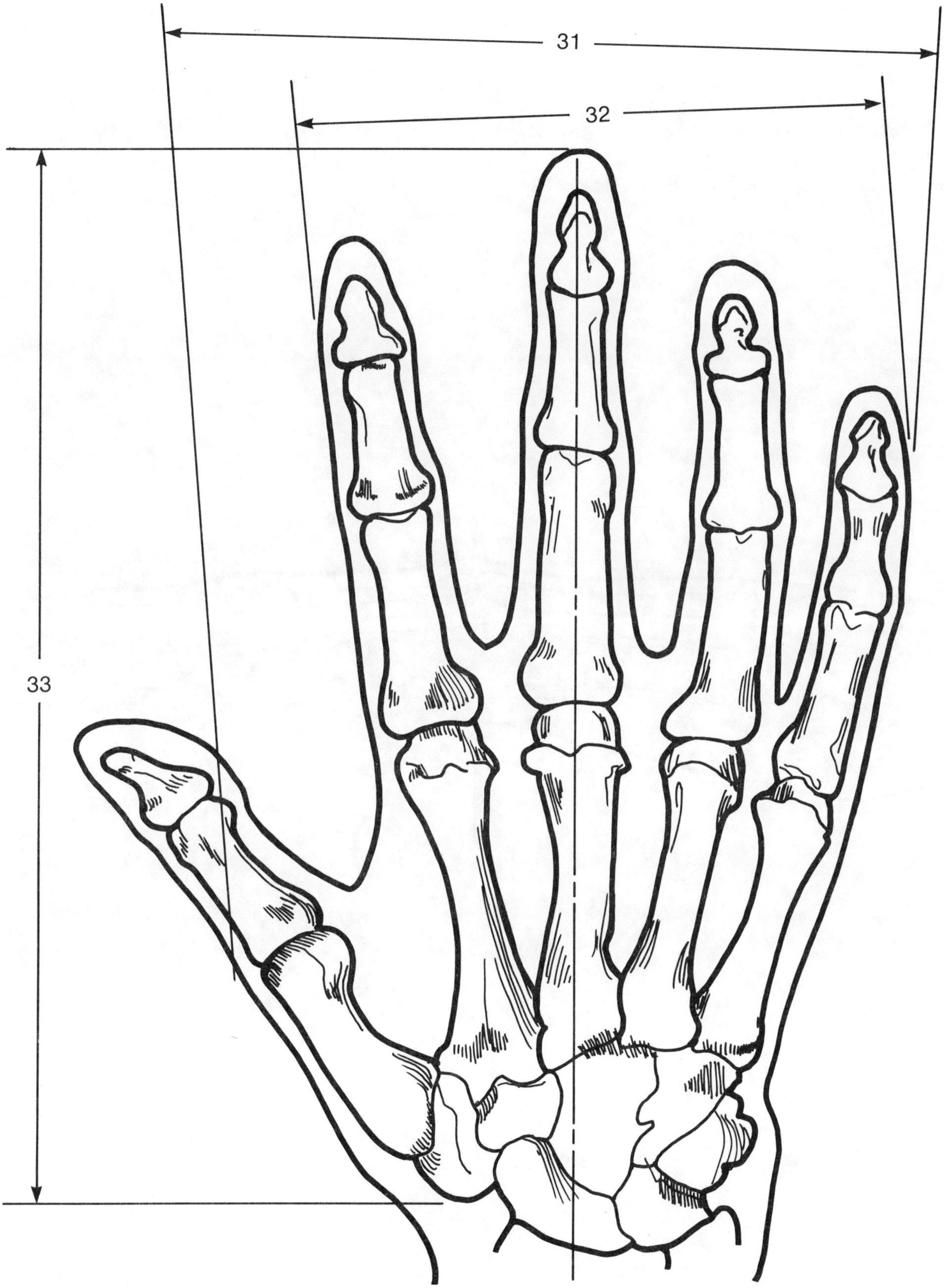

PLATE NO. V

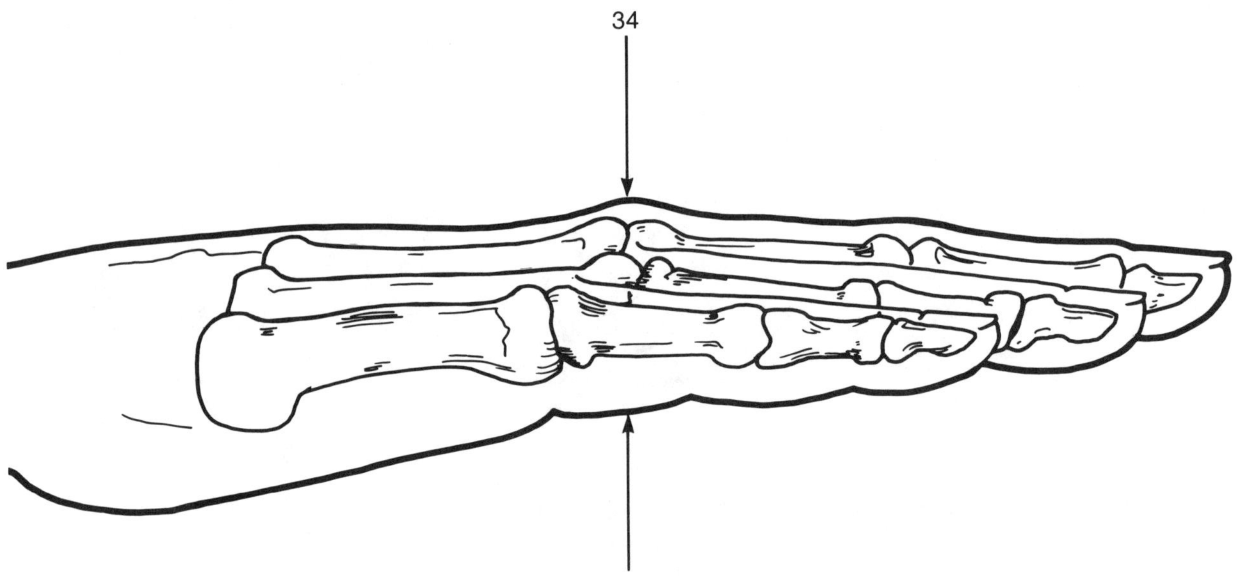

PLATE NO VI

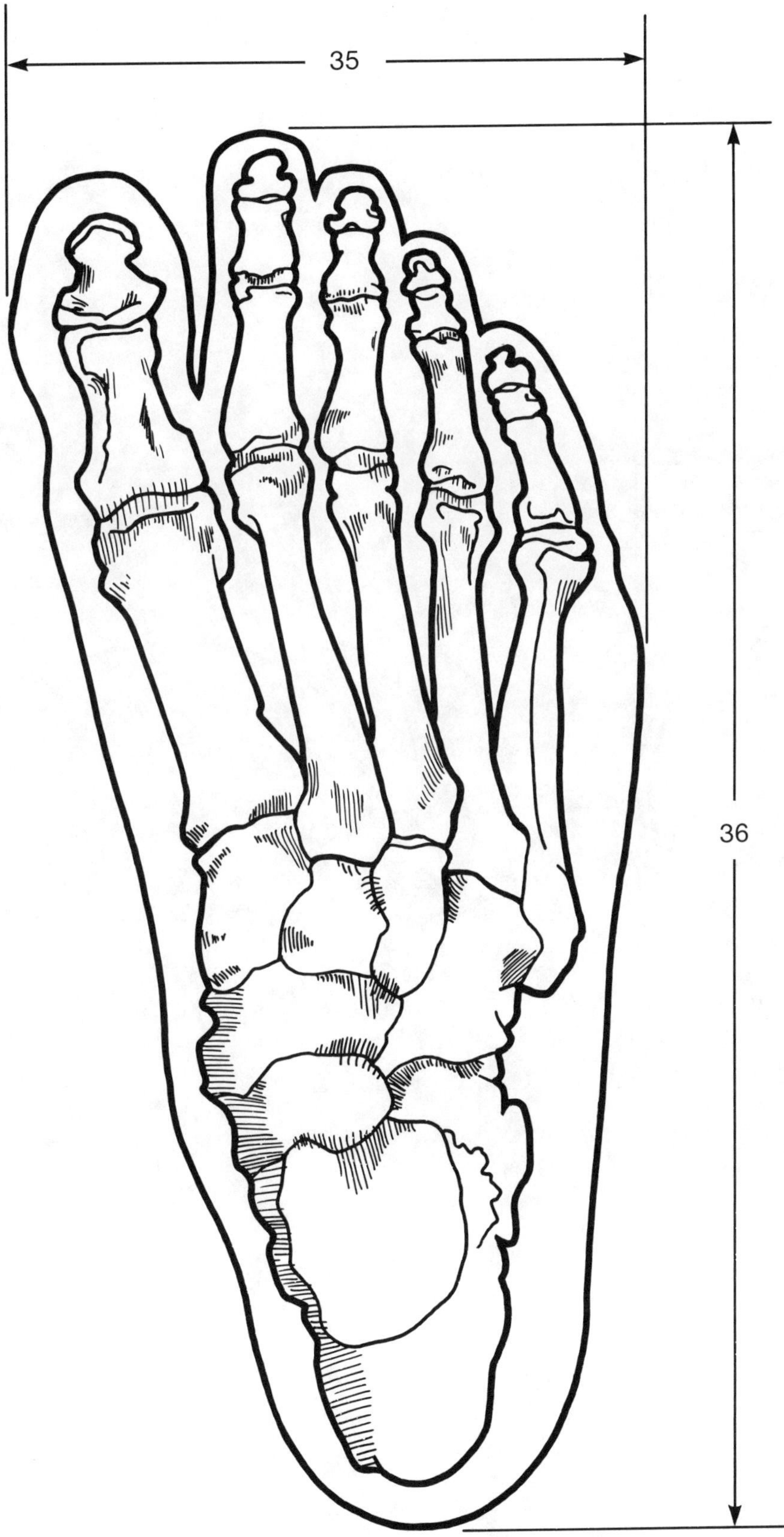

PLATE NO. VII

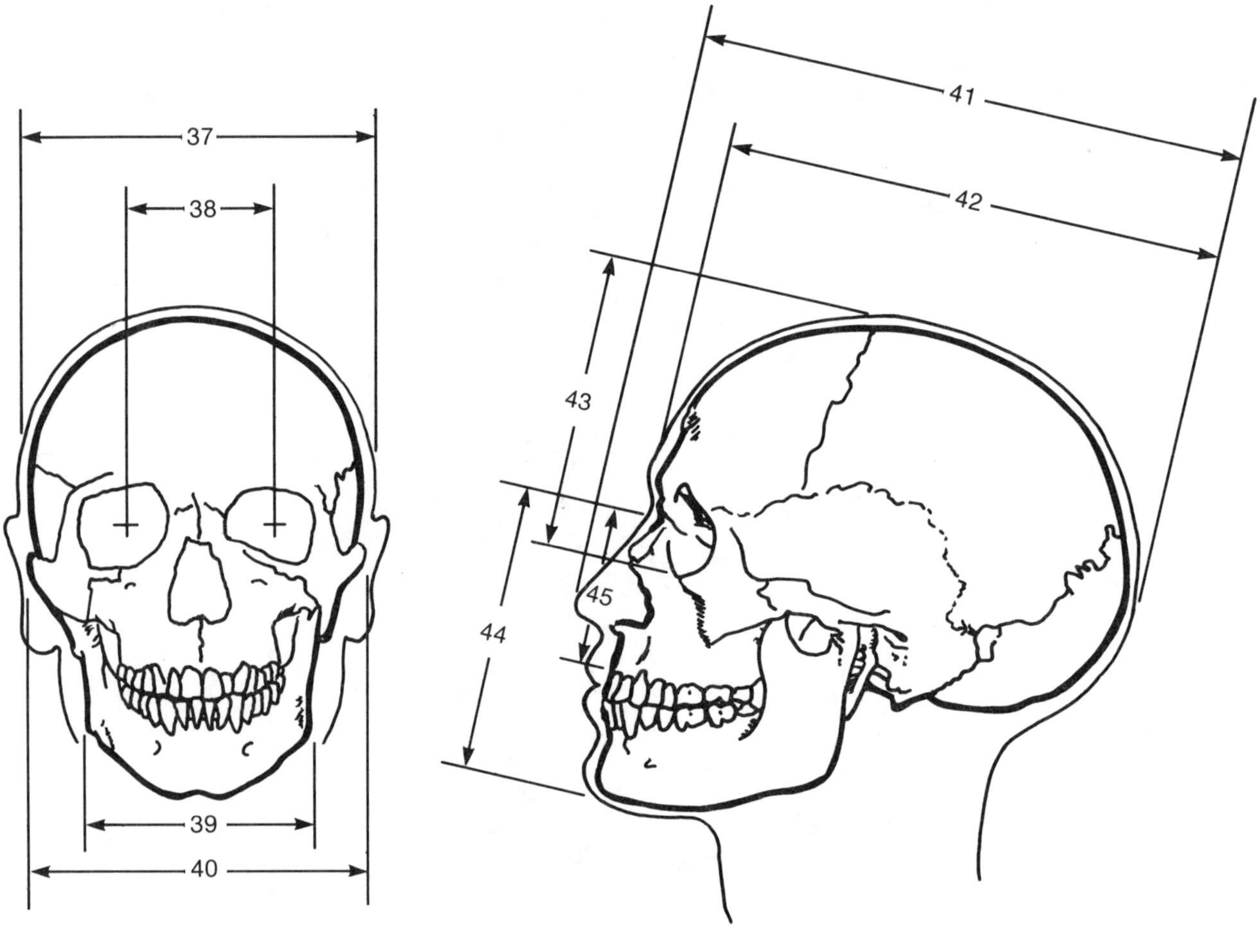

PLATE NO. VIII

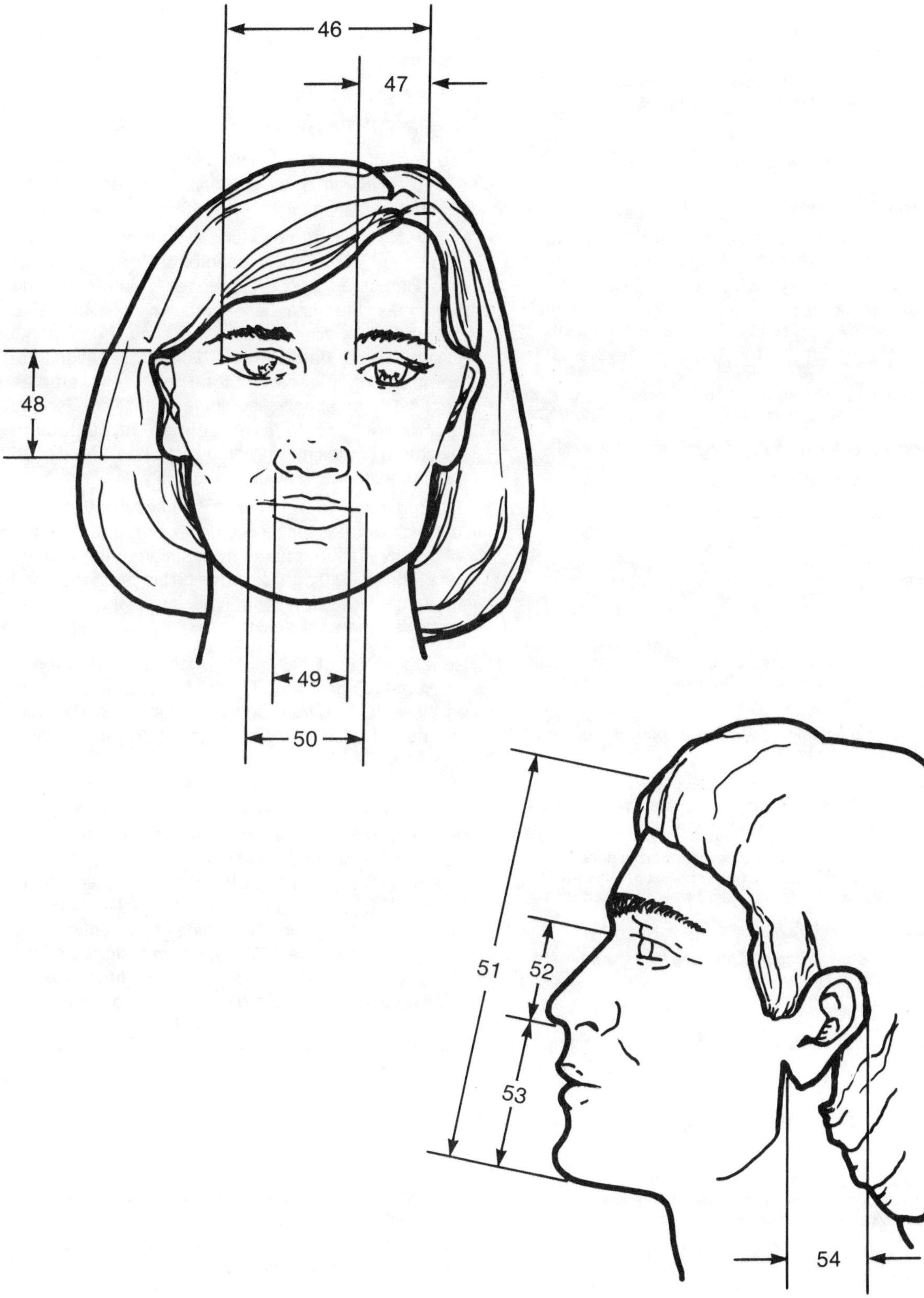

PLATE NO. IX

Glossary

anterior, Toward the front, ventral; e.g., the breast is on the anterior chest wall.

ascending ramus, Refers to that part of the mandible or jaw which forms an angle with the main body of the jaw.

coronal, Frontal; any plane dividing the body into anterior (or ventral) and posterior (or dorsal) portions at right angles to the sagittal plane.

cutaneous, Relating to the skin.

Frankfort plane, The anatomical position in which the eyes are directed straight ahead and are both parallel to the horizon.

gingiva, The gum; the dense fibrous tissue, covered by mucous membrane, that envelops the alveolar processes of the upper and lower jaws and surrounds the necks of the teeth.

hypothenar, The fleshy mass at the medial side of the palm.

hypothenar eminance, The elevation on the medial side of the palm produced by the short muscles of the little finger.

inferior, Lowermost or below; the foot is inferior to the ankle.

interpupillary, Between the pupils.

interscapulae, Between the scapulae (shoulder blade, q.v.).

lateral, Toward the side.

mandible, Jaw bone.

medial, Relating to the middle or center; nearer to the median or mid-sagittal plane.

metacarpal, Relating to the metacarpus (plural, metacarpi), which are the five bones of the hand between the carpus and the phalanges.

mid-sagittal, The plane vertically dividing the body through the midline into right and left halves.

nasal septum, The thin wall dividing the nasal cavity into halves; it is composed of a central supporting skeleton covered on each side by a mucous membrane.

nasion, A point on the skull corresponding to the middle of the nasofrontal suture; this corresponds approximately with the deepest part of the depression just below the eyebrows.

occiput, The back of the head.

palpate, To examine by feeling and pressing with the palms of the hands and the fingers.

popliteal, Relating to the posterior surface of the knee.

posterior, Toward the back, dorsal; the vertebral column is posterior to the digestive tract.

protuberance, An outgrowth, a swelling, a knob.

sagittal, Any plane parallel to the mid-sagittal line vertically dividing the body into right and left portions.

scapula, Shoulder blade; a large triangular flattened bone lying over the ribs, posteriorly on either side, articulating laterally with the clavical and the humerus.

thenar, The fleshy mass on the lateral side of the palm; the ball of the thumb.

thenar eminance, The swelling on the lateral part of the palm of the hand caused by the short muscle of the thumb.

transverse, Horizontal; any plane dividing the body into superior and inferior portions.

zygomatic arch, Cheekbone.

Primary Bibliography

Damon, Albert; Stoudt, Howard W.; and McFarland, Ross A. *The Human Body in Equipment Design.* Cambridge: Harvard University Press, 1966.

A primary source from which dimensions were chosen for the standard, the chapter on "Anthropometry and Human Engineering" presents data on 37 dimensions, the data base from which is available military and civilian studies including the National Health Survey data of 1960-1962. Explanations accompanying the dimensions include examples applicable to specific equipment design. General discussion provides guidelines for selecting minimum or maximum values and a consideration of factors affecting individual variability.

Other dimension types discussed and illustrated are circumferences, functional dimensions, and center of gravity measurements. In addition to the chapter on anthropometry are sections on biomechanics and equipment design, control design, hand and foot controls, and seat design.

Hertzeberg, H.T.E. "Chapter 11: Engineering Anthropology" in *Human Engineering Guide to Equipment Design,* edited by Harold P. Van Cott and Robert G. Kinkade. Washington, D.C.: U.S. Government Printing Office, Superintendent of Documents, 1972.

A primary source from which dimensions were chosen for inclusion in the standard, "Engineering Anthropology" presents tabular data of male and female, military and civilian studies as well as international data (where available) for 39 static dimensions. The use of anthropometry is clarified with specific guidelines for the choice of percentiles for a given design problem. Variability of military and national groups is discussed. Data for functional dimensions as well as the discussion of factors affecting muscle capacity give the reader a perspective for interpretation and use of data.

Montagu, M.F. Ashley. *A Handbook of Anthropometry.* Springfield: Charles C. Thomas, 1960.

A primary source from which definitions of head dimensions were chosen for inclusion in the standard, this volume is a detailed compendium on the theory, practice, and methods of anthropometry. It considers the cephalometry and somatometry of living humans, the craniometry of skeletal populations, postcranial osteometry, developmental assessment of the dentition, the measurement of body composition, and an unusually detailed section on references and directories for further information.

Secondary Bibliography

Abraham, Sidney. *Preliminary Findings of the First Health and Nutrition Examination Survey, United States, 1971-1972: Anthropometric and Clinical Findings.* Rockville, Maryland: Health Resources Administration, National Center for Health Statistics, April 1975.

This document presents preliminary findings collected on a probability sample of the U.S. population by age, sex, race, and income level. Data are presented on selected anthropometric measurements of children 1-17 years of age, obesity in adults 20-74 years of age, and clinical signs of possible nutritional deficiency for persons 1-74 years of age. The particular contribution to anthropometric data is the inclusion of data by yearly intervals on height, weight, triceps skinfold, and subscapular skinfold for ages 1-17.

Bass, William. *Human Osteology.* Columbia: Missouri Archeological Survey, 1971.

This volume is a primer in skeletal anatomy and in the anthropometry of the human skeleton. It gives a thorough introduction to the theory and practice of anthropometry but it emphasizes the bony landmarks and is geared to the measurement, aging, sexing, and cataloguing of skeletal material found in archeological sites.

Brozek, Josef. *Body Measurements and Human Nutrition.* Detroit: Wayne University Press, 1956.

This edited volume considers the effect that varying degrees of nutritional status has on a battery of anthropometric measurements. Most of the samples are drawn from the armed forces, yet, this volume also starts to consider the effects of nutritional status on the anthropometry of women.

Brozek, Josef and Henschel, Austin. *Techniques for Measuring Body Composition.* Washington, D.C.: National Academy of Sciences, National Research Council, 1961.

This edited volume pulls together a large portion of the available techniques for quantifying body composition. It explores the relative merits and drawbacks of different somatometric, volumetric, and biochemical approaches to the measurement of body composition and body volume. The assessment section is particularly important because it maintains that in order to progress in this field, multivariate statistical models must be adopted.

Chapanis, Alphonse (ed.). *Ethnic Variables in Human Factors Engineering.* Baltimore: The Johns Hopkins University Press, 1975.

A diverse reference on human factors, the book presents 18 essays from papers presented at a NATO symposium, three of which specifically deal with anthropometry: "Population differences in dimensions, their genetic basis and their relevance to practical problems of design" by D.F. Roberts; "Anthropometric measurements on selected populations of the world" by Robert M. White; and "International anthropometric variability and its effects on aircraft cockpit design" by Kenneth W. Kennedy. In particular, the article by White includes statistics on weight, stature, sitting height, and chest circumference for a number of populations of different countries, data important in the design of international products.

Churchill, Edmund and McConville, John T. *Sampling and Data Gathering Strategies for Future USAF Anthropometry.* Springfield, Virginia: National Technical Information Service, 1976.

A methodology is presented for tailoring a data gathering study to the specific design needs of a project as well as a discussion of the different types of variability, necessary to the determination of the data gathering approach needed. Types of variability discussed are intra-individual, inter-individual, and secular variability. The appendix includes a list of all the anthropometric dimensions available in the Aerospace Medical Research Lab Data Bank.

Clauser, C. E.; Tucker, P. E.; McConville, J. T.; Churchill, E.; Laubach, L. L.; and Reardon, J. A. *Anthropometry of Air Force Women.* Wright-Patterson Air Force Base, Ohio: Aerospace Medical Research Laboratory, 1972.

A total of 137 dimensions are included with graphs representing the actual distributions of the population surveyed. Dimension types include body lengths and heights, circumferences, weights, skinfolds, breadths and depths, body surface distances, as well as measures of the head, the hands, and the feet. Correlation coefficients which establish the ratio of certainty by which one dimension can be determined from another are developed for all variables as well as regression equations for those variables with a moderately large intercorrelation.

Dempster, Wilfred Taylor. *Space Requirements of the Seated Operator: Geometrical, Kinematic, and Mechanical Aspects of the Body with Special Reference to the Limbs.* Wright-Patterson Air Force Base, Ohio: Wright Air Development Center, United States Air Force, 1955.

This study explores the relationships between joint "centers" and joint ranges of motion based on a subject population representing a cross section of somatotypes. Links are defined, as a result of this study, as a proportion of the corresponding static segment length. Links are useful to the designer and engineer in determining workspace layout as a function of potential ranges of motion. Specifications for construction of drafting board mannequins were worked out for seated postures.

Diffrient, Niels; Tilley, Alvin R.; and Bardagjy, Joan C. *Humanscale 1/2/3.* Cambridge, The MIT Press, 1974.

Humanscale consists of three pictorial selectors with rotary dials correlating anthropometric data with age and body size. Included are body and link measurements, seat and table design measurements, and some data on wheelchair users, the handicapped and the elderly. Although the information is presented in an easily usable form, there is a drawback in this method of presentation. Since for any given dimension, there have been numerous surveys conducted, attention needs to be given to the criteria by which one selects the data of one survey over another. Humanscale presents no discussion on why or by what criteria they selected the data which is presented on the rotary selectors.

Dreyfuss, H. *The Measure of Man.* New York: Whitney Library of Design, 1967.

This portfolio of anthropometric data is accompanied by design specifications, bibliography, charts, and two life-size standing human figures. It provides male and female, seated and standing dimensions.

Hrdlicka, Ales. *Anthropometry.* Philadelphia: The Wistar Institute of Anatomy and Biology, 1920.

This volume is another detailed primer on the measurement of humans. It goes into great detail on how to measure humans, anthropometric landmarks, anthropometric instruments, as well as a section on aging and sexing skeletal material.

Lockhart, Robert Douglas; Hamilton, Gilbert Frewin; and Fyfe, Forest William. *Anatomy of the Human Body.* Philadelphia: J. B. Lippincott Co., 1959.

This volume is an authoritative source book on the structure and function of the human body. It first considers the skeleton in great detail; then, it considers each part of the human body in an integrated, functional manner.

Reid, Betty (Compiler). *An Annotated Bibliography of United States Air Force Applied Physical Anthropology January 1946-July 1976.* Wright-Patterson Air Force Base, Ohio: Aeorspace Medical Research Laboratory, 1976.

Abstracts of 153 AMRL reports on anthropometric data studies as well as discussion of anthropometry in specific design problems.

Roebuck, J. A.; Kroemer, K. H. E.; and Thomson, W. G. *Engineering Anthropometry Methods.* New York: John Wiley & Sons, 1975.

Methods for both measuring and applying data on human body dimensions and strength to engineering design are presented which the engineer and designer may use in the conceptual development of systems where applicable data is not already available. A detailed discussion of the use of statistics theory in the development of anthropometric data is overed. A section on applications methodology discusses the role of models and mock-ups of many types, for example, drawings, mannequins, models of reach envelopes, and mathematical models in the design evaluation process. A detailed bibliography documenting examples of methods presented and a correlation coefficient table relating 96 anthropometric dimensions is included in the appendices.

Sinclair, David. *Human Growth After Birth.* London: Oxford University Press, 1973.

This is a volume written for the medical scientist or the medical student on aspects of post-natal growth of humans. It covers the growth of the individual from birth until maturity with regards to height and weight, tissues, organ systems, indices of maturity, and ontogenetic changes in shape and posture. It then considers factors affecting growth, repair, and maturation.

Snyder, Richard G.; Chaffin, Don B.; and Schutz, Rodney K. *Link System of the Human Torso.* Wright-Patterson Air Force Base, Ohio: Aerospace Medical Research Laboratory, United States Air Force, 1972.

Vector prediction equations are developed by use of anthropometry, photogrammetry, radiography and cineflouroscopy of 28 male volunteer subjects for location of cervical, thoracic, and lumbar vertebrae relative to given surface landmarks. This information is useful in the development of work-space envelopes. This kind of information is extremely beneficial in the dynamic evaluation of man-machine interface problems. Included also are anthropometric dimensions and skinfolds of 28 subjects used in the study.

Snyder, Richard G.; Spencer, Martha L.; Owings, Clyde L.; and Schneider, Lawrence W. *Anthropometry of U.S.*

Infants and Children. Detroit: Society of Automotive Engineers, Inc., 1975.

This reference presents in tabular and graphic form the results of anthropometric data collection on 4024 infants and children from birth to 13 years of age, in half-year intervals. Methods of data collection and analysis are also described.

Stoudt, Howard W.; Damon, Albert; McFarland, Ross A.; and Roberts, Jean. *Skinfolds, Body Girths, Biacromial Diameter and Selected Anthropometric Indices of Adults: United States, 1960-1962.* Rockville, Maryland: Health Resources Administration, National Center for Health Statistics, 1970.

This survey includes the following data by age in 10-year intervals (except for the intervals 18-24 years and 75-79 years) and by sex: right-arm and infrascapular skinfolds; right arm, waist, and chest girths; sum of skinfolds; ponderal index; ratios of sitting height erect to stature, chest girth to stature, and biacromial diameter to stature. The equations for predicting each of the physical measurements from height, weight, and age, and the interrelation of all the measurements is summarized graphically and analyzed.

Stoudt, Howard W.; Damon, Albert; McFarland, Ross A. and Roberts, Jean. *Weight, Height, and Selected Body Dimensions of Adults: United States, 1960-1962.* Rockville, Maryland: Health Resources Administration, National Center for Health Statistics, 1965.

Data on the civilian population with a sample size of 3091 men and 3581 women, age 18 through 79, includes: height, weight, sitting height (normal and erect), knee height, popliteal height, elbow rest height, thigh clearance height, buttock-knee length, buttock-popliteal length, elbow-to-elbow breadth, and seat breadth. Results of large-scale studies done in earlier years on college populations and others are compared.

Safety

Z94.16

16

The terms and their definitions appearing in this standard arise from common usage and the terminologies available in the literature. No attempt has been made to "invent" new definitions, for there is no gain from "reinventing the wheel." However, there is a caveat in the use of these terms in the field. The safety area is undergoing a tremendous upheaval as evidenced by the passage, enforcement, and continued reappraisal of such statutes as the Occupational Safety and Health Act (OSHA) and the Consumer Product Safety Act, and the development of the Good Manufacturing Practices by the Product Safety Commission (CPSC) and the Food and Drug Administration (FDA). Both OSHA and CPSC are actively at work in continued expansion and revision.

Of more impact is the great amount of case law that is being developed in these areas of industrial and product safety. The courts, as part of their ongoing activities, have undertaken the task of defining such important terms as *safe, reasonable, defect,* and *hazardous.* Therefore, although the terms here are acceptably defined in the engineering and technical sense, when used in conjunction with legal matters, they will have to be interpreted in light of the latest court pronouncements.

The terms presented here are not limited to industrial safety but also include those from the areas of transportation, fire, automobile, and general safety, as well.

The sources used to accumulate the terms and definitions are listed below. However, of even greater importance was the expertise of the subcommittee members who contributed their time and effort to bring this matter to a "temporary" conclusion. "Temporary," because, like other standards, this one will also need revision from time to time. (As the standard is used, the Committee would appreciate a note on additions, deletions, and changes.)

CHAIRMAN:

Dr. John Mihalasky, Professor, Industrial Engineering, N.J. Institute of Technology

MEMBERS:

Jack L. Brandell
Air Products and Chemicals, Inc.

Edward S. Charkey
American Insurance Association

Donald Staudter
University of Dayton

Edward Wilczynski
Port Authority of New York and New Jersey

Bibliography

Bruer, D. B., Systems Analysis and Design for Safety, Prentice-Hall, 1976.

Firenze, R. J., *The Process of Hazard Control,* Kendall/Hunt Publishing, 1978.

Grimaldi, J. V. and Simonds, R. H., *Safety Management,* R. D. Irwin, 3rd Ed. 1975.

Hammer, W., *Handbook of System and Product Safety,* Prentice-Hall, 1972.

Hammer, W., *Occupational Safety Management and Engineering,* Prentice-Hall, 1976.

Moving People Safely, The American Public Transit Association, 3rd Ed., 1977.

Petersen, D., *Techniques of Safety Management,* McGraw-Hill, 1971.

Rapid Transit Systems: Glossary of Reliability, Availability, and Maintainability Terminology for Rapid Transit, American Public Transit Association, Feb. 1978.

Tarrants, W. E. (Editor), *Dictionary for Terms Used in the Safety Profession,* American Society of Safety Engineers, 1971.

Thorpe, J. F. and Middendorf, J. H., *What Every Engineer Should Know About Product Liability,* Marcel Dekker, Inc., 1979.

Weinstein, A. S., Twerski, A. D., Piehler, H. R., and Donaher, W. A., *Product Liability and the Reasonably Safe Product,* Wiley, 1978.

Kolb, J. and Ross, S. S., *Product Safety and Liability,* McGraw-Hill, 1980.

Journals: *Hazard Prevention, Professional Safety.*

abate, To correct under the Occupational Safety and Health Act; to come into compliance with a standard that is being violated.

abnormal use, Using the article for a purpose other than that for which it was intended to be used, and one which could not reasonably have been forseen.

abridgement of damages, The right of the court to reduce damages in certain cases.

accident, An unplanned, sometimes but not necessarily injurious or damaging event, which interrupts the completion of an activity, and is invariably preceded by an unsafe act and/or unsafe condition or some combination of unsafe acts and/or unsafe conditions.

accident cause (causal factors), One or more factors associated with an accident or a potential accident. Causal factors may be identified as time sequenced events and/or may be categorized as being related to human and/or environmental (e.g., equipment, machinery, atmospheric contaminant, temperature, etc.) influences and their interactions.

accident costs, Monetary losses associated with an accident.

accident experience, One or more indices describing accident performance according to various units of measurement (e.g., disabling injury frequency rate, number of lost-time accidents, disabling injury severity rate, number of first-aid cases, or dollar loss). A summary statement describing accident performance.

accident-free, A record of no accidents, sometimes of specified types, relating to an operation, activity, or worker performance during a specified time period.

accident hazard, A situation present in an environment or connected with a job procedure or process which has the potential for producing an accident.

accident liability, A predisposition to accident involvement. The probability associated with the likelihood of a worker becoming involved in an accident or the probability of either a set of unsafe conditions or unsafe acts producing an accident.

accident location, The exact position of the key event that produced the accident.

accident potential, A behavior (behaviors) or condition (conditions) having a probability of producing an accident.

accident prevention, The application of countermeasures designed to reduce accidents or accident potential within a system or organization. Programs directed toward accident avoidance. The reduction or elimination of behaviors and conditions having an accident potential.

accident probability, The likelihood of a worker, operation, or item of equipment becoming involved in an accident.

accident proneness, A discredited theory. Originally used to describe workers who were involved in more accidents than statistically could be expected by chance or fortuitous occurrences. Accident proneness was meant to define some personal trait as opposed to some characteristic of the environment which predisposed some to have more accidents than others in work conditions where the risk or hazard is equal to all. The term would appear to imply moreover that it is possible either (a) to differentiate clearly between two classes of people — those who are accident prone, and those who are not; or (b) at least to be able to rank the group in terms of the severity of their proneness.

accident rate, Accident experience in relation to a base unit of measure (e.g., number of disabling injuries per 1,000,000 man-hours exposure, number of accidents per 1,000,000 miles traveled, total number of accidents per 100,000 employee-days worked, number of accidents per 100 employees, etc.).

accident records, Reports and other recorded information concerning employee accident experience.

accident repeater, A worker who has been involved in two or more accidents in a predetermined period of time (regardless of fault).

accident reporting, Collecting information for, and/or preparing and submitting to a designated individual or agency, an official report of an accident.

accident and sickness benefits (sick benefits), Periodic payments to workers who are absent from work due to off-the-job disabilities through accident or sickness.

accident statistics, Descriptive or inferential data which provide information about accident occurrences.

accident susceptibility, See accident proneness.

accident type, A description of the occurrences directly related to the source of injury classification and explaining how that source produced the injury. "Accident type" answers the question "How did the injured person come in contact with the object, substance, or exposure named as the source of injury, or during what personal movement did the bodily motion occur?"

activity sampling, A measurement technique for evaluating potential accident-producing behavior. It involves the observation of worker behavior at random intervals and the instantaneous classification of these behaviors according to whether they are safe or unsafe. Calculations are then made to determine either (a) the percent of time the workers are involved in unsafe acts or (b) the percent of workers involved in unsafe acts during the observation period.

act of God, An accident causal factor generally interpreted as being beyond human control (e.g., lightning, flood, tornado, earthquake, etc.). Although the natural phenomenon itself may be beyond control, its effects can be controlled in most instances. Thus, the act of God classification should not be accepted as an excuse for not taking proper precautions or properly designing structures to reduce the severity of a so-called "act of God" occurrence.

actual exposure hours, Employee-hours of exposure taken from payroll or time-clock records, wherever possible, and including only actual straight time worked (in hours) and actual overtime hours worked.

acuity, The sharpness of a sense. Pertaining to the sensitivity of hearing, vision, smell, or touch.

acute, Severe, usually critical, often dangerous exposure in which relatively rapid changes are occuring. An acute exposure normally runs a comparatively short course and its effects are easier to reverse in contrast with a chronic exposure.

adjuster, An individual who determines the amount of loss suffered in an insurance claim.

adsorption, The taking up or condensation of a gas, liquid, vapor, or dissolved material on the surface of a solid. The adhesion of molecules of a gas, liquid, or dissolved material to a surface.

agency (agent), The principal object, substance, or premises, such as tool, machine, or equipment, involved in the accident. The term is used either to designate the object most directly causing the accident, or the object inflicting injury or property damage.

agency part, The specific hazardous part of the agency that contributed to the accident occurrence or injury.

air sampling, Determining quantities and types of atmospheric contaminants by measuring and evaluating a representative sample of air. The most numerous environmental hazards are chemical and can be conveniently divided into (a) the particulates and (b) the gases or vapors. Particulates are mixtures or dispersions of solid or liquid particles in air and include dust, smoke, mist, and similar materials.

ambient, Environmental, surrounding.

ambient noise, Total noise present in an environment. Usually a composite of sounds from many sources.

analysis, accident, A separating or breaking up of an accident into its parts so as to determine their nature, proportion, function, or relationship. A statement of the results of this process. Commonly involves the identification and evaluation of the unsafe conditions and unsafe acts that are associated with an accident. The dissection of an accident and its results to determine the basic elements and the resultant or potential damage to persons and/or things.

analysis, job hazard, See job hazard analysis.

analysis, job safety, See job safety analysis.

approval agency, An agency designated or authorized to sanction, consent to, confirm, certify, or accept as good or satisfactory for a particular purpose or use, a method, procedure, practice, tool, or item of equipment or machinery.

approved (method, equipment, etc.), A method, equipment, procedure, practice, tool, etc., which is sanctioned, consented to, confirmed, or accepted as good or satisfactory for a particular purpose or use by a person or organization authorized to make such a judgment.

arson, The burning of buildings and/or property with malicious or criminal design, generally utilizing highly flammable materials or explosives to spread the fire quickly, or deliberately placed obstructions to impede fire fighting. This action may be by the owner and/or others and is by law a crime subject to federal jurisdiction.

artificial radiation, Radioactive by-products produced by particle bombardment of radioactive substances or electromagnetic production of x-rays.

asphyxiation, Suffocation resulting from being deprived of oxygen. Simple asphyxiants act mechanically by excluding oxygen from the lungs when breathed in high concentrations (examples: nitrogen, hydrogen, carbon dioxide). Chemical asphyxiants act through chemical action preventing oxygen from reaching the tissue, or else prevent the tissue from using it even though the blood is well oxygenated (examples: carbon monoxide, hydrogen cyanide, analine).

assignable cause, A causal factor designated as having a relationship to an accident. A causal factor or several causal factors assigned to an accident based on a predetermined causal classification system.

assigned risk, Many states have "unsatisfied judgment" or "financial responsibility laws" which make the purchase of insurance mandatory. Some motorists cannot buy insurance for some reason, such as poor accident experience. To make it possible for them to be insured, there are "assigned risk" plans in which such risks are insured. These risks are rotated among the subscribing companies in proportion to the amount of automobile liability insurance each writes in the state. All companies writing this class of insurance are required to participate in this activity. A comparable system operates in some states with respect to workmen's compensation.

assumption of risk, The legal theory that a person who is aware of a danger and its extent and knowingly exposes himself to it assumes all risks and cannot recover damages, even though he is injured through no fault of his own.

atmosphere, explosive, Atmosphere containing a mixture of vapor or gas which is within the explosive or flammable range.

attention, Focusing on a task all of the personal abilities (both mental and physical) necessary for the safe accomplishment of that task. The human characteristic of giving sufficient heed or observing and perceiving with sufficient care to avoid accidents. A readiness to respond to stimuli in a safe manner.

attenuation, Reduction of intensity; to render less virulent or toxic; to reduce in strength.

attenuation (sound), The reduction, expressed in decibels, of the sound intensity at a designated first location as compared with sound intensity at a second location, which is acoustically farther from the source, or caused by an intervening material.

audible range, The frequency range over which normal ears hear — approximately 20 Hz through 20,000 Hz. Above the range of 20,000 Hz, the term ultrasonic is used. Below 20 Hz the term subsonic is used.

audiogram, A record of hearing loss or hearing level measured at several different frequencies — usually 500 to 6,000 Hz. The audiogram may be presented graphically or numerically. Hearing level is shown as a function of frequency.

audiologist, A person trained in the specialized problems of hearing and deafness.

audiometer, A signal generator or instrument for measuring objectively the sensitivity of hearing in decibels referred to audiometric zero.

automatic sprinkler system, A combination of water discharge devices (sprinklers), distribution piping to supply water to the discharge devices, one or more sources of water under pressure, water flow controlling devices (valves), and actuating devices (temperature, rate of rise, smoke or other type device). The system automatically delivers and discharges water in the fire area.

average days charged per disabling injury, This measure expresses the relationship between the total days charged to a disabling injury and the total number of disabling injuries as defined by the ANSI Z16 Standard. The average may be calculated by use of the following formula:

$$\text{Average days charged per disabling injury} = \frac{\text{Total days charged}}{\text{Total number of disabling injuries}}$$

The following alternate formula may also be used to compute this measure:

$$\text{Average days charged per disabling injury} = \frac{\text{Standard Injury Severity Rate}}{\text{Standard Injury Frequency Rate}}$$

avoidable accident, An accident which could have or can be prevented by proper behavior, environmental, or equipment modifications or controls.

background (nuclear), The total effect of natural radiation, electronic circuitry and other factors in the environment which may interfere with the radiation measurements of specific interest. The radiation coming from sources other than the radioactive material to be measured.

barrier guard, Protection for operators and other individuals from hazard points on machinery and equipment.

fixed barrier guard, A point of operation enclosure attached to the machine or equipment.

interlocked barrier guard, An enclosure attached to the machinery or equipment frame and interlocked with the power switch so that the operating cycle cannot be started unless the guard is in its proper position.

adjustable barrier guard, An enclosure attached to the frame of the machinery or equipment with front and side sections which can be adjusted.

gate or movable barrier guard, A device designed to enclose the point of operation completely before the clutch can be engaged.

bioengineering, Designing equipment, machines, and other structures to fit the characteristics of people.

biologic agents, Cause of occupational disease or infections; may be classified into four groups; viral and rickettsial, bacterial, fungal, and parasitic. These agents act on or within the body to produce disease or infection.

biological effects, Reactions of the body to biological agents.

biomechanics, Designing tools, equipment, and the workplace to meet the physiological and anthropometric characteristics of the operator. Design criteria include such factors as error reduction, fatigue reduction, improved safety, increased production quality and quantity, and work efficiency.

board of certified safety professionals (BCSP), An autonomous board established to certify the professional and technical competence of individual practitioners in the safety field.

bodily injury (BI), Injury to a human being, as opposed to injury to property.

boiler codes, Standards prescribing requirements for the design, construction, testing, and installation of boilers and unfired pressure vessels. (E.g., American Society of Mechanical Engineers, etc.)

brightness, The light intensity of a surface in a given direction per unit emissive area as projected on a plane of the same direction. (See candle, illumination.)

building code, An assembly of regulations which set forth the standards to which buildings must be constructed.

carelessness, Not paying sufficient attention to avoid an accident where attentiveness is necessary for safe performance. Not thinking about the safety aspects of a behavior before acting. Behaving in an unwary, neglectful, heedless, negligent manner with general indifference to the safety aspects of a task or situation.

casual connection, An act, agency or force occuring without design or without being forseen or expected which is a concurrent or contributing factor to the injury, but is not usually the proximate cause thereof.

casualty insurance, Insurance written by companies licensed under the casualty sections of state insurance laws as distinguished from that written under the fire or marine or life insurance sections. Concerned principally with insurance against loss due to accident or other mishap.

catastrophe, A loss of extraordinary large dimensions in terms of injury, death, damage, and destruction.

causal factors (of an accident), A combination of simultaneous or sequential circumstances directly or indirectly contributing to an accident. Modified to identify several kinds of causes such as direct, early, mediate, proximate, distal, etc.

caveat emptor, Latin for "Let the Buyer Beware." The rule of law that the purchaser buys at his own risk concerning quality and condition.

caveat venditor, Latin for "Let the Seller Beware." The doctrine of law that a seller, if he wishes to absolve himself of future responsibility, must make a specific agreement with the purchaser to that effect.

certified safety professional (CSP), An individual who has been certified by the Board of Certified Safety Professionals as having achieved professional competence by remaining abreast of the technical, administrative, and regulatory developments in his chosen field and who maintains professional integrity in his relations with clients, associates and the public that reflects the highest standard of ethics.

chain of causation, The original force is responsible for every subsequent force which it puts in motion, and for the final result.

chronic violator, The chronic or persistent violator is the individual who repeatedly violates established statutes and ordinances as in the case of traffic situations, or they may be in the form of posted rules and regulations in a particular business or industrial enterprise.

circumstances (of an accident), Any condition or action accompanying or associated with an accident.

citation, Issued by the OSHA Area Director to an employer for an alleged violation reported by the OSHA Compliance and Safety and Health Officer during a compliance visit.

claim, (1) A broad, comprehenvise term whose meanings include, but are not limited to: Cause of suit or action; judgment; right; a demand for compensation and/or for payment of medical expenses. (2) The amount which a policyholder believes he has coming from an insurance company as the result of some occurrence insured against.

claimant, One who claims or asserts a right, demand, or claim.

codes, Rules and standards which have been adopted by a governmental agency as mandatory regulations having the force and effect of Law. Also used to describe a body of standards.

collision diagram, A diagram of an intersection or section of roadway where an accident occurred. The diagram shows the manner of collision, the resting positions of vehicles and other items after the collision occurred by the use of designated symbols.

compensable injury, An occupational injury resulting in sufficient disability to require the payment of compensation as prescribed by law. A work injury for which compensation benefits are payable to the injured worker or his beneficiary under workmen's compensation laws.

compensation, Indemnity paid to an employee for disability sustained in an occupational accident.

comply, To act in accordance with the Occupational Safety and Health Standards; to follow the rules and regulations published in the CODE OF FEDERAL REGULATIONS.

conflagration, A fire extending over a considerable area, and destroying numbers of buildings and/or substantial amounts of property.

consensus standard, A standard developed according to a consensus agreement or general opinion among representatives of various interested or affected organizations and individuals.

constraint, A restriction or a compelling force affecting freedom of action. Forcing into or holding within close bounds. An operational condition which may necessitate work performance in less than an ideal safe environment (e.g., building construction).

containment, Restricting the spreading of fire. Preventing the spread of fire to surrounding structures or areas.

contest, To object to or to appeal a decision made by the OSHA Area Director.

contractual liability, Liability as set forth by agreements between people as distinguished from liability imposed by law (legal liability).

contributory negligence, The act or omission amounting to want of ordinary care on part of complaining party, which, concurring with defendant's negligence is proximate cause of injury. This is different from assumption of risk, which exists where none of fault for injury rests with plaintiff, but where plaintiff assumes consequences of injury occurring through fault of defendant, third person or fault of no one.

cost effectiveness, In this method, the cost of system changes made to increase safety are compared with either the decreased costs of fewer serious failures, or with the increased effectiveness of the system to perform its task to determine the relative value of these changes.

coverage (insurance), An insured risk or liability. That which is insured, as specified in the insurance policy.

crash safety, A system characteristic that allows the system occupants to survive the impact of a crash and to evacuate the system after potentially survivable accidents.

crashworthiness, The capacity of a vehicle to act as a protective container and energy absorber during impact conditions.

criminal negligence, Involving or relating to a legal crime due to failure to use a reasonable amount of care when such failure results in injury or death to another.

critical function, An operation or activity which is essential to the continuing survival of a system. Those function which have a major impact on system performance and safety.

critical incident technique, A set of procedures for collecting direct observations of human behavior in such a way as to facilitate their potential usefulness in solving practical problems and developing broad psychological principles. The critical incident technique outlines procedures for collecting observed incidents having special significance and meeting systematically defined criteria. A randomly selected sample of critical incidents should permit an inference to be made concerning the existence of similar incidents within the population from which the sample was taken.

damage, (1) Loss in value, usefulness, etc., to property or things. Harm causing any material loss. (2) Severity of injury or the physical, functional, or monetary loss that could result if control of a hazard is lost.

damage risk criterion, The suggested base line of noise tolerance, which if not exceeded, should result in no hearing loss due to noise. A damage risk criterion may include in its statement a specification of such factors as time of exposure, noise intensity, noise frequency, amount of hearing loss that is considered significant, percentage of the population to be protected, and method of measuring the noise.

damages, A just compensation which may be recovered in the courts by a person who has suffered loss, damage, or injury, whether to his person or property, through the unlawful act, omission, or negligence of another.

compensatory damages, That amount that will compensate the injured party for injury sustained, and nothing more, to make good or replace the loss.

punitive damages, Damages assessed as a punishment to the wrongdoer, or as an example to others, for his outrageous conduct or his gross, wanton negligence.

danger, A general term denoting liability or potential of injury, illness, damage, loss, or pain.

danger zone, A physical area or location within which a danger exists.

dangerous, Attended with risk; hazardous; unsafe. Something that if in normal use, danger or injury can be anticipated by the user. Something without adequate protection.

imminently dangerous, An article, by reason of defective construction, that threatens impending danger of injury.

inherently dangerous, An article which is usually dangerous even in its normal or non-defective state, such as explosives or poisons, and requires special precautions and warnings so as to prevent injury.

days of disability, Total full calendar days on which an injured person was unable to work as a result of injury. The total does not include the day of the injury or the day of return to work.

deaf, A term used to describe a person who has lost his hearing before the speech patterns were established.

deafened, Refers to a person who has lost his ability to hear after normal speech patterns were established.

death (accidental), An injury which terminates fatally and is causally related to an accident. Death resulting from work injuries is assigned a time charge of 6,000 days each according to ANSI Z16.1 standard.

decibel (Db), A measure of sound expressed as the logarithmic ratio of two amounts of pressure, power, or intensity between a measured quantity and a reference quantity.

decontamination, Removal of a polluting or harmful substance from air, water, earth surface, etc. For example, the process of removing radioactive contamination from objects or areas.

defect, (1) Known or unknown unsafe or unwanted physical condition of material and/or equipment due to an inherent or created weakness that may lead to an accident. (2) Anything outside of specification or standards.

defective, Lacking in some particular way which is essential to the completeness or security of the object.

deflagration, An exothermic reaction which propagates from the burning gases to the unreacted material by conduction, convection, and radiation. The combustion zone progresses through the material at a rate that is less than the velocity of sound in the unreacted material.

de minis violation, Violation of a standard that does not involve an immediate or direct relationship to the safety or health of an employee.

density (of traffic), The number of vehicles occupying a unit length of the moving lanes of a roadway at a given instant. Usually expressed in vehicles per mile.

deposition, The testimony of a witness taken upon interrogatories, not in open court, under oath, in writing and duly authenticated and intended to be used as evidence in court. (See interrogatory.)

depth perception, The ability to judge distances of nearby objects by the use of both eyes.

dermatitis, Inflammation or irritation of the skin. Industrial dermatitis is an occupational skin disease. There are two general types of skin reaction: primary irritation dermatitis and sensitization dermatitis. (See irritant and sensitizer.)

design (safety), The planning of environments, structures, and equipment, and the establishment of procedures for performing tasks, so that human exposure to injury potential will be reduced or eliminated. In product safety, design of the product for safe use.

destructive test, A procedure for quality testing whereby the material being tested is destroyed in order to obtain the desired measurements.

detonation, An exothermic reaction that is characterized by the presence of a shock wave in the material that establishes and maintains the reaction. A distinctive failure is that the reaction zone propagates at a rate greater than sound velocity in the unreacted material.

direct cause, Unsafe behaviors or unsafe conditions which contribute sequentially or concurrently in a chain of events leading to an accident.

direct damage, Damage caused by the direct action of a peril as distinguished from damage done contingently.

direct injury costs, The sum of compensation payments and medical expenses for an injury.

disability, Any injury or illness, temporary or permanent, which prevents a person from carrying on his usual activity. (See permanent and total disability.)

disabling injury, American Standard Z16.1: An injury which prevents a person from performing a regularly established job for one full day (24 hours) beyond the day of the accident.

disabling injury frequency rate, The number of disabling (lost time) injuries per million employee-hours of exposure:

$$\text{DIFR} = \frac{\text{Disabling Injuries} \times 1{,}000{,}000}{\text{Employee-hours of exposure}}$$

disabling injury severity rate, The total number of days charged per million employee-hours of exposure:

$$\text{DISR} = \frac{\text{Total days charged} \times 1{,}000{,}000}{\text{Employee-hours of exposure}}$$

disabling injury index, An index computed by multiplying the disabling injury frequency rate by the disabling injury severity rate and dividing the product by 1,000:

$$\text{DII} = \frac{\text{DIFR} \times \text{DISR}}{1{,}000}$$

This measure reflects both frequency and severity, yielding a combined index of total disabling injury (Z16.1).

disaster control, Advanced planning and established procedures for handling emergency situations.

disclaimer, The seller may insert in his contract or agreement a statement that he does not warrant at all, or that he warrants only against specified consequences or costs. Disclaimers do not release the manufacturer or defendant from liability for negligence, nor are disclaimers a defense to statutory violations.

disfigurement, A blemish, defect, or deformity which harms the appearance or attractiveness of the human body or a physical structure.

distal cause, A behavioral act or condition involved simultaneously or sequentially in the causal factors leading to an accident, but separated from the accident location by time and/or space.

dose, The amount of radiation delivered to a specified area or the whole body. A dose meter, or dosimeter is an instrument that measures radiation dose. Dose rate is the dose delivered per unit of time. The term "dose" or "dosage" is also used generally to express the amount of energy or substance absorbed in a unit volume by an organ or individual.

dry-chemical extinguisher, An extinguisher containing a chemical which extinguishes fire by interrupting the chain reaction wherein the chemicals used prevent the union of free radical particles in the combustion process so that combustion does not continue when the flame front is completely covered with the agent.

dry-powder extinguisher, A fire extinguisher designed for use on combustible metals fires, such as sodium, titanium, uranium, zirconium, lithium, magnesium, and sodium-potassium alloys.

ear defenders, Plugs or muffs designed to keep noxious noise from the ear to preserve hearing acuity.

early cause, An act on the part of some person or organization or a condition that causes or permits a proximate or mediate cause to exist (also referred to as distal cause).

eczema, A skin disease or disorder.

edema, Excessive accumulation of body fluid in the tissue space of the body, producing a swelling of body tissues.

efficient cause, The cause which originates and sets in motion a chain of causation through other causes to the result. The cause of injury attached to legal liability. This term is not as popular in use as proximate cause.

emergency alarm, A warning device, usually visual or auditory, which indicates the existence of an emergency situation requiring immediate action.

emergency procedure, A plan for action in case of emergency.

emergency stop, A switch installed in an elevator, or other similar piece of equipment, by means of which the power to the operating motor can be cut off in case of an emergency.

emergency switch, A switch placed in a convenient position for cutting off the supply of electricity to a piece of equipment or to a building, in case of emergency.

emphysema, A lung disease in which the walls of the air sacs (alveoli) have been stretched too thin and broken down.

employers liability, Legal liability imposed on an employer making him responsible for paying damages to an employee injured by the employer's negligence. Generally replaced by "Workmen's Compensation," which pays the workman whether the employer has been negligent or not.

epidemiological concept of cause, A descriptive model for use in defining causal relationships. In this framework, causation involves more than the "agent" directly concerned, e.g., the germ in disease or the loose board in a home accident. Causation is interpreted as a combination of forces from three sources at least — the host, the agent, and the environment in which both host and agent exist.

epidemiology, The study of disease as it spreads and involves large groups of people.

equivalent form, Any of two or more forms of a test that are closely parallel with respect to the nature of the content and the difficulty of the items included and that will yield very similar average scores and measures of variability for a given group.

exhaust (general), Diluting the general room atmosphere with outdoor air fast enough to keep the concentration of toxic vapor in the room air within safe limits. Also known as "general ventilation."

exhaust (local), A local exhaust system is used to carry off an air contaminant by trapping it near its source, as contrasted with general ventilation which lets the contaminant spread throughout the workroom, later to be diluted by exhausting quantities of air from the room.

exhaust ventilation, The removal of air or other gas from any space, usually by mechanical means.

experience, Data describing past events, e.g., accident experience refers to the frequency number or severity of accidents that have occurred within a specified time period.

experience rating (merit rating), Process of basing tax rates or insurance premiums on the employer's own record — as in workmen's compensation, unemployment insurance, and commercially insured health and insurance programs — so that he may benefit from a good record.

expert evidence, Testimony given in relation to some scientific, technical, or professional matter by an expert witness. An expert can reason out, infer, or conclude from hypothetically stated facts, and can offer opinions involving the subject matter under consideration. It is up to the jury to give whatever credence it wants to the testimony.

expert testimony, The opinion of a witness skilled in a particular art, trade, or profession or possessed of special knowledge derived from education or experience not within the range of common experience, education, or knowledge.

expert witness, A person possessing particular knowledge, wisdom, skill or information acquired by study, investigation, observation, practice, or experience, regarding the subject matter under consideration and not likely to be possessed by ordinary or inexperienced persons. It is up to the judge to determine if the person is qualified as an expert and if his testimony is pertinent to, or would "shed light" on the case.

explosimeter, A device which detects and measures the presence of gas or vapor in an explosive atmosphere.

explosion, A rapid increase of pressure in a confined space followed by its sudden release due to rupture of the container (vessel, structure, etc.). The increase in pressure is generally caused by an exothermic chemical reaction or over-pressurization of a system.

explosion-proof, An electrical apparatus so designed that an explosion of flammable gas inside the enclosure will not ignite flammable gas outside. (See flameproof.)

explosion venting, A means provided for the release of high pressures caused by explosions.

explosive, Limits — The minimum (lower) and maximum (upper) concentration of vapor or gas in air or oxygen below or above which explosion or propagation of flame does not occur in the presence of a source of ignition. The explosive or flammable limits are usually expressed in terms of percentage by volume of vapor or gas in air. Range — The difference between the lower and upper flammable (explosive) limits, expressed in terms of percentage by volume of vapor or gas in air.

explosive mixture, A mixture of flammable vapor or gas and air within the lower and upper limits of the explosive range.

explosives, Any chemical compound or mechanical mixture that is used or intended for the purpose of producing an explosion. Contains any oxidizing and combustive units or other ingredients in such proportions, quantities, or packing that an ignition by fire, by friction, by concussion, by percussion or by detonation of any part of the compound or mixture may cause such a sudden generation of highly heated gases that the resultant gaseous pressures are capable of producing destructive effects on contiguous objects or of destroying life or limb.

exposure, The quantity of time involved and the nature (quality) of involvement with certain types of environments possessing various degrees or types of hazards. The amount of time a worker has been exposed to certain types of job hazards.

exposure (casualty), Proximity to a condition which may produce injury, death, or damage from dusts, chemicals, high pressure, explosives, etc.

exposure (fire), The danger of loss arising from fire occurring to another risk close by.

exposure hours (employee-hours), Total number of employee-hours worked by all employees including those in operating, production, maintenance, transportation, clerical, administrative, sales, and other activities (Z-16 Standard).

extinguisher, See fire extinguisher.

extinguishing agent, Material or substance which performs a fire extinguishing function.

eye protection, A device which safeguards the eye in an eye-hazard environment.

face shield, A protective device designed to prevent hazardous substances, dust particles, sharp objects, and other materials from contacting the face.

face velocity, Air velocity at the face opening required to retain a contaminant in a hood, or booth. (Paint spray, grinding, welding operations, etc.)

factor, In mental measure a hypothetical trait, ability, or component of ability, that underlies and influences performance on two or more tests and hence carries scores on the tests to be correlated. The term "factor" strictly refers to a theoretical variable, derived by a process of factor analysis, from a table of intercorrelations among tests; but it is also commonly used to denote the psychological interpretation given to the variable, i.e., the mental trait assumed to be represented by the variable as verbal ability, numerical ability, etc.

factor analysis, Any of several methods of analyzing the intercorrelations among a set of variables such as test scores. Factor analysis attempts to account for the interrelationships in terms of some underlying "factors" preferably fewer in number than the original variables and it reveals how much of the variation in each of the original measures arises from or is associated with each of the hypothetical factors. Factor analysis has contributed to our understanding of the organization or components of intelligence, aptitudes, and personality and it has pointed the way to the development of "purer" tests of the several components.

factor of safety (FS), The ratio of ultimate strength of a material or structure to the allowable stress.

fail-safe, Design of a product or equipment, in such a manner that, when it fails or becomes inoperative it will do so in a safe position or condition.

failure mode and effect analysis (FMEA), A method of analysis used in system safety. The failure or malfunction of each component is identified, along with the mode of failure (e.g., switch jammed "on"). The effects of the failure are traced through the system, and the ultimate effect on the task performance is evaluated.

failure mode, effect and criticality analysis (FMECA), An extension of an FMEA in which each effect is assigned a criticality index which reflects both the probability of the occurrence of the effect and the seriousness of the effect in terms of loss in performance and/or safety.

failure, primary, The failure which is responsible for a system malfunction.

failure rate, The number of failures of an item per unit time (cycles, hours, miles, events, etc., as applicable for the item).

failure, secondary, A failure which occurs as the consequence of another failure (also dependent failure).

fainting, A temporary loss of consciousness as a result of a diminished supply of blood to the brain. Technically called syncope.

fatal accident, An accident resulting in the death of one or more persons.

fatality, A death resulting from an accident.

fatigue, The physical and/or mental responses to an activity which show themselves in a diminished capacity for work.

fault hazard analysis, The analysis or hazards or hazard-potential situations using fault tree methodology.

fault tree analysis, A method of analysis used in system safety. An undesired event is selected and all possible factors that can contribute to the event are diagrammed in sequence in the form of a "tree." The branches of the tree are continued until independent events are reached. Probabilities are determined for the independent events and after simplifying the tree, both the probability of the undesired event and the most likely chain of events leading up to it are computed.

final position (after an accident), The place where objects or persons involved in the accident finally come to rest without application of power. This is the position before anything is moved to help the injured or remove vehicles or equipment.

fire, Rapid oxidation with the evolution of heat and light.

fire alarm, A device or system (visual, auditory, local or transmitted to other locations, etc.) which signals the presence of a fire to those who will provide assistance and/or occupants.

fire classifications, See fire extinguishers.

fire doors, Doors rated and tested for resistance to various degrees of fire exposure and utilized to prevent the spread of fire through horizontal and vertical openings. The doors must remain closed normally or be closed automatically in the presence of fire. The degree of resistance required is determined by the anticipated fire exposure and the resistance of the structure in which it is installed.

fire extinguishers, Devices having characteristics essential to extinguish flame. Fire extinguishers may contain either liquid or dry chemicals, or gases (water, dry chemical, carbon dioxide, etc.). They are tested and rated to indicate their ability to handle specific classes and sizes of fires.

class A extinguishers: For ordinary combustibles, such as wood, paper, and textiles, where quenching-cooling effect is required.

class B extinguishers: For flammable liquid and gas fires, such as oil, gasoline, paint, and grease, where oxygen exclusion or flame interruption effect is essential.

class C extinguishers: For fires involving energized electrical wiring and equipment where the nonconductivity of the extinguishing agent is of prime importance.

class D extinguishers: For fires in combustible metals such as magnesium, potassium, powdered aluminum, zinc, sodium, titanium, zirconium, and lithium.

fail operational, A characteristic design which permits continued operation in spite of the occurrence of a discrete failure.

fail operational fail safe, A system characteristic which permits continued operation on occurrence of a failure while remaining acceptably safe. A second like failure results in the system remaining safe, but non-operational.

failure, An inability to perform an intended function.

failure analysis, The logical systematic examination of an item to identify and analyze the cause, mode, and consequence of a real failure.

failure assessment, The process by which the cause, effect, responsibility and cost of any reported problem in the system is determined and reported.

failure, critical, A failure which could result in major injury or fatality to people or which could result in major damage to any system or loss of a critical function.

failure, dependent, See failure, secondary.

failure, independent, See failure, primary.

failure management, Decisions, policies and planning which identify and eliminate or control potential failures and implement corrective or control procedures following real failures.

failure mechanism, The physics or chemistry of the failure event, i.e., the cause of the failure.

fire (flame) proof, Material incapable of burning. The term "fire-proof" is false. No material is immune to the effects of a fire possessing sufficient intensity and duration. It is commonly, although erroneously used synonomously with the term "fire resistive." Use of the term is discouraged since it is misleading.

fire prevention, Measures or actions specifically directed toward preventing the inception of fires, and minimizing the severity of fires should they occur.

fire protection, In its broadest interpretation it embraces all measures in the prevention, detection, and extinguishment of fire; relating to the safeguarding of human life and the preservation of property. In a strict interpretation, it refers to the methods of providing for fire control or fire extinguishment.

fire protection engineering, The field of engineering concerned with the safeguarding of life and property against loss from fire, explosion and related hazards. It is concerned with integrated programs involving the design and use of structures, equipment, processes and systems, including the areas of prevention, detection and alarm, and fire control and extinguishment, and gives consideration to functional, economic and operational factors.

fire resistive, Refers to properties of materials or designs to resist the effects of any fire to which the material or structure may be expected to be subjected. Fire resistive materials or structures are non-combustible, but non-combustible materials are not necessarily fire resistive. Fire resistive implies a higher degree of fire resistance than non-combustible.

fire retardant, In general denotes a substantially lower degree of fire resistance than fire resistive. The term is frequently used to refer to materials or structures which are combustible but have been subjected to treatments or surface coverings to prevent or retard ignition or the spread of fire.

fire wall, A fire resistant wall designed to prevent the horizontal spread of fire into adjacent areas, generally self-supporting and designed to maintain its integrity if the structure on either side completely collapses. If a wood roof is involved, the wall must extend through and above the roof.

first aid, The emergency care of a person who is injured or ill, to prevent death or further injury, to relieve pain, and to counteract shock, until medical aid can be obtained.

first aid injury, An injury requiring first aid treatment only.

fit for purpose intended, When the article sold is useless or so imperfect that the buyer would not have bought it had he known, the seller has not complied with statutory warranty against hidden defects and the article cannot be considered "fit for the purpose intended," unless the warranty has been specifically waived.

fitness (or fit), Satisfactoriness, suitability or appropriateness to fulfill the need or use.

flame, The visible heat rays which appear when the ignition of a material is reached. Hydrogen is one of the exceptions since the heat rays are not visible.

flame (flash) arrestor, Devices utilized on vents for flammable liquid or gas tanks, storage containers, cans, gas lines or flammable liquid pipelines to prevent flash-back (movement of flame) through the line or into the container when a flammable or explosive mixture is ignited.

flameproof, See fireproof.

flame propagation (spread), The spread of flame throughout combustible vapor area which may be in a container or across a surface, independently of the ignition source. Generally used in connection with the capability and rate of such movement.

flammable, Any substance that is easily ignited, burns intensely, or has a rapid rate of flame spread. Flammable and inflammable are identical in meaning, however, the prefix "in" indicates negative in many words and can cause confusion. Flammable therefore is the preferred term.

flammable limits, See explosive limits.

flammable liquids, Liquids having a flash point below 140°F and having a vapor pressure not exceeding 40 psia at 100°F. They are divided into two classes: Class I those with flash points below 100°F and Class II those with flash points between 100°F and 140°F. Liquids with a flash point above 140°F are known as combustible liquids. A flammable liquid does not itself burn, the vapors from the liquid burn.

flammable range, See explosive range.

flammable vapor, A concentration, by volume, of vapors in air from a flammable liquid within the lower and upper flammable limits.

flash burn, Injury or destruction of body tissue caused by exposure to a flash or sudden release of intense radiant heat.

flash point, The lowest temperature of a liquid at which it gives off sufficient vapor to form an ignitible mixture with the air near the surface of the liquid or within the vessel used. The flash point can be determined by the open cup or the closed cup method. The flash point determined by the open cup method is usually somewhat higher than the closed cup method. The latter is commonly used to determine the classification of liquids which flash in the ordinary temperature range.

foam, A fluid aggregate of gas- or air-filled bubbles formed by chemical or mechanical means that will float on the surface of flammable liquids or flow over solid surfaces. The foam functions to blanket and extinguish fires and/or to prevent ignition of the material.

foreseeability, The legal theory that a person may be held liable for actions that result in injury or damage only where he was able to foresee dangers and risks that could reasonably be anticipated.

four-to-one ratio, An arbitrary ratio frequently used in the comparison of the indirect costs of an accident to the direct costs. Generally considered obsolete since no fixed ratio exits among various types of exposures.

gamma (γ) radiation, ray, Electromagnetic radiation of short wave length omitted from the nucleus of an atom. Gamma radiation is more penetrating than alpha or beta particles. Gamma rays are very high energy X-rays.

gas mask, A face covering connected to its own purifying device which filters harmful gases from the air so uncontaminated air may be breathed. Gas masks do not add oxygen to air and cannot be used where there is oxygen deficiency.

glare, Any brightness within the field of vision of such character as to cause discomfort, annoyance, interference with vision, or eye fatigue.

goggles, Spectacles or glasses to protect the eyes against dust, impacting objects, strong light, or other harmful environmental influences.

ground fault circuit interrupter, A fast-acting circuit breaker that is sensitive to very low levels of current leakage to ground. The interrupter is designed to limit the electric shock to a current and time duration value below that which can produce serious injury.

guard, An enclosure which prevents entry into the point of operation of a machine or renders contact harmless with any substance or object.

guard, fixed barrier, A point-of-operation enclosure attached to a machine frame by fasteners.

guard, interlocking barrier, An enclosure attached to the frame of a machine and interlocked so that the machine cycle cannot be started normally unless the guard, including its hinged or movable sections, are in position. In some situations, movement of the guards will interrupt the machine cycle.

guardrail, A device consisting of posts and rail members, or of wall sections erected to mark points of major hazard, and to prevent individuals from coming in contact with the hazard.

habitual violator, An individual whose record, during a given time period, shows reports of repeated violations of laws or regulations. In traffic safety, any driver whose record during a consecutive 12-month period shows reports of more than three convictions for traffic violations or more than five times the average number of convictions for all drivers in the state, whichever is greater.

hazard, That dangerous condition, potential or inherent, which can bring about an interruption or interference with the expected orderly progress of an activity.

hazard analysis, An analysis performed to identify hazardous conditions for the purpose of their elimination or control. *(DOD) (mod.)*

hazard classification, A designation of relative accident potential based on probability of accident occurrence.

hazard control, That function in an organization directed toward the recognition, evaluation, and reduction or elimination of the destructive effects of hazards emanating from human acts of commission and omission and from the physical and environmental aspects of the workplace.

hazard level, A qualitative measure of hazards stated in relative terms. *(DOD)*

a. *Category I - Negligible*
   Will not result in personal injury or system damage.
b. *Category II - Marginal*
   Can be counteracted or controlled so that no injury to personnel or major system damage will be sustained.
c. *Category III – Critical*
   Will cause personal injury or major system damage or both.
d. *Category IV - Catastrophic*
   Will cause death to personnel.

hazard pay, Extra payments to workers to dangerous occupations or while engaged in work where the chances of injury are greater than normal.

hazard recognition, The perception of a hazardous condition.

hazardous condition, The physical condition of circumstance which is causally related to accident occurrence. The hazardous condition is related directly to both the accident type and the agency of the accident.

hazardous materials, Materials possessing a relatively high potential for harmful effects upon persons.

health and insurance plan, A program of providing financial protection to the worker and his family against death, illness, accidents, and other risks, in which the costs may be borne in whole or in part by the employer. One or more of the following major benefits may be provided for the worker and, frequently, his dependents: life insurance, accidental death and dismemberment benefits, accident and sickness benefits, hospitalization benefits, surgical and medical benefits, and major medical expense benefits.

hearing conservation, Preventing or minimizing noise-induced deafness through the use of hearing protection devices and the control of noise through engineering methods.

hearing level (hearing loss), The deviation in decibels of an individual's threshold from the zero reference of the audiometer. Formerly called "hearing loss."

heat prostration (heat exhaustion), A form of heatstroke characterized by low body temperature, collapse, and, in severe cases, coma and death.

high frequency hearing loss, A hearing deficit starting at 2000 Hz. and beyond.

hoistway, A vertical passageway designed to enclose and provide support for an elevator, platform, or other lifting device.

hoistway-door interlock, A hoistway-door interlock is a device, on an elevator shaftway, the purpose of which is, first, to prevent the operation of the elevator machine in a direction to move the car away from a landing unless the hoistway door at that landing at which the car is stopping or is at rest is locked in the closed position, and second, to prevent the opening of the hoistway door from the landing side except by special key, unless the car is at rest within the landing zone, or is coasting through the landing zone with its operating device in the stop position.

hold harmless agreement, A contract under which the legal liability of one party for damages is assumed by another party to the contract. The principal in a large construction project will frequently demand hold harmless agreements from all subcontractors in respect to claims made against him arising out of the subcontractors' negligence. The principal often stipulates the purchase of a liability policy by the subcontractor to support the hold harmless agreement.

hospitalization benefits, A plan that provides workers, and in many cases their dependents, with hospital room and board or cash allowances toward the cost of such care for a specified number of days, plus the full cost of specified services. Usually part of a more inclusive health and insurance program.

host factors, Personal characteristics of the individual. These may be grouped, as Thorndike has done in an analysis of the literature, into (a) sensory abilities, (b) psycho-motor abilities, (c) cognitive and intellectual abilities, (d) personal and emotional adjustments (including physiological adjustment), and (e) biographical facts.

housekeeping, Cleanliness, neatness, and orderliness of an area, with the designation of a proper place for everything and everything in its proper place.

illumination, The amount of light flux a surface receives per unit area. May be expressed in lumens per square foot or in foot-candles. The rate at which a source emits light energy, evaluated in terms of its visual effect, is spoken of as light flux, and is expressed in lumens.

imminent danger, An impending or threatening dangerous situation which could be expected to cause death or serious injury to persons in the immediate future unless corrective measures are taken.

implosion, A rapid expenditure of energy producing an inward burst, opposed to explosion.

incendiary, A substance causing or designed to cause fires. Also, a person who willfully destroys property by fire.

incident, An unforeseen event or occurrence which does not result in injury or property damage.

incombustible, A substance that cannot be burned. The term noncombustible is more appropriate.

indirect cause, A causal factor associated with an occupational accident other than medical costs and workmen's compensation payments.

indirect costs, Monetary losses resulting from an accident other than medical costs and workmen's compensation payments. (See accident costs.)

indirect damage, Loss resulting from a hazardous condition or incident but not caused directly thereby.

industrial dermatitis, See dermatitis.

industrial hygiene, That science and art devoted to the recognition, evaluation and control of those environmental factors or stresses, arising in or from the work, which may cause sickness, impaired health and well being, or significant discomfort and inefficiency among workers or among the citizens of the community.

industrial safety, See occupational safety.

inflammable, A general term once used to describe combustible gases, liquids or solids. Now obsolete. (See flammable.)

inflammation, The reaction of living tissues to injury, whether by infection or trauma. The affected part becoming red, hot, painful, and swollen due to hyperaemia, exudation of fluid, etc.

interrogatories, A set of formal written questions presented to a witness, usually before the trial, who must give written answers under oath.

intervening act, The negligent act of a third person to break the chain of causation in such a manner that the injury was not a probable consequence of the original wrongdoer's negligence.

intervening agency, An intermediate agency that disrupts the connection between the negligent act and the injury in such a way to supersede the original act and stand alone as the efficient cause of injury.

intervening cause, An independent cause which breaks the connection between the negligent act and the injury such that it is the immediate cause of injury.

intervening force, One which actively produces harm after the defendant's negligent act has been committed. The word "intervening" is used in a time sense to include later events, and not conditions or forces already in operation at the time of defendant's conduct. To relieve the defendant from liability for his original negligent act, it must be determined whether the intervening force was the superseding cause of injury, rather than an extraordinary cause or a cause which might reasonably have been expected.

ingestion, The act or process of taking in or putting items (food, drugs, etc.) into the body for digestion. With regard to certain cells, the act of engulfing or taking up bacteria and other foreign matter.

inhalation, The act of breathing in, or taking into the lungs.

inhibitor, A substance or an agent that slows or prevents a chemical or organic reaction, or a material used to prevent or retard rust or corrosion.

injury, Physical harm or damage to a person resulting in the marring of appearance, personal discomfort, infection, and/or bodily hurt or impairment. Note: the definition of this word is frequently determined by the government agency or other organization using it.

injury (occupational), Any hurt, harm or impairment to the body which arises out of, or in the course of, employment. Again, the government agency which has jurisdiction determined the exact wording of the term.

inrunning nip point, A rotating mechanism that can seize and wind up loose clothing, belts, hair, body parts, etc. It exists when two or more shafts or rolls rotate parallel to one another in opposite directions.

insurable interest, Refers to the relationship of the party for whom the insurance coverage is written to the peril being insured. The party must either own or have a financial stake in the subject that may suffer the damage or loss which would involve out-of-pocket expense to him if no insurance indemnity was available.

insurance, The making of a legal and enforceable contract between one party (called the insurer or underwriter) with another (called the insured) whereby in consideration of a sum of money (called the premium) the insurer agrees to pay an agreed amount of money to the insured if and when the latter may suffer some loss or may be injured by some event, the happening of which is described in the contract of insurance (which is usually a "policy").

insured, The person who has purchased a policy of insurance and is protected by it.

insured costs, Accident losses, which are covered by workmen's compensation, medical, or other insurance programs. They comprise the insured element of the total accident cost.

intensity level (sound), The sound-energy flux density level. Measured in decibels of sound. In decibels of sound, intensity level equals 10 times the logarithm to the base 10 of the ratio of the intensity of this sound to the reference intensity. (See decibel.)

interlock, A device which interacts with another device or mechanism to govern succeeding operations. For example, an interlocked machine guard will prevent the machine from operating unless the guard is in its proper place. An interlock on an elevator door will prevent the car from moving unless the door is properly closed.

intervening losses, The intrinsic value for the trouble, disadvantage, or deprivation incurred by damage to property or injury to people. Losses in addition to the material or physical losses.

intervening negligence, The negligence of an intervening person or agency between the original negligence of the wrongdoer and the injury.

intoxication, Pertains, in medicine, to poisoning as by a drug, serum, alcohol, or any poison. Also drunkenness or the acute condition produced by over-indulgence of alcohol.

irritant, Any external stimulus which produces an active response in a living organism. A primary irritant is one that has been found to produce an irritating effect at the area of the skin contact. Although they affect everyone, they do not produce the same degree of irritation.

job hazard analysis, See job safety analysis.

job safety analysis, The breaking down into its component parts of any method or procedure to determine the hazards connected therewith and the requirements or qualifications of those who are to perform it. A method for studying a job in order to (1) identify hazards or potential accidents associated with each step or task and (2) develop solutions that will eliminate, nullify, or prevent such hazards or accidents.

job safety training, Training associated with or emphasizing the safety aspects of a job and the hazards of tasks and their interrelationships within a job.

judgment, The allowance or disallowance of a claim.

jurisdiction, The power of a court to hear and decide a cause of action, having power over both the parties of the action and the subject matter thereof.

just cause, Good or fair reasons for discipline. This term is commonly used in agreement provisions safeguarding workers from unjustified discharge or a lesser punishment. When defined in agreements, it usually includes such offenses as insubordination, fighting, inefficiency, etc.

key event (of an accident), That event (or events) in a series of events in an accident that determined the exact time, place, type, and extent of consequence of the accident.

kickback, A term used to describe the reaction of a piece of material (usually wood) when being cut by a circular saw which is rotating in the opposite direction from the material as it is fed into the saw. Since the saw blade is rotating at a high rate of speed the force that can be exerted on the material is substantial and can cause serious injury to the operator.

lead intoxication, A result of lead absorption, occurring from inhalation of lead dust or fumes or from swallowing lead dust.

leave of absence, Generally, excused time (unpaid) away from work without loss of job or seniority.

legal liability, Liability imposed by law as opposed to liability arising from an agreement or contract.

lesion, Injury, damage, or abnormal change to body tissue or organ, especially one that leads to impairment or loss of function of the part involved, or, even if it produces no impairment of function, expresses a symptom or sign of disease.

level (noise), The logarithm of the ratio of the measured quantity to a reference quantity of the same kind. The base of the logrithm, the reference quantity, and the kind of level must be specified.

liability insurance, Insurance which obligates the insurance company to pay any liability for which the insured may be covered and, also at the expense of the company, to defend any damage suits brought against him to enforce such liability, thus protecting him from liability and expense of litigation growing out of claims in which he is involved.

Insurance which agrees to reimburse the policyholder for sums he may be required to pay others as the result of his negligence.

life insurance plan, Group term insurance coverage for employees, paid for in whole or in part by the employer, providing a lump-sum payment to a worker's beneficiary in the event of his death. (See health and insurance plan, death benefit.)

lifeline, A rope designed to arrest a workman's fall before he strikes a damaging obstruction.

limit switch, A switch fitted to electric lifts, traveling cranes, etc., in order to cut off the power supply if the liftcar or moving carriage travels beyond a certain specified limit.

load limit, The upper weight limit capable of safe support by a vehicle or floor. The designer of floors to meet load limit requirements should consult ANSI Standard A58.I, Building Code Requirements For Minimum Design Loads in Buildings and other Structures.

load weight or allowable load, Refers to highest weight of load to be carried by a vehicle safely.

loss, In insurance it means the amount the insurer is required to pay because of a judgment that requires the company to pay by virtue of the terms of the insurance contract. Also refers to the overall financial result of some operation, as opposed to "profit."

loss control, A program designed to minimize accident-based financial losses. The concept of total loss control is based on studies of "near misses" (noninjury accidents) and on detailed analysis of both indirect and direct accident costs. Property damage as well as injurious and potentially injurious accidents are included in the analysis.

loss prevention, A before-the-loss program designed to identify and correct potential accident problems before they result in actual financial loss or injury.

loss ratio (insurance), A fraction calculated by dividing the amount of losses by the amount of premiums. Expressed as a percentage of the premiums. Various bases are used in calculating the loss ratio, e.g., "earned premium loss ratio," "written premium loss ratio," etc.

loss reserve, An estimate of the amount an insurer expects to pay for losses incurred but not yet due for payment.

lost time accident (LT), See lost time injury.

lost time injury, A work injury which results in death or disability and in which the injured person is unable to report for duty on his next regularly scheduled shift. (See injury.)

lost workdays, See days of disability.

loudness, The intensity attribute of an auditory sensation, in terms of which sounds may be ordered on a scale extending from soft to loud. Loudness depends primarily upon the sound pressure of the stimulus, but it also depends upon the frequency and wave form of the stimulus.

loudness level, A sound, in phons, is numerically equal to the median sound pressure level, in decibels, relative to 0.0002 microbars, of a free progressive wave of frequency 1000 Hz presented to listeners facing the source, which in a number of trials is judged by the listeners to be equally loud.

lower explosive limit (LEL), The minimum concentration of combustible gas or vapor in air of flammable liquids or gases below which propagation of flame does not occur on contact with a source of ignition.

luminescent, Emitting light not due to high temperatures, usually caused by excitation by rays of a shorter wavelength.

machine guarding, The installation of equipment or devices on machines to eliminate hazards created by operation of the machines.

magnaflux, A test in which particles of iron are applied to the surfaces of a magnetized specimen; the particle pattern indicates surface or near-surface flaws or irregularities.

maintenance, preventive, A system of scheduled overhaul and replacement of key parts to forestall breakdown.

maintenance, routine and breakdown, The function of preserving facilities and equipment by making needed repairs.

major injury, An injury where there is loss of time to the injured person and a medical expense.

major medical expense benefit (catastrophe insurance), Plan designed to insure workers against the costly medical expenses resulting from catastrophic or prolonged illness or injury. If the benefit supplements the benefit payable by a basic health insurance plan (hospital, medical, or surgical), it is called a "supplementary" plan; otherwise, it is called a "comprehensive" plan.

makeup air, Clean, tempered outdoor air supplied to a workspace to replace air removed by exhaust ventilation or some industrial process.

malingerer, A person who pretends deafness, illness, physical disability, or other abnormalities, or inability to perform one's duties or service.

malpractice, Misconduct or lack of proper professional skill on the part of any professional, such as a doctor, dentist, attorney, or engineer.

man-hours, The total number of hours worked by all employees of an industrial organization or an industry. A man-hour is the equivalent of one person working for one hour.

manipulative dexterity, The degree of skill a person has in using his wrists and fingers for fine tasks.

manual control switch, An auxiliary device for manual operation of an automatic controller. Any control switch that is actuated by a person.

manual dexterity, The degree of use and facility a person has in using his hands. Includes gross arm-hand movements as well as fine wrist-finger skills.

manual rate, The compensation insurance premium in dollars per hundred dollars of payroll, for a specified classification of operation or risk as listed in an official manual for a given state or jurisdiction.

margin of safety, A border, edge, or limit beyond which a particular behavior, condition, or situation becomes hazardous or unsafe.

mask, A protective covering for the face or head, such as a wire screen, a metal shield, a respirator, or a gas mask.

masking (noise), The stimulation of one ear of a subject by controlled noise to prevent his hearing with that ear the tone or signal given to the other ear. This procedure is used where there is at least 15 to 20 dB difference in the two ears.

maximum allowable concentration (MAC), The concentration for exposures not exceeding a total of eight hours daily of an atmosphere contaminant below which it is unlikely that ill effects will be experienced by any but hypersensitive individuals.

mechanical hazards, Unsafe conditions involving machinery, equipment, tools, etc.

medical benefits, Plans which provide employees, and in most cases their dependents, with specified medical care (other than that connected with surgery) or a cash allowance toward the cost of doctors' visits. Generally part of a health insurance program.

medical evaluation, An approach based on physical examination and the use of medical and physiological techniques, concerned with the influence of physical defects, physiological malfunction, or disease on the production of accidents.

medical only (MO), An injury requiring medical attention only, without loss of time from work. In insurance terminology it describes an injury where medical costs only are paid, even when some loss of work time has been incurred.

mileage death rate, The number of deaths from vehicle accidents per 100,000,000 miles of vehicle travel in the area for which the rate is computed.

minor injury, An injury where no lost time or major medical costs are involved.

monitoring (nuclear), The periodic or continuous determination of the amount of ionizing radiation or radioactive contamination. Area monitoring is the routine monitoring of the level of radiation or of radioactive contamination of any particular area, building, room or equipment. Personnel monitoring is the monitoring of any part of an individual, his breath, or excretions, or any part of his clothing.

motor vehicle accident, Any accident involving a motor vehicle in motion that results in death, injury, or property damage. However, motion of the motor vehicle is not required in a collision between a railroad train or a street car or another motor vehicle.

motor vehicle nontraffic accident, Any motor vehicle accident which occurs entirely in any place other than a trafficway.

nature of injury, The type, or classification of the hurt, harm, or impairment received or inflicted.

nature of work, A description of the type or classification of the work actually being done by a person in connection with the work operation. Usually described in reports of occupational injuries.

near accident, A term used synonymously with "Non-Injury Accident." Also applies to an accident having a potential for property damage, but where no property damage was incurred.

nip point, The point of intersection or contact of two opposed circular surfaces, or a plane and a circular surface.

noise, Unwanted sound.

noise-induced hearing loss, The slowly progressive inner ear hearing loss that results from exposure to continuous noise over a long period of time as contrasted with acoustic trauma or immediate physical injury to the hearing function.

noise level, Sound level. For air-borne sound, unless otherwise specified, noise level is the weighted sound pressure level called sound level. The weighting must be indicated.

noise reduction, A decrease of the sound pressure level at a specified observation point, which is attributable to a designated structure. Noise reduction is also used to designate the differences in sound pressure levels existing at two different locations at a single time, when the designated structures are in position. The term "noise control" is meaningful only when noise control components and the points of observation are fully specified.

nonaccident injuries, Injuries, such as bursitis, tendosynovitis, silicosis, etc., which do not arise out of specific accidents.

nonauditory effects of noise, Refers to stress, fatigue, health, work efficiency, and performance effects of loud noise that is continuous.

noncombustible, A material or substance incapable of burning or supporting fire. (See fire resistive.)

nondisabling injury, An occupational injury which does not result in death, permanent total disability, permanent partial disability, or temporary total disability.

nondestructive testing, A test to determine the characteristics or properties of a material or substance that does not involve its destruction or deterioration (e.g., x-ray examination, ultra-high frequency sound, etc.).

nonfatal injury accident, An accident in which at least one person is injured and no injury terminates fatally.

nonflammable, A material or substance that will not burn readily or quickly.

noninjury accident, Any accident in which there is no personal injury or from which no personal injury results.

not elsewhere classified, See not otherwise classified.

not otherwise classified, A general category of items such as might appear in an accident causal classification system to permit the grouping of relatively infrequent dissimilar items.

negligence, The lack of reasonable conduct or care, characterized by "Accidental" or "Thoughtlessness," which a prudent person would ordinarily exhibit. There need not be a legal duty.

actionable negligence, The breach or non-performance of a legal duty, through neglect or carelessness, which results in damage or injury to another.

comparative negligence, Where negligence by both the Plaintiff and the Defendant is concurrent and contributes to the injury. Plaintiff's damages are diminished proportionately, provided his fault is less than the Defendant's, and that even by exercising ordinary care he could not have avoided the consequences of Defendant's negligence.

contributory negligence, Conduct by the injured person which he should have known involved an unreasonable risk to himself. Inattentiveness or carelessness when using an article known to be defective or hazardous, or disregard of warnings or instructions issued by manufacturers and sellers usually constitutes contributory negligence.

degrees of negligence, "Ordinary" negligence is based upon the fact that one ought to have known the results of his acts, while "Gross" negligence rests on the assumption that one knew the results of his acts but was recklessly or wantonly indifferent to the results. All negligence below that called "Gross" or "Ordinary" by the courts is "slight" negligence.

occupancy, The use of a building or other structure. The contents of a building or other structure.

occupational disease (OD), Bodily injury or health impairment resulting from exposure to conditions of the work and work area, materials, processes, and equipment from absorption, inhalation, or ingestion of disabling agents as distinct from trauma.

occupational illness, Any abnormal physical condition or disorder, other than one resulting from an occupational injury, caused by exposure to environmental factors associated with employment. It includes acute and chronic illnesses or disease which may be caused by inhalation, absorption, ingestion, or direct contact. Because occupational illnesses are rarely attributable to a specific incident they should be reported in the year in which the illness was first diagnosed and reported to the employer.

occupational injury, Any injury such as a cut, fracture, sprain, amputation, etc., which results from a work accident.

occupational injury and illness records, OSHA, A recording of each reportable occupational injury (including fatality) and illness required by every employer covered by the National System for Uniform Recording and Reporting of Occupational Injury and Illnesses.

ocupational injury or illness, reportable, OSHA, Any disability or permanent impairment to an employee which results from any exposure in the work environment that either: (1) results in death; or (2) prevents the employee from performing his normal assignment during the next regular or subsequent work day or shift; or (3) not causing death or loss of time, (a) results in transfer to another job or termination of employment, or (b) requires medical treatment other than first aid, or (c) results in loss of consciousness, or (d) is diagnosed as an occupational illness, or (e) results in restriction of work or motion.

occupational safety, The prevention of personnel or environmental accidents in work-related environments or situations.

occupational safety & health administration — OSHA, Responsible for regulating/enforcing workplace codes, rules, and laws.

occupational skin diseases or disorders, Contact dermatitis, eczema, or rash caused by primary irritants and sensitizers or poisonous plants; oil acne; chronic ulcers; or inflammations, etc. arising out of, or during the course of, employment.

occurrence, An incident often classified as relatively major or minor. In insurance, distinguished from accident by the fact that it is apparent or foreseen as occurring if certain activities take place.

off-the-job safety, Accident prevention activities or programs associated with non-job-related activities.

old-age survivors and disability insurance benefits (OASDI), Retirement income and survivors' and disability payments available to eligible workers covered by federal social security legislation.

oxygen deficiency, Designates an atmosphere having less than the percentage of oxygen found in normal air. Normally, air contains approximately 21 percent oxygen at sea level. When the oxygen concentration in air is reduced to approximately 16 percent, many individuals become dizzy, experience a buzzing in the ears, and have a rapid heart beat.

paid absence allowance, Payment for lost working time available to workers for various types of leave not otherwise compensated for, e.g., excused personal leave.

pathogenesis, Describes how a disease takes hold on the body, spreads, and what it does to the body.

pathology, The study of disease processes. Branch of medicine treating essential nature of disease.

pedestrian, Any person who at the time of an accident is not in a motor vehicle or non-motor vehicle. Persons on other vehicles such as coaster wagons, child's tricycle, roller skates, etc., are considered pedestrians. A person hitched onto a vehicle is considered a pedestrian unless entirely in or on the vehicle.

perception, Awareness of objects or other data through the medium of the senses. The meaning or interpretation given to stimuli received through the senses. Insight or intuition relative to information introduced through sensory reception.

permanent disability, A permanent impairment. Includes any degree of impairment from an amputation of a part of finger or a permanent impairment of vision to seriously and permanently crippling nonfatal injuries.

permanent impairment, See permanent disability.

permanent partial disability, Any injury other than death or permanent total disability which results in the loss, or complete loss of use, of any member or part of a member of the body, or any permanent impairment of functions of the body or part thereof, regardless of any pre-existing disability of the injured member or impaired body function. These cases are used in computing American Standard Z16.1 injury rates whether or not time is lost.

permanent total disability, Any injury other than death which permanently and totally incapacitates an employee from following any gainful occupation, or which results in the loss, or the complete loss of use, of any of the following in one accident: (a) both eyes (b) one eye and one hand, or arm, or leg, or foot (c) any two of the following not on the same limb: hand, arm, foot, or leg.

personal factor (unsafe), The mental or bodily characteristic which permitted or occasioned an act which contributed to an accidental occurrence.

personal protective equipment, Any material or device worn to protect the worker from exposure to, or contact with any harmful material or force.

physical factors (unsafe), Environmental factors conducive to accident occurrence. Physical or environmental hazards.

physiology, The science and study of the functions or actions of living organisms.

pinch point, Any point at which it is possible to be caught between the moving parts of a machine or between moving and stationary parts of a machine or between the material and the moving parts of a machine.

pitch (hearing), That attribute of auditory sensation in terms of which sounds may be ordered on a scale extending from low to high. Pitch depends primarily upon the frequency of the sound stimulus, but it also depends upon the sound pressure and wave form of the stimulus.

point of operation, The place of contact between the primary functional part of any machine or tool and the material or substances in the production, or process, in which the machine or tool is being used. Generally considered to be the point of greatest danger to the operator.

point of perception, The time and place at which the individual actually first perceived — that is, saw, heard, smelled or felt — the hazard, that is, the unusual or unexpected movement or condition that could be taken as a sign of the accident about to occur.

point of possible perception, The place and time at which the unusual or unexpected movement or condition could have been perceived by a normal person. This point always comes at or before the point of (actual) perception.

potential hazard, A person, situation, thing, or event having latent characteristics conducive to an accident occurrence.

power level (noise), The power level, in decibels, is 10 times the logarithm to the base 10, of the ratio of a given power to a reference power. The reference power must be indicated.

precautions, Actions taken in advance to reduce the probability of an accident.

pre-employment examination, A physical examination of a job applicant conducted prior to his employment.

preponderance of evidence, A greater weight of evidence (facts, information from all sources, past experience) to support one judgment or decision versus the opposite. A phrase frequently used to describe the basis upon which a decision is made, such as determining if an injury is occupational or non-occupational.

presbycusis, The hearing loss due to age. It is believed by some to be the degeneration of the nerve cells due to the ordinary wear and tear of the aging process.

prevention (of accident), The science and the art representing control of man performance, machine performance and environment to eliminate failures or losses. Prevention connotes correction of conditions as well as elimination or isolation of hazards.

product liability, The liability a merchant or a manufacturer may incur as the result of some defect in the product he has sold or manufactured, or the liability a contractor might incur after he has completed a job from improperly performed work. The latter part of product liability is called "completed operations."

property damage accident (PD), An accident wherein damage to or destruction of any property is the immediate and direct result. It does not include accidents resulting in loss of human life or personal injury.

protective clothing, Clothing worn to protect the worker from exposure to or contact with harmful substances.

protective coating, A thin layer of metal or organic material as paint applied to a surface primarily to protect it from oxidation, weathering, and corrosion.

protective hand cream, A cream designed to protect the hands and other parts of the skin from exposure to harmful substances.

proximate cause, The cause which directly produces the effect without the intervention of any other cause. Cause nearest to the effect in time or space.

prudent, Cautious, careful, attentive, discrete, circumspect and sensible as applied to action or conduct.

psychiatric evaluation, A method, based on case studies of accident victims and accident repeaters, which derives inferences as to the personality characteristics of persons who have accidents, and to the personal dynamics of accident causation.

psychogenic deafness, That originating in or produced by the mental reaction of an individual to his physical or social environment. It is sometimes called functional deafness or feigned deafness.

psychological evaluation, Studies of the relationships between accidents and emotional, or intellectual functions as distinct from sensory and sensori-motor factors. Intelligence tests and personality tests are examples of these measurement techniques, as are time sampling, activity analysis, and the observation of critical incidents.

psychophysical, characteristics, A combination of mental and physical qualities of humans, such as visual acuity, reaction time, hearing, depth perception, peripheral vision, manipulative dexterity, color vision, etc.

psychophysical measurement, Measurements of reaction times and various kinds of sensory-motor and psychomotor coordinations. (Laboratory experiments dealing with vision under conditions of low illumination illustrate this latter group.)

psychosocial evaluation, The evaluation of socially conditioned behavior and reactions in relation to the social environment. The measurement of attitudes, studying the effect of supervisory practices, and the use of indices of the adequacy of social adjustment, illustrate methods used in this approach to accident investigation.

psychosocial factors, Social influences that are related to or affect psychological factors of human behavior.

respirator, A protective device for the human respiratory system designed to protect the wearer from inhalation of harmful atmosphere. There are two types of respiratory protective devices: (a) air purifiers, which remove the contaminants from the air by filtering or chemical absorption before inhalation, and (b) air suppliers, which provide clean air from an outside source or oxygen from a tank.

respiratory diseases, Disease conditions due to toxic agents in the respiratory tract, e.g., pneumonitis, bronchitis, pharyngitis, rhinitis, or acute congestion due to chemicals, dusts, gases or fumes.

respiratory irritants, Irritants affecting the respiratory tract, e.g., dusts, vapors, gases.

respiratory system, A system of the body consisting of (in descending order) — the nose, mouth, nasal passages, nasal pharynx, pharynx, larynx, trachea, bronchi, bronchioles, air sacs (alveoli) of the lungs, and muscles of respiration.

rest period (coffee break; break time), Brief interruption in the workday, usually of 5 to 15 minutes' duration, during which the worker rests, smokes, or takes refreshments without loss of pay. (See relief time.)

retrograde shock amnesia, The loss or impairment of memory due to shock. It can result from shock due to head injuries, severe loss of blood, and other conditions.

right-of-way, The right of one vehicle or pedestrian to proceed in a lawful manner in preference to another vehicle or pedestrian approaching under such circumstances of direction, speed, and proximity as to give rise to danger of collision unless one grants precedence to the other.

riscident, Comes from one of the same root words as accident, and means "risk due to fortuitous change." The study of riscidents has been approached in two ways: (a) the analysis of accident records, identifying the maneuvers of the various vehicles or other agents involved, and (b) by recording near accidents.

riscutant, Refers to the behavior of a person which tends to bring about his involvement in riscident or critical situations.

risk, The chance of physical or personal loss; the amount of possible loss to the insuring company. Also used in place of insured or prospect. Hazard, danger, peril. A company protected by insurance. A subjective evaluation of relative failure potential.

safe, Relatively free from danger, injury, or damage or from the risk of danger, etc.

safety, The art of performing any activity in the most accident-free manner. Relatively free from hazard.

safety belt, A life belt worn by telephone linesmen, window washers, etc., attached to a secure object (telephone pole, window sill, etc.) to prevent falling. A seat or torso belt securing a passenger in an automobile or airplane to provide body protection during a collision, sudden stop, air turbulence, etc.

safety coupling, A friction coupling adjusted to slip at a predetermined torque, to protect the rest of the system from overload.

safety cut-out, An overload protective device in an electric circuit.

safety education (training), The transmission of knowledge, skills, attitudes, motivations, etc., concerning the safety requirements of operations, processes, environments, etc., to workers, supervisors, managers, and others. The objective of safety education and training is favorable behavior change.

safety engineering, Safety engineering is concerned with the planning, development, improvement, coordination, and evaluation of the safety component of integrated systems of men, materials, equipment, and environments to achieve optimum safety effectiveness in terms of both protection of people and protection of property.

safety factor, See factor of safety.

safety glass, Impact resistant and shatterproof glass used as eye protection and for automobile windows, large architectural windows and doors. Also includes heat-treated glass that breaks into granules instead of sharp-edged strands.

safety hats, Rigid headgear of varying materials designed to protect the workman's head — not only from impact, but from flying particles and electric shock, or any combination of the three. Safety helmets should meet the requirements of American Standard Z89, Standard for Industrial Protective Helmets.

safety lock, A lock which can be opened only by its own key. Often used to "lock out" the electrical energized sources used in equipment or machinery operation.

safety professional, An individual who, by virtue of his specialized knowledge and skill and/or educational accomplishments has achieved professional status in the safety field. He may also have been awarded the status of Certified Safety Professional by the Board of Certified Safety Professionals.

safety-prone, A predisposition toward safe behavior. A term coined by Professor Lawrence L. LeShan who discussed this concept for the first time in an article entitled, "The Safety Prone: An Approach to the Accident-Free Person," in the November, 1952, issue of Psychiatry: Journal for the Study of Interpersonal Processes.

safety rule, A rule prescribing safeguarding requirements, personal protective equipment, or safe behavior on the job.

safety rules, Codes of conduct to avoid injury and damage.

safety solvents, Solvents which are free from fire or toxicity hazards and are nondamaging to surfaces or materials being cleaned. This term is often misused. Depending on the conditions of use, none of these criteria may be met by so-called "safety solvents." For example, petroleum hydrocarbons are effective solvents, low in toxicity and inexpensive, but have relatively low flash and fire points. The following conditions must be kept in mind by users of "safety solvents": (1) The toxicological effects alone are not adequate to assess the hazard potential of a solvent. (2) The vapor pressure, ventilation, and manner of usage will determine the concentration in air. (3) Handling procedures and type of clothing will determine the degree of skin contact and absorption. (4) Ignition temperature, flash point, and other factors determining the potential for fire and explosion must be considered.

safe workplace, One in which the likelihood of all identifiable undesired events are maintained at an acceptable level.

schedule rating, A type of rating assigned under the Industrial Compensation Rating Schedule, as approved by the insurance commissioner, by which the Basic Manual Rate is modified to fit the physical conditions related to guarding of machines. It is also affected by the compliance of a safety organization to prescribe insurance standards. Not used in most states any longer.

scheduled charge, The specific charge (in days) assigned to a permanent partial, permanent total, or fatal injury. (A list of these charges is contained in the Z16.1 Table of Scheduled Charges.)

self insurance, A term used to describe the assumption of one's own financial risk.

serious injury, The classification for a work injury which includes: (1) all disabling work injuries, (2) nondisabling injuries in the following categories: (a) eye injuries from work-produced objects, corrosive materials, radiation, burns, etc., requiring treatment by a physician, (b) fractures, (c) any work injury that requires hospitalization for observation, (d) loss of consciousness (work related), (e) any other work injury (such as abrasion, physical or chemical burn, contusion, laceration, or puncture wound) which requires: 1) treatment by a medical doctor, or 2) restriction of work or motion or assignment to another regularly established job.

serious injury frequency rate, The number of serious injuries, as defined in Z16.1 per 1,000,000 employee-hours of exposure. When serious injury frequency rate is used, it should be clearly identified as serious injury frequency rate, to avoid confusion with other frequency rates. This rate relates serious injuries, as defined, to the employee-hours worked during the period and expresses the number of such injuries in terms of million-hour units by use of the following formula:

$$\text{SIFR} = \frac{\text{Number of serious injuries} \times 1{,}000{,}000}{\text{Employee-hours of exposure}}$$

severity rate, The total days charged for work injuries per 1,000,000 employee-hours exposure. Days charged include actual calendar days of disability resulting from temporary total injuries and scheduled charges for deaths and permanent disabilities. These latter charges are based on 6,000 days for a death or permanent total disability, with proportionately fewer days for permanent partial disabilities for varying degrees of seriousness. (See standard disabling injury severity rate.)

$$\text{SR} = \frac{\text{Total days charged for work injuries} \times 1{,}000{,}000}{\text{Employee-hours exposure}}$$

shock, (1) The rapid fall in blood pressure especially following injury, operation, or the administration of anesthesia. (2) A sudden disturbance of mental equilibrium.

shop rules (working rules), Either regulations established by an employer dealing with day-to-day conduct in the plant — operations, safety, hygiene, records, etc. — or working rules set forth in collective bargaining agreements and in some union constitutions.

sick leave, Period of time during which a worker may be absent without loss of job or seniority if unable to work because of illness or accident. A paid sick leave plan provides for full or partial pay for such absence, usually up to a stipulated maximum. Sick leave plans differ from accident and sickness benefits, principally in that the former cover shorter periods of absence, usually provide higher pay, and are uninsured.

single point failure, A failure of a subunit which by itself will cause a failure of the system or equipment.

skin contamination, Irritations and infections to the skin such as dermatitis and poison ivy. These contaminations are classified as work injuries if they arise out of and in the course of employment.

smoke, Carbon or soot particles less than 0.1 microns in size which result from the incomplete combustion of carbonaceous materials such as coal or oil. An air suspension (aerosol) of particles, often originating from combustion or sublimation.

sone, A unit of loudness. By definition, a simple tone of frequency 1000 Hz, 40 dB above a listener's threshold, produces a loudness of 1 sone. The loudness of any sound that is judged by the listener to be n times that of 1 sone is n sones.

sound, An oscillation in pressure, stress, particle displacement, particle velocity, etc., which is propagated in an elastic material, in a medium with internal forces (e.g., elastic, viscous), or the super-position of such propagated oscillations. Sound is also the sensation produced through the organs of hearing — usually by vibrations transmitted in a material medium, commonly air.

sound absorption, The change of sound energy into some other form, usually heat, in passing through a medium or on striking a surface. In addition, sound absorption is the property possessed by materials and objects, including air, of absorbing sound energy.

sound analyzer, A device for measuring the band-pressure level or pressure-spectrum level of a sound as a function of frequency.

sound intensity, The average rate at which sound energy is transmitted through a unit area perpendicular to a specified point.

sound level meter, An instrument for use in measuring sound pressure levels in decibels referenced to 0.0002 microbars. Readings can also be made in specific octave bands, usually beginning at 75 Hz and continuing through 10,000 Hz.

sound pressure level, The weighted sound pressure in decibels. The level, in decibels, of sound is 20 times the logarithm to the base 10 of the ratio of the pressure of this sound to the reference pressure. The reference pressure shall be explicitly stated. (See decibel.)

spasm, (1) Sudden, violent, involuntary contraction of muscle(s). (2) Sudden but transitory constriction of a passage, canal, or orifice.

speech perception test, A measurement of hearing acuity by the administration of a carefully controlled list of words. The identification of correct responses is evaluated in terms of norms established by the average performance of normal listeners.

spontaneous ignition, Ignition resulting from a chemical reaction in which there is a slow generation of heat from oxidation of organic compounds until the combustion or ignition temperature of the material (fuel) is reached. This condition is reached only where there is sufficient air for oxidation but not enough ventilation to carry away the heat as fast as it is generated.

standard disabling injury frequency rate, See disabling injury frequency.

standard disabling injury severity rate, See disabling injury severity rate.

subrogation, The legal process by which the company endeavors to recover from a third party the amount paid to an insured under an insurance policy when such third party may have been responsible for the occurrence causing the loss.

surgical benefits, Plans which provide employees, and in many cases their dependents, with specified surgical care or a cash allowance toward the cost of such care, usually in accordance with a schedule of surgeon's fees. Generally part of a health and insurance program. (See health and insurance plan; Blue Shield.)

survey, Inspection and comprehensive study or examination of an organization, environment, or activity for insurance and/or accident prevention purposes.

survivors' benefits (transition benefits: bridge benefits: widows' allowance), Payments to dependents of employees who die prior to retirement, financed in whole or in part by the employer. May be in the form of payments for a fixed period (e.g., 24 months) supplementing regular life insurance benefits, a benefit for life out of a pension program, a lump-sum payment, etc.

system safety, The application of Operating, Technical and Management techniques and principles to the safety aspects of a system throughout its life to reduce hazards to the lowest level possible through the most effective use of available resources.

system safety analysis, The safety analysis of a complex process by means of a diagram or model that provides a comprehensive, overall view of the process, including its principal elements and the ways in which they are interrelated. There are four principal methods of analysis: failure mode and effect, fault tree, THERP, and cost-effectiveness. Each has a number of variations and more than one may be combined in a single analysis. (See failure mode and effect; fault tree.)

system safety engineering, The application of scientific and engineering principles during the design, development, manufacture and operation of a system to meet or exceed established safety goals.

temporary threshold shifts (TTS), The hearing loss suffered as the result of noise exposure, all or part of which is recovered during an arbitrary period of time when one is removed from the noise. It accounts for the necessity of checking hearing acuity at least 16 hours after a noise exposure.

temporary total disability, The classification for any injury which does not result in death or permanent total or permanent partial disability, but which renders the injured person unable to perform a full day's work. This means that the injured employee cannot perform all the duties of a regularly established job which is open and available to him; or he in unable to perform such duties during the entire time interval corresponding to the hours of his regular shift on any one or more days including weekends, holidays, and other days off, or plant shutdown) subsequent to the date of the injury.

THERP, This is an acronym for Technique for Human Error Prediction, a means for quantitatively evaluating the contribution of human error to the degradation of product quality. It is often combined with failure mode and effect and fault tree methods to identify failures of the human components in a system.

threshold (hearing), The point at which a person just begins to notice the tone is becoming audible. The level where the first effects occur.

threshold limit values (TLV), Refers to airborne concentrations of substances and represents conditions under which it is believed that nearly all workers may be repeatedly exposed, day after day, without adverse effect. These values may be breathed continually for 8 hours per day without harm. Because of wide variation in individual susceptibility exposure of an occasional individual at or even below the threshold limit may not prevent discomfort, aggravation of a pre-existing condition, or occupational illness. Threshold limits should be used as guides in the control of health hazards and should not be regarded as fine lines between safe and dangerous concentrations. The American Conference of Governmental Industrial Hygienists (ACGIH) adopts a list of threshold limit values (TLV) each year for more than 450 substances.

toeboards, A guard commonly installed around flywheels and other equipment in open pits and on overhead catwalks. The installation of toeboards should conform to ANSI Standard A-12, Safety Code for Floor Openings, Railings, and Toeboards. Toeboards should be at least 4 inches high and should be made of wood, metal, or metal grille not exceeding 1-in. mesh. Toeboards at flywheels should be placed as close to the edge of the pit as possible. Wood toeboards for permanent installations should be of 1″ × 4″ stock or heavier.

tolerance, The ability of a living organism to resist the usually anticipated stress. The upper and lower limits of permissible variability. The limits of permissible inaccuracy above and below design specifications.

tongs, safety, A device for feeding objects to and removing them from a danger area.

total loss, Loss of all the insured property. Also a loss involving the maximum amount for which a policy is liable.

total reaction distance (motor vehicle), The distance traveled between the point at which the driver perceives that braking evasive action is required and the point at which he contacts the braking controls.

total reaction time (motor vehicle), The time required for a vehicle to move the total reaction distance.

total stopping distance (motor vehicle), The distance in which the vehicle comes to rest after the driver discovers a hazard which requires stopping. Includes driver reaction time, brake reaction time, and braking time.

toxemia, Poisoning by way of the blood stream.

toxic, Of, pertaining to, or caused by poison. Poisonous, harmful.

toxicity, A relative property of a chemical agent with reference to a harmful effect on some biologic mechanism and the condition under which this effect occurs. The quality of being poisonous.

toxicology, That branch of medical science which deals with the nature and effects of poisons.

toxic substance (OSHA), A substance that demonstrates the potential to induce cancer, to produce short- and long-term disease or bodily injury, to affect health adversely, to produce acute discomfort, or to endanger life of man or animal resulting from exposure via the respiratory tract, skin, eye, mouth or other routes in quantities which are reasonable for experimental animals or which have been reported to have produced toxic effects in man.

traffic accident, Any accident involving one or more motor vehicles in motion on a roadway. A traffic accident may involve more than one unit if each unit comes in contact with some other unit involved while part of either is in contact with the road or sidewalk.

transmission loss (sound), The ratio, expressed in decibels, of the sound energy incident on a structure to the sound energy that is transmitted. The term is applied both to building structures (walls, floors, etc.) and to air passages (muffler, ducts, etc.).

trauma, An injury, wound, or shock resulting from an accident or outside force.

tremor, Involuntary shaking, trembling, or quivering.

tunnel vision, Inability to see toward the sides. A narrow field of vision.

two-hand controls, Tripping devices for a machine which require simultaneous application of both hands to operate the control, so that the hands of the operator are kept out of the point-of-operation area while the machine is operating.

type of accident, The classification of an accident according to the manner of contact of the injured person with the agency. Involves movement of object, material or person and association with an agency; such as fall, struck by, struck against, etc.

under the influence of alcohol, Sufficient intoxicating liquor in the body system to cause impairment of faculties.

underwriter, An insurance company employee who accepts or rejects risks on behalf of the insurance company. More broadly, anyone who makes insurance contracts.

unsafe act, A departure from an accepted, normal, or correct procedure or practice which has in the past actually produced injury or property damage or has the potential for producing such loss in the future; an unnecessary exposure to a hazard; or conduct reducing the degree of safety normally present. Not every unsafe act produces an injury or loss but, by definition, all unsafe acts have the potential for producing future accident injuries or losses. An unsafe act may be an act of commission (doing something which is unsafe) or an act of omission (failing to do something that should have been done).

unsafe condition, Any physical state which deviates from that which is acceptable, normal, or correct in terms of its past production or potential future production of personal injury and/or damage to property or things; any physical state which results in a reduction in the degree of safety normally present. It should be noted that accidents are invariably preceded by unsafe acts and/or unsafe conditions. Thus, unsafe acts and/or unsafe conditions are essential to the existence or occurrence of an accident.

unsafe personal factor, The mental or bodily characteristic that caused or permitted the unsafe act.

upper explosive limit (UEL), The maximum proportion of vapor or gas in air above which propagation of flame does not occur. The upper limit of the flammable or explosive range.

waiting period, Certain policies which insure against loss of use or against disability provide that they will not begin payment until the interruption or the disability has equalled a specified time, which is called the "waiting period."

welding and cutting permit, Authorization for the use of open flame and spark-producing devices in areas of combustible and/or hazardous materials, where their use is normally prohibited. Frequently extended to the use of open flame devices and/or high heat-producing devices.

willful misconduct, Deliberate failure to comply with statutory regulations.

working surface, Any surface or plane on which an employee walks or works. (An agency factor used in classifying accidents.)

work injury, Any injury suffered by a person which arises out of and in the course of his employment.

workmen's compensation (WC), A system of insurance required by state law and financed by employers which provides payment to employees or their families for occupational illness, injuries, or fatalities resulting in loss of wage income incurred while at work regardless of whether or not the employee has been negligent.

wound, A break in the continuity of the skin and other body tissues.

Occupational Health and Medicine

17

Z94.17

The terms in this section were defined by the authors themselves or taken or adapted from: "The Chemical Industry Facts Book," 5th ed., published by the Manufacturing Chemists' Association, Inc., Washington, D.C.; "Occupational Diseases and Industrial Medicine," by R. T. Johnstone and S. E. Miller, published by W. B. Saunders Co., Philadelphia, 1960; "Guide for Industrial Audiometric Technicians," published by the Safety Health Services, Employers Insurance of Wausau, Wis., American National Standards S1.1, "Acoustical Terminology," and Z88.2, "Practices For Respiratory Protection"; "101 Atomic Terms and What They Mean," by the Esso Research and Engineering Company, Linden, N.J.; "Paramedical Dictionary," by J. E. Schmidt, M.D., published by Charles C. Thomas, Springfield, Ill., 1969"; "Glossary of Microbiological Terms," published by The Soap and Detergent Assn., New York, N.Y.; the "Accident Prevention Manual for Industrial Operations," 6th ed., published by the National Safety Council, 1969; "Fundamentals of Industrial Hygiene," edited by Julian B. Olishifski and Frank E. McElroy, published by the National Safety Council, 1971; "Toxicology, the Basic Science of Poisons," by Lewis J. Caserette and John Doull, published by the McMillan Publishing Company, New York, 1975; "Occupational Medicine, Principles and Practical Application," by Carl Zenz, Year Book Medical Publishers, 1975; "Industrial Noise and Hearing Conservation," by Julian B. Olishifski and Earl R. Harford, published by the National Safety Council, 1975; "Text Book of Medicine," 14th ed., edited by Paul B. Beeson and Walsh McDermott, published by W. B. Saunders Co., 1975; "Pathology, A Dynamic Introduction to Medicine and Surgery," 2nd ed., by Perry and Miller, M.D., published by Little, Brown & Co., Boston, 1971; the "Encyclopedia of Occupational Health & Safety," published by the McGraw Hill Co., 1972; "U.S. Naval Flight Surgeons Manual," U.S. Government Printing Office, Washington, 1968; "U.S. Navy Diving Manual," Navships 0994-001-9010, Navy Department, Washington, D.C., 1973; the "Physiology and Medicine of Diving and Compressed Air Work," 2nd ed., by Bennett & Elliott, Williams & Wilkins, Baltimore, 1975; "Nuclear Terms, a Brief Glossary," by the U.S. Atomic Energy Commission, Division of Technical Information, Oak Ridge, Tenn., 1969; "Radiological Health Handbook," U.S. Department of Health, Education and Welfare, Public Health Service, U.S. Government Printing Office, 1970; "Industrial Toxicology," by Hamilton & Hardy, 3rd ed., Publishing Sciences Group, Acton, Mass., 1974; "Occupational Lung Diseases," by Morgan and Seaton, W. B. Saunders Co., Philadelphia, 1975; "Chronic Obstructive Pulmonary Disease, a Manual for Physicians," 3rd ed., New York, National Tuberculosis and Respiratory Disease Association, 1975; and "Dorland's Illustrated Medical Dictionary," 24th ed., W. B. Saunders Co., Philadelphia, 1965.

The following are the persons who gave their time, effort, and expertise:

Robert Brubaker, M.D.
Medical Director, United Technologies

Jennifer Kelsey, Ph.D.
Associate Professor, Epidemiology, School of Medicine, Yale University

Leonard A. Krause, Sc.D.
Director, Environmental Hygiene and Toxicology, Olin

William C. Milroy, M.D., Ph.D.
Officer in Charge, Naval Underseas Medical Institute

Edward J. Bernacki, M.D., M.PH.
Corporate Medical Director, United Technologies
Chairman
(Note: Dr. Bernacki is now with Tenneco.)

abnormality, A defect. (See inherited abnormality.)

abrasion, An area of body surface denuded of skin or mucous membrane due to an abnormal or unusual mechanical process. A scraping away of a portion of the skin surface epithelium.

absolute risk, The probability that a person with stated characteristics will develop a given disease.

acceleration illusions, Incorrect perceptions of apparent motion resulting from acceleration-induced stimulation of the semicircular canals. This can be a significant cause of disorientation in aircraft pilots. Three specifically described illusions of frequent occurrence are:

ciriolis, A change of head position during a turning maneuver resulting in illusory motion of the aircraft.

oculogyral, A sensation of spinning in the opposite direction following cessation of a spinning or rolling maneuver,

oculogravic, A perceived tilt in the visual vertical during linear acceleration.

acceleration syndrome (G-force syndrome), Alterations in physiologic function and/or perceptual-motor performance resulting from the forces imposed on the body by changes in velocity or direction of movement. The most significant of the symptoms are related to the circulatory status of the central nervous system and progress from loss of peripheral vision through blackout to unconsciousness during increases in acceleration.

acclimatization, The state of successful physiologic adaptation of an individual to new climatic conditions.

accuracy (validity), The closeness with which a measurement approaches the true or actual value.

ACGIH, American Conference of Governmental Industrial Hygienists.

acoustic trauma, Hearing loss caused by sudden loud noise in one ear, or by sudden blow to head. In most cases, hearing loss is temporary, although there may be some permanent loss.

actinic keratonconjunctivitis, See welder's flashburn.

acute, Severe, usually crucial, often dangerous in which relatively rapid changes are occurring. An acute exposure runs a comparatively short course.

acute mountain sickness, A disorder occurring in individuals exposed to high altitudes (about 10,000 ft.) for relatively long periods (24 hrs.). It presents as malaise, headache, and vomiting attributed to cerebral and pulmonary edema. It is probably related to both hypoxia and decreased atmospheric pressure. It may be prevented by proper acclimatization.

acute radiation syndrome, The organic illness which follows exposure to relatively large acute doses of ionizing radiation. It presents as one of three separately identified dose dependent syndromes:

central nervous system syndrome, Follows doses of 2000 rem or above producing relatively prompt shock and coma, uniformly fatal within hours to days.

gastrointestinal syndrome, Follows doses of 500 to 2000 rem producing nausea, vomiting, and diarrhea within hours to days and death usually occurring within a week.

hematopoetic syndrome, Follows doses of 100 to 500 rem producing early nausea and vomiting followed by a lag period of several weeks after which infection and hemorrhage occur due to hematopoetic supression. At doses of 400 to 500 rem death will occur in about 50% of the cases within 30 days. With adequate therapy the majority of affected individuals in this category can be salvaged.

aerosols, Liquid droplets or solid particles dispersed in air, that are of fine enough particle size to remain so dispersed for a period of time.

age-adjusted death rate, (age-standardized death rate), The mortality rate of a population calculated for comparative purposes in such a way that allowance is made for the age distribution of the population. When mortality rates for two or more populations are adjusted in the same manner, they may be compared to each other as though the populations had the same age distributions.

age-specific death rate, The number of deaths occurring in a defined age group, per unit population in that age group, over a stated period of time.

AIHA, American Industrial Hygiene Association.

airborne noise, A condition when sound waves are being carried by the atmosphere.

allergy, A hypersensitive state acquired through exposure to a particular allergen, re-exposure bringing to light an altered capacity to react.

allergic alveolitis, A granulomatous pneumonitis caused by inhalation of a specific allergen. Examples are: sequosis, farmer's lung, bagassosis, suberosis, and maple bark disease.

ambient noise, Ambient noise is the characteristic or background noise of a given environment.

anechoic room, A room in which all of the sound emanating from a source is essentially absorbed at the walls. Hence, there are no reflections and the spatial sound radiation pattern of a source may be determined. An anechoic room may be described as "echoless" or acoustically "dead."

anodize, To coat a metal electrolytically with a protective film.

antagonism, The interaction of two or more physiologically active agents producing an effect which is less than the effect produced using equivalent doses of each agent presented individually.

anthracosilicosis, Pneumoconiosis caused by breathing air containing dust that has free silica as one of its components and that is generated in the various processes in mining and preparing anthracite (hard) coal and to a lesser degree bituminous coal.

AOMA, American Occupational Medical Association.

asbestos, A hydrated magnesium silicate in fibrous form.

A-scale, A filtering system that has characteristics which roughly match the response characteristic of the human ear at low sound levels (below 55 dB SPL, but frequently used to gauge levels to 85 dB). A-scale measurements are often referred to as dB(A).

association (causal), See causal association.

association (non-causal), See non-causal association.

asthma, A disease characterized by increased responsiveness of the trachea and bronchi to various stimuli, manifested by difficulty in breathing caused by generalized narrowing of the airways.

attack rate, The number of new cases of a disease per unit population occurring over a defined period of time.

attributable risk in the exposed, The rate of disease in those exposed to a given causal agent that can be attributed to the causal agent.

attributable risk percent in the exposed, The percentage of the rate of disease in those exposed to a given causal agent that can be attributed to the given causal agent.

attributable risk in the population, The rate of disease in the total population that can be attributed to a given causal agent.

attributable risk percent in the population, The percentage of the rate of disease in the total population that can be attributed to a given causal agent.

audible range, The frequency range over which normal ears hear — approximately 20 Hz through 20,000 Hz.

audiogram, A record of hearing loss or hearing level measured at several different frequencies — usually 500 to 6,000 Hz. The audiogram may be presented graphically or numerically. Hearing level is shown as a function of frequency.

avulsions, A tearing away of a portion of the skin or other body structure.

background noise, The total of all sources of interference in a system used for the production, detection, measurement or recording of a signal, independent of the presence of the signal.

background radiation, Natural radiation in the environment including cosmic rays and radiation from naturally radioactive elements; it may also mean radiation from sources other than the one directly under consideration.

band pressure level, Band pressure level of a sound for a specified frequency band is the sound pressure level for the sound contained within the restricted band. The reference pressure must be specified.

barotrauma, Tissue damage resulting from expansion or contraction in gas-filled spaces within or adjacent to the body. As external pressure varies with ascent or descent, gas volume within these spaces varies resulting in tissue distortion, edema, and hemorrhage. It is the most common occupational disease of divers.

beat knee, Bursitis of the knee joints due to friction or vibration common in mining.

Belding-Hatch index, Estimate of the body heat stress of an average man (standard man) for various degrees of activity; also relates to his sweating capacity.

bends, See decompression sickness.

benign, The property of a neoplasm, of non-invasiveness; i.e., the neoplasm remains localized.

berylliosis, Disease due to inhalation of beryllium or beryllium compounds. Acute berylliosis is characterized by bronchitis and pneumonia. The chronic disease which may follow is a systemic disease with pathologic changes not only in the lung, but in numerous other body tissues including the liver, spleen, kidneys, heart, skin, and bone.

bias, A consistent tendency for an estimate to overestimate or underestimate the true population value.

biological effects of radiation, Any alteration in a biological system resulting from exposure to radiation. Effects may or may not be hazardous or detrimental to the system.

biologic half-life, The time required for a given species, organ, or tissue to eliminate half of a substance which it takes in.

biological monitoring, The direct quantitative analysis of expired air, body fluids or tissue for the presence of the hazardous agent or its metabolites and/or evidence of biologic impairment quantified by the use of physiologic, psychometric, biochemical tests.

birth cohort analysis, The observation over time of persons born within a particular period of time, such as those born during a one-year or a five-year period.

B-scale, A filtering system that has characteristics which roughly match the response characteristics of the human ear at sound levels between 55 and 85 dB. B-scale measurements are often referred to as dB(B).

bursa, A sac or sac-like structure containing a fluid and situated at places in tissues at which friction would otherwise develop. In industry the bursae of interest are at the shoulder, the elbow, and the anterior surface of the knee.

bursitis, Inflammation of a bursa, sometimes with development of calcium deposits. Causes of bursitis are multiple, but inflammation of the shoulder bursa may be associated with repetitive or unusual motion. Bursitis of the elbow or knee may be due to prolonged pressure or blunt trauma.

byssinosis, A subacute condition characterized by chest tightness on recurrent exposure to cotton dust after a period of absence from exposure.

cancer, Any malignant neoplasm.

canister (air-purifying), A container filled with sorbents and catalysts that remove gases and vapors from air drawn through the unit. The canister may also contain an aerosol (particulate) filter to remove solid or liquid particles.

carcinogen, Any agent capable of inducing or promoting malignant neoplastic changes in living organisms.

carcinoma, A malignant neoplasm arising from epithelial cells.

carding, The process of unbinding or untangling wool, cotton, etc.

carpal tunnel, A narrow tunnel structure at the base of the palm as it joins the wrist through which pass flexor tendons to thumb and fingers, blood vessels, and the median nerve.

carpal tunnel syndrome, A condition often associated with repetitive or unusual wrist motions in which the median nerve becomes compressed in the carpal tunnel. Symptoms are pain in the palm, thumb, ring, and index fingers and numbness in the first three digits.

case-control study, A type of epidemiologic study in which persons with a given disease (the cases) and persons without the given disease (the controls) are selected; the proportions of cases and controls with certain background characteristics or exposure to possible risks are then determined and compared.

case fatality rate, The number of persons dying of a disease divided by the number of persons who contracted the disease.

CAT, Acronym for computer tomography scan. Essentially a procedure involving a computerized apparatus. (See scan, tomograph.)

causal association, A statistical association between the occurrence of a factor and a disease, in which available evidence indicates that the statistically associated factor increases the probability of occurrence of the disease and that its removal decreases the probability of occurrence of the disease.

chelating agent, Any compound which will inactivate a metallic ion with the formation of an inner ring structure in the molecule, the metal ion becoming a member of the ring. The original ion, thus chelated, is effectively trapped and unable to exert its usual effect.

chloracne, An acneiform erruption attributed to chlorinated hydrocarbon exposures. Principal agents are chlornaphthalenes, chlordiphenyls, and chlordiphenyl oxides.

chronic bronchitis, A disorder characterized by excessive mucus secretion in the bronchi manifested by a chronic or recurrent productive cough (arbitrarily for a minimum of three months per year and for at least two successive years) in persons whom other causes of productive cough have been excluded.

cluster sampling, A procedure in which clusters (e.g., city blocks) rather than individual units are first selected from population and then observations are made on all individual units (e.g., households) in each cluster.

cohort study (prospective), A type of epidemiologic study in which persons initially free from the disease under study are selected and their background characteristics and their exposure to possible risks ascertained; these individuals are then followed through time and the proportions who develop the disease among those with and without certain characteristics and among those exposed and not exposed to the risks are determined and compared.

cohort study (retrospective), A type of epidemiologic study in which background characteristics and exposure to possible risks at some time in the past are determined in a group of individuals who were at that time free of the disease under study; persons having developed the disease since the time the characteristics or exposures were measured are ascertained, usually through existing records. Comparisons are then made of the proportions having developed the disease among those who did and did not have the characteristics and among those who were and were not exposed to the risk.

combustible gas meter, Instrument which contains sensitive Wheatstone Bridge. If air containing solvent or combustibles is drawn over one leg of bridge and unbalances it, an appropriate meter indicates change. Signal change is proportional to concentration of contaminant.

comparison (or control) group, In a case-control epidemiological study, a group of individuals free of the disease under study who are believed to reflect the characteristics of the population from which the persons with the disease (the cases) were drawn.

compliance, A measure of the ease which a hollow visus (lung, urinary bladder) may be distended.

confounding variable, A variable that is associated with both the independent and dependent variables which would not otherwise exist.

congenital abnormality, Any defect in the size or function of an organism existing at birth. A congenital abnormality is distinguished from an inborn abnormality in that the change may be hereditary or due to some applied stimulus during the total period of gestation; inborn on the other hand always implies a change secondary to a hereditary influence.

contact dermatitis (allergic type), An inflamation of the skin due to contact with a sensitizing material. Certain substances may not cause a dermatitis on first contact, but may so alter the skin that it becomes inflamed after a second contact 5-10 days later. The first contact is said to have sensitized the skin. The second and subsequent contacts give rise to an allergic reaction often with a minimal exposure.

contact dermatitis (irritant type), Inflammation from a material which would cause inflammation in the skin of most people if applied in sufficient concentration over a sufficient length of time. Most industrial dermatitis is due to contact with direct irritants.

control group, See comparison group.

contusion, Injury to the body due to a blunt external force from an object, or a fall or bump. Usually accompanied by swelling and black and blue mark due to rupture of veins. Severity may vary from a small bruise to severe underlying damage to bones, vessels and nerves. Severe contusions may be accompanied by lacerations as well.

correlation coefficient, A measure of the degree of association found between two characteristics in a series of observations on the assumption that the relationship between the two characteristics is adequately described by a straight line. A positive coefficient indicates that as one variable increases in value the other also tends to increase, while a negative coefficient indicates that as one variable increases in value, the other tends to decrease. Its value must be between $+1$ and -1; either $+1$ or -1 denotes complete dependence of one characteristic on the other, and 0 denotes no association whatsoever between them.

critical organ, That organ or tissue for which radiation injury will be of greatest detriment to health and, therefore, the limiting organ for that particular circumstance. "Criticality" of an organ may be based on specific radiosensitivity, localized radiation, selective uptake of specific radioisotopes, or the importance of the organ for health.

cross-sectional study, A type of epidemiologic study in which the presence or absence of background characteristics or exposure to possible risks and also the presence or absence of the disease under study are measured at one point in time in the individuals being studied. Prevalence rates among those with and without the characteristics or risks are then compared.

C-scale, A filtering system that has characteristics which roughly match the response characteristics of the human ear at sound levels above 85 dB. C-scale readings may be referred to as dB(C).

damage risk criterion, The suggested baseline of noise tolerance which, if not exceeded, should result in no hearing loss due to noise.

decibel (dB), A unit used to express the ratio of two amounts of electric or acoustic signal power. The decibel is equal to 10 times the logarithm of the signal power ratio as expressed by the following equation:

$$n\ (\mathrm{dB}) = 10 \log_{10} \frac{(P_1)}{(P_2)}$$

decompression sickness (bends, caisson disease), A condition caused by the formation and growth of bubbles in the blood or tissue resulting from a state of supersaturation with gas. This occurs when the sum of partial pressures of gases dissolved in a tissue exceeds the ambient pressure. It occurs in divers and compressed air workers on return from hyperbaric pressures to surface pressure or in aviators going from surface pressure to hypobaric pressures at altitude. Several specific clinical syndromes are described:

- Serious Symptom or Type II
  - Cerebral
  - Spinal Cord
  - Vestibular (the staggers)
  - Pulmonary (the chokes)
- Simple or Type I
  - Pain-only bends
  - Skin bends (the niggles)

decontamination, The removal of radioactive material from a location where it is not desired. In regard to personnel it would include both removal of external contamination by washing and removal of internal contamination by the use of chelating agents or similar methods.

degenerative joint disease, See osteoarthritis.

dependent variable, A variable that is considered to take on values at least in part as a result of the particular value of the independent variable.

detector tube, A glass tube containing specific chemicals which have been impregnated on inert material granules and which will change color when chemicals in air are drawn through the tube.

disc, intervertebral, A soft tissue structure between the bodies of the vertebrae. The central portion (nucleus) is a soft, pulpy material surrounded by an annular interplacement of tough fibrous tissue.

disc, ruptured, A condition in which the central portion (nucleus) of the intervertebral disc protrudes or herniates through the annular fibrous tissue; it frequently presses on a nearby nerve. The medical term for this condition is herniation of the nucleus pulposus. Other terms commonly used are slipped disc, ruptured disc, prolapsed disc.

disease, classification of, The grouping of persons with similar signs and symptoms and diagnostic test results into a disease entity which permits them to be distinguished from people with other disease entities; also, the arrangement of these disease entities into groups with common characteristics.

drowning, Death from acute asphyxia while submerged, whether or not liquid has entered the lungs.

dummy variable, In regression analysis, a variable that is not continuously distributed, but which has two or more distinct levels; values are assigned to these levels, usually according to their presence or absence. For instance, the value 1 might be assigned to females and the value 0 to males.

dusts, Solid particles generated by handling, crushing, grinding, rapid impact, detonation, and decrepitation of organic or inorganic materials, such as rock, ore, metal, coal, wood, and grain. Dusts do not tend to flocculate except under electrostatic forces; they do not diffuse in air but settle under the influence of gravity. Dust is a term used in industry to describe airborne solid particles that range in size from 0.1 to 25 microns (1 micron = 1/10,000 cm = 0.001 mm = 1/25,000 in.). Dusts above 5 microns in size usually will not remain airborne long enough to be inhaled. (Editor's Note: Although industry may still use the "micron," it is a term outside the International System, and CIPM (International Commmittee of Weights and Measures) considers it preferable to avoid. Like such terms as "fermi," "caloric," "stere," etc., the "micron" is a nonacceptable term in ASME publications, and it should be converted to its SI counterpart, viz, 0.000 001m.)

dysbaric osteonecrosis (aseptic bone necrosis), Lesions occurring in the bones, especially in juxta-articular areas, of divers and caisson workers. Small areas of bone die probably as a result of vascular occlusion by gas bubbles. Once detected, the lesions are essentially irreversible. No accepted method of therapy currently exists.

exzema, A nonspecific term for any type of dermatitis.

electrostatic precipitator, Instrument which charges particles drawn through it. Particles are then deposited on tube for subsequent analysis.

emphysema, An enlargement of the air spaces distal to the terminal nonrespiratory bronchioles, with destruction of alveolar walls.

endemic, Usual level of disease occurrence in a population.

epicondyles (of humerus), The bony prominences at the medical and lateral side of the elbow. The extensor muscles of the wrist and digits are attached to the lateral epicondyle and the flexors to the medial epicondyle.

epicondylitis (tennis elbow), An irritation at the common muscle attachment at the epicondyles, almost always at the lateral epicondyle. The cause may be a contusion of the epicondyle area or repetitive motion involving a rotary movement at the elbow. This condition is common in tennis players who serve too vigorously for the first time in spring. Symptoms are tenderness over the external epicondyle, pain in the lateral aspects of the elbow on grasping objects.

epidemic, Unusually high level of disease occurrence in a population in the light of past experience.

epidemiology, The study of the distribution of a disease or condition in a population and of the factors that influence that distribution.

etiology, The study of causes of disease, both direct and predisposing, and of their mode of operation.

exhaust ventilation, The removal of air usually by mechanical means from any space. The flow of air between two points is due to the occurrence of a pressure difference between the two points. The pressure difference will cause air to flow from the high pressure to the low pressure zone.

explosive decompression, A sudden rapid decrease in barometric pressure. This may occur from loss of integrity of a pressurized aircraft cabin or recompression chamber or in a diver who "blows up," surfacing extremely rapidly from depth due to a loss of buoyancy control. It may result in decompression sickness or pulmonary hyperinflation and air embolism.

false negative rate, The proportion of persons with a disease for whom the screening test for that disease is negative.

false positive rate, The proportion of persons without a disease for whom the screening test for that disease is positive.

fibrosis, A reparative or reactive process characterized by the deposition of connective tissue fibers in and around the site of injury.

flash point, The lowest temperature at which a liquid gives off enough vapor to form an ignitable mixture with air and produce a flame when a source of ignition is present. Two tests are used — open cup and closed cup.

flux, Usually refers to a substance used to clean surfaces and promote fusion in soldering. However, fluxes of various chemical nature are used in the smelting of ores in the ceramic industry, in assaying silver and gold ores and in other endeavors. The most common fluxes are silica, various silicates, lime, sodium and potassium carbonate and litharge and red lead in the ceramic industry.

folliculitis, Inflammation of a hair follicle or follicles.

footcandles, A measure of illuminance produced on a surface all points of which are 1 foot from a directionally uniform point of 1 candela. That illuminance is 1 lumen/ft^2 or 1 footcandle. (Editor's Note: Under SI, illuminance is measured in terms of lux units, where 1 lx = 1m/m^2 or $m^{-2} \cdot cd \cdot sr$ where m =meter; cd = candela; sr = steradian for solid angle. To convert: 1 footcandle = 1.076 391 lx.)

forced expiratory flow (FEF$^{200-200}$), The average rate of flow for a specified portion of the forced expiratory volume, usually between 200 and 1200 ml.

forced expiratory volume (qualified by subscript indicating the time interval in seconds, FEVT), Volume of gas exhaled over a given time interval with expiration as forceful as possible.

forced expiratory volume percentage expired ($FEV_{T\%}$), Forced expiratory volume expressed as a percentage of the forced vital capacity $(\frac{FEVT}{FVC} \times 100)$.

forced mid-expiratory flow ($FEF_{25\text{-}75\%}$), The average rate of flow during the middle half of the forced expiratory volume.

forced vital capacity (FVC), The vital capacity performed with expiration as forceful and rapid as possible.

frequency (in cycles per second, or cps, Hertz or Hz), The time rate of repetition of a periodic phenomenon. The frequency is the reciprocal of the period. It "defines" pitch or the highness or lowness of sound.

frostbite, A freezing injury of the skin due to exposure to extreme cold. It may be recognized by whitening of the skin and loss of sensation. It is treated by rapid thawing. Deep frostbite with freezing of tissues deep to the skin generally results in dry gangrene.

fume, Solid particles generated by condensation from the gaseous state, generally after volatilization from molten metals. A fume is formed when a volatized solid, such as metal, condenses in cool air. The solid particles that make up a fume are extremely fine, usually less than 1.0μ (0.001 mm). In most cases, the hot material reacts with the air to form an oxide.

functional disease, Disease in which some change in function of the body or its parts occurs without changes in structure. Usually refers to psychiatric diseases.

gases, Normally formless fluids which occupy the space or enclosure and which can be changed to the liquid or solid state only by the combined effect of increased pressure and decreased temperature.

gas mask, See respirators.

genetic effects of radiation, Any changes in the genetic material following irradiation which are hereditarily transmitted and expressed in the progeny of the irradiated subject.

genetic mutation, A permanent, genetically transmitted variation in which there is an alteration in the molecular arrangement of the inherited DNA in the chromosomes of the germ cells. These changes may occur spontaneously or be induced by the application of an external stimulus. (Note: DNA = Deoxyribonucleic Acid = a nucleic acid originally isolated from fish sperm and thymus gland, but later found in all living cells; on hydrolisis it yields adenine, guanine, cytosine, thymine, deoxyribose, and phosphoric acid. It is the carrier of genetic information for all organisms except RNA viruses. The Watson-Crick Helix is a double helix, each chain of which contains information completely specifying the other chain, representing a structural formulation of the mechanism by which the genetic information of the DNA reproduces itself.)

G-force syndrome, See acceleration syndrome.

glare, The sensation produced by luminance within the visual field that is sufficiently greater than the luminance to which eyes are adapted to cause annoyance, discomfort, or loss in visual performance or ability.

glassblowers' cataract, Posterior polar lenticular opacities occurring in occupations entailing long exposures to intense heat and glare. It is probably due to focusing of infrared radiation at the posterior pole of the lens.

hard metal disease, A condition of fibrosis of the lungs thought to be due to inhalation of dust of tungsten carbide and cobalt used as a binder.

health physicist, A health professional trained in radiation physics and concerned with problems of radiation damage and protection.

heat cramps (miners' or stokers' cramps), Painful spasms of the voluntary muscles due to salt depletion. These occur in healthy, heat acclimatized individuals and are due to excessive sweating without salt replacement rather than directly to high temperature exposure. They are easily relieved with fluid and salt replacement.

heat exhaustion (heat prostration, heat collapse), A state of peripheral vascular collapse in an unacclimatized individual attributable to exposure to a high temperature environment. Treatment consists of removal to a cool environment, rest, and salt and water replacement. It is prevented by adequate acclimatization and maintaining adequate salt and water intake.

heat pyrexia (heat stroke), A very serious and often fatal condition resulting from breakdown of thermoregulatory mechanisms during exposure to high temperature environments. It is characterized by extremely high deep-body temperature and an absence of sweating. Treatment consists of rapid cooling in an ice bath.

hematoma, A collection of blood, usually clotted, in an organ, space, or tissue, resulting from a break in the wall of a blood vessel. Frequently occurs with a contusing wound.

hertz, The frequency measured in cycles per second. 1 cps = 1 Hz.

high pressure nervous syndrome, A disorder occurring at extremely deep depths (in excess of 600 feet or about 20 atmospheres pressure) characterized by tremor, nausea, and decreased psychomotor ability. It appears to be due to both direct hydrostatic pressure and to compression rate. It is counteracted to some extent by addition of a narcotic agent to the breathing mixture. It usually resolves completely with decrease in pressure.

hyperbaric oxygen therapy, A treatment modality using pure oxygen at high pressure in a recompression chamber. It is useful in the treatment of diving casualties but is also valuable in treatment of medical diseases such as carbon monoxide intoxication and gas gangrene. Hyperbaric oxygen cannot be used at pressures exceeding 3 atmospheres.

hypercapnia, An increased carbon dioxide level in the blood. It results from carbon dioxide accumulation due to inadequate respiratory exchange or through breathing elevated concentrations. Although not acutely toxic below relatively high concentrations (5%), it is implicated as a synergist in most other high pressure gas-related casualties.

hypersensitization, An antigen-antibody mediated response characterized by an immediate vascular exudative reaction in previously sensitized tissue following exposure to a specific exiting agent.

hypersusceptibility, A markedly greater quantitative response to a physiologically active agent qualitatively identical to its usual response.

hypothermia, Lowered core temperature. Drop in core temperature to 35°C. It produces shivering and discomfort sufficient to adversely affect performance; at about 25°C. hypothermia is ordinarily fatal. Cold water immersion produces hypothermia very rapidly whereas exposure in cold air environments is tolerable for much longer periods. Treatment for hypothermia is rapid rewarming in a warm bath.

hypothesis, A conjecture subject to verification or proof.

immersion foot (trench foot), A painful condition of the feet resulting from relatively long wet exposures at temperatures above freezing. Nerve, muscle, and blood vessel injury occurs as a result of cooling. Prevention and treatment consist of proper foot care including regular drying of the feet, shoes, and socks.

impinger, Glass device for drawing contaminated air through a liquid to entrap particles, solvents, aerosols for later analysis.

incidence rate, The number of new cases, per unit population occurring during a stated period of time.

incubation time, The elapsed time between exposure to an environmental agent or microorganism and the appearance of biologic alteration.

independent variable, A variable that can either be set to a desired value by the investigator or observed as it naturally occurs.

industrial accident, See occupational accident.

industrial disease, See occupational disease.

industrial hygiene, Industrial hygiene is that science and art devoted to the recognition, evaluation, and control of those environmental factors or stresses arising in or from the workplace which may cause sickness, impaired health and well-being, or significant discomfort and inefficiency among workers or among the citizens of the community.

industrial hygienist, A person with the training and ability to: (a) recognize the environmental factors and stresses associated with work and work operations to understand their effect on man and his well-being; (b) evaluate, on the basis of experience and with the aid of quantitative measurement techniques, the magnitude of these stresses in terms of ability to impair man's health and well-being; and (c) prescribe methods to eliminate, control, or reduce such stresses when necessary to alleviate their effects.

industrial medicine, See occupational medicine.

inert gas narcosis (nitrogen narcosis, "rapture of the depths"), The toxic effect of the carrier or inert gas in a breathing mixture at increased pressures. It is characterized by euphoria, impaired coordination, and diminished cognitive function. Breathing air, nitrogen begins to become toxic at about 4 atmospheres absolute pressure and becomes disabling at pressures exceeding 8 atmospheres absolute. It resolves completely with decreasing pressure.

infant mortality rate, The number of deaths under 1 year of age occurring during a stated period of time divided by the total number of live births occurring during that period of time.

inflammation, The reaction of body tissue to injury, whether by infection or trauma. The inflamed area is red, swollen, hot and usually painful.

inherited abnormality, A defective character or quality transmitted from parent to offspring.

intervention study, A type of epidemiologic study in which the investigator assigns, usually at random, individuals to be exposed or not to be exposed to a given factor, such as a treatment or a vaccine. These individuals are then followed through time and the proportion developing or dying from a disease among those exposed to the factor is compared to the proportion developing or dying from the disease among those not exposed, the exposure having been determined by initial assignment.

interstitial, (1) Pertaining to the small spaces between cells or structures; (2) occupying the interstices of a tissue or organ; (3) designating connective tissue occupying spaces between the functional units of an organ or a structure.

inversion, Phenomenon of a layer of cool air trapped by a layer of warmer air above it so that the bottom layer cannot rise.

irradiation, Exposure to radiation.

laceration, A term applied to cleanly cut incised wounds as well as jagged, irregular, blunt breaks or tears through the skin. Severity extends from small cuts which can be taped together to severe dirty wounds with damage to underlying structures. The term is also applied to wounds of the mucous membranes and the surface of the eyes.

laser, An acronym for Light Amplification by Stimulated Emission of Radiation.

laser burn, A punctate tissue injury caused by a beam of coherent light. Laser beams are particularly hazardous to the eye, where they may cause burns of the cornea, lens, or retina with consequent effects on visual acuity. A laser burn of the retinal fovea may destroy central vision sufficiently to produce legal blindness.

late effects of radiation, Effects which occur with a long delay time following sublethal doses of radiation or following relatively low dose rates carried over a long period of time. Examples are carcinogenesis and premature aging.

LD_{50}, The dose which is required to produce in 50 percent of the exposed species.

local effects of radiation, Effects limited to a specific organ or system due to specific irradiation of that organ or increased sensitivity of the organ.

loudness level, A subjective method for rating loudness in which a 1000 Hz tone is varied in intensity until it is judged by listeners to be equally as loud as a given sound sample. The loudness level in phons is taken as the sound pressure level, in decibels, of the 1000 Hz tone.

lumen, Under the "common" system, it was defined as the flux on one square foot of a sphere, one foot in radius, with a light source of one candle at the center that radiates uniformly in all directions. Under SI, luminous flux is measured in lumen units (symbols: lm) and has its formula cd • sr which are the SI Base Units of candela and steradian for solid angle. Thus, the lumen is the luminous flux emitted in a solid angle of one lumen uniformly distributed in a solid angle of one steradian by a point source having a uniform intensity of one candela. (See lux.)

luminaire, A complete lighting unit consisting of a lamp or lamps together with parts designed to distribute the light, position and protect the lamps and connect the lamps to the power supply.

lung volumes:

inspiratory capacity (IC), Maximal volume inspired from resting expiratory level.

expiratory reserve volume (ERV), Maximal volume inspired from resting expiratory level.

vital capacity (VC), Maximal volume expelled by complete expiration after a maximal inspiration.

functional residual capacity (FRC), Volume of gas in the lungs at the resting expiratory level.

residual volume (RV), Volume of gas in the lungs after maximal expiration.

total lung capacity (TLC), Volume of gas in the lungs after a maximal inspiration; that is, the VC + RV.

tidal volume (TV), Volume of gas inspired or expired during each breathing cycle.

lux, The illuminance produced by a luminous flux of one lumen uniformly distributed over a surface of one square meter.

malformation, A deformity. (See congenital abnormality.)

malignancy, The property of a neoplasm, marginal invasion, relentless growth, and distal spread with a lethal effect.

malignant neoplasm, See cancer.

matching, In an epidemiological case-control study, the selection of controls such that the control group has the same distribution of certain variables as the case group; in a cohort study, the selection of those not exposed to a given factor such that they have the same distribution of certain variables as those exposed to the given factor.

matching frequency (stratified matching), In a case-control epidemiological study, an attempt to make cases and controls comparable by subdividing the population of potential controls according to certain variables and sampling such that the proportion of controls in each subgroup is the same as the proportion of cases in each subgroup. In a cohort study, the population not exposed to the risk under study would be subdivided and sampled such that the proportion of nonexposed in each subgroup was the same as the proportion of exposed in each subgroup.

matching individual, In a case-control epidemiological study, an attempt to make cases and controls comparable by selecting for each case a control who matches the case with respect to specific criteria. In a cohort study, each person exposed to a possible adverse factor would be matched to a person not exposed, according to specific criteria.

maximum permissible body burden (MPBB), That amount of an internally deposited radioisotope which will result in a maximum permissible dose to the individual or a critical organ.

maximum permissible concentration (MPC), That amount of radioactive material in air, water, or food which will result in accumulation of a maximum permissible body burden when assimilated at a standard rate for that medium.

maximum permissible dose (MPD), A dose established by competent authority as the highest allowable dose which is not expected to cause injury to a person during his lifetime. A wide safety margin below the dose at which detrimental effects are known to occur is normally provided in establishing MPD's.

maximum voluntary ventilation (MVV), Volume of air which a subject can breathe with voluntary maximal effort for a given time.

measurement, The assigning of numbers to observations in such a way that the numbers are amenable to analysis according to certain rules.

median lethal dose (LD-50), The acute dose of radiation required to kill 50% of the population within a specified time, usually 30 days.

medical radiation, Radiation from diagnostic or therapeutic radiological procedures, e.g., x-rays, radioactive tracers, cobalt therapy.

medical monitoring, See biological monitoring.

medical surveillance, See biological monitoring.

membrane filter, A thin, pliable material of varying porosity designed to be placed in filter holder and have air contaminants drawn across it entrapping particles for gravimetric or instrumental/chemical analysis.

metal fume fever, An acute illness characterized by the sudden onset of chills and fever several hours after exposure to metal fumes. It is most common in brass foundries and in the smelting of zinc. The condition is thought to be due to inhalation of metal oxides. Zinc, copper, iron, and magnesium have been documented as causing the syndrome, but other metals may behave in a similar manner under certain circumstances.

metastatic tumor, A secondary malignant growth that originates from a primary tumor and is separated from the primary site.

microwave bioeffects, Alteration in biological systems resulting from exposure to microwave radiation. Two broad types of bioeffects are generally described:

thermal effects, Effects produced by the heating effect of absorbed microwave energy. These effects are fairly well documented and most are accompanied by a detectable rise in tissue temperature. Some effects in which a temperature rise is not detected by available instrumentation are termed microthermal.

nonthermal effects, Effects which are thought to be due to a mechanism other than heating. Certain experimentally produced field effects fall in this category but are generally biologically insignificant. The existence of any true biologically significant nonthermal effects has not been conclusively demonstrated and is highly controversial.

microwave cataracts, Posterior subcapsular lenticular opacities attributed to exposure to microwave radiation. Such opacities have been experimentally produced and several cases involving human exposures have been reported. The occurrence of microwave cataracts in occupationally exposed personnel in normal industrial circumstances has not been demonstrated.

microwave dosimetry, Measurement of the amount of microwave energy to which a system has been exposed. The term microwave is usually used to cover radiations in the radio and radar frequencies (kilohertz through gigahertz). Absorbed doses cannot currently be measured and can only be roughly estimated from measurements of temperature increase. Time integrating instruments are also lacking so that exposures are usually expressed in terms of power flux density (watts per square centimeter) or field strength (volts per meter).

microwave hearing effect, The auditory sensation evoked in man exposed to low level pulsed microwave radiation. It is perceived as an audible buzz or click. It was long thought to demonstrate a direct, nonthermal effect but recent evidence indicates that it may be due to rapid transient thermal expansion of fluid. No hazard has ever been linked with this effect.

mists, Suspended liquid droplets generated by condensation from the gaseous to the liquid state or by breaking up a liquid into a dispersed state, such as by splashing, foaming, or atomizing. Mist is formed when a finely divided liquid is suspended in the atmosphere.

morbidity, Sickness.

morbidity rate, The number of cases of a specific disease occuring per unit population over a stated period of time.

mortality, Death.

mortality rate, The number of deaths occurring per unit population over a stated period of time.

mutagen, Any agent capable of producing genetic or somatic mutation. (See genetic mutation, somatic mutation.)

near-drowning, Submersion with asphyxia in which the victim survives the acute episode. Liquid may or may not have entered the lungs. Many near-drowning victims succumb to pulmonary edema hours to days following the acute episode.

neoplasm, A local autonomous new growth having no useful function.

neuropathy (peripheral), Any disease of the peripheral nerves. Principal types are: (a) demyelinating — a peripheral neuropathy in which the prominent pathologic alteration is an inflammatory destruction of the myelin sheath (example — Guillian Barré syndrome). (b) Vascular — a peripheral neuropathy secondary to small vessel occlusion; destruction involves both the axon and myelin sheath (example — diabetes mellitus mononeuropathy multiplex). (c) Axonal, a peripheral neuropathy in which the primary pathologic alteration is the retro-grade destruction of the axon (examples — toxic neuropathy caused by exposure to arsenic, n-hexane, acrylamide, methyl N-butyl ketone, tri-orthocresyl phosphate, carbon disulfide).

nickel itch, A dermatitis due to sensitization to nickel and an allergic reaction on contact with nickel or nickel compounds. It is seen in nickel-platers and in persons wearing jewelry containing this metal.

NIOSH, National Institute for Occupational Safety and Health.

nitrogen narcosis, See inert gas narcosis.

noise, Any undesired sounds usually resulting in an objectionable or irritating sensation.

non-causal asociation, A statistical association between the occurrence of a factor and a disease in which the factor is not a cause of the disease.

nuisance dust, Generally innocuous dust not recognized as the direct cause of a serious pathological condition.

occupational disease, A disease arising out of, and in the cause of, employment — resulting from exposure to, the absorption of or intoxication from harmful chemical, biologic, or physical agents to which the general public would not normally be exposed.

occupational injury, An injury arising out of, and in the course of, employment — resulting from the action of traumatizing physical or chemical agents in the workplace.

occupational medicine (health), The promotion and maintenance of the highest degree of physical, mental, and social well-being of workers in all occupations. The prevention among workers of departures from health caused by their working conditions. The protection of workers in their employment from risks resulting from factors adverse to health. The placing and maintenance of the worker in an occupational environment adapted to his physiological and psycological equipment.

occupational radiation, Radiation to which an individual is exposed in the cause of his employment. For instance, x-ray technicians, radioisotope technicians, uranium miners, nuclear reactor technicians are occupationally exposed to radiation.

octave band, A range of frequency where the highest frequency of the band is double the lowest frequency of the band. The band is usually specified by the center frequency.

odds ratio, In an epidemiological study, the ratio of affected individuals among those exposed to a given factor divided by the ratio of affected to unaffected individuals among those not exposed to the factor. (For a rare disease, the odds ratio is approximately equal to the relative risk.)

oil acne, A skin condition which affects those surfaces in contact with insoluble oils or oil-soaked clothing. Comedones, raised papules, or an infection about the hair follicles (oil boils) occur.

oncology, The study of neoplasia.

organic disease, Disease in which some change in the structure of body tissue could either be visualized or positively inferred from indirect evidence.

OSHA, Occupational Safety and Health Act of 1970.

osteoarthritis, Chronic multiple degenerative joint disease.

otitis externa (swimmer's ear), An inflammatory condition of the external ear canal. It occurs in swimmers and divers and in long-term compression chamber operators, in whom it is one of the most regularly occurring disorders. It is a result of excess moisture in the ear canal resulting in tissue maceration and infection.

otosclerosis, A condition caused by a growth of bony tissue about the foot plate of the stapes and the oval window of the inner ear. It results in a gradual loss of hearing.

oxygen toxicity, A disorder associated with increased partial pressures of oxygen. There are two types of oxygen toxicity:

pulmonary, Occurring with exposures of about 48 hours to high oxygen tensions at normal barometric pressures. It is characterized by chest pain and decreased vital capacity. It may not be fully reversible upon removal from exposure.

central nervous system, Occurring within minutes to hours with exposure to oxygen at increased barometric pressures (greater than 2.5 atmospheres). It is characterized by twitching, nausea, tunnel vision, and ultimately grand mal convulsions. It is fully reversible on return to normoxic conditions.

particulate, A particle of solid or liquid matter.

patch test, A test in which a diluted sample of a material suspected of causing an allergic dermatitis is applied to the patient's skin (usually the back) with an adhesive patch and removed in 48 hours. If the patient is sensitive, a small area of dermatitis will appear at the point of contact of the material.

pesticides, General term for that group of chemicals used to control or kill such pests as rats, insects, fungi, bacteria, weeds, etc., that prey on man or agricultural products. Among these are insecticides, herbicides, fungicides, rodenticides, miticides, fumigants, and repellents.

peripheral neuropath, See neuropathy (peripheral).

phon, The unit of loudness level.

pitch, That attribute of auditory sensation in terms of which sounds may be ordered on a scale extending from low to high. Pitch depends primarily upon the frequency of the sound stimulus, but it also depends upon the sound pressure and wave form of the stimulus.

pneumoconiosis, A fibrous induration of the lungs attributable to the irritation caused by the inhalation of dust incident to exposures such as antimony, asbestos, barium, bauxite, beryllium, coal, graphite, hematite, silicon dioxide, tin, titanium, and tungsten.

pneumonitis, An inflammation of the lung.

presbycusis, Hearing loss associated with increasing age.

prevalence rate, The number of new and existing cases, per unit population, occurring during a stated period of time.

probability sampling, Sampling in which each sampling unit has a known non-zero probability of being included in the sample.

prospective study, See cohort study; prospective and retrospective.

pulmonary hyperinflation syndrome (burst lung), Pulmonary barotrauma of ascent. It occurs during decreases in pressure from overdistension and rupture of the lungs by expanding gases. It is a result of breath-holding or air-trapping during ascent. The clinical presentations are pneumothorax, mediastinal or subcutaneous emphysema, and air embolism.

pure tone, A pure tone has a unique pitch and is characterized by a sinusoidal variation in sound pressure with time. The frequency spectrum of a pure tone shows a single line at a discrete frequency.

quality factor (QF), The factor by which absorbed dose in rads (q.v.) is multiplied to obtain effective or equivalent dose in rem. Sometimes used interchangeably with relative biological effectiveness (RBE), a term which should be limited to research applications. Typical QFs are: x-ray = 1, beta = 1, alpha = 20, protons = 10, fast neutrons = 10.

rad, Radiation absorbed dose. Under the common system it is the basic unit of absorbed dose of any type radiation equivalent to absorption of 100 ergs per gram of any material. However, under SI, the unit is the gray (Gy) = 0.01 rad. The gray is the energy imparted by ionizing radiation to a mass of matter corresponding to one joule per kilogram.

radiation, The emission and propagation of energy through space or a material medium in the form of electromagnetic waves (gamma or x-rays) or particles (alpha and beta). Electromagnetic radiation may be ionizing or nonionizing; all particulate radiation is ionizing.

radiation dosimetry, Measurement of the amount of radiation delivered to or absorbed at a specific place. Personnel dosimetry is accomplished with such devices as the film badge, thermoluminescent dosimeter, or pocket ionization chamber. In this way continuous recording of cumulative radiation dose can be maintained.

radiation monitoring, Measurement of the amount of radiation present in a given area. Radiation exposures can be limited by monitoring and limiting access and stay time in areas in which high levels of radiation are present.

radioactive contamination, The deposition or presence of radioactive material in a place where it is not desired and may be harmful. In regard to personnel:

external contamination, The presence of radioactive material on the skin.

internal contamination, The presence of radioactive material within the body due to ingestion, inhalation, or absorption.

random noise, A sound or electrical wave whose instantaneous amplitudes occur, as a function of time according to a normal (Gaussian) distribution curve. Random noise is an oscillation whose instantaneous magnitude is not specified for any given instant of time. The instantaneous magnitudes of a random noise are specified only by probability functions giving the fraction of the total time that the magnitude, or some sequence of the magnitudes, lies within a specific range.

random sampling, A procedure in which each sampling unit in the population has an equal chance of being included in the sample.

rate, An expression of the speed or frequency with which a certain event or circumstance occurs in relation to a certain period of time, a specific population, or some other fixed standard. (E.g., mortality rate, pulse rate, etc.)

Raynaud's phenomenon (dead hand or white hand syndrome), A vascular disorder of the extremities, usually the hand, resulting from the use of vibrating tools, particularly pneumatic tools. Intermittent cyanosis or pallor of the part, coldness, numbness, and paresthesias are typical. Once the disorder develops it generally does not improve even with cessation of work with vibrating tools.

recompression therapy, Treatment of decompression sickness or air embolism by returning the victim to a hyperbaric environment, usually in a recompression chamber. Standard tables for treatment of diving casualties are available, the most frequently used being those developed by the U.S. Navy.

regression coefficient, In popular terms, it is the amount of change that will on the average take place in one characteristic when the other characteristic changes by one unit. However, in more rigorous (mathematical) terms:

regression, This term was originally used by Galton to indicate certain relationships in the theory of heredity but it has come to mean the statistical method developed to investigate those relationships.

If a variate y consists of two components, a variate and a systematic element f (X) depending on a variable X; i.e., if

$$y = f(X) + \epsilon$$

then the regression of y on X is the equation

$$Y = f(X)$$

where it is supposed that ϵ has zero expectation. The definition remains valid if X, instead of being a single variable, refers to a set of variables X_1, X_2, etc.

In particular, X itself may be given as the values of a variate, in which case the regression of y on x may be regarded as expressing the dependence of the mean of y (for given x) on the corresponding x:

$$E(y \mid x) = f(x).$$

The most frequently considered form of f (x) is a polynomial, particularly a linear function, giving the regression of y on X

$$Y = \beta_0 + \beta_1 X$$

or, for p variables

$$Y = \beta_0 + \beta_1 X_1 + \ldots + \beta p X p$$

Such expressions are called regression equations. The X's are called "independent," "predicated" variables, "predictors" or "regressors." y is called the "dependent variate," "predictand" or "regressand." Hence regression coefficient is the coefficient of an independent variable in a regression equation.

relative risk, In an epidemiological study, the rate of disease among those exposed to a given factor divided by the rate of disease among those not exposed to the given factor. (For a rare disease, the relative risk may be approximated by the odds ratio.)

reliability, See reproducibility.

rem, Roentgen equivalent man. Unit of dose of any type radiation which produces in man an effect equivalent to exposure to one rad or roentgen of x- or gamma rays.

reproducibility (reliability, precision), The extent to which a measurement produces the same results when repeatedly applied to the same situation.

respirator, A device designed to protect the wearer from inhalation of harmful atmospheres.

retrospective study, See case-control study.

reverberation, The persistence of echoing of previously generated sound caused by reflection of acoustic waves from the surfaces of enclosed spaces.

risk factor, A characteristic or agent whose presence increases the probability of occurrence of a disease.

roentgen (r), The common unit of exposure to x- or gamma rays equivalent to the absorption of 83 ergs per gram of air. Under SI, it is equal to 0.000 258 coulombs per kilogram (c/kg).

sample, A number of sampling units from the population of units eligible to be included.

sampling (general), The collection of samples of contaminants in air in the general work environment for a specified length of time.

sampling (grab), The collection of samples of contaminants in air in a work environment for small time periods (usually 5 min. or less).

sampling (multistate), A procedure in which primary sampling units (e.g., municipalities) are first selected from a population, and then secondary units (e.g., city blocks) are sampled from within each chosen primary unit. This may be extended so that tertiary units (e.g., households) or further units (individuals) are selected within the secondary units.

sampling (personnel), The collection of samples of contaminants in air for the duration of the specific part of an operation by attaching an instrument to the worker.

sampling unit, The basic unit (e.g., person, household, etc.) around which a sampling procedure is planned.

sarcoma, A mlignant neoplasm arising from connective tissue cells.

scan, A shortened form of scintiscan — variously designated according to the organ under examination (e.g., brain scan, lung scan, etc.). Scanning refers to the visual examination of a small area in some detail.

screening test, A test or series of tests to which an individual submits to determine whether enough evidence of a disease exists to warrant a further diagnostic examination.

secondary attack rate, The number of new cases of a disease which occur over a relatively short period of time within households in which there is a first or primary case of the disease per unit susceptible population in households with primary cases.

secular trends, Changes in incidence rates, mortality rates, and other indicators of disease frequency that occur gradually over relatively long periods of time.

sensitivity, The extent to which a test identifies as positive all individuals who have a given disease.

Shaver's disease, A condition of pulmonary fibrosis and emphysema caused by the inhalation of a dust of alumina with silicia.

siderosis, A pneumoconiosis resulting from chronic exposure to iron oxide and silicon dioxide.

silicates, Compounds made up of silicon, oxygen, and one or more metals with or without hydrogen. These dusts cause nonspecific dust reactions, but generally do not interfere with pulmonary function.

silicones, A unique group of compounds made by molecular combinations of the element silicon or certain of its compounds with organic chemicals. Produced in variety of forms, including silicone fluids, resins, and rubber. Silicones have special properties such as water repellency, wide temperature resistance, high durabillity, and great dielectric strength.

silicosis, A fibrotic disease of the lungs caused by the chronic inhalation of dust containing silicon dioxide.

sintering, Process of mak ng coherent powder of earthy substances by heating but without melting.

slot velocity, Linear flow rate through the openings in a slot-type hood.

smoke, An air suspension (aerosol) of particles, often originating from combustion or sublimation. Carbon or soot particles less than 0.1μ in size result from the incomplete combustion of carbonaceous material such as coal or oil. Smoke generally contains droplets as well as dry particles. Tobacco, for instance, produces a wet smoke composed of minute tarry droplets.

somatic mutation, An alteration in the molecular arrangement of the inherited DNA of the chromosomes of the general body cells (as opposed to the germ cells) occurring spontaneously or by some abnormal form of external stimulus.

sonarography, Ultrasonic scan (q.v.) that provides a two-dimensional image corresponding to clear sections of acoustic interfaces in tissue.

sone, A unit used in judging the loudness of sounds. One sone is defined as the level, at 1000 Hz, that is 40 dB above the subjects' thresholds of hearing. The loudness of a given sound is rated by the listener as some multiple of the sone.

sonometer, An appliance for testing acuteness in hearing.

sorbent, A material which removes toxic gases and vapors from air inhaled through a canister or cartridge.

sound, An oscillation in pressure, stress, particle displacement, particle velocity, etc., which is propagated in an elastic material, in a medium with internal forces (e.g., elastic viscous), or the super-position of such propagated oscillations. Sough is also the sensation produced through the organs of hearing — usually by vibrations transmitted in a material medium, commonly air.

sound-level meter, An instrument for the direct measurement of sound pressure level. Sound-level meters often are made with various filtering networks that measure the sound directly on A, B, C, etc., scales. Sound-level meters may also incorporate octave-band filters for measuring sound directly in octave bands.

specificity, The extent to which a test identifies as negative all individuals who are free of a given disease.

sprain, A joint injury in which some of the fibers of a supporting ligament are ruptured, but the continuity of the ligament remains intact.

squeeze, This occurs during increases in pressure from the failure or inability to equalize pressure in gas-filled spaces. It occurs most commonly in the middle ear (aerotitis media) due to eustachian tube dysfunction or the nasal sinuses (aerosinusitis) due to blockage of the ostia. Squeeze may also occur in carious or poorly restored teeth, in the lung in breath-hold diving, or in the mask, suit, or helmet.

standard man, A theoretical physically fit man of standard (average) height, weight, dimensions, and other parameters (blood composition, percentage of water, mass of salivary glands, to name a few); used in studies of how heat or ionizing radiation affects humans.

standardized mortality ratio, The ratio of the number of deaths observed in a group of individuals to the number of deaths that would have been expected if some selected age (or sex) specific mortality rates had been applied to that group. The ratio is usually multiplied by 100, and is most frequently used in comparisons of the mortality experience of a component part of a population with the mortality that would be expected on the basis of the experience of the whole population.

stannosis, A condition due to inhalation of tin or tin oxide dust. It is characterized by the appearance of nodules in the lungs on chest x-ray, but autopsy studies have shown no evidence of fibrosis. The condition at present is thought to be benign.

stenosing tenosynovitis (trigger or snapping finger), A localized inflammation of the tendon sheath with localized swelling. The swollen tendon snaps back and forth through the narrowed sheath during flexion and extension. The condition frequently occurs at the base of the thumb or fingers in workers who repeatedly grasp hard or push with the distal palm, causing points of pressure over the tendon sheaths.

strain, Overstretching or overexertion of some part of the body musculature.

stratified sampling, A procedure in which the population is divided into strata, or groups of units having certain characteristics in common, and a sample of units drawn from each stratum.

supplied-air suit, A one or two-piece suit that is impermeable to most particulate and gaseous contaminants and is provided with an adequate supply of respirable air.

synergism, The interaction of two or more physiologically active agents producing an effect which is equal to (additive) or greater than (potentiation) the effect produced using equivalent doses of each agent presented individually.

systematic sampling, A procedure in which the selected sampling units are spaced regularly throughout the population; that is, every nth unit is selected.

temporary threshold shift (TTS), The hearing loss suffered as the result of noise exposure, all or part of which is recovered during an arbitrary period of time when one is removed from the noise.

tenosynovitis, Inflammation of a tendon sheath. Causes are varied, but commonly seen in industry associated with repetitive motion in an awkward or strained position. Commonly occurs about the wrist, the extensor tendons of the thumb, or in the fingers.

tennis elbow, See epicondylitis.

teratogen, Any agent capable of producing a permanent alteration in the structure and/or function of cells, tissue and organs occurring during the period of fetal organogenesis.

theory, A system of assumptions, accepted principles and rules of procedure devised to analyze, predict, or otherwise explain a specified set of phenomena.

threshold dose, The minimum dose of radiation which will produce a detectable biological effect.

tinnitus, A ringing sound in the ears.

TLV (threshold limit value), An exposure level under which most people can work consistently for 8 hours a day, day after day, with no harmful effects.

tolerance (drug), The progressive diminution of susceptibility to the effects of a drug resulting from its continued administration.

toxicant, A substance that causes a degenerative alteration in any anatomic, physiologic, or biochemical system of a formed organism.

trigger finger, See stenosing tenosynovitis.

TSCA, Toxic Substances Control Act.

tumor, Any neoplasm.

TWA (time-weighted average), An average over a given (working) period of a person's exposure as determined by sampling at given times during the period.

vapor, The gaseous form of substances which are normally in the solid or liquid state (at room temperature and pressure). The vapor can be changed back to the solid or liquid state either by increasing the pressure or decreasing the temperature alone.

variable, A quantity capable of assuming two or more different values.

ventilation, The process causing fresh air to circulate to replace foul air simultaneously removed.

vertigo, Dizziness due to vestibular dysfunction. In true vestibular vertigo the victim experiences a sensation of rotary motion of either himself or the environment. Vertigo may be experienced in aviators as a function of acceleration. In divers, vertigo may occur as a result of caloric stimulation or unequal middle ear pressurization (alternobaric vertigo). Permanent vestibular injury and vertigo may result from vestibular decompression sickness or round window rupture.

vesicant, Any substance that produces blisters.

vibration, An oscillation wherein the quantity is a parameter that defines the motion of a mechanical system.

wavelength, The distance between a point of a given phase of one wave and a point of the same phase of an adjacent wave.

welders' flashburn (actinic keratoconjunctivitis), An inflammatory condition of the corneal and conjunctival epithelium resulting from exposure to intense ultraviolet radiation as from a welder's arc. It is characterized by corneal edema, pain, a gritty sensation, tearing, and photophobia occurring several hours after exposure. It is generally self-limited, resolving spontaneously within 24 to 48 hours and is fully preventable by training and the use of proper goggles.

white noise, Noise of wide frequency range in which the amplitude of the noise is essentially the same in all frequency bands.

workmen's compensation, A system in which employers have a statutory obligation to pay medical expenses, disability, and other benefits for work-related injuries or illnesses. Administrative aspects and extent of coverage of workmen's compensation laws vary from state to state.

Workmen's Compensation Insurance, A form of insurance imposed on employers by legislative statute which insures coverage for medical expenses and disability and other benefits for work-related injuries or illnesses to be paid by the employer. Depending on the state, an employer may be self-insured or may pay premiums to a carrier to administrate the insurance program, or the state may run the program itself for all employers.

Index
(Referenced by section number)

B

C

D

E

F

H

J

K

L

M

O

P

Q

R

T

U

X

Y

Z